Green Energy and Technology

More information about this series at http://www.springer.com/series/8059

Medhat A. Nemitallah ·
Mohamed A. Habib · Hassan M. Badr

Oxyfuel Combustion for Clean Energy Applications

Springer

Medhat A. Nemitallah
TIC in CCS and Mechanical
Engineering Department
King Fahd University of Petroleum
and Minerals
Dhahran, Saudi Arabia

Mohamed A. Habib
TIC in CCS and Mechanical
Engineering Department
King Fahd University of Petroleum
and Minerals
Dhahran, Saudi Arabia

Hassan M. Badr
TIC in CCS and Mechanical
Engineering Department
King Fahd University of Petroleum
and Minerals
Dhahran, Saudi Arabia

ISSN 1865-3529 ISSN 1865-3537 (electronic)
Green Energy and Technology
ISBN 978-3-030-10587-7 ISBN 978-3-030-10588-4 (eBook)
https://doi.org/10.1007/978-3-030-10588-4

Library of Congress Control Number: 2018965458

This Springer imprint is published by the registered company Springer Nature Switzerland AG
The registered company address is: Gewerbestrasse 11, 6330 Cham, Switzerland

Preface

The problem of global warming is becoming one of the most important problems facing mankind because of its direct effect on the entire planet (coastal flooding, heat waves, rainfalls, wildfires, food production, and many others). The emission of greenhouse gases resulting from the burning of fossil fuels has been identified as the main cause of current climatic changes. Currently, about 80% of the global energy demand comes from the burning of fossil fuel, resulting in the emission of a huge amount of CO_2 to the atmosphere. Also, the burning of coal, natural gas, and oil for electricity and heat is the largest single source of global greenhouse gas emissions. Researchers and scientists are currently striving to find different means for tackling this problem either by increasing the efficiency of all equipment involved in the processes of energy production or energy consumption. Also, increasing the utilization of clean energy sources such as solar energy, hydroelectric power, and geothermal energy represents another way to reduce CO_2 emissions. The third option is to achieve clean combustion through the modification of various combustion processes in order to enable carbon capture and its utilization in other industries or its sequestration in underground aquifers.

This book is intended to be a basic reference for graduate students, practicing engineers, and young researchers in the area of clean combustion. The motivation for writing this book originates from the current international thrust for reducing greenhouse gas emission to the atmosphere for the sake of reducing global warming. As a result, very many industries worldwide start modifying their existing processes/equipment to comply with the Paris Agreement (Paris Climate Conference, December 2015) adopted by 195 countries. Accordingly, it becomes essential for engineers and scientists to develop green combustion systems that are friendly to the environment. Currently, gas turbines used for power generation, boilers used for steam generation, and cogeneration plants are the largest sources of greenhouse gas emissions. This book contains an extensive review of different carbon capture methodologies associated with fuel combustion. Novel approaches for clean combustion are introduced including design and performance analysis of burners. The feasibilities of different combustion technologies are also presented and discussed. Special emphasis is given to basic formulation of various combustion processes and computational modeling of conventional combustors together with applications to gas turbines and boilers supported by numerical results and detailed discussions for a number of case studies.

This book consists of six chapters: The first chapter is an overview of the greenhouse gas emission problem and brief presentation of the current carbon capture and sequestration technologies. The second chapter introduces oxy-fuel combustion technologies with emphasis on system efficiency, combustion and emission characteristics, applications, and related challenges. The third chapter focuses on the recent developments in ion transport membranes and their performance in oxygen separation units and oxygen transport reactors. The fourth chapter presents novel approaches for clean combustion in gas turbines. The fifth chapter presents the computational modeling and optimization of combustion in gas turbine combustors with some numerical results and detailed analyses. The sixth chapter presents the replacement of conventional combustion systems by oxygen transport reactors of distinctive designs together with applications in gas turbine combustors and furnaces of fire tube boilers.

The authors wish to acknowledge the support received from King Fahd University of Petroleum & Minerals under Grant # BW171005 for the preparation of this book.

Dhahran, Saudi Arabia

Medhat A. Nemitallah
Mohamed A. Habib
Hassan M. Badr

The original version of the book was revised: Incorrect grant number has been corrected. The correction to the book is available at https://doi.org/10.1007/978-3-030-10588-4_7

Contents

Chapter 1
Introduction

1.1 Global Warming

Greenhouse gas (GHG) anthropogenic emission in the atmosphere has been the ultimate cause of the current climatic change [1]. Among the main sources of anthropogenic greenhouse gas emissions, burning of fossil fuels has been identified as the main concern in the current century [2]. As reported by the International Energy Agency (IEA) [3], the global energy consumption based on fossil fuel amounts to about 80% of the total global energy demand. This resulted in the emission of 32.3 Gt of CO_2 to the atmosphere in the year 2014 [3]. Recent findings indicated that about 40% of the global CO_2 emission is a direct result of electricity generation, with more than 30% coming from fossil fuels [4]. Several routes for lowering CO_2 emissions can be applied including increasing of plant efficiency (provides reduction of CO_2 emission by 2–3% for increasing plant efficiency by 1%), decreasing of carbon content in the fuel by utilizing less carbon fuels, reducing unnecessary fuel consumption, and employing carbon capture and storage technologies [2, 5]. Carbon capture and sequestration (CCS) is the process of capturing, purifying, compressing, transporting, and storing of unwanted CO_2 emitted from the above-mentioned sources. The CCS technologies include post-, pre-, and oxy-combustion [6–9]. In the pre-combustion technique, the fuel carbon is converted into syngas at initial stage before the combustion process. In the post-combustion process, carbon dioxide emission is mitigated by separation and removal of CO_2 from the flue gases via absorption using solvents (basically amines), adsorption using metal–organic framework (MOF), as well as CO_2 selective membrane separation. In oxy-fuel combustion process, pure oxygen is being used for the combustion of fuel resulting in only H_2O and CO_2 as products of combustion. Oxy-fuel combustion process results in high combustion temperature. Part of the flue gas is usually recycled into the combustion zone to lower the combustion temperature [10]. To maintain a stable flame under oxy-combustion conditions, there should be certain minimum amount of oxygen concentration in the oxidizer

M. A. Nemitallah et al., *Oxyfuel Combustion for Clean Energy Applications*,
Green Energy and Technology, https://doi.org/10.1007/978-3-030-10588-4_1

(O_2 + CO_2) to enable sufficient temperature levels within the combustor to sustain the combustion reactions. Carbon dioxide addition, as a diluent, affects the oxy-combustion characteristics in terms of variations in the following parameters within the combustor: (a) adiabatic flame temperature of the mixture, (b) radiative heat transfer, (c) transport properties including viscosity, mass diffusivity, and thermal conductivity, (d) specific heat of the mixture, (e) chemical kinetics, and (f) flame structure. Therefore, combustion in an environment of high CO_2 concentrations provides new combustion challenges and opportunities, which is the subject of the book.

1.2 Carbon Budget for the 2 °C Limit

In December 2015, a new global treaty to battle climate change (Paris Agreement [11]) was adopted under the United Nations Framework Convention on Climate Change (UNFCCC). In preparation of this treaty, countries submitted national plans that spell out their aims for addressing the climate change challenge after 2020. These Intended Nationally Determined Contributions (INDCs) address a range of matters, which can relate to avoiding, adapting, or coping with climate change, among other things. However, targets and actions for reducing greenhouse gas (GHG) emissions are core components. At this point, the INDCs are not final and can be modified up until the time the Paris Agreement is ratified. However, for the time being they represent our best understanding of the climate actions countries intend to pursue after 2020.

The main climate goal of the Paris Agreement is to hold "the increase in the global average temperature to well below 2 °C above pre-industrial levels and to pursue efforts to limit the temperature increase to 1.5 °C above pre-industrial levels" [11]. This climate goal represents the level of climate change that governments agree would prevent dangerous interference with the climate system, while ensuring sustainable food production and economic development [12, 13], and is the result of international discussions over multiple decades [14]. Figure 1.1 shows the global greenhouse gas emissions as implied by INDCs compared to no-policy baseline, current policy, and 2 °C scenarios. White lines display the median of each range. The white dashed line demonstrates the median evaluation of what the INDCs would bring if all conditions are met. The 20–80% ranges are shown for the scenarios of the no-policy baseline and 2 °C. For current policy and INDC scenarios, the minimum–maximum and 10th–90th percentile range across all evaluated studies are given, respectively. Symbols characterize single studies and are offset slightly as to increase readability. Dashed brown lines connect data points for each study [15].

Limiting warming to a particular level implies that the total amount of carbon dioxide (CO_2) that can ever be emitted into the atmosphere is finite [16]. From a geophysical standpoint, global CO_2 emissions thus need to become net zero [17, 18]. Around two thirds of the available budget for setting warming to below 2 °C

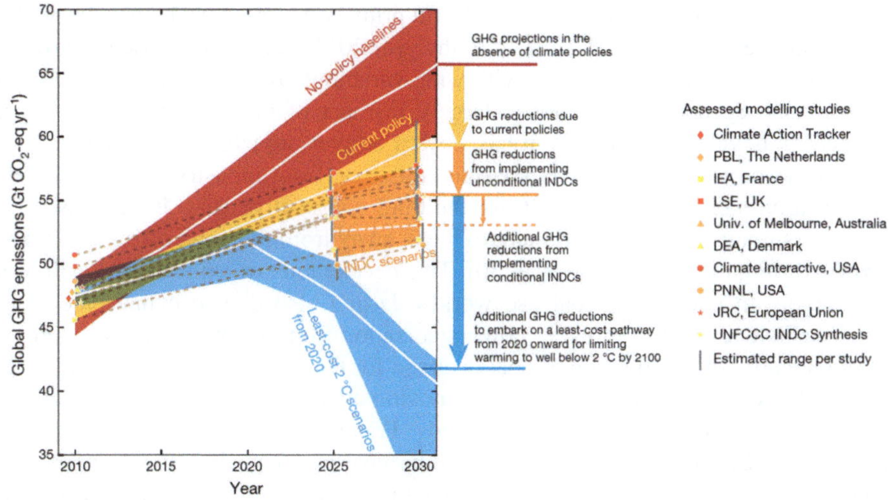

Fig. 1.1 Global greenhouse gas emissions as implied by INDCs compared to no-policy baseline, current policy, and 2 °C scenarios [15]

have already been emitted [19–21], and growing trends in CO_2 emissions [22] indicate that global emissions urgently need to start to decline so as to not foreclose the possibility of holding warming to well below 2 °C [23, 24]. Total anthropogenic emissions of one trillion tons of carbon (3.67 trillion tons of CO_2), about half of which has already been emitted since industrialization began, results in a most likely peak carbon dioxide-induced warming of 2 °C above pre-industrial temperatures, with a 5–95% confidence interval of 1.3–3.9 °C [25]. The window for limiting warming to below 1.5 °C with high probability and without temporarily exceeding that level already seems to have closed [26]. The Paris Agreement implicitly acknowledges these insights and has the aim to reach a global peak in GHG emissions as soon as possible together with achieving a balance between the removals of GHGs and the anthropogenic emissions in the second half of this century. Both targets are in principle consistent with the temperature objective of the Agreement [27, 28], but request the broader question of whether the current Intended Nationally Determined Contributions are already putting the world on a path toward achieving them.

1.3 Status of Renewable Energies

During 2016, several developments and ongoing trends were made. These developments have a bearing on renewable energy, including the continuation of relatively low global fossil fuel prices; dramatic price reductions of several renewable energy technologies (especially solar photovoltaic and wind power); and a continued growth in attention to energy storage. World primary energy demand has grown by an annual average of around 1.8% from the year 2011, although the pace of growth has slowed in the past few years, with wide variations by country [29]. Growth in primary energy demand has arisen largely in most of the developing countries, whereas in developed countries, it has slowed or even sometimes declined.

Global energy-related carbon dioxide emissions from industry and fossil fuels were nearly even in the year 2016 for the third consecutive year. It was raised by only an estimate of 0.2%, continuing to break away from the trend of 2.2% average growth during the previous decade [30]. This decelerating of emissions growth was due largely to decreasing coal use worldwide and also due to enhancements in energy efficiency and to growing power generation from renewable energy sources [31]. Globally, the production of coal declined for the second year in a row [32]. In 2016, additional countries committed to moving away from or phasing out coal for electricity generation (e.g., the Netherlands, France, Finland, and the US state of Oregon and Canada) or to no longer funding coal use (e.g., Brazil's development bank) countering this trend. However, a number of countries declared tactics to expand coal production and usage [33].

Starting 2015, renewable energy provided a projected percentage of 19.3% [29] of global final energy consumption. Out of this total share, traditional biomass that is used primarily for heating and cooking in remote and rural areas of developing countries, accounted for about 9.1%. Modern renewables, that do not include traditional biomass, increased their share relative to 2014 to approximately 10.2%. During the year 2015, hydropower accounted for a likely 3.6% of total final energy consumption. Other renewable power sources comprised 1.6%; renewable heat energy accounted for approximately 4.2%, and transport biofuels provided about 0.8% [34], as per Fig. 1.2. Despite the overall decline in coal production, relatively low global prices for oil and natural gas during much of the year continued to contest renewable energy markets, especially in the heating and transport sectors [35]. Fossil fuel subsidies, which remained significantly higher than subsidies for renewable energy, have continued to affect renewable energy growth [36].

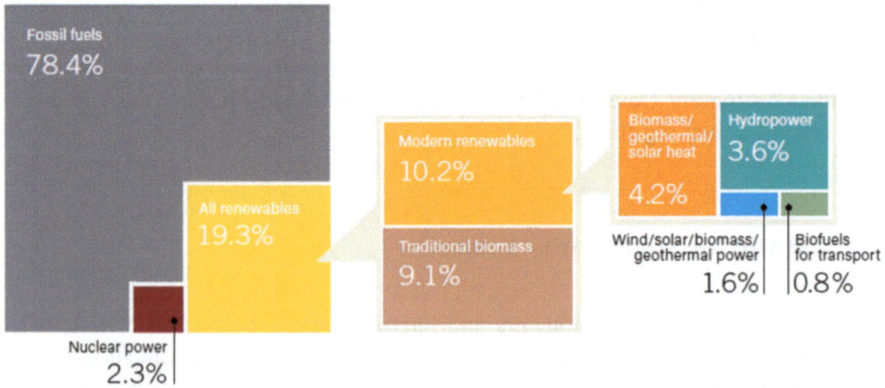

Fig. 1.2 Estimated renewable energy share of total final energy consumption, 2015 [37]

1.3.1 Market and Industry Trends

Greenhouse gas emissions can be further reduced by the growing utilization of renewable energy sources. Mature technologies are now available for the utilization of these sources that includes biomass energy, geothermal power and heat, hydropower, ocean energy, solar photovoltaics (PV), and wind power. In the following, we give a brief about the market and industry trends of such renewable energies.

Biomass feedstocks can be converted into useful renewable energy by many ways. A broad range of residues, wastes, and crops grown for energy purposes can be used directly as fuels for heating and cooling or for electricity production. As well, they can be converted into gaseous or liquid fuels to be used in transportation or as substitutes for petrochemicals [38]. Many bioenergy technologies and conversion procedures are now well-established and fully marketable. A further set of conversion processes, in particular to produce advanced liquid fuels, is maturing rapidly [39]. In the year 2016, global environmental concerns in addition to the growing energy demand and energy security continued to drive resulting in amassed production and the use of bioenergy. Bioenergy consumption and investment in new capacities are supported by policies in many countries. The low prices of fossil fuel during the year 2016 have discouraged, in some countries, the investment in bioenergy-based heating. Unlike transport use of biofuels, bio-heat is not sheltered by blending mandates from changes in fossil fuel prices. Increased competition from other low-cost renewable sources of electricity acted as a barrier to bio-power production during that year [40]. The continuing discussion about the sustainability of some forms of bioenergy has led to regulatory and policy uncertainty in some markets created a more difficult investment climate [41]. Bioenergy is the largest contributor to global renewable energy supply in traditional and modern uses [42]. The supply of biomass for energy has been growing at around 2.5% per year since 2010 [43] and reached approximately 62.5 exajoules (EJ) in the year 2016. The share of bioenergy in total global primary energy consumption has remained relatively stable since 2005, at around 10.5%, despite a 21% growth in overall global energy demand over the last decade [43]. The global bio-power capacity increased at an estimated rate of 6% in 2016, thus reaching 112 GW [44] and resulting in an increase of energy generation to 504 terawatt hours (TWh) [45]. In 2016, global biofuels' production, which closely tracks demand, increased around 2% compared to the year 2015, reaching 135 billion liters [46]. The increase was due largely to a rebound in biodiesel production after a decline in 2015. An estimated 72% of biofuel production (in energy terms) was ethanol, 23% was biodiesel, and 4% was hydrotreated vegetable oil (HVO), as per Fig. 1.3.

On the other hand, the use of geothermal resources was mainly for electric power generation and thermal energy services (heating and cooling). In the year 2016, the estimated electricity and thermal output from geothermal sources was 157 TWh, with each providing approximately the same share [47]. Some geothermal plants produce both thermal output and electricity for various heat applications.

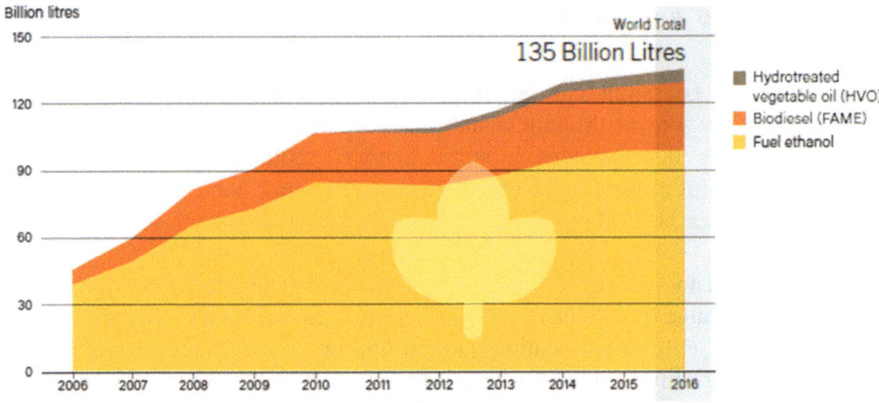

Fig. 1.3 Global trends in ethanol, biodiesel, and HVO production, 2006–2016

The geothermal industry continued to face challenges in 2016. It is burdened by (1) the inherent high risk of geothermal exploration and project development, (2) the associated lack of risk mitigation, and (3) the constraints of financing and competitive disadvantage relative to low-cost natural gas. Yet the industry achieved progress with new project development in key markets, and industry leaders cemented partnerships to tackle new opportunities.

Global hydropower generation was estimated to be 4102 TWh in 2016, up about 3.2% over 2015 record [48]. As well, global hydropower capacity additions in the year 2016 are estimated to be at least 25 GW, with total capacity reaching approximately 1096 GW [49]. Global pumped storage capacity (being counted independently) was estimated as 150 GW at year's end, with about 6.4 GW additions in 2016 [48]. In addition to ongoing improvements to mechanical equipment such as turbines, plant operators also continued to implement advanced control technologies and data analytics for digitally enhanced power generation. It is expected that these steps will aid to optimize plant management for greater reliability, efficiency, and lower cost. They will also allow for more flexible integration with other grid resources, including variable renewable energy [50]. Ocean energy is defined as any energy harnessed from the ocean. These are generated by means of ocean waves, tidal waves, ocean permanent currents, temperature gradients, and salinity gradients. Very few commercial ocean energy facilities have been built with operating capacities reaching approximately 536 MW at the end of 2016 [51]. The character of 2016 was similar to that of the previous year for the ocean energy industry, with a rising number of companies around the world advancing their technologies and installing new and improved devices. However, commercial success for ocean energy technologies continued to be in check due to perennial challenges. These include financing obstacles in an industry characterized by relatively high risk and high upfront costs and the need for improved planning, consenting, and licensing procedures [52].

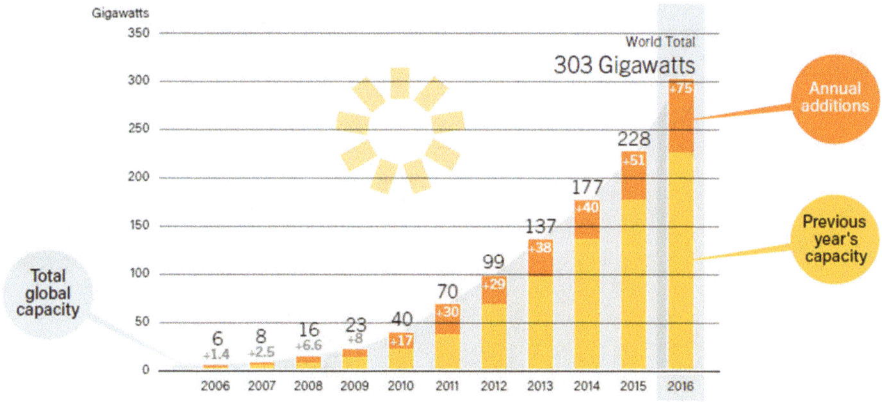

Fig. 1.4 Solar PV global capacity and annual additions, 2006–2016

During 2016, at least 75 GW of solar photovoltaic (PV) capacity was added worldwide, equivalent to the installation of more than 31,000 solar panels every hour [53]. More solar PV capacity was installed in 2016 (up 48% over 2015) than the cumulative world capacity five years earlier [54]. By year's end, global solar PV capacity totaled at least 303 GW [55], as per Fig. 1.4. Despite tremendous demand growth in 2016, the year brought unprecedented price reductions for modules, inverters, and structural balance of systems [56]. Due to even greater increases in production capacity, as well as to lower market expectations for 2017, module prices plummeted [57]. Average module prices went down by an estimated 29%, to USD 0.41 per watt between the fourth quarter of 2015 and a year later, reducing to historic lows [58].

In 2015, wind power was the second largest annual market in the renewable energy sector. In addition, about 55 GW of wind power capacity was added during 2016, increasing the global total by about 12% to reach nearly 487 GW [59]. Gross additions were 14% below the record high in 2015 [60], as per Fig. 1.5. At the end of 2016, over 90 countries had seen commercial wind power activity, and 29 countries, representing every region, had more than 1 GW in operation [61].

1.3.2 Renewables for Global Warming Control

Renewable energy, often distributed units or part of a larger diversified energy system, can ensure the delivery of energy services in direct response to climate change impacts. In order to ensure a reliable supply of energy to energy consuming facilities, future energy schemes need to be resilient and to maintain service even under extreme, varying, or unpredictable conditions by being robust, yet flexible and adaptive [62]. However, discussion about the specific role of renewables in energy system resilience, and in adaptation activities more generally, is still

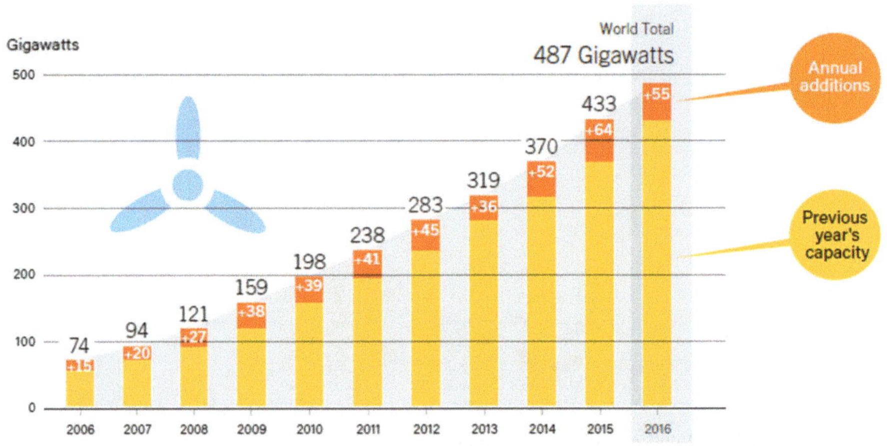

Fig. 1.5 Wind power global capacity and annual additions, 2006–2016

comparatively limited. Most literature focuses on power infrastructure and looks primarily at how renewable energy can contribute to disaster recovery, as well as at the backup functions that renewables can provide in cases of increased demand or grid failure. Although the impact of climate variability on energy systems is being discussed increasingly in various research forums and in an expanding number of documents and studies (at the national and local levels through regional initiatives and international bodies such as the UNFCCC), the focus primarily is on identifying the impacts currently being witnessed and anticipating future impacts [63]. Little is printed about the proactive role that renewable energy can show in increasing energy system resilience, and how these technologies can provide services as part of larger adaptation activities.

The role of renewables has already been established in climate mitigation. As the effects of extreme weather are felt progressively, more attention will need to be paid to how renewable energy can support adaptation activities so that energy services can be assured. Mitigation and adaptation responses to climate change are closely dependent on each other. Both responses need to occur simultaneously, illustrating their balancing nature and their collective role in meeting climate change challenges [64].

In conclusion and based on the above discussion, renewable sources of energy share about 24.5% of the global electricity production as per Fig. 1.6. Such technologies have great potential for application; however, the tremendous increase in global energy demand necessitates the use of non-renewable fossil fuel sources of energy. To date, non-renewable sources of energy share three-quarters of the energy market through the tremendous reduction in oil prices. This situation forced many countries to invest in clean combustion technologies for the control of global warming. Carbon capture and storage (CCS) technologies (including pre-combustion, oxy-combustion, and post-combustion) are considered as very promising technologies for the control of carbon emissions.

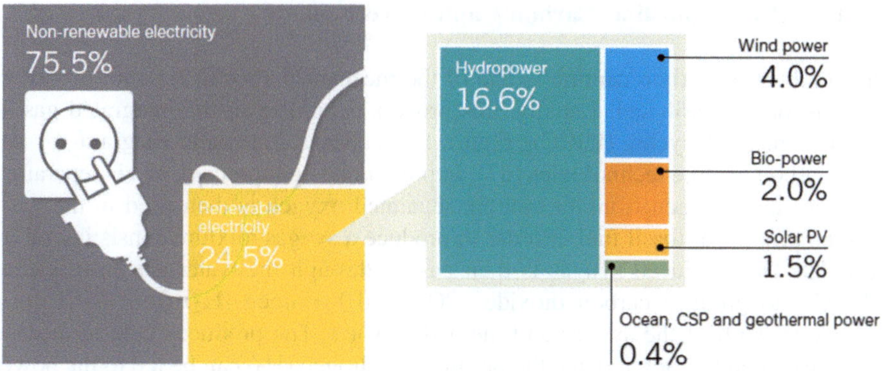

Fig. 1.6 Estimated renewable energy share of global electricity production, end 2016 [37]

1.4 Carbon Capture and Storage (CCS) Techniques and Limitations

1.4.1 Carbon Capture Technologies

The increasing level of CO_2 emission in the atmosphere due to fossil fuel combustion and the dissolved CO_2 in ocean waters is developing critical environmental concerns in terms of global warming and ocean acidification [65]. Electrical power plants using fossil fuel are the major contributor to greenhouse gas emissions with 41% [66]. As a solution for such problem, intensive research works on renewable and nuclear energies are being conducted in many developed and even in developing countries. However, renewable energies still need more work to make them economically competitive to the oil price. There are also many safety issues and technical problems associated with the uncontrolled spread of nuclear energy [67]. As well, the world's energy demand is tremendously increasing which requires the use of fossil fuels at least at the present time. This necessitates the handling of the combustion process to capture CO_2 before it influences the atmosphere. There are three carbon capture technologies (CCTs) which can be applied in order to capture CO_2. The difference among these technologies depends on the order of the capture process with respect to the combustion process. These technologies include: (1) pre-combustion carbon capture in which CO_2 is being captured before the combustion process; (2) oxy-combustion carbon capture in which the fuel is being oxidized using pure oxygen instead of air, and thus, the exhaust gases are highly CO_2-concentrated which facilitates CO_2 capture after H_2O condensation; and (3) post-combustion carbon capture in which CO_2 is being captured after the combustion process.

1.4.1.1 Pre-combustion Carbon Capture Technology

Pre-combustion carbon capture is one of the mechanisms for CO_2 capture prior to combustion of fossil fuel. Usually, this process is applied in the integrated gasification combined cycles (IGCC). Figure 1.7 shows a schematic diagram for the three carbon capture technologies [67]. In pre-combustion process, an air separation unit separates oxygen from air and the separated oxygen is then used in the gasification process of fossil fuel in order to produce a syngas mixture consisting of H_2 and CO. The produced syngas is then passed through the water-gas shift reactor (WGSR) to produce carbon dioxide (CO_2) and hydrogen (H_2) from the carbon monoxide (CO) of the syngas and the added water. The produced carbon dioxide (CO_2) can then be captured and the produced hydrogen (H_2) can be fed to the power generation device as the working fuel. The high concentrations of carbon dioxide (CO_2) in the produced mixture (CO_2 plus H_2) leaving the WGSR facilitate the capture of CO_2 as compared to the process of CO_2 capture from normal exhaust flue gas (a mixture of NO_x, SO_x, O_2, N_2, and unburned hydrocarbons) out of a conventional combustor.

There are different mechanisms for CO_2 separation from the gas mixture leaving the WGSR including absorption, adsorption, cryogenic separation, and membrane separation. Each of these methods has its own merits and limitations [68, 69]. In the absorption process, solvents are used to selectively absorb CO_2 from the exhaust stream. The absorption process can be classified as chemical absorption, which can be applied in pre-combustion and post-combustion CO_2 capture technologies, and physical absorption, which can only be applied in pre-combustion CO_2 capture technology. The application of pre-combustion CO_2 capture technology in IGCC plants, using dimethyl ethers of polyethylene glycol, results in capital cost increase of 19.55% for 70% CO_2 capture [70]. There are several problems associated with the application of the absorption process using selexol or sectisol solvents for CO_2 capture in power plants. These problems include solvent degradation, efficiency of

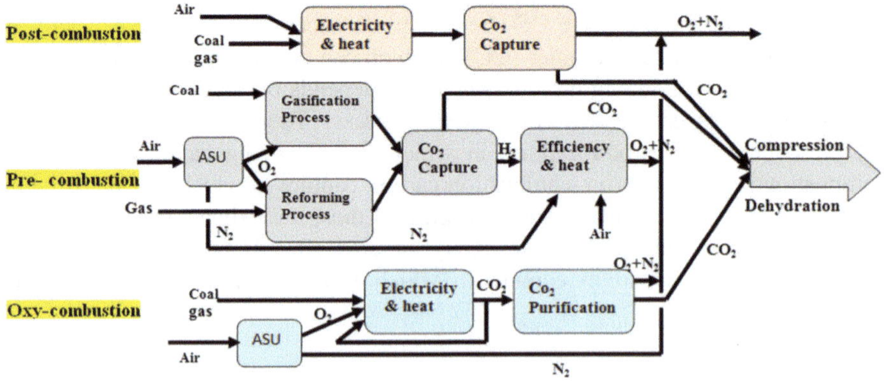

Fig. 1.7 A schematic diagram showing the three technologies of carbon capture [67]

the solvent regeneration, and corrosion of the vessel [69]. Among the potential replacement of amine solvents, ionic liquids are promising, as they are characterized by high CO_2 absorption potential, but they do not evaporate and degrade at temperatures higher than 250 °C [71]. Other technologies that have potential for pre-combustion capture such as the use of separation membranes, adsorption, and absorption processes that use ionic liquids are still at the phase of laboratory testing.

Adsorption process is used to separate CO_2 from the exhaust flow of methane reforming and gasification after the water-gas shift reaction. Both pressure swing adsorption (PSA) and temperature swing adsorption (TSA) can be used to desorb CO_2 and regenerate the adsorbent [72]. In PSA, the gas mixture flows through a packed bed of adsorbents at an elevated pressure, until the concentration of the required gas reaches equilibrium. The bed is then regenerated by lowering the pressure. In TSA, the adsorbent is regenerated through increasing the temperature. Both processes are commercially available for CO_2 separation [73]. In the last two decades, metal–organic frameworks (MOFs) have attracted great attention in the field of gas storage and separation due to their high selectivity and capacity [74]. Among porous materials, MOFs are characterized by having the highest surface area [75], hydrogen uptake [76], and methane and CO_2 storage [77]. There are many different types of gas separation membranes that can be utilized in the separation process of CO_2 before the combustion process, including porous inorganic membranes, polymeric membranes, palladium membranes, zeolites microporous silica, and ceramic membranes [78]. However, to make this technology cost-effective for application in IGCC, the membranes should have high selectivity, high permeability, and high physical/chemical resistance [79]. The process that is based on hydrate gas separation is a novel promising technology for CO_2 separation before the combustion process [80, 81]. In this technology, the basic principle of separation is the selective partition of the carbon dioxide component from the fuel or the mixture of flue gases between the solid hydrate crystal and the gaseous phases upon the formation of hydrate crystal. Intensive research is currently ongoing in the field of pre-combustion carbon capture; however, rigorous technical devices are needed for gasification and water-gas shift reaction [69]. A summary of the characteristics of different techniques for CO_2 separation before the combustion process is presented in Table 1.1.

1.4.1.2 Post-combustion Carbon Capture Technology

In post-combustion technology, the fuel is burned with normal air and the carbon dioxide is separated from the flue gases after the combustion process. Therefore, the main feature of this technology is that it does not require modifications in the combustion system. Accordingly, post-combustion technology is retrofittable to the majority of the existing plants that utilize liquid, gaseous, or coal-fired fuels in their combustion chambers. The concentration of CO_2 in the exhaust flue gases of these plants is normally low and it depends on the type of the plant and air-to-fuel ratio [82]. It is very low in simple gas turbines and rises significantly in other

Table 1.1 Characteristics of different pre-combustion carbon capture technology [68]

Pre-combustion technology	Characteristics
Adsorption	• It is a surface phenomenon • Requires high surface area to volume ratio • Performs at a much higher feed gas temperature than is possible for other gas separation technologies • Requires low energy to regenerate sorbent material
Physical absorption	• Absorption occurs at low temperature and high partial pressure of CO_2 • Solvents are regenerated by either heating, pressure reduction, or a combination of both • Has high CO_2 capture capacity • Selexol: low toxicity, less corrosive, and low vapor pressure solvent • Rectisol: more stable and less corrosive absorbent • Purisol: low energy consumption
Membranes	• Widely used for removal of CO_2 from natural gas and H_2 recovery • Polymer-based high-temperature membrane: PBI_Combining membranes with chemical solvents: ILs in supported liquid membranes
Gas hydrate crystallization	• Non-stoichiometric crystalline compounds consisting of a lattice of water molecules that physically encage CO_2 • Additives are used like TBAF and TBAB to reduce hydrate formation pressures to feasible industrial conditions • Necessity of reliable phase equilibrium data for the relevant CO_2 hydrate systems
Cryogenic	• Cryogenic processes are widely used to separate gases into high-purity streams • Results in significant energy penalty • Water must be removed before the cooling process

applications such as boilers. Therefore, separation of CO_2 from other gases is very important for reducing the cost of compression before sequestration. The most important methods of CO_2 separation from flue gases include absorption, adsorption, cryogenic processes and membranes [83]. These methods differ from each other based on their simplicity, economics and technology readiness level. Gas–liquid absorption is one of the mature CO_2 capture technologies economically and technologically in the chemical sector such as ammonia and fertilizer production [84]. Different chemical solvents were tested in the industrial field including methyl-diethanol- amine (MDEA), diethanol-amine (DEA), and mono-ethanol-amine (MEA) [85]. Recently, new classes of porous materials having high surface area, known as metal–organic framework (MOFs), are developed. MOFs are better absorbent compared to liquid solvents in terms of energy consumption and ease of regeneration [86].

Hermosillalara et al. [87] studied the thermal effects of charging a tank packed with activated carbon to store hydrogen at 10 MPa. The results showed that about 22% of the observed heating effects are due to adsorption. Richard et al. [88] used

Dubinin–Astakhov (D-A) model to get adsorption isotherm for hydrogen, nitrogen, and methane over activated carbon at higher pressure and superficial temperatures. They concluded that the D-A model best fits the experimental data. Xiao et al. [89] investigated heat and mass transfer phenomena during adsorption of hydrogen in a stainless steel tank with activated carbon packing. The results showed that the tank central region temperature was high as compared to those values in the areas near the walls and at the entrance. The absolute adsorption of hydrogen was found to have maximum values at the entrance of the tank. Ye et al. [90] simulated the charging and discharging processes of hydrogen in a stainless steel storage tank with activated carbon packing at 302 K and 10 MPa. They reported that the amount of adsorbed hydrogen was greater than that of compressed gaseous hydrogen. Xiao et al. [91] simulated charging–discharging cycles of hydrogen in two samples of MOF-5 (compacted tablet and powder) using Comsol Multiphysics. They compared the results with the case of activated carbon. They showed that the maximum pressure was the highest in the tank containing the MOF-5 powder, followed by that tank of activated carbon, then that of the tank packed with MOF-5 compacted tablet. In conclusion, high energy is required in case of post-combustion CO_2 capture. This can be attributed to regeneration and loss of the solvent during the absorption. This motivates the need to develop improved solvents to reduce the cost.

1.4.1.3 Oxy-fuel Combustion Carbon Capture Technology

Oxy-fuel combustion technology depends on the concept of using pure oxygen, instead of air, to burn the fuel. The flue gases of the combustion process consist primarily from water vapor and carbon dioxide. The high concentration of CO_2 in the exhaust gases facilitates its capture after the condensation of water vapor. Due to the absence of nitrogen, emissions of nitrogen oxides are fully eliminated unless there is a leak from outside air or impurities in the fuel. As well, the absence of nitrogen results in high CO_2 contents in the exhaust gases. It also leads to smaller volume of constituents in the combustion chamber and results in different combustion characteristics when compared to combustion using normal air. The oxygen needed for combustion is currently is produced from air through cryogenic processes; however, such conventional methods for oxygen production are costly. New promising technologies that use membranes to separate oxygen from air are under investigation and attracted the attention of many researchers. Membranes can be made from polymers and/or ceramics. Polymer membranes provide high oxygen flux across the membrane, however, at high concentration of nitrogen in the permeate side. On the contrary, ceramic membranes produce highly pure oxygen but with very low flux. Membrane development is receiving significant attention by many investigators in the recent decades. The objective is to provide high flux of oxygen and more stability in the case of ceramic membranes and to achieve high purity of oxygen in the case of polymer membranes. Nowadays, most of the oxy-combustion research work is forced toward the use of mixed ionic and electronic conducting ceramic membranes. Also, major scientific and industrial efforts are recently focused on developing ceramic membranes for syngas production via partial oxidation of hydrocarbons. In particular, the process of syngas production through partial oxidation of methane is heavily investigated [92]. Among of membrane materials, lanthanum cobaltite perovskite type ceramics are extensively studied [93, 94]. New materials of ceramic-based mixed-conducting membranes have been developed for membrane-based reactor applications including $SrFeCo_{0.5}O_x$ modified perovskite ceramics [95], $Sr_{1.4}La_{0.6}GaFeO_{3-\delta}$ brownmillerite-structured ceramic [96], $Sr_{0.2}La_{0.8}Fe_{0.69}Co_{0.1}Cr_{0.2}Mg_{0.01}O_3$ + 50Ag/50Pd ceramic–metal-based dual-phase membranes [97] and chemically stable yttria-stabilized zirconia (YSZ)-Pd thin dual-phase membranes [98]. In general, the capabilities of such oxygen semi-permeable membranes to produce high oxygen fluxes increase when it is exposed to air from feed side and a hydrocarbon fuel (such as methane) in the permeate side.

1.4.1.4 Comparison Between Different CCS Technologies

Figure 1.8 shows various CO_2 separation techniques used in different carbon capture technologies. Excluding cryogenic separation method, all separation techniques require some materials as carriers [68]. Table 1.2 summarizes the merits and demerits of the existing technologies for CO_2 capture or separation. Pre-combustion (high pressure, mainly CO_2/H_2 separation), post-combustion (low-pressure, mostly CO_2/N_2), and oxy-fuel combustion (predominantly CO_2/H_2O separation) technologies as well as processes and new materials are presented and analyzed in this

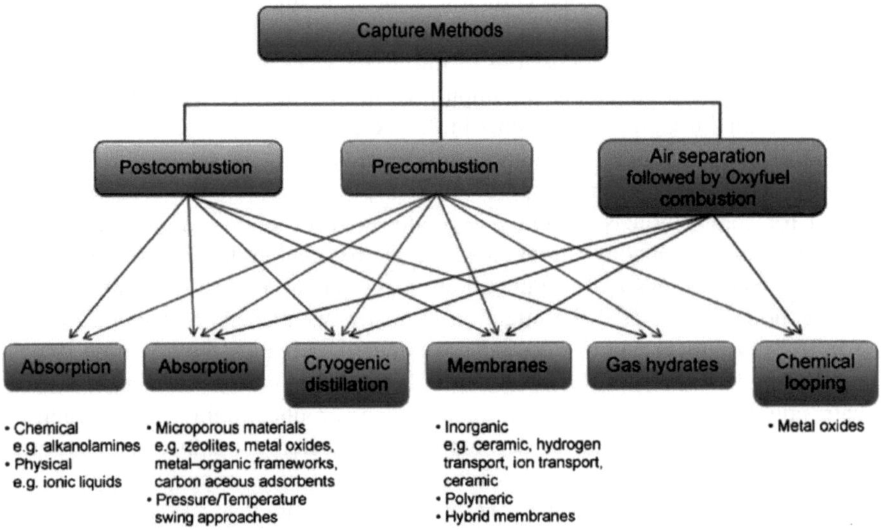

Fig. 1.8 Different CO_2 separation techniques used in different carbon capture technologies [68]

Table 1.2 Comparison between different existing technologies for CO_2 capture or separation [68]

Technology	Examples	Advantages	Disadvantages
Physical absorption	– Selexol process – Rectisol process – Purisol process	– Toxicity is low – Low corrosion – Low energy consumption	– Low capacity – High operational and capital costs
Chemical absorption	– MEA, DEA, MDEA – Sterically hindered amine (AMP)	– The technology is well received and is widely used in various industries – Suitable for retrofit – Suitable for CO_2 separation at low concentrations – Product purity > 99 vol% – Recovery rates of up to 95%	– Significant energy requirement because of solvent regeneration – Degradation and equipment corrosion – Solvent loss – Solvent emissions have negative impacts on the environment – Large absorber volume
	– Ionic liquid	– Low vapor pressure – Not toxic – Thermal stability	– High viscosity – Energy required for regeneration is high – Unit cost is high
Physical adsorption	– Activated carbon – Zeolite – Mesoporous silica – Metal–organic frameworks (MOFs)	– Regeneration and CO_2 recovery has less energy consumption – CO_2 and H_2S capture can be combined – High pore size and tunable pore structure (Mesoporous silica and MOFs)	– Solid handling is difficult – Adsorption kinetics are low – CO_2 selectivity is low – Cycling is questionable to be thermally, chemically, and mechanically unstable
Chemical adsorption	– Amine-based adsorbent – Alkaline earth metal adsorbent	– Adsorption capacity is high – Low cost in natural minerals – Exothermal reaction	– Loss of sorption capacity over multiple cycles – CO_2 selectivity is low – Diffusion resistance issue
Membrane technology		– No regeneration processes – Simple modular system – Has no waste streams	– Plug of membranes by impurities in the gas stream – Not proven industrially

(continued)

Table 1.2 (continued)

Technology	Examples	Advantages	Disadvantages
Oxy-fuel		– Relatively simple technology – Suitable for retrofit – NO_x are not significant	– Significant energy requirement for separation of O_2 from air
CLC		– The technology is well known – No thermal formation of NO_x – Suitable for retrofitting – Cheap and abundant sorbent (limestone) – Exhaust gas streams are not harmful – Low energy penalty and operational costs	– Large-scale demonstration are not available – Decay in sorbent's capture capacity

section. The suitability of each of the carbon capture techniques depends on the system that they are used for. Accordingly, the carbon capture method should be compatible with the combustion system for each application. For example, the pre-combustion capture method is shown to be compatible with IGCC systems [99]. Physical absorption is used to separate CO_2 in high-pressure applications. Post-combustion and oxy-fuel combustion capture techniques can be adapted to pulverized coal (PC) combustion. On the other hand, any of the three capture technologies can be implemented for the systems of natural gas combined cycles (NGCC). However, the pre-combustion carbon capture option is expensive in comparison with the other methods. In such a method, methane reforming is performed, and then carbon dioxide in the synthetic gas is captured after conversion of CO to CO_2. The reported plant efficiencies in the literature (defined as a percentage of lower heating value of the fuel) are as high as 50% with CO_2 post-combustion capture for NGCC in comparison with 60% [100] for cases without capture as shown in Fig. 1.9. The oxy-combustion in PC presents the next highest potential, with 35% efficiency in comparison with 45% for the case without capture. The IGCC Puertollano with pre-combustion capture provides efficiency of the order of 33.5% compared with 44% with no capture. An efficiency reduction of 15% (efficiency of 30%) is reported for post-combustion capture in PC with respect to that of PC without capture (if MEA is used) [100]. Skipping the economic aspects while considering the energy aspects, one would prefer applying pre-combustion capture in IGCC, oxy-combustion for PC, and post-combustion with NGCC. Table 1.3 shows the average efficiency of the plant (in reference to fuel calorific

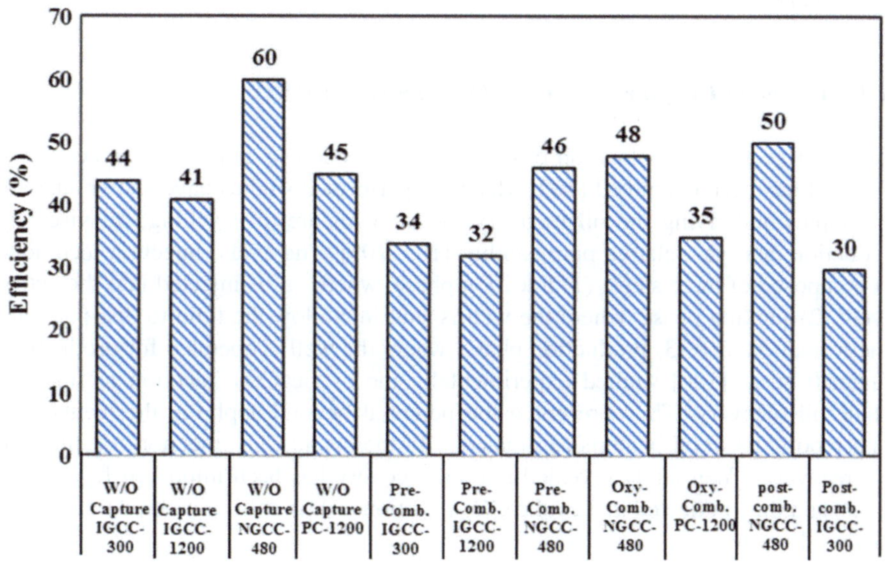

Fig. 1.9 Comparison between the efficiency of different power systems with and without including one of the carbon capture technologies [100]

Table 1.3 Plant efficiency and fuel use and fuel penalty for the three CCS technologies for different plant types [101]

CO_2 capture technology	Type of power plant	Plant efficiency	Fuel use [MJ/kWh]	Fuel penalty [%]
No CO_2 capture	IGCC	42	8.6	0
	NGCC	57	6.4	0
	PC	40	9.1	0
Post-combustion	NGCC	49	7.4	18
	PC	31	11.8	37
Pre-combustion	IGCC	35	10.4	24
Oxy-fuel	NGCC	46	7.9	25
	PC	33	10.9	30

value), fuel use, and penalty per carbon capture technology. The data in Table 1.3 tell that the CCS power plants with the highest average efficiencies have the lowest fuel uses and fuel penalties.

1.4.2 Carbon Storage Techniques

Several techniques have been adopted for storing CO_2 after its capture. These techniques include enhanced oil recovery (EOR), depleted oil/gas fields, and deep saline aquifers.

1.4.2.1 Using CO_2 for Enhanced Oil Recovery (EOR)

In recent years, the application of cyclic CO_2 injection to enhance the recovery of light oil has been examined [102]. Recently performed studies have shown that the principles underlying the oil recovery mechanisms are oil swelling, oil viscosity reduction, and gas relative permeability [103–105]. The cyclic injection technique is composed of three steps: (1) injection phase, where gas is injected into the active well; (2) soaking phase, where the well is shut into allow the fluid to dissipate into the formation; and (3) production phase, where the well is operated for production. Gamadi et al. [106] studied experimentally the cyclic CO_2 injection to improve shale oil recovery. They presented the potential beyond applying the technology and reported several benefits associated with the cyclic CO_2 injection for effective oil recovery. Their study revealed that oil recovery has been improved from 33 to 85%, and they indicated that cyclic CO_2 injection is a promising method to improve shale oil recovery.

1.4.2.2 CO_2 Injection in Depleted Oil/Gas Fields and Deep Saline Aquifers

Capture and geological storage of CO_2 provide a way to avoid emitting CO_2 into the atmosphere, by capturing CO_2 from major stationary sources, transporting it usually by pipeline and injecting it into suitable deep rock formations. The subsurface is the Earth's largest carbon reservoir, where the vast majority of the world's carbon is held in coals, oil, gas organic-rich shales, and carbonate rocks. Geological storage of CO_2 has been a natural process in the Earth's upper crust for hundreds of millions of years. Carbon dioxide derived from biological activity, igneous activity, and chemical reactions between rocks and fluids accumulates in the natural subsurface environment as carbonate minerals, in solution or in a gaseous or supercritical form, either as a gas mixture or as pure CO_2.

Geological storage of anthropogenic CO_2 as a greenhouse gas mitigation option was first proposed in the 1970s, but little research was done until the early 1990s, when the idea gained credibility through the work of individuals and research groups [107–110]. Recently, geological storage of CO_2 has grown from a concept of limited interest to one that is quite widely regarded as a potentially important mitigation option; see Fig. 1.10. There are several reasons for this. First, the level of confidence in the technology has increased as research has progressed and as demonstration and commercial projects have been successfully undertaken. Second,

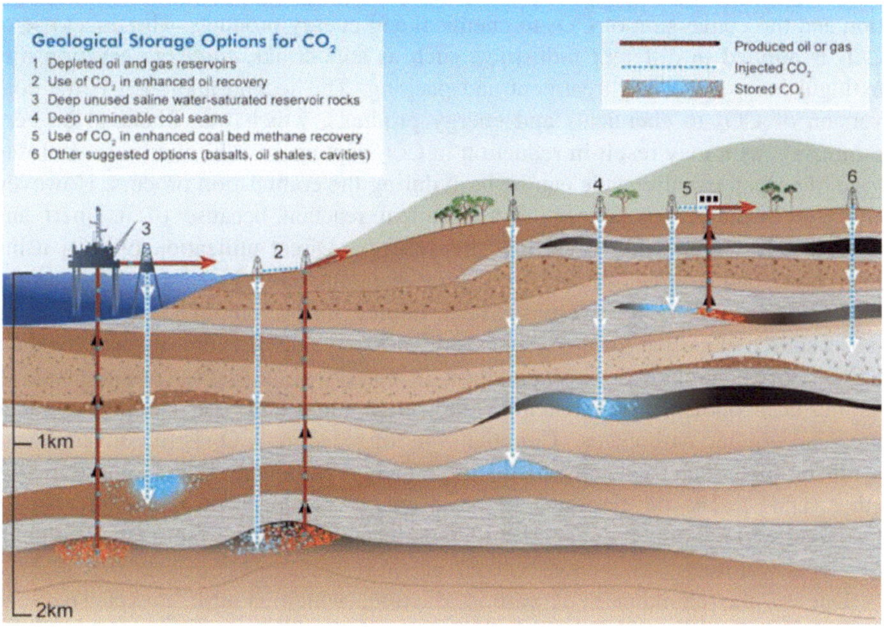

Fig. 1.10 Options for storing CO_2 in deep underground geological formations [111]

there is a consensus that a broad portfolio of mitigation options is needed. Third, geological storage (in conjunction with CO_2 capture) could help to make significant cuts in atmospheric CO_2 concentration.

In the Intergovernmental Panel on Climate Change (IPCC) report [111], it was suggested that underground saline aquifers have a storage capacity of around 2×10^3 gigaton (Gt) of carbon dioxide which is about two orders of magnitude higher than the total annual worldwide emissions, making it a potential disposal option. In order to make disposal in underground aquifers a viable option to mitigate climate change, we should be able to sequester large quantities of CO_2 with scales of 10–30 Mt/year per injection site. Currently, the typical injection rates used in research studies and in field projects are around one megaton (Mt) per year. Larger, by order of magnitude, volumes of CO_2 injected within a short period of time (50–100 years) increase the reservoir pressure extremely fast which may lead to loss in reservoir integrity [112]. Therefore, the projected capacity of a reservoir should be evaluated not by available pore space but by injection capacity, defined by how much carbon dioxide can be injected within a given period and within particular injection area.

1.4.3 Carbon Utilization Techniques

Two approaches have been adopted for CO_2 utilization, namely the direct utilization and the conversion of CO_2 to chemical and energy products. The direct use of CO_2 is applied in different industries, such as soft drinks, food preservation, fire extinguishers, and water treatment and packing. The second approach is the conversion of CO_2 to chemicals and energy products, which was found to be very promising, as it may result in reduction in CO_2 capture cost. Furthermore, a closed loop of carbon capture cycle can be built during the combustion process. However, CO_2 has some disadvantages as a chemical reactant because of its inert and non-reactive nature with low Gibbs free energy. Direct utilization of CO_2 using microalgae can be very promising knowing that cultivating 1 ton of microalgae can fix 1.8 tons of CO_2 from the environment. Indirect utilization of CO_2 includes using it as a source of carbon for the synthesis of various valuable chemicals and fuels via CO_2 hydrogenation, CO_2 cycloaddition to epoxides, and CO_2 carbonylation of amines or alcohols. Existing chemical industries utilize CO_2 conversion to produce urea and organic carbonates. The other way of recycling CO_2 is to convert it into synthetic fuels such as syngas, methanol, dimethyl carbonate (DMC), and dimethyl ether (DME). The development of these energy products via renewable energy resources will not only reduce the burden on fossil fuels but also mitigate the threats of global warming.

Due to the abundance of carbon dioxide, which is also inexpensive and non-toxic, it is considered an attractive raw material for incorporation into important industrial processes. The increase in fossil fuel cost, coupled with the need for cheap plastics, is forcing the industry to reduce the production cost of plastics by using CO_2 as a cheap bio-renewable resource that can potentially solve several problems related to plastics production. The catalytic bonding of CO_2 and epoxies to create carbonates or polycarbonate has proven to be very promising techniques in the implementation of CO_2 as a major component in a wide variety of plastic products. Carbon dioxide is a significant C_1 source of carbon both as gas and in-bound carbonate and has been considered as a possible carbon source for synthesizing essential chemical precursors. Therefore, its utilization as a renewable chemical feedstock has become one of the great concerns and prime challenges for scientists in the twenty-first century. The development of an efficient catalyst-based technology for the use of CO_2 would provide a highly desirable, renewable, and sustainable C_1 carbon resource that would revolutionize the energy sector while addressing climate change concerns. Among these energy products are the conversion of CO_2 to methanol. To produce methanol, CO_2 could be combined with hydrogen (separated or electrolyzed from water), compressed, then reacted to produce methanol and water. Estimations of the possible yield range from 30 MMt (million metric tons) of methanol produced per year to over 300 MMt of CO_2 per year, with the amount of CO_2 per tones of methanol ranging from 3.1 to 14 tones. The potential market of CO_2 conversion to methanol will be significantly increasing if methanol consumption increases and if methanol is able to replace methane for energy production.

Nowadays, membrane technology has a rising prominent role in various process engineering applications in the chemical and petrochemical sectors, in addition to applications in water treatment, gas separation, syngas production, biotechnology, methane steam reforming, energy production, etc. Mixed ionic–electronic conducting (MIEC) perovskite-based ceramic membranes have an increasingly eminent role in membrane engineering for oxygen separation from air on large scale for different applications. The role of these membranes can be extended for the separation of oxygen from feeding captured CO_2. The membrane allows oxygen to migrate from a CO_2 stream through the membrane to the other side of the membrane, thus producing carbon monoxide. The produced CO during this process can be utilized as a fuel by itself or can be mixed with H_2 and/or H_2O to provide other different hydrocarbon fuels. Other chemicals including methanol (an important fuel for motorized vehicles), syngas, fertilizers, and others can be produced. This process could become an important part of the carbon capture/utilization and sequestration technologies (CCUS). If this approach is applied in the electric power generation sector, it could reduce the impact of using fossil fuel on global warming. The perovskite membrane, 100% selective for oxygen, allows only CO_2 atoms to pass. The separation process is driven by hot temperatures of up to 900 °C. The key for making the process working is to maintain high flux of the separated oxygen from CO_2 across the membrane. This could be made through creating a vacuum on the sweep side of the membrane. This would require a lot of energy to maintain the process. As well, a stream of fuel, such as H_2 or CH_4, can be used in the sweep side in place of vacuum. These substances are so readily oxidized that they will actually attract the oxygen atoms across the membrane without requiring a significant pressure difference. The membrane also stops the oxygen from drifting back to avoid recombining with the CO and forming CO_2 in a reversible process. In order to reduce the energy required for driving the process and producing a useful product, a combination of vacuum and fuel can be used. The combination depends on the concerned specific application. The energy input required to keep the process working is heat, which might be provided by solar energy or by waste heat. The waste heat could be obtained from the power plant itself or from other sources. Basically, the process makes it possible to store that heat in the form of chemical energy to be used whenever needed. Chemical energy storage is characterized by high energy density (energy per unit weight) as compared to other energy storage techniques. In a natural gas power plant, the inward natural gas could be divided into two streams. One stream would be burned in a conventional process to produce electricity while capturing CO_2. The second stream would be driven to the fuel side of the membrane reactor to react with the separated oxygen. This stream would produce a second output from the plant, which is a Syngas mixture of H_2 and CO. Syngas is extensively used as an industrial fuel and a feedstock. The syngas can also be mixed with the natural gas distribution network. This technique may, thus, result in another potential revenue stream to help defray its costs in addition to cut greenhouse emissions. Using ceramic membranes for oxygen separation can significantly reduce the combustion temperature, from 1500 °C to below 1000 °C, resulting in great energy saving compared to the conventional CO_2 decomposition method. It is believed that this work is important to the fields of membrane processes and sustainable energy and environment.

1.5 Bio-energy with CCS (BECCS) for Negative CO_2 Emissions

Interest in bio-energy with CCS has developed quickly as it has the potential to offer deep reductions in atmospheric carbon dioxide concentrations. BECCS also appears to be feasible and economical. The IPCC Special Report on Renewable Energy Sources and Climate Change Mitigation addresses the technical and economic feasibility BECCS in some detail [113]. It states that successful deployment of CCS in combination with biomass conversion could result in the elimination of greenhouse gases from the atmosphere at attractive mitigation cost levels. BECCS offers the potential to achieve long-term reductions in GHG emissions necessary to stabilize atmospheric CO_2 concentrations and can be applied to a widespread range of biomass-related technologies [114].

1.5.1 Concept of BECCS

Bioenergy with carbon capture and storage (BECCS) is a carbon reduction technology offering permanent net removal of CO_2 from the atmosphere. This has been expressed as "negative carbon emissions." It offers a significant advantage over other mitigation alternatives, which only decrease the amount of emissions to the atmosphere. BECCS can achieve this because it uses biomass that has removed atmospheric carbon while it was growing, and then stores the carbon emissions resulting from combustion permanently underground. It has been proposed that BECCS can be applied to a wide range of biomass-related technologies and may also be attractive from a relative cost perspective. However, up to date, the combination of bioenergy and carbon capture and storage (CCS) has not been fully recognized or realized. Incentive policies to support BECCS need to be based on an assessment of the net impact on emissions that the technology can achieve.

1.5.2 Status of BECCS

There is a potential drawback that could undermine the positive opportunities provided by BECCS. The biomass used during conversion into energy may or may not come from sustainable sources. The use of unsustainable biomass in BECCS could negate any carbon benefits and may even cause net positive CO_2 emissions rather than reductions. The IPCC has noted that both direct land use changes that include conversion and forest management and leads to a loss of carbon stocks and indirect land use changes can diminish and possibly more than counteract any of the net positive GHG mitigation impacts deriving from BECCS. Activity boundaries are, therefore, critical when assessing the pros and cons of BECCS projects. Inclusion of the impacts on land use and land use changes may change the amount of total avoided emissions significantly and are likely to depend heavily on the specific cases and circumstances. To assume that BECCS is beneficial in all cases would be simplistic. The potential to reduce atmospheric CO_2 levels offered by BECCS is unlikely to be realized unless there is an incentive to deploy it. An appropriate incentive policy for BECCS needs to be based on an assessment of the emissions reduction that the technology can deliver.

1.6 Approaches for Oxy-fuel Combustion Technology

Oxy-combustion technology is among the foremost technologies that are being considered recently for CO_2 capture from power plants. Oxy-fuel combustion is the process of combusting/burning fuels with pure O_2 rather than air. It is characterized by high flame temperatures which necessitate the recycling of some part of the flue gases back into the combustor to control the flame temperature. There has been significant progress in the development of oxy-fuel combustion technology after the International Panel for Climate Change [111] publication of its special report on CO_2 capture and storage [115]. Most of the existing power plants utilize air as the oxidizer for combustion; meanwhile, oxy-fuel combustion concept requires the use of pure oxygen. This necessitates the oxygen separation from air using various techniques to fit into the conventional as well as the future power plants. The three main air separation processes are cryogenic, membrane, and solid sorbent processes [116, 117]. The most established technology among the three is the cryogenic distillation which allows for large-scale production of O_2 at very high purities (up to 99%) at low temperature [117, 118]. Despite the production of high purity of O_2 by this technique, the design complexity, high energy demand as well as low exergetic efficiencies (15–24%) necessitate the investigation of alternative air separation concepts [119, 120]. The solid sorbent (adsorption) process, despite having O_2 purity of up to 95%, has not been fully implemented due to excessive capital requirements of the solvent [118, 121]. The most recent separation technology is the membrane separation which includes the use of polymeric membranes (PM) and high-temperature ion transport ceramic membrane (ITM). The polymer-based membrane can provide O_2-enriched air at higher concentrations (60–80%) with low energy consumption in comparison with the cryogenic and adsorption techniques [117, 122]. Whereas the PM can enrich the oxygen content in air up to 80% and ion transport membranes can achieve near 100% oxygen purity, even though both types of membranes are still under development. The use of ITMs for O_2 separation has been given significant attention much research due to its potential of reduction in the cost of energy requirement to produce O_2 by about 35%. This can be ascertained by assessing the file of patents and research publications on the matter over the last decade. There are some proposals to combine the above membrane technologies to produce high purity O_2; however, the ideas have not been yet fully exploited.

Based on that, there are two existing approaches for the application of oxy-combustion technology. The first approach is applied using air separation unit (ASU) to separate oxygen from air, and then, the separated oxygen is used in oxidizing the fuel in conventional combustion systems. The second approach is applied using oxygen transport reactors (OTRs). Within OTRs, ceramic membranes, selective to oxygen, are used to separate oxygen from the feeding air from one side of the membrane to be burned with a sweeping fuel stream on the other side of the membrane. Below, we give briefs about the two approaches to be discussed in detail in the next chapters.

1.6.1 Conventional Combustion Systems

In conventional combustion systems, the required oxygen is supplied by an air separation unit where the nitrogen is separated from air. A great portion of the flue gases must be recycled to substitute the removed nitrogen. A key component of the oxy-fuel process with high-temperature membrane air separation unit (HTM-ASU), which is in the stage of development, is a dense membrane made of ceramic materials. The separated oxygen is then used in the combustion process in conventional combustion systems such as gas turbines and boilers. The process takes place in a nitrogen-free (or low-nitrogen) environment resulting in a flue gas composed mainly of CO_2 and H_2O, as well as a low concentration of impurities such as argon and oxygen. Therefore, a simplified flue gas processing by means of condensation of H_2O to capture CO_2, without using costly separation methods such as chemical absorption, becomes possible.

The oxy-fuel combustion technology is an effective approach to eliminate NO_x emissions out of conventional combustion systems [123, 124]. In this technology, N_2 is removed from air using an air separation unit. The remaining gas (mainly O_2) is used as oxidizer. However, the combustion of fuel using pure oxygen results in excessively high flame temperature [3]. For the sake of preventing this, some of the CO_2 in the exhaust stream is captured and recirculated to be mixed with the incoming oxygen [125]; a technology that is termed exhaust gas recirculation (EGR). This modern technology makes the exhaust gases mainly consist of CO_2 and H_2O and, hence, facilitates the capture and sequestration of CO_2 to eliminate its release into the atmosphere. This technology also reduces the mass and volume of the exhaust gases significantly with further benefits of reducing the amount of heat losses and reducing the size of the treatment equipment of the flue gases [126]. A detailed description of this approach for oxy-fuel combustion application along with design and performance analyses is presented in the next chapters.

1.6.2 Oxygen Transport Reactors (OTRs)

The other approach for the application of oxy-fuel combustion technology is through using OTRs within which ceramic membranes are used for separation of oxygen from feeding air stream to be used for burning fuel stream on the other side of the membrane within the same unit. These membranes are capable of oxygen–air separation at 700–900 °C and can be formed in tubular or planar shapes, thus enabling the formation of compact OTRs. Nowadays, most of the research works are focused on the application of the oxy-fuel combustion in the permeate side of the membrane. Those studies aimed at the understanding the oxygen permeation and the oxy-fuel combustion characteristics inside OTRs. Nemitallah et al. [127] performed numerical investigations on a LSCF OTR trying to characterize the membrane performance under oxy-fuel combustion conditions using a modified two-step oxy-combustion reaction kinetic model for CH_4. They reported sharp increase in the oxygen permeation flux when the reactions were activated in the permeate side of the membrane. Comparable results were reported for the same membrane material by Ben-Mansour et al. [128, 129] using BSCF membrane material. An experimental study coupled with mathematical formulations has been conducted by Akin and Jerry [130] to examine the influences of sweep gas flux and reactivity in the permeate side on the oxygen flux. They showed a reduction in the oxygen permeation flux when the sweep gas flux was increased due to the lowered reaction rates of the fuel at these conditions.

Recently, very few number of research works have been conducted to apply the OTR technology into large-scale power plants aiming at a real application of the zero-emission power plant (ZEPP) concept. The OTR technology is supposed to replace the present cryogenics and decrease the cost of oxygen production by about 35% [131, 132]. This cost reduction can result in a 50% reduction in the energy required for CO_2 capture, when the OTR is integrated into power plants. For ZEPP applications, high oxygen mass flow rates are required which forced the research in this direction to maximize the ratio between the membrane surface area and the total volume of the reactor. Nemitallah et al. [5] presented a design for an ion transport membrane reactor for the substitution of a conventional gas turbine combustor. The OTR in their work is monolith structure design using LSCF-1991 membrane type. Optimizations for mass flow rates, channel geometry, and percentage of the fuel in the sweep gas have been performed. They came up with an OTR with a height of 3.35 m, membrane surface area of overall 2700 m^2, total reactor volume of 10 $m^{3,}$ and output power of 5–8 MWe. Another design of an OTR was introduced by Mancini and Mitsos [133] to produce the required oxygen for combustion inside the reactor for power plant applications. The OTR in their study can deliver a power of 300:500 MWe based on the cycle first law efficiency. The subject of using membranes for OTR applications is discussed in detail in the next chapters.

1.7 Why Oxy-combustion

Most of today's combustion applications and operational gas turbines utilize air for combustion process. The use of air as oxidizer generates an exhaust flue gas stream that contains mainly CO_2, H_2O, O_2, and N_2; however, CO_2 is a greenhouse gas that is primarily responsible for the escalating global warming problem. Currently, the world is largely dependent on fossil fuels for energy production and it is expected that this will continue for many decades to come [134], but CO_2 is inevitably created as a product of burning fossil fuels. The Paris Agreement [135] has declared that limiting future temperature rise to 2 °C will be very difficult to achieve without the implementation of carbon capture and sequestration. Separating CO_2 from the other products of air combustion, N_2 in particular, is a costly and difficult process [136]. However, burning the fuel in pure oxygen instead of air, i.e., oxy-fuel combustion, the flue gas products will contain mainly of CO_2 and H_2O. Carbon capture utilizing simple processes to condense the water vapor can, thus, be implemented at the lowest cost. In the absence of air-based nitrogen, the volume of exhaust stream can be also significantly smaller, thus reducing the size and increasing the efficiency of the treatment equipment. One of the other benefits of employing oxy-combustion is the inherent drastic reduction of emitted nitrogen oxides, which result from the reaction of air-based oxygen and nitrogen. The ability to retrofit existing air-based gas turbines and other combustion applications to implement oxy-combustion without much hardware changes can offer one of the essential motivations for most clients to switch to oxy-combustion.

1.8 Oxy-combustion in Gas Turbines

Another method of specific interest is the use of oxy-fuel combustion in gas turbine applications. For oxy-fuel gas turbine cycles, researchers to date have mainly focused on thermodynamic studies of system performance. However, the combustion behavior, e.g., the reaction zone structures and flame dynamics in the gas turbine combustors, is less addressed. Swirl-stabilized flames are used extensively in real-world combustion systems because they enable high energy conversion in a small volume and exhibit good ignition and stabilization conditions over a wide operating range. In order to reduce the flame temperature and thereby the remaining oxygen in the flue gas, it is beneficial to premix the oxygen and CO_2 or steam before introducing them to the combustor. Also, to generate a stable combustion in a gas turbine combustor through oxy-fuel combustion, certain minimal oxygen level in the oxidizer has to be maintained. This is mainly due to the need to have the required high-temperature environment inside the reaction zones for the chain reactions to proceed. Heil et al. [137] reported that poor burnout and lifted dark flames were shown when the oxygen mole fraction in the O_2/CO_2 stream was set to 21%. When the oxygen volume fraction was increased to 27 and 34%, full fuel burnout and stable flames were obtained. In order to burn the fuel with lower oxygen level in the oxidizer (O_2/CO_2) stream, the burner had to be modified to allow for recirculation of hot gases to the flame. With this high oxygen concentration in the oxidizer mixture, the combustion products, however, become hot. This may lead to high concentration of oxygen in the flue gas due to the dissociation reactions that occur at high temperatures. There is an optimal "window" of oxygen/diluent ratio in the oxidizer stream.

1.8.1 Required System Modifications

The gas turbine processes using oxy-fuel combustion technology have several combined-cycle concepts such as O_2/CO_2 [138–140], COOLENERG [141], COOPERATE [142], and MATIANT [143] cycles. These cycles are called semi-closed oxy-fuel combustion cycles (SCOF-CC) and are likely to require modified design of the turbo-machineries along with high-temperature turbines. In general, any conventional combustion system adapting oxy-combustion requires a set of additional units to be fixed along with the system. Those units include air separation unit (ASU) for oxygen separation from air, flue gas recirculation system to control combustion temperature, and carbon dioxide purification unit to remove impurities in the exhaust stream for CO_2 sequestration. The application of oxy-fuel combustion technology in gas turbines is very specific as the control of combustion temperature becomes a critical parameter to protect the turbine blades against high-temperature oxidation and corrosion. Also, the operation of the turbine should be kept around stoichiometric combustion to make the oxy-combustion process economically viable. The range of stable flame operation is expected to be changed while adapting oxy-combustion technology as compared to normal air combustion. The significant changes in the properties of the reacting mixture (highly concentrated CO_2) necessitate modifications in the inlet section design of the gas turbines in order to obtain stable flame operation over considerable ranges of operating conditions. The required system modifications for conventional systems to work under oxy-combustion conditions are discussed in detail in the next chapters.

1.8.2 Gas Turbine Performance Under Oxy-combustion

Stable combustion and low turbine inlet temperature can be obtained simultaneously by optimizing the ratio of oxygen to CO_2 in the oxidizer mixture fed to the combustion chamber. Liu et al. [144] reported that the primary oxidizer which is supplied in the upstream through the dome of the combustion chamber should have minimal oxygen level of 24% under the oxidizer temperature 520 K condition. Kutne et al. [145] have checked the stability of a swirl-stabilized oxy-fuel/CH_4 flames for O_2/CO_2 percentages of 20/80–40/60%, equivalence ratios of 0.5–1 and thermal powers of 10–30 kW. They reported that attempts of operating the burner with less than 22% O_2 were unsuccessful even at stoichiometric conditions. Syngas and methane flames for premixed swirl-stabilized conditions for two different oxidizers of air and $O_2/CO_2/N_2$ were studied by Williams et al. [146]. They reported lower nitrogen oxides concentrations (NO_x) for the quasi-oxy-fuel flames and higher carbon monoxide (CO) concentrations, suggesting stoichiometric operation at 20–24% O_2 as ideal for low emissions. Ditaranto and Hals [147] studied the influence of stoichiometric operation and high oxygen content in the oxidizer mixture on thermo-acoustic oscillations in sudden expansion jet configuration. They reported occurrence of thermo-acoustic instabilities as O_2 content in the oxidizer was increased, characterizing different instability modes dependent on flow velocity and flame speed variations. Anderson et al. [148] have performed experiments on a 100-kW test unit for air and two O_2/CO_2 test cases with different recycled feed gas mixture concentrations of O_2 (OF = 21 at 21 vol% O_2, 79 vol% CO_2 and OF = 27 at 27 vol% O_2, 73 vol% CO_2). They showed that the fuel burnout is delayed for the OF = 21 case compared to air-fired conditions because of reduced temperature levels. Instead, the OF = 27 case results in more similar combustion behavior as compared to the reference conditions in terms of in-flame temperature and gas concentration levels, but with significantly increased flame radiation intensity. Nemitallah and Habib [149] performed experimental and numerical investigations of an atmospheric diffusion oxy-combustion flame in a gas turbine model combustor. They investigated oxy-combustion and emission characteristics and stabilization behavior of oxy-methane non-premixed flames in a swirl-stabilized gas turbine combustor. The experimental and numerical results showed that the stability of the oxy-combustion flame is affected when the operating percentage of oxygen in the oxidizer mixture is reduced below 25%. In all cases, flame was extinct for conditions of less than 21% oxygen in the oxidizer mixture.

1.8.3 Combustion and Emission Characteristics

As a promising CCS technology, oxy-fuel combustion can be used for existing power plants and also for new ones. Oxy-fuel combustion gives different characteristics of heat transfer, ignition, burnout, as well as NO_x emission [150]. During an oxy-fuel combustion process, a mixture of oxygen and recycled flue gases is used for the combustion of the fuel. The exhaust gases consist of a mixture of CO_2 and H_2O. A large portion of the flue gases should be recycled in order to substitute the removed nitrogen to ensure that there is enough gas to carry the heat through the boiler and in order to control the flame temperature.

Oxy-fuel and air–fuel combustion technologies have different degrees of freedom that confine the operational space. Oxy-fuel flames are more likely to be operated close to stoichiometry, in order to effectively utilize both fuel and O_2 with controlled EGR (i.e., controlled ratio of O_2/CO_2) to maintain the combustion temperature within desired limits. Fuel-rich combustion increases fuel consumption and results in incomplete combustion as well as generation of excessive CO emissions while fuel-lean operation is associated with unutilized O_2 in the exhaust stream. The characteristics of oxy-fuel combustion also differ from those of air–fuel combustion. This may be attributed to significant differences in the physical properties of CO_2 and N_2 [151–153]. The replacement of N_2 by CO_2 in the oxidizer mixture affects the flame from different aspects, namely the temperature distribution and changes in oxidizer mixture density, volumetric heat capacity, and transport properties, including thermal conductivity, mass diffusivity, and dynamic viscosity. Table 1.4 compares selected properties of air to those of O_2/CO_2 mixtures (with the oxygen fraction values OF = 25% and OF = 50%, by volume) at 298 K.

Table 1.4 Select properties of air and different O_2/CO_2 mixtures at 298 K [154]

	Air	O_2/CO_2 mixture 25% OF	O_2/CO_2 mixture 50% OF	% change 25% → 50% OF
Density [kg/m^3]	1.17	1.65	1.53	−7%
Dynamic viscosity [$\times 10^{-6}$ Pa s]	18.6	16.2	17.5	8%
Kinematic viscosity [$\times 10^{-6}$ m^2/s]	15.9	9.82	11.4	16%
Vol. heat capacity [kJ/m^3/K]	1.18	1.44	1.35	−6%

The increased concentration of CO_2 in the oxidizer mixture also results in reduced rates of chemical kinetics, which, in turn, decrease the laminar burning velocity and combustion efficiency [155, 156]. Consequently, combustion in a CO_2-diluted atmosphere needs higher minimum oxygen threshold, i.e., more than 21% by volume, in order to obtain stable flame at the same level of operating equivalence ratio [157, 158]. Rashwan et al. [159–161] illustrated the effect of carbon dioxide addition. CO_2 has higher density than N_2, which affects gas density, jet velocity, flame shape, and pressure drop. The higher density of CO_2 also leads to higher heat capacity of the O_2/CO_2 mixtures as compared to air, which directly reduces the flame temperature level and results in lower flame speeds and reduced flame stability. Flame speed is also affected by gas transport properties [162].

1.8.4 Flame Stability

Most gas turbine combustors used in jet engines and power plants utilize non-premixed flames because of their inherent stability under wide ranges of operating conditions. However, diffusion flames result in spots of high-temperature values [3, 163, 164]. Consequently, generation of high levels of nitric oxides, NO_x, *can result from* this temperature increase [165]. Public awareness and legislation have led to strict policies for the reduction of the pollutants. Therefore, alternatives such as lean premixed flames (*LPF*) have been proposed and their application is currently expanding. In this case, the fuel and oxidizer are mixed upstream in order to prevent the formation of stoichiometric zones and, hence, reduce the combustion temperature and, accordingly, reduce the NO_x emissions [166]. Unfortunately, lean premixed flames are subjected to combustion instabilities [167, 168]. Combustion instabilities are resonant phenomena that occur when a positive feedback is established between the acoustic environment and heat release. Resulting pressure fluctuations can reach critical values at which the engine operation can be affected leading to failure [169].

The geometry of combustion systems and the associated flame anchoring mechanism are some of the most significant parameters affecting the combustion stability. The combustor geometry controls the size and structure of the recirculation zone that is formed in order to stabilize a flame [170]. Li and Gutmark [171] studied the flame stability with and without center body recess in dump combustor utilizing bluff body for stabilization. The results showed that the flame is stabilized, and the oscillations are reduced when the center body is recessed. Speth and Ghoniem [172] studied the combustion instabilities of a syngas–air premixed flame in a swirl-stabilized combustor over wide ranges of operating parameters. Their results showed strong dependence of the combustion instabilities on the combustor geometry [173, 174], operating conditions, and fuel compositions. Altay et al. [175] studied the flame–vortex interaction driven combustion dynamics of a premixed flame in a backward-facing step combustor under different fuel compositions and operating conditions. They observed unstable flames at high equivalence ratio, quasi-stable flames at intermediate equivalence ratio, and long stable flame near the lean blowout limit. Hong et al. [176] studied the impact of fuel composition (C_3H_8/H_2) on the structure of the recirculation zone and its role in lean premixed flame anchoring in a backward-facing step combustor. Their results demonstrated a complex coupling between the size and the structure of the recirculation zone and the flame anchoring. Two counter-rotating eddies, a primary eddy (PE) and a secondary eddy (SE), were observed in the recirculation zone at relatively low equivalence ratio. Shrinkage of the SE size was observed while increasing the equivalence ratio until this zone completely disappeared. Adding hydrogen to the fuel resulted in higher temperatures and the motion of the flame tip toward the reactor step [176].

Details of the dynamics and phenomenology of near blow-off flames were explained by Shanbhogue et al. [177]. They showed that temporally localized extinction, like holes in the flame structure, occurs close to the blow-off conditions. The number of holes increases as the conditions of blow-off approach. Kedia and Ghoniem [178] investigated the mechanism of a laminar premixed flame anchoring close to a heat-conducting bluff body. They used a fully resolved unsteady two-dimensional simulations coupled with detailed chemical kinetics for methane–air combustion. Their results showed a shear layer-stabilized flame in the vicinity and downstream of the bluff body, where favorable ignition conditions are established, and a recirculation zone was formed by the combustion products. Altay et al. [179] investigated the effect of oscillations in the equivalence ratio on the dynamics of combustion of a lean premixed propane–air flame in a backward-facing step combustor. Equivalence ratio oscillations were performed by altering the location of the fuel injector. They reported that flame–vortex interactions are the primary source of the combustion dynamics and the oscillations in the equivalence ratio have secondary effects.

The effects of the enthalpy of reaction and fuel composition on combustion dynamics were examined by Ferguson et al. [180]. They utilized two different combustors, laboratory scale and atmospheric pressure combustors. Different fuel blends of ethane, propane, and natural gas were considered for the combustion in an air environment. They observed different dynamic response with increased mole fraction of propane. Fritsche et al. [181] performed an experimental study of thermo-acoustic instabilities in a premixed flame on a swirl-stabilized combustor under different inlet temperature and air-to-fuel ratio. The results exhibited the existence of two stable flames, one is lean and the other is rich, separated by a range of unstable flames. The unstable flames exhibited different shapes as well as pressure oscillations. Seo [182] studied the effect of each of the combustion chamber pressure, the operating temperature, and equivalence ratio on combustion dynamics of a lean premixed flame on single-element swirl injector using gaseous fuel. Unstable flames were recorded when the equivalence ratio was in the range between 0.5 and 0.7. Also, unstable flames appeared when the inlet temperature was greater than 650 K. Venkataraman et al. [183] studied the effects of inlet Reynolds numbers, swirl number, and equivalence ratio on combustion instabilities of a premixed natural gas–air flame in a coaxial dump combustor stabilized using a bluff body. Unstable flames were recorded close to stoichiometric conditions and near the lean blowout limit. Combustion stability was affected negatively when the inlet velocity was raised.

1.9 Conclusions

In this chapter, we introduced the oxy-fuel combustion technology and discussed the alternative technologies for clean and sustainable environment applications. The global warming issue is discussed and the agreement to keep the average global temperature rise below 2 °C is introduced. About two thirds of the available budget for keeping the temperature rise below 2 °C has already been released. Along with that, fossil fuels continue to provide most of the world's energy demand and expecting 78.4% more CO_2 to be emitted. In addition, there is a set of mature renewable energies including biomass energy, geothermal energy, hydropower, ocean energy, solar photovoltaics (PV), and wind power. However, reduced oil price is the main obstacle toward rapid and wide application of renewables. This necessitates the application of carbon capture and storage (CCS) technologies parallel to investment in renewables in order to have effective control of CO_2 emissions. Different techniques for carbon capture, storage, and utilization have been presented and discussed. We also discussed the combination of the use of bio-energy and CCS technologies (BECCES) for negative CO_2 emissions. The potential to reduce atmospheric CO_2 levels offered by BECCS is unlikely to be realized unless there is an incentive to deploy it. An appropriate incentive policy for BECCS needs to be based on an assessment of the emissions reduction that the technology can deliver. Among CCS technologies, oxy-fuel combustion technology is considered as one of the most promising foremost technologies that are being considered recently for CO_2 capture from power plants. This technology can be applied through two main approaches. The first approach is applied using air separation unit (ASU) to separate oxygen from air, and then, the separated oxygen is used in oxidizing the fuel in conventional combustion systems. The second approach is applied using oxygen transport reactors (OTRs). Both approaches are discussed in detail in the following chapters. The application of oxy-fuel combustion technology results in the lowest cost for carbon capture after simple condensation of the water vapor from the exhaust stream containing mainly CO_2 and H_2O. The volume of exhaust stream is significantly reduced resulting in reduction the size and increase in the efficiency of the treatment equipment. The ability to retrofit existing air-based gas turbines and other combustion applications to implement oxy-combustion with no significant hardware changes will offer one of the essential motivations for most customers to switch to oxy-combustion. Also, NO_x emissions are completely controlled due to the absence of nitrogen in the combustion zone.

References

1. Nemitallah MA, Ben-Mansour R, Habib MA, Ahmed WH, Toor IH, Gasem ZM, Badr HM (2015) Solid particle erosion downstream of an orifice. ASME J Fluids Eng 137(021302): 1–11
2. Habib MA, Tahir F, Nemitallah MA, Ahmed WH, Badr HM (2015) Experimental and numerical analysis of oxy-fuel combustion in a porous plate reactor. Int J Energy Res 39:1229–1240
3. Habib MA, Nemitallah MA, Sharqawy MH, Perveez A, Guhlam I, Yaqub M (2015) Experimental analysis of oxygen-methane combustion inside a gas turbine reactor under various operating conditions. Energy 86:105–114
4. Binash IA, Habib MA, Nemitallah MA, Jamal Aqil (2015) Investigation of liquid ethanol evaporation and combustion in air and oxygen environments inside a 25-kW vertical reactor. Proc Inst Mech Eng Part A J Power Energy 229(6):647–661
5. Nemitallah MA, Habib MA, Ghoniem AF (2014) Design of an ion transport membrane reactor for gas turbine combustion application. J Membr Sci 450:60–71
6. Habib MA, Nemitallah MA, El-Nakla M (2014) Current status of CHF predictions using CFD modeling technique and review of other techniques especially for non-uniform axial and circumferential heating profiles. Ann Nucl Energy 70:188–207
7. Ben-Mansour R, Habib MA, Nemitallah MA, Rajhi M, Suara KA (2014) Characteristics of oxy-fuel air fuel combustion in an industrial water tube boiler. J Heat Transf Eng 35:1394–1404
8. Rajhi MA, Ben-Mansour R, Habib MA, Nemitallah MA, Andersson K (2014) Evaluation of gas radiation models in CFD modeling of oxy-combustion. Energy Convers Manag 81:83–97
9. Habib MA, Mokhiemer EM, Sofihullahi Y, Sanusi Y, Nemitallah MA (2014) Numerical investigations of combustion and emissions of syngas as compared to methane in a 200 MW Packaged Boiler. Energy Convers Manag 83:296–305
10. Habib MA, Said SA, Khalifa A, Nemitallah MA, Ayinde T (2014) Experimental investigations of the flow maldistribution inside an air-cooled heat exchanger. Arab J Sci Eng 39:8187–8198
11. UNFCCC. Adoption of the Paris Agreement. Report No. FCCC/CP/2015/L.9/Rev.1. http://unfccc.int/resource/docs/2015/cop21/eng/l09r01.pdf (UNFCCC, 2015)
12. UNFCCC. United Nations Framework Convention on Climate Change. Report No. FCCC/INFORMAL/84. https://unfccc.int/resource/docs/convkp/conveng.pdf. (UNFCCC, 1992)
13. Knutti R, Rogelj J, Sedláčˇek J, Fischer EM (2016) A scientific critique of the two-degree climate change target. Nat Geosci 9:13–18
14. Randalls S (2010) History of the 2 °C climate target. Wiley Interdiscip Rev Clim Change 1:598–605
15. Rogelj J, den Elzen M, Höhne N, Fransen T, Fekete H, Winkler H, Schaeffer R, Sha F, Riahi K, Meinshausen M (2016) Paris Agreement climate proposals need a boost to keep warming well below 2 °C. Nature 534. https://doi.org/10.1038/nature18307
16. Collins M et al (2013) Stocker TF et al. (eds) In: Climate change 2013: the physical science basis. Contribution of Working Group I to the fifth assessment report of the Intergovernmental Panel on Climate Change. Cambridge University Press, Cambridge, pp 1029–1136
17. Matthews HD, Caldeira K (2008) Stabilizing climate requires near-zero emissions. Geophys Res Lett 35:L04705
18. Knutti R, Rogelj J (2015) The legacy of our CO_2 emissions: a clash of scientific facts, politics and ethics. Clim Change 133:361–373
19. IPCC (2014) Climate change 2014: synthesis report. Contribution of Working Groups I, II and III to the fifth assessment report of the Intergovernmental Panel on Climate Change
20. Meinshausen M et al (2009) Greenhouse-gas emission targets for limiting global warming to 2°C. Nature 458:1158–1162

21. Rogelj J et al (2016) Differences between carbon budget estimates unravelled. Nat Clim Change 6:245–252
22. Friedlingstein P et al (2014) Persistent growth of CO_2 emissions and implications for reaching climate targets. Nat Geosci 7:709–715
23. Rogelj J, McCollum DL, Reisinger A, Meinshausen M, Riahi K (2013) Probabilistic cost estimates for climate change mitigation. Nature 493:79–83
24. Luderer G et al (2013) Economic mitigation challenges: how further delay closes the door for achieving climate targets. Environ Res Lett 8:034033
25. Allen MR, Frame DJ, Huntingford C, Jones CD, Lowe AJ, Meinshausen M, Meinshausen N (2009) Warming caused by cumulative carbon emissions towards the trillionth tonne. Nature 458. https://doi.org/10.1038/nature08019
26. Rogelj J et al (2015) Energy system transformations for limiting end-of-century warming to below 1.5 °C. Nat Clim Change 5:519–527
27. Clarke L et al (2014) In: Edenhofer O et al (eds) Climate change 2014: mitigation of climate change. Contribution of Working Group III to the Fifth Assessment Report of the Intergovernmental Panel on Climate Change, 413–510. Cambridge University Press, Cambridge
28. Rogelj J et al (2015) Zero emission targets as long-term global goals for climate protection. Environ Res Lett 10:105007
29. International Energy Agency (IEA) (2016) World energy outlook 2016. Paris. http://www.worldenergyoutlook.org/publications/weo-2016/
30. Le Quere C et al (2016) Global carbon project 2016. Earth Syst Sci Data 8:605–649
31. Waldman S (2016 Nov 14) Global carbon emissions have now been flat for 3 years. E&E News. http://www.eenews.net/climatewire/2016/11/14/stories/1060045682
32. Enerdata, Global energy statistical yearbook 2016—coal and lignite production. https://yearbook.enerdata.net/coal-and-lignite-production.html, viewed 21 Mar 2017
33. Neslen A (2016 Sept 23) Dutch parliament votes to close down country's coal industry. The Guardian (UK). https://www.theguardian.com/environment/2016/sep/23/dutch-parliament-votes-toclose-down-countrys-coal-industry
34. International Energy Agency (IEA) (2016) World energy statistics and balances, 2016 edition. OECD/IEA, Paris
35. IEA (2016) Medium-term renewable energy market report 2016. Paris. https://www.iea.org/newsroom/news/2016/october/mediumterm-renewable-energy-market-eport-2016.html
36. Coady David et al (2015) How large are global energy subsidies?. IMF, Washington, DC
37. Renewables 2017, Global status report
38. International Energy Agency (IEA) (2017) Bioenergy How2Guide. Paris. https://www.iea.org/publications/freepublications/publication/technology-roadmap-how2guide-for-bioenergy.html
39. International Renewable Energy Agency (IRENA), Innovation Outlook, Advanced Biofuels (Abu Dhabi: 2016). http://www.irena.org/DocumentDownloads/Publications/IRENA_Innovation_Outlook_Advanced_Biofuels_2016_summary.pdf
40. IEA (2016) Medium-term renewable energy market report 2016. Paris. https://www.iea.org/newsroom/news/2016/october/medium-term-renewable-energy-market-report-2016.html
41. European Commission (2017) Final report from special group on advanced biofuels: building up the future. Brussels (forthcoming)
42. IEA (2016) Renewables information. Paris. http://wds.iea.org/wds/pdf/Ren_documentation.pdf
43. World Energy Outlook (2016) Projections for 2015 and 2016 are from a linear extrapolation based on data for 2010–14 from IEA. Paris. http://www.worldenergyoutlook.org/publications/weo-2016/
44. UK Department for Business, Energy and Industrial Strategy, National Statistics, Energy Trends Section 6: Renewables, Table 6.1, updated 3 Apr 2017. https://www.gov.uk/government/statistics/energy-trends-section-6-renewables

45. US data (corrected for difference between net and gross electricity generation) from US Energy Information Administration (EIA), Electric Power Monthly, 24 March 2017. http://www.eia.gov/electricity/monthly/epm_table_grapher.cfm?t=epmt_1_01_a
46. Licht FO (2017) Fuel ethanol: world production by country. Where provisional data have been replaced in the source, these have been used
47. International Energy Agency (IEA) (2017 Feb) Geothermal, annual report 2015. Taupo, New Zealand. http://iea-gia.org/wp-content/uploads/2017/02/2015-IEA-Geothermal-Annual-Report.pdf
48. IHA, op. cit. note 1, and on IHA, Hydropower Status Report 2016 (London: May 2016). http://www.hydropower.org
49. International Hydropower Association (IHA) (2017 Apr) Key trends in hydropower. London
50. GE (2016) Powering the digital transformation of electricity Boston. https://www.ge.com/digital/sites/default/files/Power%20Digital%20Solutions%20Product%20Catalog.pdf
51. International Renewable Energy Agency (IRENA) (2017 Apr) Renewable capacity statistics 2017. Abu Dhabi. http://www.irena.org/DocumentDownloads/Publications/IRENA_RE_Capacity_Statistics_2017.pdf
52. European Commission (EC), Ocean energy forum, Ocean energy strategic roadmap—Building ocean energy for Europe. Brussels (Nov 2016), https://webgate.ec.europa.eu/maritimeforum/en/node/3962
53. International Energy Agency (IEA) Photovoltaic Power Systems Programme (PVPS) (Apr 2017) Snapshot of global photovoltaic markets 2016. Paris, p. 4
54. IEA PVPS (Apr 2016) Snapshot of global photovoltaic markets 2015, Paris
55. IEA PVPS (2016) Trends in photovoltaic applications, 2016: survey report of selected IEA countries between 1992 and 2015, Paris
56. GTM Research and SEIA, op. cit. note 26, p. 15
57. Innovations in solar plant assembly drive costs towards $1 per watt in 2017", PV Insider, 12 Oct 2016. http://analysis.pv-insider.com/innovations-solar-plant-assembly-drive-costs-towards-1-watt-2017
58. GTM Research, personal communication with REN21, April 2017; historic lows from Mints, op. cit. note 12
59. Global Wind Energy Council (GWEC) (2017 Apr) Global wind report—annual market update 2016. Brussels
60. GWEC, op. cit. note 1; FTI Intelligence, "Vestas returns to no. 1 spot in global wind turbine supplier ranking in 2016", press release (London: 20 Feb 2017)
61. Shruti Shukla, GWEC, personal communication with Renewable Energy Policy Network for the 21st century (REN21), 13 Apr 2017
62. Lily Riahi et al (2015) District energy in cities. Unlocking the potential for renewable energy and energy efficiency. UNEP, Paris
63. Gerdes J (2012, Oct 24) Copenhagen's seawater cooling delivers energy and carbon savings. Forbes
64. Commonwealth of Australia, op. cit. note 31
65. Babu P, Linga P, Kumar R, Englezos P (2015) A review of the hydrate-based gas separation (HBGS) process for carbon dioxide pre-combustion capture. Energy 85:261–279
66. International Energy Agency (2011) World energy outlook
67. Habib MA, Nemitallah MA, Ben-Mansour R (2012) Recent development in oxy-combustion technology and its applications to gas turbine combustors and ITM reactors. Energy Fuels 27:2–19
68. Kenarsari SD, Yang D, Jiang G, Zhang S, Wang J, Russell AG et al (2013) Review of recent advances in carbon dioxide separation and capture. RSC Adv 3(45):22739–22773
69. Mondal MK, Balsora HK, Varshney P (2012) Progress and trends in CO_2 capture/separation technologies: a review. Energy 46(1):431–441
70. Padurean A, Cormos CC, Agachi PS (2012) Pre-combustion carbon dioxide capture by gas-liquid absorption for integrated gasification combined cycle plant. Int J Greenh Gas Control 7:1–11

71. Hasib-ur-Rahman M, Siaj M, Larachi F (2010) Ionic liquids for CO_2 capture-development and progress. Chem Eng Process 49:313–322
72. Wang Q, Luo J, Zhong Z, Borgna A (2011) CO_2 capture by solid adsorbents and their applicaions: current status and new trends. Energy Environ Sci 4:42–55
73. Mart′ın CF, Stöckel E, Clowes R, Adams DJ, Cooper AI, Pis JJ, Rubiera F, Pevida C (2011) Hypercrosslinked organic polymer networks as potential adsorbents for pre-combustion CO_2 capture. J Mater Chem 21:5475–5483
74. Long JR, Yaghi OM (2009) The pervasive chemistry of metal–organic frameworks. Chem Soc Rev 38:1213–1214
75. Furukawa H, Ko N, Go YB, Aratani N, Choi SB, Choi E, Yazaydin AO, Snurr RQ, O'Keeffe M, Kim J (2010) Ultrahigh porosity in metal-organic frameworks. Science 329:424–428
76. Sculley J, Yuan D, Zhou HC (2011) The current status of hydrogen storage in metal–organic frameworks—updated. Energy Environ Sci 4:2721–2735
77. Li JR, Ma Y, McCarthy MC, Sculley J, Yu J, Jeong HK, Balbuena PB, Zhou HC (2011) Carbon dioxide capture-related gas adsorption and separation in metal-organic frameworks. Coord Chem Rev 255:1791–1823
78. Kovvali AS, Sirkar K (2001) Dendrimer liquid membranes: CO_2 separation from gas mixtures. Ind Eng Chem Res 40:2502–2511
79. Brunetti A, Scura F, Barbieri G, Drioli E (2010) Membrane technologies for CO_2 separation. J Membr Sci 359:115–125
80. Linga P, Kumar R, Englezos P (2007) Gas hydrate formation from hydrogen/carbon dioxide and nitrogen/carbon dioxide gas mixtures. Chem Eng Sci 62(16):4268–4276
81. Xu CG, Li XS (2014) Research progress of hydrate-based CO_2 separation and capture from gas mixtures. RSC Adv 4(35):18301–18316
82. Meisen A, Jim Lim C (2006) Kinetics of carbon dioxide absorption and desorption in aqueous alkanolamine solutions using a novel hemispherical contactor-II: experimental results and parameter estimation. Chem Eng Sci 61(19):6590–6603
83. Jamal A, Meisen A, Jim Lim C (2006) Kinetics of carbon dioxide absorption and desorption in aqueous alkanolamine solutions using a novel hemispherical contactor-I. Experimental apparatus and mathematical modeling. Chem Eng Sci 61(19):6571–6589
84. Kohl A, Nielsen R (1997) Gas purification, 5th edn. Gulf Publishing Company
85. Singh D, Croiset E, Douglas P, Douglas M (2003) Techno-economic study of CO_2 capture from an existing coal-fired power plant: MEA scrubbing vs. O_2/CO_2 recycle combustion. Energy Convers Manag 44(19):3073–3091
86. Folger P (2013) Carbon capture: a technology assessment
87. Hermosillalara G, Momen G, Marty P, Leneindre B, Hassouni K (2007) Hydrogen storage by adsorption on activated carbon: investigation of the thermal effects during the charging process. Int J Hydrog Energy 32:1542–1553
88. Richard MA, Bénard P, Chahine R (2009) Gas adsorption process in activated carbon over a wide temperature range above the critical point. Part 1: Modified Dubinin-Astakhov model. Adsorption 15:43–51
89. Xiao J, Tong L, Deng C, Bénard P, Chahine R (2010) Simulation of heat and mass transfer in activated carbon tank for hydrogen storage. Int J Hydrog Energy 35:8106–8116
90. Ye F, Xiao J, Hu B, Benard P, Chahine R (2012) Implementation for model of adsoptive hydrogen storage using UDF in fluent. Phys Procedia 24:793–800
91. Xiao J, Hu M, Bénard P, Chahine R (2013) Simulation of hydrogen storage tank packed with metal-organic framework. Int J Hydrog Energy 38(29):13000–13010
92. Dyer P, Richards R, Russek S, Taylor D (2000) Ion transport membrane technology for oxygen separation and syngas production. Solid State Ion 134:21
93. Lin Y (2001) Microporous and dense inorganic membranes: current status and prospective. Sep Purif Technol 25:39–55

94. Zeng Y, Lin Y, Swartz S (1998) Perovskite type ceramic membranes: synthesis, oxygen permeation and membrane reactor performance for oxidative coupling of methane. J Membr Sci 150:87–98
95. Balachandran U, Kleefisch M, Kobylinski T, Morissette S, Pei S (1997) Oxygen ion-conducting dense ceramic membranes. US Patent 5:639,437
96. Schwartz M, White J, Sammels A (2000) Solid state oxygen anion and electron mediating membrane and catalytic membrane reactors containing them. US Patent 6:033,632
97. Chen C, Prasad R, Gottzmann C (1999) Solid electrolyte membrane with porous catalytically-enhancing constituents (Assigned to Praxair Technology). US Patent 5:938,822
98. Kim J, Lin Y (2000) Synthesis and oxygen permeation properties of thin YSZ/Pd composite membranes. AIChE J 46:1521
99. Descamps C (2004) CO_2 capture study by physical absorption in power production systems based on Integrated Gasification Combined Cycle. Ph.D. thesis, Ecole des Mines, Paris
100. Kanniche M, Bonnivard R, Jaud P, Marcos J, Amann J, Bouallou C (2010) Pre-combustion, post-combustion and oxy-combustion in thermal power plant for CO_2 capture. Appl Therm Eng 30:53–62
101. IEA Greenhouse Gas R&D Programme (IEAGHG) report. Emissions of substances other than CO_2 from power plants with CCS. March 2012 (03)
102. Thomas GA, Mcclure TGM (1991) Feasibility of cyclic CO_2 injection for light-oil recovery, 179–84
103. Torabi F, Qazvini Firouz A, Kavousi A, Asghari K (2012) Comparative evaluation of immiscible, near miscible and miscible CO_2 huff-n-puff to enhance oil recovery from a single matrix-fracture system (experimental and simulation studies). Fuel 93:443–453. https://doi.org/10.1016/j.fuel.2011.08.037
104. Li JH, Bao R, Qin B, Jiang T (2013) Numerical simulation of foamy oil stability using natural gas huff and puff for ultra-deep heavy oil reservoir, vol 318. doi:10.4028/www.scientific.net/AMM.318.405
105. Li G, Li X (2011) Numerical simulation for gas production from hydrate accumulated in Shenhu Area, South China Sea, using huff and puff method. Huagong Xuebao/CIESC J 62:458–468
106. Gamadi TD, Sheng JJ, Soliman MY, Menouar H, Watson MC, Emadibaladehi H (2014) An experimental study of cyclic CO_2 injection to improve shale oil recovery. SPE Improv Oil Recover Symp, pp 1–9. https://doi.org/10.2118/169142-ms
107. Marchetti C (1977) On geoengineering and the CO_2 problem. Clim Change 1:59–68
108. Baes CF, Beall SE, Lee DW, Marland G (1980) The collection, disposal and storage of carbon dioxide. In: Bach W, Pankrath J, William J (eds) Interaction of energy and climate. D. Reidel Publishing Co, pp 495–519
109. Koide H, Yamazaki K (2001) Subsurface CO_2 disposal with enhanced gas recovery and biogeochemical carbon recycling. Environ Geosci 8(3):218–224
110. Gunter WD, Bachu S, Benson S (2004) The role of hydrogeological and geochemical trapping in sedimentary basins for secure geological storage for carbon dioxide. Geological Society, London, Special Publications, 233, pp 129–145, 1 January 2004
111. IPCC (Intergovernmental Panel on Climate Change) (2005) Special report on carbon capture and storage. Cambridge University Press, Cambridge
112. Robinson M, Leonenko Y (2017) Analytical approach for modeling of multi well CO_2 injection. Energy Procedia 114:3406–3416
113. IPCC (2011) Special report on renewable energy sources and climate change mitigation, Working Group III Mitigation of Climate Change, Potsdam Institute for Climate Impact Research (PIK), Potsdam, Germany. http://srren.ipcc-wg3.de/report
114. Karlsson H, Byström L, Biorecro AB (2010) Global status of BECCS Projects. Global CCS Institute, Stockholm Sweden

115. Stanger R, Wall T, Spörl R, Paneru M, Grathwohl S, Weidmann M, Scheffknecht G, McDonald D, Myöhänen K, Ritvanen J (2015) Oxyfuel combustion for CO_2 capture in power plants. Int J Greenh Gas Control 40:55–125
116. Chen W-Y (2012) Handbook of climate change mitigation. Springer, Berlin
117. Burdyny T, Struchtrup H (2010) Hybrid membrane/cryogenic separation of oxygen from air for use in the oxy-fuel process. Energy 35:1884–1897
118. Banaszkiewicz T, Chorowski M, Gizicki W (2014) Comparative analysis of oxygen production for oxy-combustion application. Energy Procedia 51:127–134
119. Simpson AP, Simon A (2007) Second law comparison of oxy-fuel combustion and post-combustion carbon dioxide separation. Energy Convers Manag 48:3034–3045
120. Kakaras E, Koumanakos A, Doukelis A, Giannakopoulos D, Vorrias I (2007) Oxyfuel boiler design in a lignite-fired power plant. Fuel 86:2144–2150
121. Athayde DD, Souza DF, Silva AM, Vasconcelos D, Nunes EH, da Costa JCD, Vasconcelos WL (2016) Review of perovskite ceramic synthesis and membrane preparation methods. Ceram Int 42:6555–6571
122. Murali RS, Sankarshana T, Sridhar S (2013) Air separation by polymer-based membrane technology. Sep Purif Rev 42:130–186
123. Scheffknecht G, Al-Makhadmeh L, Schnell U, Maier J (2011) Oxy-fuel coal combustion—a review of the current state-of-the-art. Int J Greenh Gas Control 5:16–35
124. Habib MA, Nemitallah MA (2015) Design of an ion transport membrane reactor for application in fire tube boilers. Energy 81:787–801
125. Granados DA, Chejne F, Mejía JM (2015) Oxy-fuel combustion as an alternative for increasing lime production in rotary kilns. Appl Energy 158:107–117
126. Aneke M, Wang M (2015) Process analysis of pressurized oxy-coal power cycle for carbon capture application integrated with liquid air power generation and binary cycle engines. Appl Energy 154:556–566
127. Nemitallah MA, Habib MA, Ben Mansour R (2012) Investigations of oxy-fuel combustion and oxygen permeation in an ITM reactor using a two-step oxy-combustion reaction kinetics model. J Membr Sci 432:1–12
128. Ben-Mansour R, Habib MA, Badr HM, Nemitallah M (2012) Characteristics of oxyfuel combustion in an oxygen transport reactor. Energy Fuels 26:4599–4606
129. Ben Mansour R, Nemitallah MA, Habib MA (2013) Numerical Investigations of oxygen permeation and methane-oxycombustion in a stagnation flow ion transport membrane reactor. Energy 1:1–11
130. Akin F, Jerry Y, Lin YS (2004) Oxygen permeation through oxygen ionic or mixed-conducting ceramic membranes with chemical reactions. J Membr Sci 231:133–146
131. Habib MA, Nemitallah MA, Ben-Mansour R (2013) Recent development in oxy-combustion technology and its applications to gas turbine combustors and ITM reactors. Energy Fuels 27:2–19
132. Zhang K, Sunarso J, Shao Z, Zhou W, Sun C, Wang S, Liu S (2011) Research progress and materials selection guidelines on mixed conducting perovskite type ceramic membranes for oxygen production. RSC Adv 1:1661–1676
133. Mancini ND, Mitsos A (2011) Ion transport membrane reactors for oxy-combustion. Part II: Analysis and comparison of alternatives. Energy 36:4721–4739
134. IEA (2016) 20 years of carbon capture and storage. [Online]. Available: https://www.iea.org/publications/freepublications/publication/20YearsofCarbonCaptureandStorage_WEB.pdf
135. UN, "Paris Agreement," 2015. [Online]. Available: https://unfccc.int/files/essential_background/convention/application/pdf/english_paris_agreement.pdf
136. Li H, Yan J, Yan J, Anheden M (2009) Impurity impacts on the purification process in oxy-fuel combustion-based CO_2 capture and storage system. Appl Energy 86(2):202–213
137. Heil P, Torporov D, Stadler H, Tschunko S, Forster M, Kneer R (2009) Development of an oxycoal swirl burner operating at low O_2 concentrations. Fuel 88:1269–1274

138. Kvamsdal H, Jordal K, Bolland O (2007) A quantitative comparison of gas turbine cycles with CO_2 capture. Energy 32:10–24
139. Bolland O, Mathieu P (1998) Comparison of two CO_2 removal options in combined cycle power plants. Energy Convers Manag 39:1653–1663
140. Dillon D, Panesar R, Wall R, Allam R, White V, Gibbins J et al, Oxy-combustion processes for CO_2 capture from advanced supercritical PF and NGCC power plant. In: Proceedings of the seventh international conference on greenhouse gas control technologies-GHGT7 2004; Vancouver, Canada
141. Staicovici M (2002) Further research zero CO_2 emission power production: the coolenerg process. Energy 27:831–844
142. Yantovski E (1996) Stack downward zero emission fuel-fired power plants concept. Energy Convers Manag 37:867–877
143. Mathieu P, Nihart R (1999) Sensitivity analysis of the MATIANT cycle. Energy Convers Manag 40:1687–1700
144. Liu C, Chen G, Sipocz N, Assadi M, Bai X (2012) Characteristics of oxy-fuel combustion in gas turbines. Appl Energy 89:387–394
145. Kutne P, Kapadia B, Meier W, Aigner M (2010) Experimental analysis of the combustion behavior of oxyfuel flames in a gas turbine model combustor. In: Proceedings of the combustion institute. https://doi.org/10.1016/j.proci.2010.07.008
146. Williams T, Shaddix C, Schefer R (2008) Effect of syngas composition and CO_2-diluted oxygen on performance of a premixed swirl stabilized combustor. Combust Sci Technol 180:64–88
147. Ditaranto M, Hals (2006) Combustion instabilities in sudden expansion oxy–fuel flames. J Combust Flame 146:493–512
148. Andersson K, Johnsson F (2007) Flame and radiation characteristics of gas-fired O_2/CO_2 combustion. Fuel 86:656–668
149. Nemitallah MA, Habib MA (2013) Experimental and numerical investigations of an atmospheric diffusion oxy-combustion flame in a gas turbine model combustor. Appl Energy 111:401–415
150. Wall T (2007) Combustion processes for carbon capture. Proc Combust Inst 31:31–47
151. Nemitallah MA (2016) A study of methane oxy-combustion characteristics inside a modified design button-cell membrane reactor utilizing a modified oxygen permeation model for reacting flows. J Nat Gas Sci Eng 28:61–73
152. Nemitallah MA, Habib MA, Mezghani K (2015) Experimental and numerical study of oxygen separation and oxy-combustion characteristics inside a button-cell LNO-ITM reactor. Energy 84:600–611
153. Habib MA, Salaudeen SA, Nemitallah MA, Ben-Mansour R, Mokheimer EMA (2016) Numerical investigation of syngas oxy-combustion inside a LSCF-6428 oxygen transport membrane reactor. Energy 96:654–665
154. http://navier.engr.colostate.edu/~dandy/code/code-2/
155. Williams TC, Shaddix CR, Schefer RW (2007) Effect of Syngas Composition and CO_2-Diluted Oxygen on Performance of a Premixed Swirl-Stabilized Combustor. Combust Sci Technol 180:64–88
156. Wicksall DM, Agrawal AK, Schefer RW, Keller JO (2005) The interaction of flame and flow field in a lean premixed swirl-stabilized combustor operated on H_2/CH_4/air. Proc Combust Inst 30:2875–2883
157. Xie Y, Wang J, Zhang M, Gong J, Jin W, Huang Z (2013) Experimental and numerical study on laminar flame characteristics of methane oxy-fuel mixtures highly diluted with CO_2. Energy Fuels 27:6231–6237
158. Joo PH, Charest MRJ, Groth CPT, Gülder ÖL (2013) Comparison of structures of laminar methane-oxygen and methane-air diffusion flames from atmospheric to 60 atm. Combust Flame 160:1990–1998

159. Rashwan SS, Ibrahim AH, Abou-Arab TW, Nemitallah MA, Habib MA (2016) Experimental investigation of partially premixed methane-air and methane-oxygen flames stabilized over a perforated-plate burner. Appl Energy 169:126–137
160. Ramadan IA, Ibrahim AH, Abou-Arab TW, Rashwan SS, Nemitallah MA, Habib MA (2016) Effects of oxidizer flexibility and bluff-body blockage ratio on flammability limits of diffusion flames. Appl Energy 178:19–28
161. Rashwan SS, Ibrahim AH, Abou-arab TW (2015) Experimental investigation of oxy-fuel combustion of CNG flames stabilized over a perforated-plate burner. In: 18th international flame research foundation. Friesing, Munich, 1–11
162. Taamallah S, Vogiatzaki K, Alzahrani FM, Mokheimer EMA, Habib MA, Ghoniem AF (2015) Fuel flexibility, stability and emissions in premixed hydrogen-rich gas turbine combustion: technology, fundamentals, and numerical simulations. Appl Energy 154:1020–1047
163. Kashir B, Tabejamaat S, Jalalatian N (2015) A numerical study on combustion characteristics of blended methane-hydrogen bluff-body stabilized swirl diffusion flames. Int J Hydrog Energy 40:6243–6258
164. Yu B, Lee S, Lee CE (2015) Study of NO_x emission characteristics in CH_4/air non-premixed flames with exhaust gas recirculation. Energy 91:119–127
165. Gao X, Duan F, Lim SC, Yip MS (2013) NO_x formation in hydrogen–methane turbulent diffusion flame under the moderate or intense low-oxygen dilution conditions. Energy 59:559–569
166. Li YH, Chen GB, Lin YC, Chao YC (2015) Effects of flue gas recirculation on the premixed oxy-methane flames in atmospheric condition. Energy 89:845–857
167. Altay HM, Hudgins DE, Speth RL, Annaswamy AM, Ghoniem AF (2010) Mitigation of thermoacoustic instability utilizing steady air injection near the flame anchoring zone. Combust Flame 157:686–700
168. Ghoniem AF, Park S, Wachsman A, Annaswamy A, Wee D, Altay HM (2005) Mechanism of combustion dynamics in a backward-facing step stabilized premixed flame. Proc Combust Inst 30:1783–1790
169. Lee K, Kim H, Park P, Yang S, Ko Y (2013) CO_2 radiation heat loss effects on NO_x emissions and combustion instabilities in lean premixed flames. Fuel 106:682–689
170. Yadav NP, Kushari A (2009) Visualization of recirculation in low aspect ratio dump combustor. J Flow Vis Image Process 16:127–136
171. Li G, Gutmark EJ (2005) Effect of nozzle geometry on combustion flow field and combustion characteristics. Proc Combust Inst 30:2893–2901
172. Speth RL, Ghoniem AF (2009) Using a strained flame model to collapse dynamic mode data in a swirl-stabilized syngas combustor. Proc Combust Inst 32:2993–3000
173. Speth RL, Hong S, Shanbhogue SJ, Ghoniem AF (2011) Mode selection in flame-vortex driven combustion instabilities. In: 49th AIAA aerospace sciences meeting; 236
174. Speth RL (2010) Fundamental studies in hydrogen-rich combustion: in-stability mechanisms and dynamic mode selection. Ph.D. thesis, Massachusetts Institute of Technology
175. Altay HM, Speth RL, Hudgins DE, Ghoniem AF (2009) Flame-vortex interaction driven combustion dynamics in a backward-facing step combustor. Combust Flame 156:1111–1125
176. Hong S, Shanbhogue SJ, Ghoniem AF (2015) Impact of fuel composition on the recirculation zone structure and its role in lean premixed flame anchoring. Proc Combust Inst 35:1493–1500
177. Shanbhogue SJ, Husain S, Lieuwen T (2009) Lean blow-off of bluff body stabilized flames: scaling and dynamics. Prog Energy Combust Sci 35:98–120
178. Kedia KS, Ghoniem AF (2014) The anchoring mechanism of a bluff-body stabilized laminar premixed flame. Combust Flame 161:2327–2339
179. Altay HM, Speth RL, Hudgins DE, Ghoniem AF (2009) The impact of equivalence ratio oscillations on combustion dynamics in a backward-facing step combustor. Combust Flame 156:2106–2116

180. Ferguson D, Straub D, Richards G, Robey E (2007) Impact of fuel variability on dynamic instabilities in gas turbine combustion. In: 5th US combustion meeting
181. Fritsche D, Furi M, Boulouchos K (2007) An experimental investigation of thermoacoustic instabilities in a premixed swirl-stabilized flame. Combust Flame 151:29–36
182. Seo S (1999) Parametric study of lean-premixed combustion instability in a pressurized model gas turbine combustor. Ph.D. thesis; Department of Mechanical and Nuclear Engineering, the Pennsylvania State University: University Park, PA
183. Venkataraman KK, Preston LH, Simons DW, Lee BJ, Lee JG, Santavicca DA (1999) Mechanisms of combustion instability in a lean premixed dump combustor. J Propul Power 15:909–918

Chapter 2
Application of Oxy-fuel Combustion Technology into Conventional Combustors

2.1 Introduction

Nowadays, emission of greenhouse gases (mainly CO_2) is a critical challenge facing the world due to the associated global warming. By the year 2050, the emission levels of CO_2 are expected to be increased by about 70% compared to the present levels [1]. The huge world's energy demand forces the governments to continue using fossil fuels and the researchers to continue developing new technologies that can reduce the emissions of greenhouse gases while burning fossil fuels [2]. Carbon capture technologies (CCTs) including pre-combustion, oxy-combustion, and post-combustion capture are good tools toward the control of CO_2 emissions [3]. Among these carbon capture technologies, oxy-combustion technology is considered as one of the most promising CCTs [4]. This technology can be integrated either with currently existing power plants with slight modifications or with the new power plants [5]. In this process, fuel is burned using oxidizer mixture consisting of high purity oxygen and recycled exhaust gases (consists mainly of CO_2). Portion of the exhaust stream is recycled to the combustor to control the combustion temperature. Theoretically, if a hydrocarbon fuel is to be burned with pure oxygen under perfect complete combustion, the combustion products should be mainly CO_2 plus H_2O; however, some other species usually exist in the exhaust stream. This can be attributed to the dissociation of different species within the flame core, impurities in fuel or oxidizer mixture, air leak, combustion instabilities, and improper mixing and incomplete combustion. However, oxy-combustion process results in highly CO_2-concentrated exhaust stream which facilitates the capture process of CO_2 after H_2O condensation [6].

There are two existing approaches for the application of oxy-combustion technology. The first approach is applied through the use of air separation unit (ASU) to separate oxygen from air, and then, the separated oxygen is used in oxidizing the fuel in conventional combustion systems, as shown in Fig. 2.1. However, the ASU requires additional energy for the operation either in cryogenic separation or in

M. A. Nemitallah et al., *Oxyfuel Combustion for Clean Energy Applications*,
Green Energy and Technology, https://doi.org/10.1007/978-3-030-10588-4_2

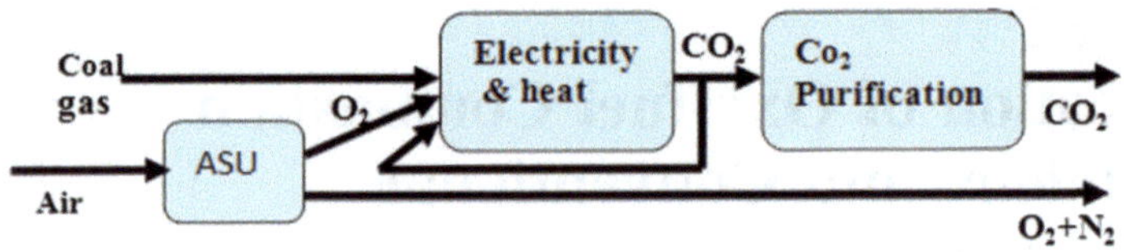

Fig. 2.1 Oxy-combustion system for conventional combustors

terms of gas heating (to activate the oxygen separation membrane) in case of membrane separation [7]. The second approach is applied through the use of what is known as ion transport membrane reactor (ITMR). In ITMR, the two processes of oxygen separation and oxy-combustion are performed simultaneously within the same unit, as shown in Fig. 2.2. An ion transport membrane (ITM) is utilized for the separation of oxygen from the flowing air in the feed side of the membrane. In the permeation side, the permeated oxygen across the membrane is burned with the fuel in a highly concentrated medium of recirculated CO_2 [8]. The design of the ITMR results in a compact size of the system, due to the integration of both the separation and combustion processes in one common unit. In addition, the resultant ITMR system will not require any additional powering for oxygen separation because the membrane is heated using part of the heat released due to combustion in permeate side of the membrane. On the other hand, integrating oxy-combustion technology to existing regular combustion systems via ASU requires part of the output power of the combustion system to separate oxygen and, as a result, the overall system efficiency is reduced [9]. Accordingly, utilizing oxygen transport membrane reactors (OTMRs) is recently suggested as an application of

Fig. 2.2 Stagnation flow oxy-combustion system utilizing oxygen transport reactor (OTR)

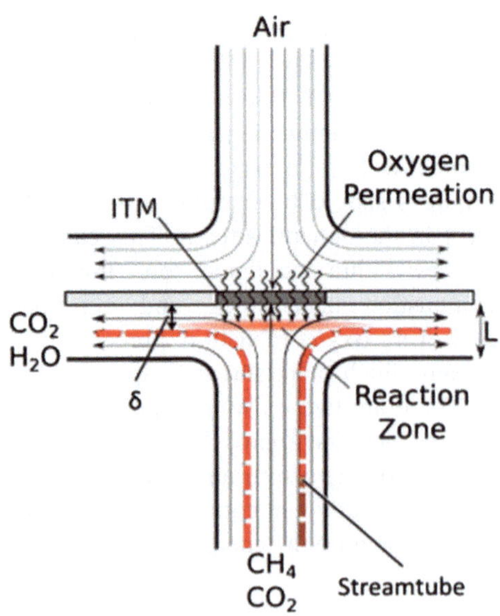

oxy-combustion technology. In such kind of reactors, OTMRs, oxygen is being separated inside the combustion system using ITM. These membranes are activated for oxygen permeation at temperatures ranging from 650 to 950 °C [10]. There are variety of membrane materials that can be used in such applications including lanthanum cobaltite perovskite ceramics, modified perovskite ceramics [11], brownmillerite structured ceramics [12], ceramic metal dual-phase membranes [13], in addition to thin dual-phase membranes which consist of chemically stable yttria-stabilized zirconia (YSZ) [14]. In this chapter, the focus is made on the first approach for the application oxy-combustion technology in conventional combustion systems, mainly in gas turbines. The second approach of using OTMRs is investigated in detail in the next chapter considering the applications in gas turbine combustors and fire tube boiler furnaces.

In the industrial field, the use of liquid fuels is preferred due to their high energy density and low volume as compared to gaseous fuels, which facilitate handling and transportation processes. This forces the research toward the design of new combustion systems that can easily handle liquid fuels. However, care should be taken while designing a combustion system that can handle liquid fuels to reduce soot formation, reduce NO_x emissions, and properly vaporize the fuel [15]. Liquid fuel combustion is utilized in many industrial devices including diesel engines and gas turbine engines. The subject of liquid fuel evaporation and combustion has been studied numerically using 2D and 3D direct numerical simulations (DNSs) [16] and large eddy simulations (LESs) [17]. Yin [18] performed numerical modeling of *n*-heptane droplets. They were able to develop a computational code for the heating and vaporization of liquid droplet taking into consideration droplet dynamics and the droplet/free stream interaction. Kitano et al. [19] numerically explored the influence of fuel composition on the droplet evaporation and combustion. They reported that the evaporation rate becomes slower for multi-component fuel as compared to single fuel. Targeting reduced emission and clean combustion, Jiang et al. [20] experimentally investigated the combustion process of different liquid fuels including diesel, biodiesel, and straight vegetable oil (VO) using a novel design flow burning (FB) injector. The results showed the capability of the FB injector to produce clean blue flames indicating mainly premixed combustion for all the tested fuels.

Emission control, especially soot formation, is another critical issue in case of liquid fuel combustion. Park and Yoon [21] proposed a two-stage injection strategy to simultaneously reduce the NO_x and soot formation in dimethyl ether (DME)-fueled engine. The results showed low NO_x, HC, CO, and soot emissions using their injection strategy. Speth et al. [22] conducted a study on the effect of using alternative low-aromatic-content jet fuels on the reduction of gas turbines black carbon emissions. They established a correlation between the reduction in the gas turbines black carbon emissions and the alternative fuel aromatic volume fraction. Choi et al. [23] investigated the effect of liquid fuel doping on the formation of soot and polycyclic aromatic hydrocarbon in counterflow ethylene diffusion flames. The results showed that adding liquid fuel could lead to the formation of higher levels of aromatic ring. Higher formation of aromatic ring in the toluene mixture compared with that in case of benzene doping flame was predicted.

In this chapter, the available CCTs are studied considering different applications and limitations of use. Oxy-combustion technology is studied in detail in terms of its applications in existing conventional combustion systems. This is followed by discussing the possibility of integrating oxy-combustion technology with conventional combustion systems in terms of required modifications to existing systems and system performance considering gaseous, liquid, and coal fuels. The recent advances in oxy-combustion and its technology readiness level are introduced for both coal-fired power plants and gas turbines. Status and future trends of oxy-combustion technology are described and analyzed. A techno-economic analysis of oxy-combustion integrated power systems is presented.

2.2 Oxy-fuel Combustion Characteristics

The characteristics of oxy-fuel combustion with recycled flue gas differ with air combustion in several aspects primarily related to the higher CO_2 levels and system effects due to the recirculated flow, including the following [24]: (1) To attain a similar adiabatic flame temperature (AFT), the O_2 proportion of the gases passing through the burner is higher, typically 30%, than that for air (of 21%), necessitating that about 60% of the flue gas is recycled. (2) The high proportions of CO_2 and H_2O in the furnace gases result in higher gas emissivities, so that similar radiative heat transfer for a boiler retrofitted to oxy-fuel will be attained when the O_2 proportion of the gases passing through the burner is less than the 30% required for the same AFT. (3) The volume of gases flowing through the furnace is reduced somewhat, and the volume of flue gas (after recycling) is reduced by about 80%. (4) The density of the flue gas is increased, as the molecular weight of CO_2 is 44, compared to 28 for N_2. (5) Typically, when air-firing coal, 20% excess air is used. Oxy-fuel requires a percent excess O_2 (defined as the O_2 supplied in excess of that required for stoichiometric combustion of the coal supply) to achieve a similar O_2 fraction in the flue gas as air-firing, in the range of 3–5% [25]. (6) Without removal in the recycle stream, species (including corrosive sulfur gases) have higher concentrations than in air-firing. (7) As oxy-fuel combustion combined with sequestration must provide power to several significant unit operations, such as flue gas compression, that are not required in a conventional plant without sequestration, oxy-fuel combustion/sequestration is less efficient per unit of energy produced.

2.2.1 *Reactions and Emission Characteristics*

The combustion of fuel in a mixture of recirculated flue gas (RFG) and oxygen, however, presents new challenges to combustion specialists. Several experimental investigations with oxy-firing pulverized coal burners report that flame temperature and stability are strongly affected [26, 27]. This work focuses on the investigation of the oxy-combustion of methane to see the effect of CO_2 recirculation on combustion characteristics. The substitution of N_2 with CO_2 in the oxidizer will lead to a reduction of the flame speed as reported by Zhu et al. [28]. This causes poor combustion performance and a modified distribution of temperature and species in the combustion chamber. Liu et al. [29] have performed numerical investigations on the chemical effects of CO_2. A comparison between numerical and experimental data showed that the decrease in burning velocity for the oxy-fuel combustion cannot entirely be described by only considering the material properties of CO_2. CO_2 affects the combustion reactions especially by the reaction $CO + OH \rightarrow CO_2 + H$. The competition of CO_2 for H radical through the above reverse reaction with the single most important chain branching reaction $H + O_2 \rightarrow O + OH$ significantly reduces the concentrations of important radicals, i.e., O, H, and OH, leading to significant reduction of fuel burning rate. This hypothesis is supported by a comparison of the burning velocity of methane flames and hydrogen flames in a CO_2/O_2 gas mixture.

The influence of CO_2 on the burning velocity of hydrogen flames is less significant because the concentration of hydrogen radicals is much higher. Finally, it was summarized that the chemical effect of CO_2 significantly reduces the burning velocity of methane, whereby the relative importance of this chemical effect increases with increasing CO_2 concentration in the oxidizing mixture. Andersson et al. [30] have performed experiments in a 100 kW test unit which facilitates O_2/CO_2 combustion with real flue gas recycle. The tests comprise a reference test in air and two O_2/CO_2 test cases with different recycled feed gas mixture concentrations of O_2 (OF21 @ 21% O_2, 79% CO_2 and OF27 @ 27% O_2, 73 vol% CO_2). The results show that the fuel burnout is delayed for the OF21 case compared to air-fired conditions as a consequence of reduced temperature levels. Instead, the OF27 case results in more similar combustion behavior compared to the reference conditions in terms of in-flame temperature and gas concentration levels, but with significantly increased flame radiation intensity.

During oxy-fuel combustion, the amount of NO_x exhausted from the system can be reduced to less than one-third of that with combustion in air [31–33]. The NO_x reduction is thought to be the result of several mechanisms [34]: Decrease of thermal NO_x due to the very low concentration of N_2 from air in the combustor, the reduction of recycled NO_x as it is reburn in the volatile matter release region of the flame, and the reaction between recycled NO_x and char. Okazaki and Ando used a bench-scale reactor to examine the effects of the latter two factors during oxy-fuel combustion with an O_2 concentration of 21% (i.e., recycling ratio as high as 80%) at a maximum flame temperature of 1450 K [34]. They concluded that the reduction of recycled NO_x is the dominant mechanism for the reduction in NO_x emissions. They estimated that more than 50% of the recycled NO_x was reduced when 80% of the flue is recycled. It has also been found that oxy-fuel combustion can decrease the So_2 emissions compared to that in air combustion [32–35].

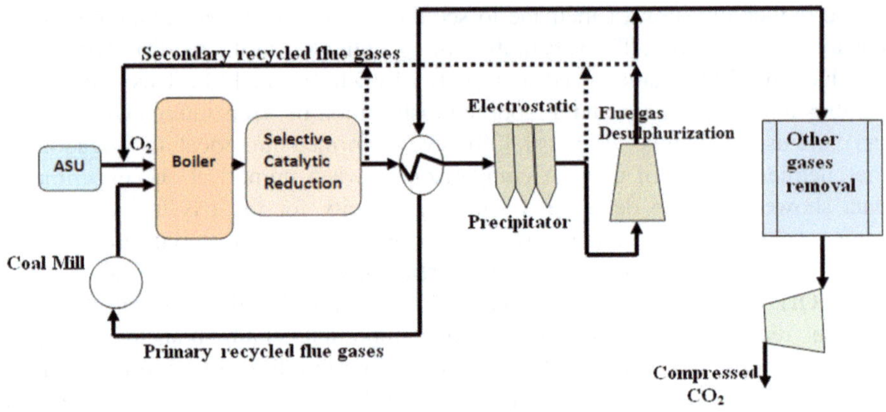

Fig. 2.3 Atmospheric oxy-coal combustion system with flue gas recycle proposed for carbon capture in coal power plants based on the work in [36, 46]

2.2.2 *Oxy-combustion Systems*

The concept of oxy-combustion involves the burning of fuel in pure oxygen in addition to some recycled flow gases or steam in order to control the flame temperature. The aim is to obtain a carbon dioxide-rich stream that is ready for sequestration, after removing water vapor and other impurities. Various oxy-combustion systems have been introduced in the literature [36–40]. The first version is the atmospheric pressure oxy-combustion system where part of the flue gases is recycled in order to control the flame temperatures. Another alternative to using recycled flue gases is to inject steam in order to control the flame temperature [38]. To further increase the performance of these systems, pressurized systems have been proposed for both systems: oxy-combustion with recycled flue gases [39, 41–45] and oxy-syngas combustion in combination with solid fuel gasification technology [40]. There is also a recent technology which is the ion transport membrane reactor technology that can also be applied, and it is discussed in detail later. The atmospheric oxy-coal combustion system shown in Fig. 2.3 was introduced as a short-term solution to retrofit existing coal-fired power plant to include the option of carbon capture and sequestration. The additional required equipment as compared with air-fired systems is considered as discussed in the following sections.

2.2.2.1 Air Separation Unit

When retrofitting existing power plants to be used with oxy-combustion, the system uses the same equipment used in the conventional combustion in addition to an ASU used to produce an oxygen-rich stream for combustion. Currently, the only ASU technology that can meet the volume and purity demand of a large-scale coal-fired utility boiler is based on cryogenic distillation. Air is compressed, cooled, and cleaned prior to being introduced into the distillation column to separate air into an oxygen-rich stream and a nitrogen-rich stream [47]. Cryogenic air separation is consuming about 0.24 kW h/kg O_2 with 95% oxygen purity [48, 49]. The oxygen purity requirement for oxy-coal combustion (85–98%) is lower than that needed in the process industry (99.5–99.6%) [50]. The ASU can consume more than 15% of the gross power output [48, 51–53].

2.2.2.2 Carbon Dioxide Purification Unit

A carbon purification unit consists of gas cleanup units in order to remove water and other gases from the flue gas before being compressed for the sequestration process. Because oxy-combustion is compatible with retrofits, selective catalytic reduction (SCR), electrostatic precipitator (ESP), and flue gas desulfurization (FGD) are typically retained as means of NO_x, particulate matter, and SO_x removal from the flue gases. These pollutants control devices are also suitable for use in conjunction with amine-type absorbents for post-combustion capture plants [47].

After the removal of acid gases such as SO_x and NO_x, non-condensable N_2, O_2, and Ar should also be purged using a non-condensable gas purification unit. This unit is made of multistage compression units with inter-stage cooling in order to separate out the inert gases.

2.2.2.3 Flue Gas Recirculation System

Recycled flue gases are required for replacement of nitrogen in order to control the combustion temperature. These flue gases can be recycled at different locations downstream of the economizer in the form of wet or dry recycles. Since SO_2 concentration in the flue gas may accumulate due to flue gas recycle, resulting in two or three times higher concentration than in conventional air-firing systems, the primary recycle has to be at least partially desulfurized for medium and high sulfur coal, to avoid corrosion in the coal mill and flue gas pipes.

2.3 Oxy-combustion Alternatives

Capture of CO_2 from large point sources such as power plants with subsequent geological storage offers the possibility of a significant and relatively quick response to climate change at a reasonable cost. Successful commercialization of such technology could therefore provide a transition to a future during which energy production from non-fossil energy sources can grow over time. At present, there are no power plants with CO_2 capture available on a commercial scale, but long-time aquifer storage is currently applied and evaluated in the North Sea and show promising results [54]. The CO_2 could also be stored in connection to enhanced oil recovery (EOR). Such storage has been closely monitored in the Weyburn project in Canada; see, e.g., [55]. The highest cost is, however, on the capture side, and to reduce the specific costs for capture, different concepts are discussed. To recover and store carbon dioxide from flue gases of fossil fuel power plants, processes based on oxy-combustion appear to be promising. Concept of the technology is the combustion with commercially pure oxygen to achieve high CO_2 concentrations in the flue gases for the final CO_2 separation. The required oxygen is supplied by an air separation unit where the nitrogen is separated from the air. A great portion of the flue gases has to be recycled to substitute the removed nitrogen. This measure is inevitable to maintain the temperature level in the combustion chamber and in particular not to increase the heat transferred to the membrane walls of the steam generator which is limited by material parameters [56].

In the past decades, intense research efforts have been directed to the development and improvement of ceramic-based membranes for oxygen separation from air at high-temperature operations. Ceramic-based membranes for oxygen separation systems can be categorized into pure oxygen-conducting membranes and mixed ionic–electronic-conducting membranes. Mixed ionic and electronic-conducting ceramic membranes have received increasing interest from academia and industry. A major industrial effort is currently devoted to the development of the mixed-conducting ceramic membrane reactor technology for partial oxidation of hydrocarbons, in particular, partial oxidation of methane to syngas [57, 58].

2.3.1 Using Air Separation Unit and Conventional Combustion Chamber

The required oxygen in this case is supplied by an air separation unit where the nitrogen is separated from the air. A great portion of the flue gases has to be recycled to substitute the removed nitrogen. Key component of the oxy-fuel process with high-temperature membrane air separation unit (HTM-ASU), which is in the stage of development, is a dense membrane made of ceramic materials. These materials begin to conduct oxygen ions above a material-dependent temperature (usually above 700 °C). Driving force for the mass transport is the differential oxygen partial pressure across the membrane, while the oxygen flux is enhanced with decreasing membrane thickness and rising temperature. As only oxygen permeates the membrane, 100% pure oxygen could be produced provided that air leakage within the membrane module is avoided. For further details regarding membrane materials, references [59, 60] are recommended. The basic idea of the HTM-ASU, as illustrated schematically in Fig. 2.4, is the elevation of the oxygen partial pressure on the air side with an air compressor. The partial pressure difference across the membrane can be further enhanced by lowering the oxygen partial pressure on the oxygen-receiving side of the membrane by sweeping with

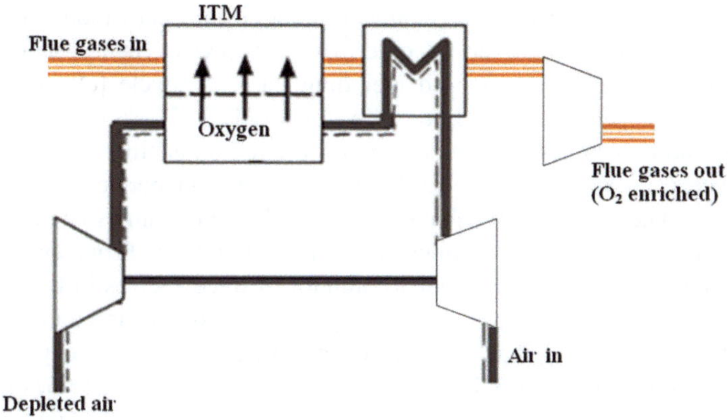

Fig. 2.4 Scheme of an air separation unit based on high-temperature membranes (exemplary flue gas swept)

flue gas, which contains only a small amount of oxygen. As temperatures at the compressor outlet are not sufficient to activate the membrane material's conduction mechanism, the air needs to be preheated with counter-current oxygen-enriched flue gas. To recover parts of the spent energy for compression, the oxygen-depleted air is expanded in a turbine. As the off-gas leaves the HTM-ASU at still elevated temperatures, the heat can be recovered in the power plant cycle. The energy demand of the HTM-ASU is determined by the required high-temperature heat. In addition, mechanical driving power is needed or produced depending on the ASU process parameter design.

2.3.1.1 Applications of Oxy-fuel Combustion in Gas Turbines

Typical Characteristics

As an option to get O_2 required for combustion, oxygen may be obtained via air separation units, e.g., cryogenic or membrane-based processes. The combustion process takes place in nitrogen-free or low-nitrogen environment, resulting in a flue gas composed mainly of CO_2 and H_2O, as well as a low concentration of impurities such as argon and oxygen. Therefore, a simplified flue gas processing by means of condensation of H_2O to capture CO_2, without using costly separation methods such as chemical absorption, can be possible. There are several proposed combined-cycle concepts in oxy-fuel gas turbine processes with natural gas combustion in oxygen and CO_2, for example, the O_2/CO_2 cycle [61–63], the COOLENERG cycle [64], the COOPERATE cycle [65], and the MATIANT cycle [66]. These cycles belong to the group also known as semi-closed oxy-fuel combustion combined cycles (SCOF-CC). Recent studies within the European Union-funded research project ENCAP (Enhanced Capture of CO_2) have concluded that SCOF-CC has good potential with limited techno-economical hinders for realization [64, 67, 68]. Besides SCOF-CC, a number of other oxy-fuel cycles using steam/water as working fluids have been proposed including the Graz cycle [69] and the water cycle [70] developed by Clean Energy Systems (CES). These cycles may require high-temperature turbines and new design for the turbomachinery. For oxy-fuel gas turbine cycles, researches have been focused on thermodynamic studies of system performance. The combustion behavior, e.g., the flame dynamics and reaction zone structures in the gas turbine combustors, is less addressed. From thermodynamic studies, it has been shown that a small amount of trace species in the combustion products can have a great impact on the CO_2 capture, storage, and transportation. Li et al. [71] demonstrated that the purification process of the flue gas stream of oxy-fuel combustion is highly influenced by the existence of impurities such as the small amount of N_2 resulted from the air separation units and the remaining O_2 in the flue gas due to incomplete combustion. The presence of non-condensable gases results in increased condensation duty for the recovery of the CO_2. This in turn leads to lower system efficiency and increased cost for separation. To minimize the oxygen concentration in the flue gas and meanwhile achieve complete combustion of fuel, stoichiometric mixture is preferred in oxy-fuel combustion. CO_2 and/or steam are used to control the flue gas temperature. Jericha and Gottlich [72] outlined a burner and combustor configuration, in which fuel, oxygen, and steam were supplied separately in different inlets. The steam was supplied through an annular outer swirler inlet to form a swirling flow motion to wrap the flames and to cool down the flue gases.

Such combustor configuration would likely generate rather high flame temperature locally in the reaction zones that will enhance the dissociation of H_2O and CO_2 and thus affect the composition of the flue gas such that the unconsumed oxygen can be high in the flue gas. To reduce the flame temperature and thereby the remaining oxygen in the flue gas, it can be beneficial to premix the oxygen and CO_2 or steam before injecting them to the combustor. There are several possibilities that need to be explored, for example, different levels of premixing of the fuel/oxygen/steam/CO_2 prior to their injection into the combustor, and different mixing patterns inside the combustor. The thermodynamic studies will give the same answer for the flue gas in the post-flame zone if the inlet temperature, combustor pressure, and overall mass flows of fuel, oxygen, steam, and CO_2 streams are kept the same. However, the flame dynamics and reaction zone structures are dependent on combustor configurations as they are dictated by the detailed inflow conditions for the fuel/oxygen/steam and CO_2 supplies.

Optimal Supply of Oxygen and Diluent to Oxy-fuel Combustion

To generate stable combustion in gas turbine combustion chambers with oxy-fuel combustion, certain minimal oxygen level in the oxidizer or elevated oxidizer temperature must be maintained. The fundamental reason for this is the need to have sufficiently elevated temperature in the reaction zones for the chain reactions to proceed. Flame instability and poor burnout have been experienced when oxygen/CO_2 are premixed and supplied together to the flame as the oxidizer [73]. For example, in the recent experiments of Heil et al. [74], it was shown that poor burnout and lifted dark flames appeared when the oxygen mole fraction in the O_2/CO_2 stream was set to 21%; when the oxygen volume fraction was increased to 27 and 34%, full burnout and stable flames were obtained. To burn the fuel with lower oxygen level in the oxidizer (O_2/CO_2) stream, the burner had to be modified to allow for recirculation of hot gases to the flame [74]. To improve the recirculation, the inlet design of the oxidizer mixture should be modified in order to mix the hot burned gases in the flame zone with the incoming fresh cold gases in order to stabilize the flame.

In the study done by Kutne et al. [75, 76] the burner considered was a modified version of a practical gas turbine combustor with an air blast nozzle for liquid fuels [77]. Co-swirling oxidizer mixture was supplied to the flame through a central nozzle and an annular nozzle. The radial swirlers consisted of 8 channels for the central nozzle and 12 channels for the annular nozzle. The overall flow field of the flames is characterized by a conically shaped inflow of fresh gas, an inner recirculation zone (IRZ), and outer recirculation zone (ORZ) as sketched in Fig. 2.5.

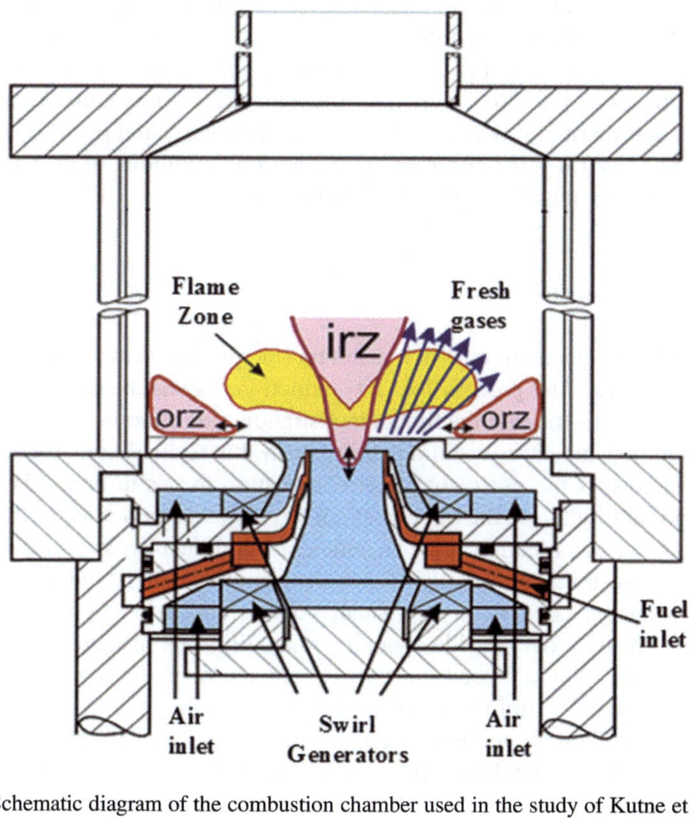

Fig. 2.5 Schematic diagram of the combustion chamber used in the study of Kutne et al. [75, 76]

In the shear layer formed between the inflow and the IRZ, the mixing of hot combustion products with fresh gas leads to a continuous ignition and stabilization of the flame. Same idea of conically shaped inlet for the oxidizer mixture is applied by our research group to our running system of gas turbine model combustor in order to improve the mixing process and so stabilizing the flame. However, it was found by our research group that the aspect ratio between the diameter of the inlet nozzle and the combustor diameter is playing an important role in calculating the amount of the oxidizer mixture that will be available for combustion and the amount that will escape between the flame zone and the reactor walls.

With high level oxygen in the oxidizer, the combustion products become hot and this may lead to high level of oxygen in the flue gas due to the dissociation reactions at high temperatures. There is an optimal "window" of oxygen/diluent ratio in the oxidizer stream [80]. In the work done by Liu et al. [78], they reported that the primary oxidizer which is supplied in the upstream through the dome of the combustion chamber should have minimal oxygen level of 24% under the oxidizer temperature 520 K condition. The excessive CO_2 shall be supplied through the liner holes downstream of the primary reaction zones to have a suitable temperature when the flue gas enters the turbines. This will cool down the combustion products generated in the primary reaction zones. Stable combustion and low turbine inlet temperature can be obtained simultaneously by optimizing the oxygen and CO_2 supplies to the combustion chamber.

The stability of swirl-stabilized oxy-fuel/CH_4 flames was studied in the work done by Kutne et al. [75, 76] for O_2 mol fractions of 20–40%, equivalence ratios of $\Phi = 0.5$–1, and thermal powers of 10–30 kW. However, attempts of operating the burner with <22% O_2 were unsuccessful even with conditions of $\Phi = 1$ at 20 and 30 kW resulting in unstable operation and blowout.

Reactions Characteristics

The oxy-fuel combustion of coal in a steam turbine process is regarded as a possible way to use the oxy-fuel process for CO_2 reduction. Research on this field is very active with the outcome that the first demonstration plants are in operation, and the power generation industry is willing to invest in this technology. Another way of particular interest is the use of oxy-fuel combustion in gas turbines. This process offers the possibility to use the same post-combustion techniques as for the oxy-fuel coal process, in combination with an efficient combined-cycle process [75, 76]. Swirl flames are used extensively in practical combustion systems because they enable high energy conversion in a small volume and exhibit good ignition and stabilization behavior over a wide operating range [79–82]. In stationary gas turbine (GT) combustors, they are used mostly as premixed or partially premixed flames, and in aero-engines, as diffusion flames. To reduce pollutant emissions, especially NO_x, today the flames are operated generally very lean [83–85]. Under these conditions, the flames tend to exhibit undesired instabilities, e.g., in the form of unsteady flame stabilization or thermo-acoustic oscillations. The underlying mechanisms of the instabilities are based on the complex interaction between flow field, pressure, mixing, and chemical reactions, and are not well enough understood to date. Detailed measurements in full-scale combustors are hardly possible and very expensive, and numerical tools have not yet reached a sufficient level of confidence to solve the problems. A promising strategy lies therefore in the establishment of a laboratory-scale "standard combustor" with practical relevance and detailed, comprehensive measurements using non-intrusive techniques with high accuracy. The gained data set will be used for validation and optimization of numerical combustion simulation codes which then can be applied to simulate the behavior of technical combustors. Intrusive probe measurements are less suited for these applications as they disturb the local flow field and change the conditions for stabilization and for reaction—locally or even in general [86, 87]. In turbulent reacting flows, the use of optical measurement techniques is therefore essential for reliable information. Laser-based tools are the method of choice offering the potential to measure most of the important quantities with high temporal and spatial resolution, often as one- or two-dimensional images, and the ability to perform the simultaneous detection of several quantities [88–91].

In recent years, a variety of laser-based investigations in GT model combustors have been reported that, besides feasibility studies, concentrated on certain aspects of the combustion process or model validation. For example, Kaaling et al. [92] performed temperature measurements with CARS in a rich-quench-lean (RQL) combustor, and Kampmann et al. [93] used CARS simultaneously with 2D Rayleigh scattering to characterize the temperature distribution in a double-cone burner. In the same combustor, Dinkelacker et al. [94] studied the flame front structures and flame lift. Their experiments have been conducted at bluff body-stabilized premixed methane/air flames, where flow and flame parameters have been varied systematically over a broad range of exit velocities and stoichiometries. They found that for this burner configuration not only one, but two different liftoff criteria must be met. For very lean mixtures, the chemically dominated ignition delay is found to be the rate-determining step. For other cases, the liftoff height can be determined by a newly described turbulent mixing dominated model [94]. Fink et al. [95] investigated the influence of pressure on the combustion process by applying PLIF of OH and NO in a lean pre-evaporized premixed (LPP) model combustor. With respect to NO_x reduction strategies, Cooper and Laurendeau [96, 97] performed quantitative NO LIF measurements in a lean direct injection spray flame at elevated pressures. They have performed excitation scans and calibration comparisons to assess the background contribution for PLIF detection. Also, they presented and analyzed quantitative radial NO profiles measured by LIF so as to correct the PLIF measurements to within the accuracy bars of the LIF measurements via a single-point scaling of the PLIF image. Shih et al. [98] applied PLIF of OH and seeded acetone in a lean premixed GT model combustor, and Deguchi et al. [99] used PLIF of OH and NO in a large practical GT combustor. They found that the automated LIBS unit is capable of monitoring trace element concentration fluctuations at ppb levels with a 1 min detection time under actual plant conditions. In addition, real-time measurement of O_2 and CO concentrations in a commercial incinerator furnace was performed using TDLAS to improve the combustion control. They reported also, using the multi-point laser measurement results to control secondary air allocation, higher secondary combustion efficiency was achieved, and CO concentration was reduced. Hedman and Warren [100] used PLIF of OH, CARS, and LDV for the characterization of a GT-like combustor fired with propane in order to achieve a better understanding of the fundamentals of GT combustion. PLIF of OH was also applied by Lee et al. [101] to study flame structures and instabilities in a lean premixed GT combustor, by Arnold et al. [102] to visualize flame fronts in a GT combustor flame of 400 kW, and by Fritz et al. [103] for revealing details of flashback. Lofstrom et al. [104] performed a feasibility study of two-photon LIF of CO and 2D temperature mapping by LIF of seeded indium in a low-emission GT combustor. Four different laser diagnostic techniques were investigated in their work. The two more mature techniques, planar Mie scattering/laser-induced fluorescence and planar laser-induced fluorescence of OH, were used for fuel- and OH-visualization, respectively. In addition, the applicability of some novel techniques in harsh industrial environments was investigated: two-line atomic fluorescence (TLAF) to obtain two-dimensional temperature distributions and two-photon LIF for the detection of CO. A comparison of two different laser excitation schemes for major species concentration measurements with laser Raman scattering was performed by Gittins et al. [105] in a GT combustion simulator. At a high-pressure test rig of the DLR, various laser techniques (LDV, CARS, PLIF of OH and kerosene, and 2D temperature imaging via OH PLIF) were applied to GT combustors under technical operating conditions to achieve a better understanding of combustor behavior and to validate CFD codes [106–109].

Williams et al. [110] investigated syngas and methane flames for premixed swirl-stabilized conditions for two different oxidizers of air and $O_2/CO_2/N_2$. Simple flame images for different conditions have been presented along with exhaust gas emissions. They report lower nitrogen oxide concentrations (NO_x) for the quasi-oxy-fuel flames and higher carbon monoxide concentrations (CO), suggesting stoichiometric operation at 20–24% O_2 as ideal for low emissions. Sautet et al. [111] studied the length of natural gas/oxygen diffusion flames in a jet burner for free and confined configurations. Fuel jet Reynolds numbers were varied from 8362 to 16,300 for five flames of which two were buoyancy controlled. The flame lengths were calculated from OH-chemiluminescence and indicated flames to be 2–3 times shorter than air flames with adiabatic flame temperatures in the region of 3050 K. Ditaranto and Hals [112] discussed the effect of stoichiometric operation and high O_2 content in oxidizer on thermo-acoustic oscillations in sudden expansion jet configuration. They reported occurrence of thermo-acoustic instabilities as O_2 content in the oxidizer was increased, characterizing different instability modes dependent on flow velocity and flame speed variations.

2.3.2 Using Membrane Reactor Technology

The other approach for the application of oxy-fuel combustion technology is through using ion transport membranes in order to separate oxygen from air and then this oxygen that permeates through the membrane is used in the combustion process in the other side of the membrane as shown in Fig. 2.6. More recently, strong demand for tonnage quantities of oxygen is encouraged by the steady growth in chemical process operations. Examples of such processes are the oxy-fuel combustion and the oxygen-blown gasification processes. The process of oxygen-blown gasification is used to convert coal and natural gas into an intermediate synthesis gas that can be further processed to produce electricity, chemicals, and transportation fuels [61]. There have been two fundamental approaches to air separation, which are cryogenic and non-cryogenic distillation. The cryogenic distillation is typically reserved for applications that require tonnage quantity of oxygen at ultra-low temperature. The latter involves the separation of air at ambient temperatures using either molecular sieve adsorbents via pressure swing adsorption (PSA) or membrane separation process using the polymeric membranes. Recently, a third category of air separation has emerged, which is based on specialized ceramic membranes that separate oxygen from air at elevated temperatures, in contrast to the supercooled temperatures required by conventional cryogenic distillation. This novel technique is based on dense ceramic membranes, which carry out the separation of oxygen from air at elevated temperatures, typically 800–900 °C. Mixed ionic–electronic-conducting (MIEC) membranes, ion transport membranes (ITM), and oxygen transport membranes (OTM) are acronyms that used to refer to high-temperature ceramic membranes [114]. These terms will be used throughout

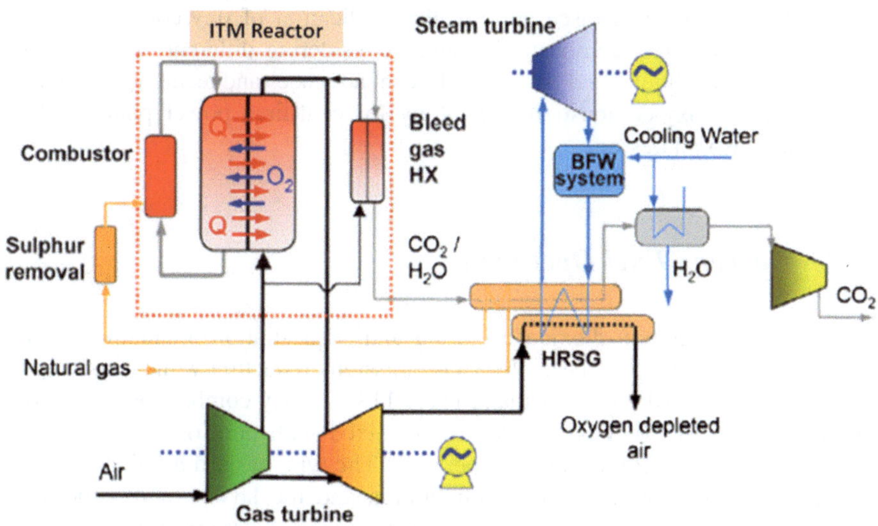

Fig. 2.6 Illustrative flow sheet for oxy-fuel combustion process using membrane reactor technology, with additional unit operations for carbon capture [113]

this work. In this chapter, we are focusing on oxy-fuel combustion in conventional combustion systems. More details about the topic of application of oxygen transport reactors (OTRs) are presented in a separate chapter in this book.

2.4 Oxy-fuel Combustion in Conventional Combustion Systems

In fact, oxy-combustion technology can be applied in conventional combustion systems including gas turbines and boilers with slight modifications to the system [115, 116]. In this approach, oxygen is separated in a device called air separation unit (ASU) which separates oxygen from the flowing air through membranes and, then, oxygen is used in the combustion process inside another conventional combustion device. Usually, the membranes used in the ASU are of ceramic type and are activated to separate oxygen from air at temperatures ranging from 650 to 950 °C [8], which requires additional energy for heating the gases. Also, energy is required for pumping air through the ASU which results in lowering the efficiency of the conventional combustion systems. The other approach for the application of oxy-combustion technology is through the use of the novel ion transport membrane reactors (ITMRs). In this reactor, both oxygen separation and the oxy-combustion processes are carried out within the same unit resulting in more compact size of the ITMR. The membrane in the ITMR is heated to be activated for oxygen separation utilizing part of the heat of combustion, resulting in lower energy consumption and higher efficiency of the reactor. In this section of the present review, the application of oxy-combustion technology in conventional combustion systems is discussed. The required system modifications for oxy-combustion application will be first addressed followed by a discussion on the application of oxy-combustion technology in conventional combustion systems considering different operating fuels including gaseous, liquid, and coal fuels. Recent advances and technology readiness level (TRL) for oxy-combustion technology for coal-fired power plants and gas turbines are also addressed.

2.4.1 Gaseous Fuel Operation

Technologies for carbon dioxide capturing and sequestration, including oxy-fuel combustion, are important means for CO_2 emissions reduction while satisfying the ever-increasing global energy demand [117, 118]. In oxy-combustion technology, part of the exhaust gases is recycled to be mixed with pure oxygen to act as a diluent to control the combustion temperature. The type of used dilution substance may lead to modification on the configuration and the layout of oxy-fuel combustion systems. When water is the dilution substance, water cycle is usually used [119]. A mixture of water vapor and carbon dioxide is used as dilution in Graz

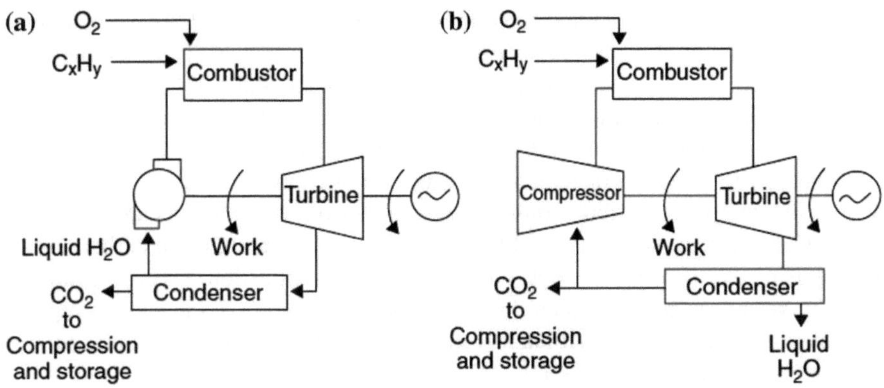

Fig. 2.7 Oxy-fuel power cycle with H_2O as a diluent (left) and CO_2 as a diluent (right) [122]

cycle [120]. In semi-closed oxy-fuel combustion combined cycles, the water vapor is separated by condensation and only carbon dioxide is recycled to the combustor [121]. Figure 2.7 shows different cycle configurations using H_2O or CO_2 as diluent. Figure 2.7a presents the case of burning clean hydrocarbon (C_xH_y) with oxygen in high-pressure combustion. The injection of water is aimed at controlling the maximum temperature. This steam and the produced steam during the combustion processes expand in the turbine along the produced CO_2 in the turbine as shown in Fig. 2.7a. At exit of the turbine, the gases are cooled to separate H_2O from CO_2. Water vapor is recirculated back to the combustor for temperature control. The remaining CO_2 is to be compressed for storage. Figure 2.7b shows another oxy-combustion power cycle. Under high pressure, hydrocarbon is being burned again with oxygen. Carbon dioxide, instead of water, is recycled to control the flame temperature. The working fluid in this configuration is mostly CO_2 with a relatively low level of steam of normally less than 10% by volume. Combustion products expand through a turbine for power generation. For the separation of the produced CO_2, exhaust gases leaving the turbine are cooled in order to remove water. The remainder of the CO_2 is compressed and recycled again to the burner [122]. Specifically, the technology of implementing carbon capture and sequestration to natural gas cycles can compete with carbon-free renewable energy sources from economic prospective [123]. Increasing the shale gas production along with the application of reducing CO_2-emissions policies would make this trend continue in future. Chakroun and Ghoniem [124] conducted a study on high efficiency, low levelized cost of electricity (LCOE) combined cycles utilizing sour gas for oxy-combustion with capture of CO_2 while using CO_2 as a diluent. Their study showed a better technical and economic performance of the sour gas-based-oxy-combustion combined cycles.

Oxy-fuel gas turbine combustors operate differently from air-based gas turbine combustors. The difference in the properties of the O_2/CO_2 mixture, used as oxidizer for oxy-fuel combustion in comparison with that of air, leads to changes in the kinetics and the dynamics of the combustion process as well as the heat transfer characteristics of the flame and the combustion products flow. Using a combustor designed originally for air–methane combustion, Kutne et al. [125] showed that oxygen concentration in the oxidizer significantly affects the flame stability. Due to higher atomic mass of carbon dioxide compared to that of nitrogen, higher oxygen concentration is required in the O_2/CO_2 oxidizer mixture for oxy-fuel combustion as compared to oxygen concentration in air combustion in order to obtain similar laminar flame speed and temperature. Thus, higher oxygen contents are required in the O_2/CO_2 oxidizer to achieve stable flame with a temperature close to that of air in a combustor that is originally designed for air–fuel combustion. Oxygen content of 22% (by volume) was not enough for stable oxy-combustion of methane as reported by Kutne et al. [125] and Williams et al. [126]. In another work, Andersson et al. [127] reported a flame suppression and lower reaction rate for oxy-combustion with oxygen content of 21% (by volume) in CO_2 while 27% (by volume) of oxygen could achieve similar combustion characteristics compared with that of air–fuel combustion in 100-kW gas-fired test ring. However, Williams et al. [126] could obtain a stable flame with lower oxygen contents but for syngas with high hydrogen concentrations.

About 200% increase in the CO equilibrium concentration in the exhaust products was reported by Amato et al. [128] for atmospheric pressure oxy-fuel combustion compared to that for the air–fuel combustion. Despite favoring low CO and O_2 at low flame temperature, kinetics analysis showed that the equilibrium concentrations are not achievable for residence time up to 40 ms. However, high pressures would reduce the carbon monoxide concentration at equilibrium as per Le Chatelier's principle and would lead to higher rates of reaction. Richards et al. [129] used an oxy-fuel gas turbine combustor with a combination of stirred and a plug-flow reactor (PSR/PFR) to explore the CO emissions. Recycling a CO_2 stream at 10 bar and 1650 K results in a flame temperature that could achieve 200 ppm equilibrium concentration in 20 ms at equivalence ratio (ϕ) of 0.98 [130]. CO emissions were experimentally reported to be insignificant from an ideal laboratory-scale atmospheric pressure gas turbine combustor for $\phi < 0.95$. However, CO emissions showed rapid increase for higher values of the equivalence ratio compared with air combustion [126]. Glarborg and Bentzen [130] attributed the considerable increase of CO concentrations in the flame zone to the reaction of CO_2 with atomic hydrogen leading to near-burner corrosion. Cooling rates of the combustion chambers of typical power plant assure no problem anticipation due to the increased CO emissions. Sundkvist et al. [131] predicted about 500 ppm CO concentrations using reactor network calculations for 0.5% excess oxygen at the end of the combustor leading to undesirable further reaction in the turbine section reducing the emissions to 400 ppm. On the other hand, the effect of equivalence ratio was found to be insignificant. However, oxy-combustion with equivalence ratio close to 1 with 30% oxygen contents in the oxidizer mixture leads to a very high temperature compared to the material limitations of the first stages of the gas turbine blades. Liu et al. [132] studied numerically the characteristics of oxy-fuel combustion in gas turbines. The results showed increase in the adiabatic flame temperature in all cases increases as the O_2 concentration in the oxidizer increases,

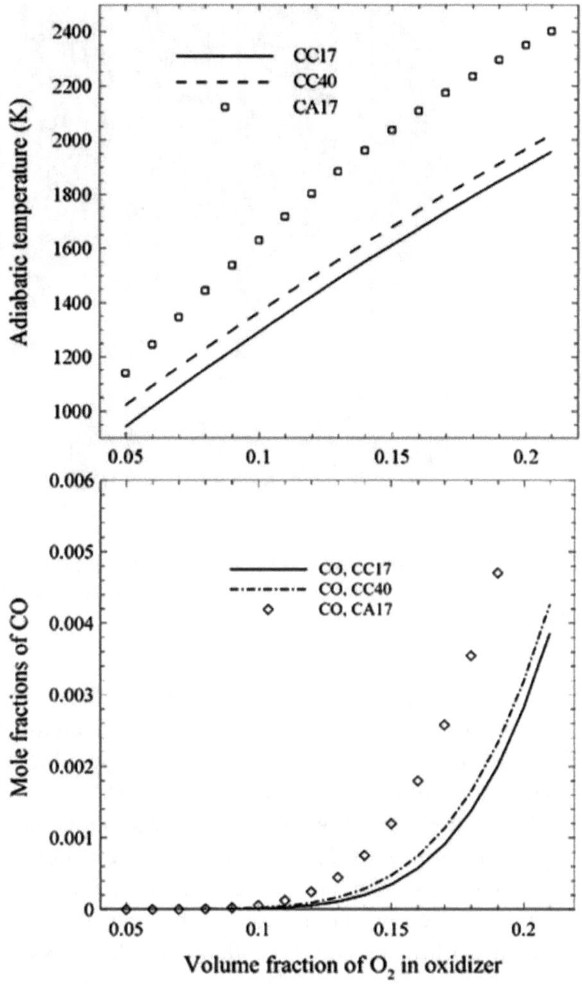

Fig. 2.8 Effect of O_2 concentration in the oxidizer mixture on adiabatic temperature and mole fraction of CO for: CC17: oxy-combustion at 17 bar; CC40: oxy-combustion at 40 bar; CA17: $CH_4/N_2/O_2$ combustion at 17 bar [132]

as shown in Fig. 2.8, due to the decrease of diluents in the mixture. Also, the equilibrium concentrations of CO are increased while increasing O_2 in the oxidizer mixture, due to the dissociation reaction of CO at high temperatures as shown in Fig. 2.8. Recently, Aliyu et al. [133] performed combined experimental and numerical study to characterize hydrogen-enriched oxy-methane flames in swirl-stabilized gas turbine combustor. The results showed that the hydrogen enrichment significantly improves the flame stability. As shown in Fig. 2.9, increasing hydrogen concentration in the inlet fuel mixture (CH_4 plus H_2) resulted in extension of the flame length. This may be attributed to high reactivity, diffusivity, and combustibility nature of hydrogen which results in more stable flame.

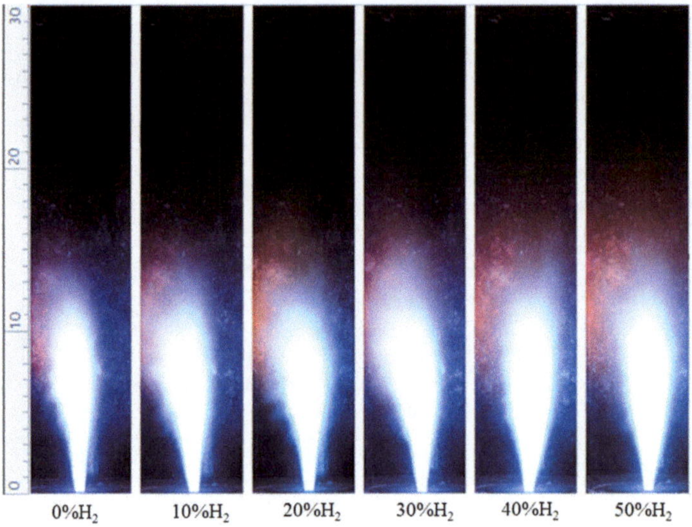

Fig. 2.9 Comparison of different flame shapes using photons captured at steady-state conditions over a range of hydrogen enrichment level and at oxidizer mixture composition of 50%O_2/50% CO_2 [133]

Increasing oxygen concentration in the oxidizer mixture resulted in higher combustion temperature due to the associated increase in flame speed, as shown in Fig. 2.10. However, the operation was not possible under any fuel composition for oxygen concentrations below 22% [7, 133].

One of the main problems toward safe application of oxy-combustion technology in gas turbines is the excessive increase in combustion temperature. Premixing of the reactants upstream of the combustor inlet section seems to be a reasonable solution. Recently, premixed combustion dominated the field of combustion research worldwide. However, flame instability is one of the major resisting issues for proper application of lean premixed combustion (LPM) technology in gas turbines [134, 135]. Rashwan et al. [136] performed an experimental study on premixed oxy-methane flames stabilized over a perforated plate burner trying to identify a range of equivalence ratio for stable flame operation. The results showed that air–fuel combustion flames have larger stability limits over all considered oxygen fractions as shown in Fig. 2.11. However, using a mixture of O_2/CO_2 with oxygen fraction of 36% achieves about 79–82% of the stability limits range of air–fuel combustion flames. Hudak et al. [137] and Amato et al. [138] studied the blow-off measurements and modeling of methane oxy-combustion for low CO cycles. They calculated the chemical timescale depending on the adiabatic flame temperature. The results demonstrated adverse effect of CO_2 dilution on the combustion process. A summary of some review studies on combustion instabilities in gas turbines is presented in Table 2.1.

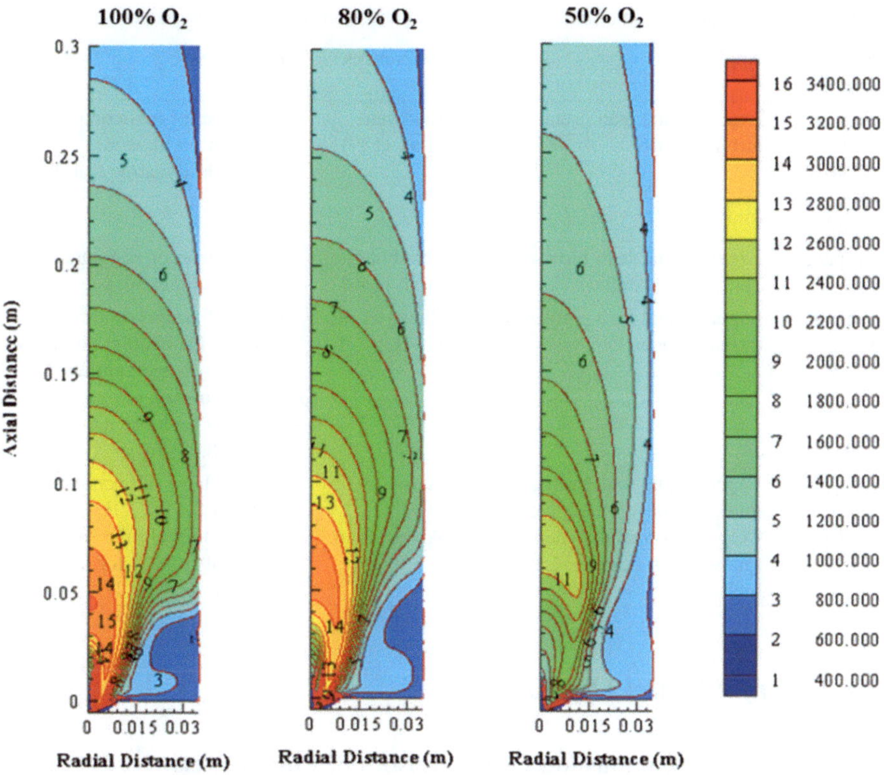

Fig. 2.10 Comparison of temperature distributions over a range of oxygen concentration in the oxidizer mixture and at fuel composition of 20%H_2/80%CH_4 [133]

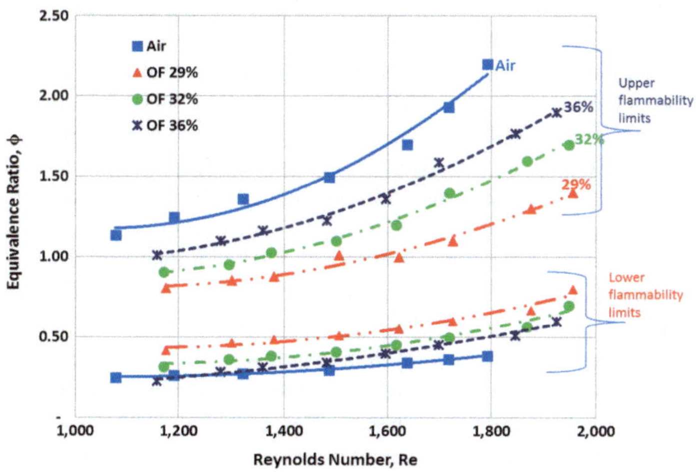

Fig. 2.11 Stability limits of air and oxy-combustion flames over a range of equivalence ratio [136]

Table 2.1 Summary of some review studies on combustion instabilities in gas turbines

Reacting mixture	Application	References	Findings
Syngas fuel $CO/H_2/CH_4$ blends	Experimental lean premixed gas turbine model	Lieuwen et al. [139]	–The effect of fuel composition/ mixtures on blowout, flashback, dynamic stability
Propane/ air	Experimental lean blow-off of bluff body stabilizer	Shanbhogue et al. [140]	–Comparison of computed chemical timescales
Syngas fuel $CO/H_2/CH_4$ blends	Experimental lean premixed gas turbine model	Zhang et al. [141]	–Blow-off occurs at almost constant Damköhler number of 0.6 for hydrogen enriched from 0 to 60% –Blow-off limit reduction was about 28.5%
CH_4/air and $CH_4/O_2/CO_2$	Experimental and Numerical investigation on the effect of CO_2 dilution in the chemical timescale as compared to air–fuel	Hudak et al. [137] Amato et al. [138]	–Higher flame temperature is required for the $CH_4/O_2/CO_2$ mixture to yield the same chemical time as for the CH_4/ air mixture

2.4.2 *Liquid Fuel Operation*

2.4.2.1 General Combustion Characteristics of Liquid Fuels

Liquid fuels are very acceptable and widely used sources of energy due to the advantages of high mass and energy densities (which makes them easy to handle and reduce their transportation economics) and low explosion risks. Liquid fuels are heavily used in generating power in spark and compression ignition internal combustion engines, gas turbine, propulsion rocket, etc. However, these fuels have major disadvantages such as difficult vaporization and formation of soot. Combustion takes place only in gaseous phase, after complete vaporization of the liquid fuel and reaching its flash point temperature. Thus, proper atomization and vaporization are the key concerns while designing a liquid fuel reactor especially for systems which are designed to work under oxy-combustion conditions [142].

An experimental investigation, by Li et al. [143] on evaporation characteristics and droplet size distribution in a spray of fuel atomized by a swirler, showed that the spray inner zone has smaller diameter droplets than that of outer zone, which can be attributed to the effect of spray-induced flow of the ambient air. The decrease in ambient pressure resulted in increase in the air entrainment, and thereby, the evaporation rate of the droplets increased. Ghassemi et al. [144] have also observed similar trend in their experimental vaporization of Kerosene droplets. Negeed et al. [145] reported that the maximum diameter droplet spread increases with the increase of Weber number, size or impinging velocity of the droplet.

Bhattacharya et al. [146] compared the inter-diffusion model and rapid mixing model in predicting the droplets vaporization of two-component fuel. They reported that the inter-diffusion model is more successful in predicting the experimental data, as it assumes that more volatile component of the fuel firstly vaporizes out of the droplet surface while the internal composition remains same. Kamal and Mohamad [147] investigated the combustion characteristics of liquid fuel in a cross-flow swirl combustor. They found that introducing the swirl resulted in mixing enhancement which consequently increased the reaction kinetics and significantly reduced the formation of NO_x and CO. They reported an increase of 1.9 order of magnitude in the radiation from the flame in the swirl air combustor compared with that in the experimentally tested one. Heyes et al. [148] experimentally investigated the combustion characteristics of liquid fuels in a small-size gas turbine burner. They reported an increase in the arithmetic and Sauter mean diameters with the increase in air-to-fuel ratio and a decrease with the increase of preheat temperature due to the fast droplets vaporization and combustion. They reported higher efficiency of combustion at lower air-to-fuel ratio and higher preheat temperature. However, higher unburned hydrocarbon was recorded at lower preheat temperature. An experimental investigation by Lacas et al. [149] on ethanol oxy-combustion showed an increase in the stability of the flame and its extinction velocity with the increase of the mixture oxygen contents.

The subject of liquid fuel evaporation and combustion has been studied numerically using 2D and 3D direct numerical simulations (DNSs) [150] or large eddy simulations (LESs) [151]. Yin [152] performed numerical modeling of *n*-heptane droplets heating and evaporation aiming at the development of a general model for the conversion of fuel droplet. They were able to develop a simulation for droplet heating and vaporization taking into account the droplet dynamics and the droplet/free stream interaction. Kitano et al. [153] numerically investigated the influence of difference in fuel composition on the droplet evaporation and combustion. They reported that the evaporation rate becomes slower for multi-component fuel as compared to single fuel. Targeting reduced emission and clean combustion, Jiang et al. [154] experimentally investigated the combustion process of different liquid fuels including diesel, biodiesel, and straight vegetable oil (VO) using a novel design flow burning (FB) injector. The results showed the capability of the FB injector to produce clean blue flames indicating mainly premixed combustion for all the tested fuels.

Emission control, especially soot formation, is another critical issue in case of liquid fuel combustion. Park and Yoon [155] proposed a two-stage injection strategy to reduce the emission of NO_x and soot simultaneously in dimethyl ether (DME)-fueled engine. The results showed low emissions of NO_x, hydrocarbons, carbon monoxide, and soot when they used the pilot injection strategy with advanced main injection. Speth et al. [156] conducted a study on the effect of using low-aromatic-content alternative jet fuels on the reduction of gas turbines' black carbon emissions. They developed a correlation between the black carbon emissions reduction and the aromatic volume fraction in the alternative fuel. Choi et al. [157] investigated the effect of liquid fuel doping on the formation of soot and polycyclic aromatic hydrocarbon in counterflow ethylene diffusion flames. The results showed an increase in the aromatic ring formation due to addition of liquid fuel. In addition, they predicted an increase in the aromatic ring formation in a mixture of toluene compared with that in case of benzene doping flame.

Oxy-combustion Characteristics of Liquid Fuels

The characteristics of oxy-combustion of liquid fuels depend on many factors including the chemical composition of the fuel, the design and performance of the fuel atomizer, the composition of the oxidizer gas, the furnace design, and the flow field characteristics. Imteyaz et al. [142] investigated numerically air- and oxy-combustion of liquid ethanol mediums inside a 25 kW reactor. Different oxidizer compositions were considered including air, oxygen-enriched air, and tow mixtures of oxygen and carbon dioxide mixture (OF21: O_2/CO_2 as 21%/79% and OF29: O_2/CO_2 as 29%/71%). Figure 2.12 shows the flame shapes and the temperature distributions for three combustion cases including air, OF21, and OF29 [142]. A flame lift and reduction in temperature were observed for the oxy-combustion using the OF21 oxidizer in comparison with that when air is the oxidizer. On the other hand, the OF29 oxidizer could achieve very close

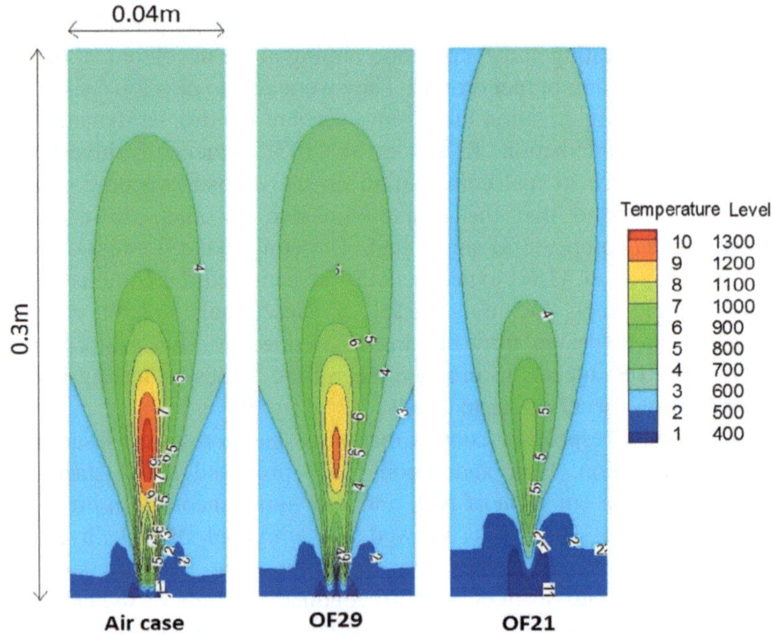

Fig. 2.12 Flame shapes and temperature distributions for the three combustion cases of air, OF21, and OF29 [142]

performance to that of the air as shown in the figure. This indicates that oxy-combustion of liquid fuels is applicable at low concentrations of oxygen in the oxidizer mixture, i.e., at reasonable cost.

The oxy-combustion of liquid ethanol was investigated experimentally by Lacas et al. [158]. They emphasized the effect of oxidizer dilution on the flame stability utilizing an air-assisted injector that is placed concentrically within a vertical combustion chamber. The oxygen dilution was carried out using air while the oxidizer velocity was kept unchanged. It was found that the dilution results in a larger diameter of the flame. In addition, the oxidizer dilution led to a less stable flame that was attributed to the laminar flame speed reduction with increase in diluent contents in the oxidizer mixture. Chaillou et al. [159] studied experimentally the characterization of heavy fuel oil atomizers at industrial scale. The main aspect of this study was to evaluate and optimize the design of an air-assisted atomizer for the particular application on oxy-combustion. The results indicated that the atomizer B at lowest atomizing air flow rate in comparison to high air flow rate for atomizer A produced a spray with smaller droplet sizes than atomizer A, thus achieving a better performance in oxy-combustion as compared to atomizer A. Furthermore, it was observed that the atomizer B demonstrates a reduction of the air atomizing flow rate by three times in comparison with the atomizing flow rate for the atomizer A. Therefore, an increase of at least 1 point in furnace thermal efficiency and a reduction in NO_x by 20% were achieved by using atomizer B. Yi and Axelbaum [160] conducted experiments on low volatile fuels combustion with high water contents present under oxy-fuel conditions. Mixtures of glycerol with ethanol and butanol were used and their concentrations were varied in order to achieve stable combustion with 100 and 85% swirl operations. Two fuel mixtures, 8.3% butanol plus 30% glycerol and 10% ethanol plus 30% glycerol, were found to have stable flame and better combustion. It was found that the swirl intensity affects the flame width and the flame becomes thinner with the decrease in intensity. To overcome the difficulties of burning fuels with high concentrations of water, Yi [161] used the concept of oxy-fuel combustion. Flame stability and flame characteristics for a variety of fuels that have high water contents were investigated. Also, the vaporization of wet fuels was modeled in addition to studying the mechanisms of mass and heat transfer within the droplets in light of the flame characteristics. Having the coal particles insurable in water, Yi [161] investigated the procedures to prepare stable water–coal slurry. Moreover, three different fuel delivery systems were used in order to produce stable flames to properly study the flame stability conditions for each of the three types of fuels considered. Another experimental study of oxy-oil combustion was reported by Chi and Lin [162] utilizing a multi-fuel combustion test furnace modified to suit oxy-fuel combustion. They investigated the impact of circulating the exhaust gas while enriching it with different oxygen concentrations on the combustion of heavy oil in the modified furnace, taking into consideration the effect of operating pressure. They reported the variation of temperature and the composition of the exhaust gases under different operating conditions. It was found that CO_2 concentration increases by oxygen enrichment from 13 to 34.4% in sub-atmospheric operation and from 14.7 to 61.1%

in overatmospheric operation with an estimate of 30% air leakage of total exhaust gas in the first case and 10% in the second case. It was also found that the change of combustion from air/oil to oxy/oil has insignificant effects on flame stability and operating characteristics.

2.4.3 Coal Fuel Operation

2.4.3.1 Oxy-combustion Characteristics of Coal

The cheap price and abundance of world's coal reserves make it a viable fossil fuel to be used in power generation in the near future [163]. Buhre et al. [164] reviewed different environment-friendly technologies related to coal-fired power plants available in the literature. They mainly focused on oxy-fuel combustion as it came out to be a favorable option as compared to post-combustion. They have compared numerical laboratory-scale and pilot-scale studies in order to compare different performance parameters like heat transfer, NO_x and SO_x emissions, ash production, combustion characteristics, flame, and ignition stability. It was indicated that the ratio of oxygen concentration and CO_2 recycling needs to be set for same heat transfer and stable flame for retrofitting of existing plants and new plants. For SO_x reduction, gas cleaning is required to reduce hazardous emissions to the environment. It was concluded that the oxy-fuel combustion is the feasibly economic method for capturing CO_2 among the current available technologies. Another review of carbon capture and storage technologies relevant to coal combustion was reported by Wall [165]. These technologies are namely post-combustion CO_2 capturing via scrubbing removal of CO_2 from the flue gases, integrating the coal gasification with the water shift reaction to remove CO_2 before the combustion, hence called pre-combustion CO_2 and oxy-fuel combustion (fuel combustion in oxygen instead of air). Out of his review, Wall [165] recommended the planning to include fundamental studies as a component of practical-scale pilot plant and demonstrations. A third review on the most recent advances in oxy-fuel coal combustion with focus on pulverized coal was reported by Scheffknecht et al. [166]. Their review revealed that the conventional oxy-fuel combustion along with carbon dioxide capturing and sequestration could lead to an efficiency reduction of 8–12%, which is equivalent to a 21–35% extra fuel consumption. They also reported that the use of ion transport membranes for oxygen/air separation for oxy-fuel combustion purposes has high potential as a more energy-efficient alternative technology for cryogenic air separation. A comprehensive review of various aspects of oxy-combustion of pulverized coal was reported by Chen et al. [167]. The authors discussed the recent developments in the oxy-combustion technique including the experimental and numerical studies. All of the three proposed carbon-capturing techniques for coal-fired power plants, namely pre-combustion capturing or IGCC, oxy-combustion capturing, and post-combustion capturing, have been reviewed emphasizing power generation, capital costs and cost of electricity. It is found that the post-combustion capture is comparatively disadvantageous in terms of capital costs and COE, while IGCC offers slightly better efficiency and less cost compared with that of oxy-combustion capturing. However, oxy-combustion capturing is the most promising technology for retrofitting existing coal-fired power plants. Aspects of atmospheric oxy-coal combustion system and pressurized oxy-coal combustion have been discussed in detail, and it was suggested that pressurized systems have an advantage over conventional systems in terms of overall processing efficiency. Their experimental work on coal-fired oxy-fuel flames showed that temperature distribution in the flame zones in oxy-fuel combustion with 25% oxygen in the oxidizer gas is almost the same as that of conventional system; however, the gas radiation is significantly higher in oxy-coal combustion compared to that with air combustion which is attributed to the high concentration of carbon dioxide for the case of oxy-coal combustion. However, the difference in particle radiation is insignificant. It is also concluded that there exists a considerable operating range of the percentage of recycling the exhaust gases which would make the convective heat transfer in the oxy-fuel combustion comparable to that of the conventional air-based fuel combustion. Thus, it can be indicated that conventional coal-based power plants can be retrofitted to be operated as oxy-fuel combustion power plants [168, 169].

An experimental demonstration of stable operation of dry lignite oxy-combustion in a 0.5 MW_{th} test plant was reported by Kab et al. [170]. They emphasized the influences of oxidizer composition on the time and rate of burning the fuel. Their results indicated that the combustion rate under oxy-fuel conditions differs significantly from that of air combustion and that a substitution of nitrogen by CO_2 yields a reduction of combustion time. The combustion time in an O_2/CO_2 environment is less than that in air for temperature greater than 1100 K in spite of having the same O_2 concentrations in the oxidizing gas. The results also indicated that the distribution oxygen at the furnace entrance has a high influence on the formation of NO_x and the addition of calcium carbonate ($CaCO_3$) as a desulfurization additive results in a reduction of SO_2 by a factor of four during the combustion. Experimental investigation of oxy-combustion of coal and the associated heat transfer characteristics was also carried out by Miklaszewski et al. [171]. Pilot-scale experiments utilized a boiler modified for oxy-combustion to operate at a temperature range above 3000 K. Different operating conditions were investigated resulting in peak temperatures ranging from 2400 to 3200 K. It was found that the flame speed and spectral intensity increase while increasing oxygen and volatile contents in the used coal, decreasing particle sizes and using diluents with lower specific heats. The flame speed varied between 0.8 and 4.1 m/s in all experiments.

A pilot-scale experimental facility for oxy-coal combustion was constructed by Jia et al. [172]. The commissioning tests indicated that the operation of the unit was stable and reliable and results in a stream of exhaust gases comprising 80–90% CO_2 and very low amount of NO_x emissions as compared to air combustion for the same operating fuel. The dependence of operating pressure on the efficiency of pressurized power cycle plant based on oxy-fuel combustion technology was studied by Hong et al. [173]. The study focused on the recovered thermal energy, gross power, and compression power required for air separation and CO_2 compression and purification. The combustor operating pressure was varied from 1.238 to 30 bars. It was found that increasing the operating pressure increases the energy available for recovery, so that less steam bleeding is required. The ASU compression work increases as the pressure is increased. The maximum net efficiency of the plant was occurring near 10 bar operating pressure that reduces the size and economics of the plant. The overall performance of the plant was found to strongly depend on exhaust gas recycling and energy recovery. Chen and Wu [174] performed a study on efficiency improvement of high-pressure oxy-coal power plant with heat integration. The results showed that the high-pressure combustion resulted in reduction of the electricity demand for the CCS system. The waste heat of the high-pressure exhaust gases is also utilized by the heat integration technique to improve the output power of steam turbines. Tumsa et al. [175] examined the effects of coal characteristics on the operation of an efficient high-pressure oxy-coal thermal power generation system. The results indicated that the pressurized system efficiency, at pressure of 30 bars, was higher than that of the atmospheric system by 6.02% when burning lignite coal. Also, the efficiency enhancements in cases of subbituminous and bituminous coals are 3 and 2.61%, respectively. The optimization of the oxy-coal combustion process is very important because of the many variables involved. Zebian et al. [176] simultaneously optimized a 300 MWe oxy-coal combustion system using the gradient-based multivariable optimization approach focusing on multi-start for pressurized wet recycling plant integrated with carbon. Previous published models conducted the sensitivity analysis for one variable which usually fails to capture the complexity of the cycle interactions. The multi-veritable optimization model considers realistic behavior including leaks of steam, losses in heat and drops in pressure, irreversibility in the cycle, and other techno-economic considerations. Using such model indicated that the optimal solution ranges between 3.75 and 6.25 bar reflecting a favorable behavior of the system toward modeling approximations and/or operation fluctuations. Furthermore, the optimization of process of regeneration in the cycle resulted in enhancement of cycle performance for a wide range of operating pressure. A computational model for multi-phase reacting flow simulating the combustion of pulverized coal using LES turbulence model was presented by Donato et al. [177]. They proposed numerical treatments for the dispersed phase equations and implemented sub-models to simulate coal evolution in the burner. Their simulations showed that underestimation of the dispersed phase motion small-scale fluctuations as well as that of particle front velocity is mainly attributed to the difficulties in resembling the evolution of inertial particles under dilute condition. Another computational model was used by Hong et al. [178] to predict the homogenous fuel conversion processes while taking into considerations all the details of the transportation and chemistry characteristics of the gas phase. This modeling was carried out for examining the important processes related to the fuel conversion processes which are the temperature, location, structure and flame thickness, the oxidation rate and the oxygen permeation rate relation, hydrocarbon pyrolysis, and soot formation precursors. The distribution of the measured heat extraction in a 3-MW_{th}

pilot-scale boiler was also plotted at different heat exchanger sections. Measured peak radiative and convective heat fluxes and calculated adiabatic flame temperature under oxy-combustion conditions are compared to the values under air combustion conditions. Hydrogen, carbon monoxide, and methane concentrations during lignite coal pyrolysis in nitrogen and carbon dioxide environments were obtained. The results showed that as compared to the case of N_2-diluted sweep gas, high concentration of CO_2 in the sweep gas stream leads to the suppression of the CH_4 kinetics and a lower flame temperature with a larger thickness. Suppression of H radical chemistry and enhanced OH-driven reactions resulted in the formation of products with smaller H_2 and higher H_2O and CO concentrations. In addition, the dilution of CO_2 suppresses the soot precursors' formation and reduces CH_3 formation. Furthermore, it was deduced that the flame location affects both the diffusion of all species and heat transfer to the membrane from the flame zone, which affects the membrane temperature and, accordingly, the oxygen permeation flux.

Focusing on achieving high efficiency for coal-fired power plants with carbon dioxide capture technologies, Stadler et al. [179] investigated the three-end and four-end integration of oxygen transport membrane (OTM). The investigation proved that maximum efficiencies of 40.1 and 40.7% can be achieved with a three-end and with a four-end membrane integration, respectively, while achieving CO_2 compression with 90% capture for a coal-fired power plant of 1210-MW_{th} compared to 45.9% efficiency for the reference power plant without carbon capturing. A feasibility study for the use of oil shale in oxy-fuel combustion was reported by Al-Makhadmeh et al. [180] using a 20-kW once-through reactor. Both oxy-fuel and air-firing combustion of oil shale were investigated with emphasis on combustion characteristics. It was found that oxy-combustion of oil shale is feasible with 100% oil shale burnout in both cases. It was also found that the SO_2 emission during oil shale oxy-combustion is less than that of air combustion by about 30%. Also, the emissions of NO_x were found to be lower and can be reduced effectively through the adoption of staged oxy-combustion technology as well as air-firing. However, the investigations considering oxy-combustion conditions were performed without the recirculation of exhaust flue gases. A comparison of oxy-combustion and air combustion of solid particles was reported by Marek and Swiatkowski [181] who carried out experiments for single-particle combustion using 2 mm particles as test objects in different gas mixtures to gain basic combustion and ignition knowledge. They focused mainly on comparing the combustion characteristics under air–fuel, oxy-fuel dry, and oxy-fuel wet operating conditions. The experimental results indicated that particles in O_2/CO_2 mixture are burned at lower temperature than in N_2-diluent mixture and that the addition of water vapor under oxy-fuel combustion increased the particle temperature during combustion. They attributed this behavior to the lower molar specific heat of water than that of CO_2 and four times higher reaction rate for char-H_2O gasification reaction than char-CO_2 reaction.

Application of Coal Oxy-combustion in Power Cycles

In most of oxy-fuel combustion research works, natural gas is being used as the working fuel because the combustion exhaust gases are used as the working fluid in turbomachinery. For the application in gas turbines, the working fluid should be ash-free and free of corrosive impurities that normally exist in untreated coal combustion exhaust gases. In order to obtain sufficiently clean coal combustion products, pressurized fluidized bed designs with hot gas purification should be used [182]. However, the more common proposal is being applied through gasification of coal to produce syngas which should be purified before using it the power cycle. In their research, Chiesa and Consonni [183] and Chiesa and Lozza [184, 185] compared three different technologies of integrating carbon capturing with coal syngas. The first approach is to integrate water-gas shift with syngas for hydrogen production and CO_2 separation before the combustion. The second is recycling part of the combustion gases leaving the IGCC in order to increase the percentage CO_2 contents in the exhaust and remove the CO_2 using an exhaust flue gas scrubber. The third approach is to supply syngas to a semi-closed oxy-combustion combined cycle (SCOC-CC). The comparative analysis revealed that the efficiency of the three approaches differed insignificantly for high levels of CO_2 capture. Accordingly, it was concluded that the choice of technology development might be needed based on other factors besides the efficiency. In their investigation on coal syngas with coal gasification, Sanz et al. [186] described the thermo-economic performance of Graz cycle. They compared their findings to those of an IGCC system using water-gas shift for hydrogen production from syngas integrated with pressurized amine scrubbing for carbon dioxide capturing. It was concluded that they overestimated the efficiencies. This was attributed to the neglect of some losses including losses during the purification process of syngas. Comparative analysis based on the cycle configuration under the same conditions revealed higher efficiency for the Graz cycle as compared to the IGCC with CO_2 capture. Pronske et al. [187] presented a primary analysis for a coal syngas oxy-combustion power cycle with water vapor recirculation and suggested that it is possible to obtain a good efficiency compared with that of conventional IGCC with CO_2 capture. Further detailed analysis might benefit from the available data and the gained experience in the currently progressing development studies considering natural gas.

Research was focused, over the last few decades, on the oxy-combustion of pulverized coal (PC) boilers due to its dominant use for electric power generation by most of the utility companies. The way of introducing the oxygen and the recycled part of the exhaust is the main controller of the oxy-combustion in the boiler. Figure 2.13 illustrates various oxygen-pulverized coal combustion power plant layouts indicating the locations of exhaust gas and oxygen [188]. It is worth noting that determining the location of injecting the recycled exhaust gases depends on the fuel water and sulfur contents as well as the sulfur dioxide and water contents in the exhaust gases. Establishing a stable flame under oxy-fuel combustion conditions is imperative but it imposes considerable challenges to the combustor control. Challenges are originated from the fact that having the carbon dioxide as

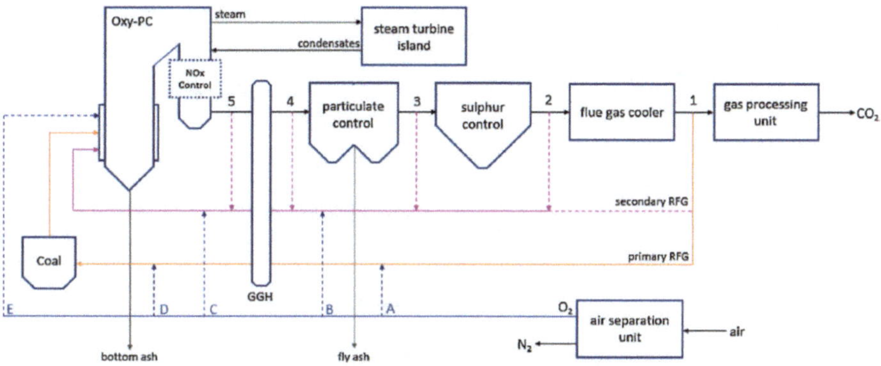

Fig. 2.13 Layout of oxygen-pulverized coal power plants with CO_2 capture, indicating possible locations for exhaust and O_2 addition [188]

the oxidizer (oxygen) diluent results in low adiabatic flame temperature, low flame speed and delays coal particles ignition. Many of the available studies revealed that a rich environment of CO_2 suppresses coal ignition. Liu et al. [189] observed an increasing delay in coal ignition in 20-kW-scale O_2/CO_2 coal combustion tests. Kiga et al. [190] conducted microgravity measurements of coal clouds combustion in 40% (by volume) O_2 in addition to the use of N_2, CO_2 or argon as a balancing diluent. They observed a flame speed reduction in the order: Ar, N_2, and CO_2. Molina and Shaddix [191] conducted experiments on single particle and could record a considerable ignition time delay in low oxygen contents in O_2/CO_2 environment but with insignificant effects on the time needed to completely burn the volatiles. Moreover, it was observed that particle ignition, adiabatic flame temperature, and devolatilization characteristics in a mixture of 30% (by volume) O_2 in CO_2 are close to those in air. Pilot-scale experiments indicated that ignition and flame stability are dependent on the oxygen contents in the oxidizer and become sensitive to changes in the volume flow in oxy-fuel operations [192].

2.4.4 Recent Advances and Technology Readiness Level (TRL)

2.4.4.1 Oxy-combustion for Coal-Fired Power Plants

The recent developments in oxy-combustion technologies and its application in coal-fired power plants are presented in Fig. 2.14 for scales from pilot facilities to demonstration level [193]. A breakthrough in the application of integrated large-scale pilot plant has been encountered during the last decade [193]. These pilot plants include the Vattenfall's Schwarze Pumpe 30-MW_{th} Plant [194], Total's Lacq 30-MW_{th} Project [195], CIUDEN's Technology Development Platform [196], and HUST's 35-MW_{th} pilot plant facility [197]. Demonstrations of such

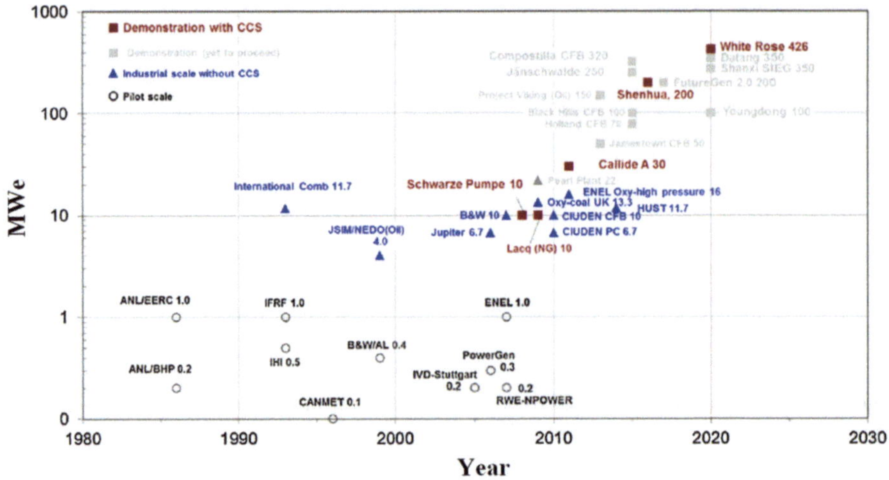

Fig. 2.14 Recent developments of application of oxy-combustion projects from pilot to demonstration size [193]

technologies are needed to commercialize it on utility scale. A few oxy-combustion demonstration projects are current examples of success stories such as Callide Oxy-fuel Project in Australia and the White Rose Project in Europe [193].

The technology readiness level (TRL) system, developed by NASA, as depicted in Table 2.2 [193] can be used to assess the readiness of oxy-fuel technologies for coal-fired power plants. The Electric Power Research Institute (EPRI) has adopted this system to compare different CCS technologies. Based on the TRL, Table 2.2 depicts ordering from 1 for basic principles level to 9 for commercial deployment level. According to Table 2.2 and Fig. 2.14, most of the oxy-fuel combustion systems developed before 2005 (which are of sizes less than 5-MWe) are of TRL-5 or lower. It is worth noting here that most of the research before 2005 was focused on combustion performance assessment. However, large-size combustor testing facilities established by different original equipment manufacturers (OEMs) such as the CEDF test rig (30-MW_{th}) at Ohio, Doosan Babcock (40-MW_{th}), Oxy-Coal UK test rig at Renfrew, and Alstom's (15-MW_{th}) boiler simulator furnace (BSF) test rig at Connecticut can be ranked between TRL-5 and TRL-6; since these facilities are basically partially integrated systems such as boiler testing system, exhaust gas filtration, and oxygen supply systems. On the other hand, pilot plants of large scale such as Vattenfall's Schwarze Pumpe, Total's Lacq Pilot Project, and CIUDEN's oxy-CFB Pilot can be ranked as of TRL-6 while Callide oxy-fuel facility is classified as a large-scale pilot plant of TRL-7 demonstrating the application of oxy-combustion for electric power generation. White Rose Project and any similar large-scale project can be classified as TRL-8. However, technologies that would be classified as TRL-9 (commercial deployment) are yet to be economically viable. Currently, oxy-coal power plant technologies are under trials to establish TRL-8 during the period from 2016 to 2020.

Table 2.2 Technology readiness level of oxy-combustion technology for coal-fired power plants with carbon capture [193]

TRL	TRL phase name for R&D initiatives	Phase for facility development	Oxy-fuel projects
9	Full-scale commercial deployment	Commercial	
8	Sub-scale commercial demonstration plant	Demonstration	White Rose Project
7	Pilot plant	Industrial-scale pilot	Callide oxy-fuel project
6	Component prototype demonstration	Industrial-scale pilot	Schwarze Pumpe pilot plant, Lacq pilot plant, CIUDEN's TDP pilot plants
5	Component prototype development	Industrial-scale pilot	Various large-scale burner test facilities (i.e., B&W, Alstom, Doosan Babcock)
4	Laboratory component testing	Bench	
3	Analytical "proof of concept"	Bench	
2	Application formulation	Bench	
1	Basic principles observed	Bench	

2.4.4.2 Oxy-combustion for Gas Turbines

Implementation of oxy-combustion for gas turbines is more complex than those coal combustion power plants. Operation of gas turbine would involve oxy-fuel combustion at high pressure while the oxidizer and the combustion products go through compression and expansion, respectively, in turbomachine devices following Joule or Brayton cycles. Various cycles are being proposed in the literature or reported as being under development. Only Allam Cycle and the Clean Energy System Water Cycle could achieve TRL-5 or 6. Up to date, a CES Cycle that could achieve a TRL-5/6 demonstrated technologies of the gas turbine main components of a nominal scale of 12–220 MWe in California, USA. On the other hand, Allam Cycle development is led by Net Power. A TRL-5 of such technology would be achieved in its 25-MWe prototype startup which is under construction [193].

2.5 Trends of Oxy-combustion Technology

2.5.1 Oxy-combustion Integrated Power Plants

One of the main problems in using the conventional oxy-fuel combustion technology is the energy penalty arising from the production of pure oxygen using cryogenic air separation in addition to the energy required for CO_2 compression and sequestration. The utilization of waste energy from oxy-fuel combustion for improving the system efficiency was recently investigated by a number of researchers. Ekstrom et al. [198] have evaluated the most promising approach for power plants with CO_2 capturing, developed in enhanced capture of CO_2 projects (ENCAP). Their evaluation included comparison and benchmark regarding the technical performance costs and the technical maturity level and the required future R&D. The comparison with the corresponding reference conventional power plant without CO_2 capturing indicated a reduction of about 6–9% points reduction in net electric efficiencies for the IGCC pre-combustion technologies, oxy-fuel pulverized fuel (PF), and circulating fluidized bed (CFB) technologies and about 15% points reduction for natural-fired integrated reforming combined-cycle (IRCC) pre-combustion technologies. The electric power generation costs for the studied technologies increased by 30–60% compared with the baseline power plants of the evaluated technologies. Chemical looping combustion (CLC) for coal, pet coke, and natural gas proved to have potential to be of high efficiency and lower cost. Colombo et al. [199] analyzed as the full- and part-load operation of a membrane Co-based combined-cycle power plant (CCPP). They applied two load-control strategies (with/without variable guide vanes) in their analysis. The variable guide vanes (VGVs) are utilized to control the mass flow of air in the gas turbine compressor. The results of the analysis indicated that using VGVs improved the combined-cycle efficiencies and increased load reduction capability. Also, the catalytic combustion in the membrane reactor, operating at near stoichiometric conditions, improved as well.

Hong et al. [200] analyzed an oxy-combustion power cycle utilizing a pressurized coal combustor. They showed that this approach could recover more thermal energy from the exhaust gases, and such system will be able to eliminate bleeding the steam at high and low pressure which is typically used in conventional steam cycles by recovering the waste heat and, hence, more thermal energy is utilized achieving higher efficiency. The pressurized combustion process provides the purification and compression unit with a concentrated carbon dioxide stream. A comparison between the case of 1.1 bar combustor with a base case with a 10 bar combustor is carried out by the authors. The results indicated about 3% increase in the net efficiency for the pressurized case and that the proposed approach has lower parasitic power demand. Pak et al. [201] evaluated the characteristics of power generation, economics, and CO_2 reduction effects of a proposed CO_2-capturing repowering system. Their proposed system utilizes low-pressure steam (LPS) from a combined-cycle power generation system (CCPS) to increase generated power and capturing the produced CO_2 based on the oxy-combustion approach. Their results indicated a generation of 2.03 times greater electric power compared with the conventional steam cycle of the same LPS, with an energy efficiency of 54.2%. Also, the net CO_2 reduction amount of the proposed system was 2.03 times larger than that of the conventional one. Their results also indicated that the proposed system is economically feasible and it will surpass the conventional system if the price of captured CO_2 emission allowances goes higher than 30 \$/ton-CO_2. Their analyses indicated that retrofitting the proposed system into the CCPS increases the net generated power by 27.9% and reduces the CO_2 emission amount by 21.8% with a 2.41% degradation of the net power generation efficiency.

A thermodynamic and economic study for using liquid petroleum fuels with the capability of CO_2 capture was carried out by Jamal et al. [202]. In their study, detailed analyses for nine different options for producing H_2 and electricity from liquid petroleum fuels at refueling stations, with CO_2 capture to be used for enhanced oil recovery (EOR), were presented. The study identified two most promising options; the first one is an on-site H_2 production utilizing a high-temperature H_2 selective membrane reactor in combination with selective polymeric membranes for CO_2 capture, and the second is a centralized H_2 production with a carrier for delivery to refueling stations. The study also found that the co-production of electricity at individual refueling stations is not a cost-effective option and that the cost incurred due to CO_2 avoidance makes up almost half of the cost of H_2 production. The feasibility of integrating ITM-based oxy-fuel combustion with thermal desalination technologies was studied by Zak et al. [203]. The thermodynamic analysis showed the feasibility of using this technology to desalinate water on industrial scale. However, they recommended further economic and environmental studies along with numerical optimization to reach at the optimal configuration of the ITM-based oxy-fuel combustion integrated thermal desalination power plants compared with conventional dual-purpose cycles. Gunasekaran et al. [204] optimized the design and operation of advanced zero-emission power cycles that are based on the use of ITM oxy-fuel combustion and carbon capture technologies. The simulation revealed the existence of an optimal membrane temperature of 850 °C operation that results in the highest efficiency. On the other hand, Zebian et al. [205] investigated the optimization of a 300-MWe wet recycling pressurized oxy-coal combustion process with carbon capture and sequestration. They used a simultaneous multivariable gradient-based optimization. Their investigation revealed that cycle configuration design modifications and controlling the operating parameters could lead to an optimal and economic operation of the cycle.

One of the main problems in using the conventional oxy-fuel combustion technology is the energy penalty arising from the production of pure oxygen using cryogenic air separation in addition to the energy required to compress the captured CO_2. This problem motivated Job et al. [206] to analyze, thermodynamically, different compression configurations that use the waste energy from the oxy-combustion unit to replace the heat regeneration in a steam cycle oxy-combustion power plant. On the other hand, the use of ion transport membrane (ITM) to separate oxygen from air makes oxy-fuel combustion a feasible and attractive CO_2-capturing technology. In this regard, Park et al. [207] proposed and analyzed the integration of an ITM oxygen separation unit along with an oxy-fuel combustion system to a solid oxide fuel cell–gas turbine hybrid cycle. The parametric analysis revealed that a properly designed integrated system that comprises oxy-fuel combustion and CO_2 capturing systems can achieve an efficiency equivalent to that of the solid oxide fuel cell–gas turbine hybrid simple cycle. Skorek-Osikowska et al. [208] presented the main results from a thermodynamic analysis of a supercritical unit (steam operation of 600 °C/30 MPa) operating on oxy-combustion technology and fueled with pulverized coal with a power output of 460-MW. The analyses were conducted numerically by building models of individual components, including an oxygen production installation (ASU), a boiler, a steam cycle, and a flue gas conditioning system (CPU). The models, utilizing waste heat from inter-stage cooling of compressors, were built in the commercial programs Gate Cycle and Aspen and then integrated into the Excel environment. The use of waste heat was considered as the methods for counteracting the efficiency decrease resulting from the introduction of ASU and CPU. The auxiliary power rates were determined for the individual installations. The analyses indicated that the auxiliary power rates are highest (15.7–19.1%) for cryogenic ASU, while reaching 35% for the whole installations. The efficiency decreased by 3.5% by increasing the number of multi-section compressors and the use of waste heat. The results of the economic analyses indicated that the oxy-combustion power plant will be economically competitive to the conventional plant without carbon dioxide capture when the price of emission allowances are between 34 and 41 €/ton.

A number of technologies related to oxy-fuel combustion (e.g., air separation, CO_2 recycling, supercritical cycle) were examined and compared for gas-fired and coal-fired power plants [209]. It was found that for the advanced supercritical pulverized fuel power plant, the net efficiency reduces from 44.3 to 35.4% by incorporating CO_2 capture, and for the natural gas combined-cycle (NGCC) power plant, the net efficiency reduces from 56 to 44.7% by incorporating CO_2 capture. The cost analysis was carried out and difference of 2.3 US cents/kW h was found for advanced supercritical pulverized fuel power plant with and without carbon capture and 2.8 US cents/kW h was found for the natural gas combined-cycle (NGCC) power plant. The overall efficiency is strongly affected by air separation unit and CO_2 compression for transportation or sequestration. The plant performance can be enhanced by putting more research efforts in optimizing the plant startup and control systems, and combustion characteristics. Also, new materials have to be developed to sustain high temperatures and to have higher oxygen permeation for air separation.

2.5.2 Third-Generation Technologies for CO_2 Capture

New technologies are being developed targeting the reduction of CO_2 capture costs. Emerging technologies include processes that show (either in the laboratory or in the field) potential to significantly reduce the CO_2 capture cost. However, technologies with high potential for cost reduction also require more time to commercialization. Next to amines and physical solvents considered for CO_2 capture, second-generation technologies for CO_2 capture using advanced amine systems for CO_2 separation are being developed. Also, third-generation technologies for CO_2

Table 2.3 Status of technology development for CO_2 capture and application to different power plants (✓✓: second generation and ✓✓✓: third generation) [210]

Technique		USC PC	IGCC	Oxy-fuel	NGCC
Membranes	O_2/N_2		✓✓✓	✓✓✓	
	CO_2/N_2	✓✓✓			
	CO_2/H_2		✓✓✓		
Solvents	Amine-based solvents	✓✓			✓✓
	Aminoacid-based solvents	✓✓			✓✓
Sorbents		✓✓✓	✓✓✓		

capture are being developed including membrane systems and solid sorbents. Table 2.3 shows the most relevant technologies that are being under development and its application in different power plants including Ultra-Supercritical Pulverized Coal (USC PC), IGCC, Oxy-fuel, and NGCC.

Currently, different technology suppliers and process developers are developing amine systems for post-combustion capture of CO_2. There are different techniques that can be applied to improve the current amine technology including the use of modified packing materials that reduce pressure drop and improve contacting, increasing heat integration to reduce energy requirements, using of additives that reduce corrosion and allow higher amine concentrations and using amine systems or mixture of amines that exhibit higher CO_2 absorption capacity. Solvent absorption is the current technology option for capturing carbon dioxide from syngas. However, membrane technology offers advantages with respect to current technology. Membranes have in general a good prospective and high potential to increase the efficiency of IGCC plants with pre-combustion capture. Membranes are also more environmental friendly than solvent applications and are easy to scale-up. There are many strategies for the application of membranes in IGCC. A recent review by Scholes and co-workers [211] summarizes the implementation possibilities in the following paths including stand-alone membrane technology (considering retention of hydrogen and retention of carbon dioxide) and integrated with water-gas shift reactor. IGCC capture systems with membranes are being investigated in different projects worldwide. There are different materials being investigated for pre-combustion application, such as metallic membranes, ceramic membranes, and polymeric membranes. One other possibility is the oxy-combustion of the sweep stream containing fuel in the permeate side of the membrane (ceramic type) to generate power; then, the stream is condensed to separate water and capture CO_2. This technology is still under development with great potential for applications in power plants; however, this depends on the development of membrane materials that can withstand high temperature and separate enough amount of oxygen for complete combustion of the fuel.

2.6 Conclusions

In this chapter, the current status of oxy-fuel combustion technology and its applications in conventional combustion systems have been reviewed. The different carbon capture technologies have been presented, and their thermal and economic performances have been evaluated considering their limitations for different applications. The operational characteristics of different combustion systems handling different fuels, including gaseous, liquid, and solid (coal) fuels have been studied. The technology status and technology readiness level of oxy-fuel combustion have been discussed considering applications in conventional coal-fired power plants and conventional gas turbines. The new technologies that are being under development toward the capture of CO_2 capture at reduced costs have been itemized and discussed in the present study.

References

1. Lupion M, Diego R, Loubeau L, Navarrete B (2011) CIUDEN CCS project: status of the CO_2 capture technology development plant in power generation. Energy Procedia 4: 5639–5646
2. Amponsah NY, Troldborg M, Kington B, Aalders I, Hough RL (2014) Greenhouse gas emissions from renewable energy sources: a review of lifecycle considerations. Renew Sustain Energy Rev 39:461–475
3. Habib MA, Nemitallah MA (2015) Design of an ion transport membrane reactor for application in fire tube boilers. Energy 81:787–801
4. Gonzalez-Salazar MA (2015) Recent developments in carbon dioxide capture technologies for gas turbine power generation. Int J Greenhouse Gas Control 34:106–116
5. Habib MA, Nemitalah MA, Ben-Mansour R (2013) Recent development in oxy-combustion technology and its applications to gas turbine combustors and ITM reactors. Energy Fuels 27:2–19
6. Guo J, Liu Z, Wang P, Huang X, Li J, Xu P, Zheng C (2015) Numerical investigation on oxy-combustion characteristics of a 200 MWe tangentially fired boiler. Fuel 140:660–668
7. Nemitallah MA, Habib MA (2013) Experimental and numerical investigations of an atmospheric diffusion oxy-combustion flame in a gas turbine model combustor. Appl Energy 111:401–415
8. Nemitallah MA, Habib MA, Ben-mansour R, Ghoniem AF (2014) Design of an ion transport membrane reactor for gas turbine combustion application. J Membr Sci 450:60–71
9. Haslbeck J, Capicotto P, Juehn N, Lewis E, Rutkowski M, Woods M et al (2007) Bituminous coal to electricity. Washington D.C. (1:DOE/NETL-1291)
10. Ahmed P, Habib MA, Ben-Mansour R, Kirchen P, Ghoniem AF (2014) CFD (computational fluid dynamics) analysis of a novel reactor design using ion transport membranes for oxy-fuel combustion. Energy 77:932–944
11. Balachandran U, Kleefisch MS, Kobylinski TP, Morissette SL, Pei S (1997) Oxygen ion-conducting dense ceramic membranes. US patent 5:639-437
12. Schwartz M, White JH, Sammels AF (2000) Solid state oxygen anion and electron mediating membrane and catalytic membrane reactors containing them. US patent 6:033-632
13. Chen CC, Prasad R, Gottzmann CF (1999) Solid electrolyte membrane with porous catalytically-enhancing constituents (assigned to praxair technology). US patent 5:938-822

14. Kim J, Lin YS (2000) Synthesis and oxygen permeation properties of thin YSZ/Pd composite membranes. AIChE J 46:1521
15. Vijaykant S, Agrawal AK (2007) Liquid fuel combustion within silicon-carbide coated carbon foam. Exp Therm Fluid Sci 32:117–125
16. Wang H, Luo K, Fan J (2014) Effects of turbulent intensity and droplet diameter on spray combustion using direct numerical simulation. Fuel 121:311–318
17. Moriai H, Kurose R, Watanabe H, Yano Y, Akamatsu F, Komori S (2013) Large-eddy simulation of turbulent spray combustion in a subscale aircraft jet engine combustor-predictions of NO and soot concentrations. J Eng Gas Turbines Power 135:091503
18. Yin C (2015) Modeling of heating and evaporation of n-heptane droplets: towards a generic model for fuel droplet/particle conversion. Fuel 141:64–73
19. Kitano T, Nishio J, Kurose R, Komori S (2014) Evaporation and combustion of multicomponent fuel droplets. Fuel 136:219–225
20. Jiang L, Agrawal AK, Taylor RP (2014) Clean combustion of different liquid fuels using a novel injector. Exp Therm Fluid Sci 57:275–284
21. Park SH, Yoon SH (2015) Injection strategy for simultaneous reduction of NO_x and soot emissions using two-stage injection in DME fueled engine. Appl Energy 143:262–270
22. Speth RL, Rojo C, Malina R, Barrett SRH (2015) Black carbon emissions reductions from combustion of alternative jet fuels. Atmos Environ 105:37–42
23. Choi SK, Choi BC, Lee SM, Choi JH (2015) The effect of liquid fuel doping on PAH and soot formation in counter flow ethylene diffusion flames. Exp Therm Fluid Sci 60:123–131
24. Wall TF (2007) Combustion processes for carbon capture. Proc Combust Inst 31:31–47
25. Khare S, Wall TF, Gupta RP, Elliott L, Buhre B (2005) Oxy-fuel (O_2/CO_2, O_2/RFG) technology for sequestration-ready CO_2 and emission compliance. The Clearwater coal conference: the 30th international technical conference on coal utilisation & fuel systems, coal technology: yesterday-today-tomorrow, Clearwater, USA, April 17–21
26. Croiset E, Thambimuthu KV (2007) NO_x and SO_2 emissions from O_2/CO_2 recycle coal combustion. Fuel 80:2117–2121
27. Rohan S, Wall T (2011) Sulphur impacts during pulverised coal combustion in oxy-fuel technology for carbon capture and storage. Prog Energy Combust Sci 37:69–88
28. Zhu DL, Egolfopoulos FN, Law CK (1998) Propagation and extinction of stretched premixed flames. Symp (Int) Combust 21(1):1419–1426
29. Liu F, Guo H, Smallwood G (2003) The chemical effect of CO_2 replacement of N_2 in air on the burning velocity of CH_4 and H_2 premixed flames. J Combust Flame 133(4):495–497
30. Andersson K, Johnsson F (2007) Flame and radiation characteristics of gas-fired O_2/CO_2 combustion. Fuel 86:656–668
31. Croiset E, Thambimuthu KV (2007) NO_x and SO_2 emissions from O_2/CO_2 recycle coal combustio. Fuel 80:2117–2121
32. Jyh-Cherng C, Zhen-Shu L, Jian-Sheng H (2007) Emission characteristics of coal combustion in different O_2/N_2, O_2/CO_2 and O_2/RFG atmosphere. J Hazard Mater 142:266–271
33. Kimura K, Omata K, Kiga T, Takano S, Shikisima S (1995) Characteristics of pulverized coal combustion in O_2/CO_2 mixtures for CO_2 recovery. Energy Convers Manag 36:805–808
34. Okazaki K, Ando T (1997) NO_x reduction mechanism in coal combustion with recycled CO_2. Energy 22:207–215
35. Hu Y, Naito S, Kobayashi N, Hasatani M (2000) CO_2, NO_x and SO_2 emissions from the combustion of coal with high oxygen concentration gases. Fuel 79:1925–1932
36. Buhre BJP, Elliott LK, Sheng CD, Gupta RP, Wall TF (2005) Oxy-fuel combustion technology for coal-fired power generation. Prog Energy Combust Sci 31:283–307
37. Seepana S, Jayanti S (2008) In: ASME international mechanical engineering congress and exposition, Boston, MA, pp 435–444
38. Seepana S, Jayanti S (2010) Steam-moderated oxy-fuel combustion. Energy Convers Manag 51:1981–1988

39. Hong J, Chaudhry G, Brisson J, Field R, Gazzino M, Ghoniem A (2009) Analysis of oxy-fuel combustion power cycle utilizing a pressurized coal combustor. Energy 34: 1332–1340
40. Andersen R, MacAdam S, Viteri F, Davies D, Downs J, Paliszewski A (2008) Adapting gas turbines to zeroemission oxy-fuel power plants. ASME turbo expo 2008: power for land, sea, and air. Volume 2: controls, diagnostics and instrumentation; cycle innovations; electric power, Berlin, Germany, June 9–13, 2008. Paper No. GT2008-51377, 781–791; https://doi.org/10.1115/GT2008-51377
41. Zheng L, Pomalis R, Clements B, Herage T (2010) In: The 35th international technical conference on clean coal & fuel systems, June 6–10, Clearwater, Florida, USA
42. Fassbender A, Henry R, Tao L (2009) AEA report, AEA grant number—AEA 07-014
43. Hong J (2009) Ph.D. Thesis, Mechanical Engineering Department, Massachusetts Institute of Technology, Cambridge
44. Hong J, Field R, Gazzino M, Ghoniem A (2010) Operating pressure dependence of the pressurized oxy-fuel combustion power cycle. Energy 35:5391–5399
45. Yang T, Hunt P, Lisauskas R, Ballas E, Vitalis B (2010) In: The 35th international technical conference on clean coal & fuel systems, June 6–10, Clearwater, Florida, USA
46. Toftegaard M, Brix J, Jensen P, Glarborg P, Jensen A (2010) Oxy-fuel combustion of solid fuels. Prog Energy Combust Sci 36:581–625
47. Chen L, Zheng S, Yong, Ghoniem A (2012) Modeling the slag behavior in three dimensional CFD simulation of a vertically-oriented oxy-coal combuston. Prog Energy Combust Sci 38:156–214
48. Haslbeck J, Capicotto P, Juehn N, Lewis E, Rutkowski M, Woods M et al (2010) Bituminous coal to electricity, vol. 1. Washington D.C. (DOE/NETL-1291)
49. BERR, AEA Energy & Environment (2007) Coal R309 BERR/Pub URN 07/1251
50. Darde A, Prabhakar R, Tranier JP, Perrin N (2009) Air separation and flue gas compression and purification units for oxy-coal combustion systems. Energy Procedia 1(1):527–534
51. Andersson K, Johnsson F (2006) Process evaluation of an 865 MWe lignite fired O_2/CO_2 power plant. Energy Convers Manag 47:3487–3498
52. Okawa M, Kimura N, Kiga T, Takano S, Arai K, Kato M (1997) Trial design for a CO_2 recovery power plant by burning pulverized coal in O_2/CO_2. Energy Convers Manag 38: S123–S127
53. Varagani R, Chatel F, Pranda P, Rostam M, Lu Y, Bose A (2005) Performance simulation and cost assessment of oxy-combustion process for CO_2 capture from coal-fired power plants. In: The 4th annual conference on carbon sequestration, May 2–5, Alexandria, VA
54. Torp TA, Gale J (2004) Demostrating storage of CO_2 in geological reservoirs: the Sleipner and SACS projects. Energy 29(1361):1369
55. Emberly S, Hutcheon I, Shevalier M, Durocher K, Gunter WD, Perkins EH (2004) Geochemical monitoring of fluid-rock interaction and CO_2 storage at the Weyburn CO_2-injection enhanced oil recovery site. Energy 29(1393):1401
56. Pfaff I, Kather A (2009) Comparative thermodynamic analysis and integration issues of CCS steam power plants based on oxy-combustion with cryogenic or membrane based air separation. Energy Procedia 1:495–502
57. Dyer PN, Richards RE, Russek SL, Taylor DM (2000) Ion transport membrane technology for oxygen separation and syngas production. Solid State Ionics 134:21
58. Lin YS (2001) Microporous and dense inorganic membranes: current status and prospective. Sep Purif Technol 25:39–55
59. Li K (2007) Ceramic membranes for separation and reaction. Wiley, West Sussex
60. Sirman J (2006) Chap 6. In: Nonporous inorganic membranes. Wiley-VCH Verlag GmbH & Co.KG, Weinheim, pp 165–184
61. Kvamsdal HM, Jordal K, Bolland O (2007) A quantitative comparison of gas turbine cycles with CO_2 capture. Energy 32:10–24
62. Bolland O, Mathieu P (1998) Comparison of two CO_2 removal options in combined cycle power plants. Energy Convers Manag 39(16–18):1653–1663

63. Dillon DJ, Panesar RS, Wall RA, Allam RJ, White V, Gibbins J et al (2004) Oxy-combustion processes for CO_2 capture from advanced supercritical PF and NGCC power plant. In: Proceedings of the seventh international conference on greenhouse gas control technologies—GHGT7, Vancouver, Canada, Sept 2004
64. Staicovici MD (2002) Further research zero CO_2 emission power production: the COOLENERG process. Energy 27:831–844
65. Yantovski EI (1996) Stack downward zero emission fuel-fired power plants concept. Energy Convers Manage 37:867–877
66. Mathieu P, Nihart R (1999) Sensitivity analysis of the MATIANT cycle. Energy Convers Manag 40:1687–1700
67. The European Technology Platform for Zero Emission Fossil Fuel Power Plants (ZEP) (2006) The final report from working group 1 power plant and carbon dioxide capture, 13 Oct 2006
68. Sanz W, Jericha H, Bauer B, Göttlich E (2007) Qualitative and quantitative comparison of two promising oxy-fuel power cycles for CO_2 capture. Paper GT2007-27375, ASME turbo expo, Montreal, Canada
69. Jericha H, Sanz W, Göttlich E (2006) Design concept for large output Graz cycle gas turbines. ASME paper GT2006-90032, ASME turbo expo 2006, Barcelona, Spain
70. Anderson RE, MacAdam S, Viteri F, Davies DO, Downs JP, Paliszewski A (2008) Adapting gas turbines to zero emission oxy-fuel power plants. Paper GT 2008-51377, ASME turbo expo, Berlin, Germany
71. Li H., Yan J., Yan J., Anheden M., Impurity impacts on the purification process on oxy-fuel combustion based CO2 capture and storage system, Appl Energy 2009;86:202–13.
72. Jericha H, Göttlich E (2002) Conceptual design for an industrial prototype Graz cycle power plant. Paper GT 2002-30118, ASME turbo expo, Amsterdam, Netherlands
73. Woycenko D, van de Kamp W, Roberts P (1997) European Commission Joule II clean coal technology program 1992–1995, vol II. Powder coal combustion projects final reports. ISBN:92-9-828-006-7
74. Heil P, Torporov D, Stadler H, Tschunko S, Forster M, Kneer R (2009) Development of an oxycoal swirl burner operating at low O_2 concentrations. Fuel 88:1269–1274
75. Kutne P, Kapadia BK, Meier W, Aigner M (2010) Experimental analysis of the combustion behavior of oxyfuel flames in a gas turbine model combustor. Proc Combust Inst https://doi.org/10.1016/j.proci.2010.07.008
76. Kutne P, Kapadia BK, Meier W, Aigner M (2011) Experimental analysis of the combustion behavior of oxyfuel flames in a gas turbine model combustor. Proc Combust Inst 33 (2):3383–3390
77. Cao M, Eickhoff H, Joos F, Simon B (1987) In: ASME propulsion and energetics, 70th symposium, AGARD conference proceedings 422, 8.1
78. Liu C, Chen G, Sipöcz N, Assadi M, Bai X (2012) Characteristics of oxy-fuel combustion in gas turbines. Appl Energy 89:387–394
79. Gupta A, Lilley D, Syred N (1984) Swirl flows. Abacus Press, Kent
80. Syred N, Chigier NA, Beér JM (1971) Flame stabilization in recirculation zones of jets with swirl. In: Thirteenth symposium on combustion, University of Utah, Salt Lake City, 1970. Thirteenth Int Symp Combust Combust Inst 13:617–624
81. Syred N, Beér JM (1974) Combustion in swirling flows: a review. Combust Flame 23: 143–201
82. Weber R, Dugué J (1992) Combustion accelerated swirling flows in high confinements. Prog Energy Combust Sci 18:349–367
83. Correa SM (1998) Power generation and aeropropulsion gas turbines: from combustion science to combustion technology. Combust Inst 27:1793–1807
84. Lefebvre AH (1999) Gas turbine combustion. Taylor & Francis, Philadelphia
85. Bauer HJ (2004) New low emission strategies and combustor designs for civil aeroengine Applications. Prog Comput Fluid Dyn 4:130–142

86. Meier W, Duan XR, Weigand P, Lehmann B (2004) Temperatur-Messungen in turbulenten Drallflammen: Thermoelemente im Vergleich zu Laser-Raman-Streuung. Gaswärme Int 53:153–158
87. Stricker W, Kohse-Höinghaus K, Jeffries J (2002) Applied combustion diagnostics. Taylor & Francis, New York, pp 155–193
88. Eckbreth AC (1996) Laser diagnostic for combustion temperature and species. Gordon & Breach
89. Kohse-Höinghaus K, Jeffries J (2002) Applied combustion diagnostics. Taylor & Francis, New York
90. Masri AR, Dibble RW, Barlow RS (1996) The structure of turbulent nonpremixed flames revealed by Raman-Rayleigh-LIF measurements. Prog Energy Combust Sci 22:307–362
91. Wolfrum J (1998) Lasers in combustion: from basic theory to practical devices. Symp (Int) On Combust 27(1):1–41
92. Kaaling H, Ryden R, Bouchie Y, Ansart D, Magre P, Guin C (1997) In: 13th international symposium on air breathing engines (ISABE), Chattanooga, TN (USA)
93. Kampmann S, Seeger T, Leipertz A (1995) Simultaneous CARS and 2D laser Ragleigh thermometry in a contained swirl combustor. Appl Opt 34:2780–2786
94. Dinkelacker F, Soika A, Most D, Hofmann D, Leipertz A, Polifke W, Döbbeling K (1998) Structure of locally quenched highly turbulent lean premixed flames. Symp (Int) On Combust 27(1):857–865
95. Fink R, Hupfer A, Rist D (2002) In: Proceedings, ASME turbo expo, GT-2002-30078
96. Cooper CS, Laurendeau NM (2000) Comparison of laser-induced and planar laser-induced fluorescence measurements of nitric oxide in a high-pressure, swirl-stabilized, spray flame. Appl Phys B 70:903–910
97. Cooper CS, Laurendeau NM (2000) Quantitative measurements of nitric oxide in high-pressure (2–5 atm), swirl-stabilized spray flames via laser-induced fluorescence. Combust Flame 123:175–188
98. Shih WP, Lee JG, Santavicca DA (1996) Stability and emissions characteristics of a lean premixed gas turbine combustor. Proc Combust Inst 26:2771–2778
99. Deguchi Y, Noda M, Fukuda Y, Ichinose Y, Endo Y, Inida M, Abe Y, Iwasaki S (2002) Industrial applications of temperature and species concentration monitoring using laser diagnostic. Meas Sci Technol 13:R103–R115
100. Hedman PO, Warren DL (1995) Turbulent velocity and temperature measurements from a gas-fueled technology combustor with a practical fuel injector. Combust Flame 100:185–192
101. Lee SY, Seo S, Broda JC, Pal S, Santoro RJ (2000) An experimental estimation of mean reaction rate and flame structure during combustion instability in a lean premixed gas turbine combustor. Proc Combust Inst 28:775–782
102. Arnold A, Bombach R, Hubschmid W, Käppeli B (2000) Fuel-oil concentration in a gas turbine burner measured with laser-induced fluorescence. Exp Fluids 29(5):468–477
103. Fritz J, Kröner M, Sattelmayer (2001) In: Proceedings, ASME turbo expo, 2001-GT-0054
104. Löfström C, Engström J, Richter M, Kaminsky CF, Johansson P, Nyholm K, Nygren J, AldénM (2000) In: Proceedings, ASME turbo expo, 2000-GT-0124
105. Gittins C, Shenoy S, Aldag H, Pacheco D, Miller M, Allen M (2000) Measurements of major species in a high pressure gas turbine combustion simulator using Raman scattering. 38th Aerospace Sciences Meeting and Exhibit, Andover, MA, USA
106. Meier UE, Wolff-Gaßmann D, Heinze J, Frodermann M, Magnusson I, Josefsson G (1999) In: 18th international congress on instrumentation in aerospace simulation facilities (ICIASF 99), Toulouse, pp 7.1–7.7
107. Meier UE, Wolff-Gaßmann D, Stricker W (2000) LIF imaging and 2D temperature mapping in a model combustor at elevated pressure. Aerosp Sci Technol 4:403–414
108. Carl M, Behrendt T, Fleing C, Frodermann M, Heinze J, Hassa C, Meier U, Wolff-Gaßmann D, Hohmann S, Zarzalis N (2001) Experimental and numerical investigation of a planar combustor sector at realistic operating conditions. J Eng Gas Turbines Power 123:810–816

109. Kunz O, Noll B, Lückerath R, Aigner M, Hohmann S (2001) In: 37th AIAA/ASME/ SAE/ ASEE joint propulsion conference and exhibition, Salt Lake City, UT, AIAA 2001-3706
110. Williams TC, Shaddix CR, Schefer RW (2008) Effect of syngas composition and CO_2-diluted oxygen on performance of a premixed swirl stabilized combustor. Combust Sci Technol 180:64–88
111. Sautet JC, Salentey L, DitarantoM, Samaniego JM (2001) Length of natural gas-oxygen non-premixed flames. Combust Sci Technol 166:131–150
112. Ditaranto M, Hals J (2006) Combustion instabilities in sudden expansion oxy–fuel flames. Combust Flame 146:493–512. https://doi.org/10.1021/ef300539c. Energy Fuels 2012, 26: 4599–4606
113. Habib MA, Badr HM, Ahmed SF, Ben-Mansour R, Mezghani K, Imashuku S, Lao GJ, Shao-Horn Y, Mancini ND, Mitsos A, Kirchen P, Ghoniem AF (2011) A review of recent developments in carbon capture utilizing oxy-fuel combustion in conventional and ion transport membrane systems. Int J Energy Res 35:741–764
114. Hashim SM, Mohamed A, Bhatia S (2010) Current status of ceramic-based membranes for oxygen separation from air. Adv Coll Interface Sci 160:88–100
115. Habib MA, Tahir F, Nemitallah MA, Ahmed WH, Badr HM (2015) Experimental and numerical analysis of oxy-fuel combustion in a porous plate reactor. Int J Energy Res 39:1229–1240
116. Habib MA, Nemitallah MA, Ahmed P, Sharqawy MH, Badr HM, Muhammad I, Yaqub M (2015) Experimental analysis of oxygen-methane combustion inside a gas turbine reactor under various operating conditions. Energy 86:105–114
117. Figueroa JD, Fout T, Plasynski S, McIlvried H, Srivastava RD (2008) Advances in CO_2 capture technology-the U.S. Department of Energy's carbon sequestration program. Int J Greenhouse Gas Control 2(1):9–20
118. Gielen D (2003) The future role of CO_2 capture and storage: results of the IEA-ETP model. Report no EET/2003/04. http://www.iea.org/textbase/papers/2003/eet04.pdf
119. Gou C, Cai R, Hong H (2006) An advanced oxy-fuel power cycle with high efficiency. Proc Inst Mech Eng A J Power Energy 220(4):315–325
120. Sanz W, Jericha H, Luckel F, Heitmeir F (2015) A further step towards a Graz cycle power plant for CO_2 capture. ASME paper GT2005-68456, ASME turbo expo
121. Dahlquist A, Genrup M, Sjoedin M, Jonshagen K (2013) Optimization of an oxy-fuel combined cycle regarding performance and complexity level. In: ASME turbo expo 2013: turbine technical conference and exposition. American Society of Mechanical Engineers, V002T07A011-V002T07A011
122. Richards G, Williams M, Casleton K (2012) Novel cycles: oxy-combustion turbine cycle Systems. In: Combined cycle systems for near-zero emission power generation. Woodhead publishing series 32, pp 186–219
123. U.S. Energy Information Administration (2014) Estimated levelized cost of electricity (LCOE) for new generation resources-2019
124. Chakroun NW, Ghoniem AF (2015) High-efficiency low LCOE combined cycles for sour gas oxy-combustion with CO_2 capture. Int J Greenhouse Gas Control 41:163–173
125. Kutne P, Kapadia BK, Meier W, Aigner M (2011) Experimental analysis of the combustion behaviour of oxy fuel flames in a gas turbine model combustor. Proc Combust Inst 33 (2):3383–3390
126. Williams TC, Shaddix CR, Schefer RW (2007) Effect of syngas composition and CO_2-diluted oxygen on performance of a premixed swirl-stabilized combustor. Combust Sci Technol 180(1):64–88
127. Andersson K, Mönckert P, Maier J, Scheffknecht G, Johnsson F (2006) Combustion and flame characteristics of oxy-fuel combustion. In: 8th international conference on greenhouse gas control technologies, Trondheim, Norway, pp 9–22

128. Amato A, Hudak R, Noble DR, Scarborough D, Peter A, Seitzman JM, Lieuwen TC (2010) Methane oxy-combustion for low CO_2 cycles: measurements and modeling of CO and O_2 emissions. In: ASME turbo expo: power for land, sea, and air. American Society of Mechanical Engineers, pp 213–222
129. Richards GA, Casleton KH, Chorpening BT (2005) CO_2 and H_2O diluted oxy-fuel combustion for zero-emission power. Proc Inst Mech Eng A 219(2):121–126
130. Glarborg P, Bentzen LLB (2008) Chemical effects of a high CO_2 concentration in oxy-fuel combustion of methane. Energy Fuels 22(1):291–296
131. Sundkvist SG, Dahlquist A, Janczewski J, Sjödin M, Bysveen M, Ditaranto M, Langörgen y, Seljeskog M, Siljan M (2014) Concept for a combustion system in oxy-fuel gas turbine combined cycles. J Eng Gas Turbines Power 136(10):101513–101513
132. Liu CY, Chen G, Sipöcz N, Assadi M, Bai XS (2012) Characteristics of oxy-fuel combustion in gas turbines. Appl Energy 89:387–394
133. Aliyu M, Nemitallah MA, Said SA, Habib MA (2016) Characteristics of H_2-enriched CH_4-O_2 diffusion flames in a swirl-stabilized gas turbine combustor: experimental and numerical study. Int J Hydrogen Energy 41(44):20418–20432
134. Mattingly JD, Von OH (2006) Elements of propulsion : gas turbines and rockets. https://doi.org/10.2514/4.861789
135. Verhoek FH (1969) Thermodynamics and rocket propulsion. J Chem Educ 46:140. https://doi.org/10.1021/ed046p140
136. Rashwan SS, Ibrahim AH, Abou-Arab TW, Nemitallah MA, Habib MA (2016) Experimental investigation of partially premixed methane-air and methane-oxygen flames stabilized over a perforated-plate burner. Appl Energy 169(1):126–137
137. Hudak R, Noble DR, Scarborough D, Carlo PAD (2010) GT2010-22302:1-11
138. Amato A, Hudak B, D'Souza P, D'Carlo P, Noble D, Scarborough D, et al (2011) Measurements and analysis of CO and O2 emissions in CH4/CO2/O2 flames. Proc Combust Inst 33:3399–3405
139. Lieuwen T, McDonell V, Petersen E, Santavicca D (2008) Fuel flexibility influences on premixed combustor blowout, flashback, autoignition, and stability. J Eng Gas Turbines Power 130:011506
140. Shanbhogue SJ, Husain S, Lieuwen T (2009) Lean blowoff of bluff body stabilized flames: scaling and dynamics. Prog Energy Combust Sci 35:98–120
141. Zhang Q, Noble DR, Meyers A, Xu K, Lieuwen T (2005) Characterization of fuel composition effects in H_2/CO/CH_4 mixtures upon lean Blowout. ASME/IGTI turbo expo 129:1–13
142. Imteyaz B, Habib MA, Nemitallah MA, Jamal A (2015) Investigation of liquid ethanol evaporation and combustion in air and oxygen environments inside a 25-kW vertical reactor. Proc Inst Mech Eng Part A J Power Energy:1–15
143. Li T, Nishida K, Hiroyasu H (2011) Droplet size distribution and evaporation characteristics of fuel spray by a swirl type atomizer. Fuel 90(7):2367–2376
144. Ghassemi H, Baek SW, Khan QS (2006) Experimental study on evaporation of kerosene droplets at elevated pressures and temperatures. Combust Sci Technol 178(9):1669–1684
145. Negeed ESR, Ishihara N, Tagashira K, Hidaka S, Kohno M, Takata Y (2010) Experimental study on the effect of surface conditions on evaporation of sprayed liquid droplet. Int J Therm Sci 49(12):2250–2271
146. Bhattacharya P, Ghosal S, Som SK (1996) Evaporation of multicomponent liquid fuel droplets. Int J Energy Res 20(5):385–398
147. Kamal MM, Mohamad AA (2007) Investigation of liquid fuel combustion in a cross-flow burner. Proc Inst Mech Eng Part A J Power Energy 221(3):371–385
148. Heyes AL, Jelercic D, Whitelaw JH (1998) Experiments in a small gas-turbine combustor with gas and liquid fuels. RTO MP 14:12–16
149. Lacas F, Leroux B, Darabiha N (2005) Experimental study of air dilution in oxy-liquid fuel flames. Proc Combust Inst 30(2):2037–2045

150. Wang H, Luo K, Fan J (2014) Effects of turbulent intensity and droplet diameter on spray combustion using direct numerical simulation. Fuel 121:311–318
151. Moriai H, Kurose R, Watanabe H, Yano Y, Akamatsu F, Komori S (2013) Large-eddy simulation of turbulent spray combustion in a subscale aircraft jet engine combustor-predictions of NO and soot concentrations. J Eng Gas Turbines Power 135:091503
152. Yin C (2015) Modeling of heating and evaporation of n-heptane droplets: towards a generic model for fuel droplet/particle conversion. Fuel 141:64–73
153. Kitano T, Nishio J, Kurose R, Komori S (2014) Evaporation and combustion of multicomponent fuel droplets. Fuel 136:219–225
154. Jiang L, Agrawal AK, Taylor RP (2014) Clean combustion of different liquid fuels using a novel injector. Exp Therm Fluid Sci 57:275–284
155. Park SH, Yoon SH (2015) Injection strategy for simultaneous reduction of NO_x and soot emissions using two-stage injection in DME fueled engine. Appl Energy 143:262–270
156. Speth RL, Rojo C, Malina R, Barrett SRH (2015) Black carbon emissions reductions from combustion of alternative jet fuels. Atmos Environ 105:37–42
157. Choi SK, Choi BC, Lee SM, Choi JH (2015) The effect of liquid fuel doping on PAH and soot formation in counter flow ethylene diffusion flames. Exp Therm Fluid Sci 60:123–131
158. Lacas F, Leroux B, Darabiha N (2005) Experimental study of air dilution in oxy-liquid fuel flames. Proc Combust Inst 30:2037–2045
159. Chaillou D, Alaterre V, Paubel X, Marcano N, Leroux B, Tsiava R (2010) Characterization of fuel oil atomizers at industrial scale, pp 1–8
160. Yi F, Axelbaum RL (2013) Oxy-combustion of low-volatility fuel with high water content. Department of Energy, Environmental & Chemical Engineering, St. Louis, Washington University, USA, pp 1–8
161. Yi F (2013) Oxy-combustion of high water content fuels. Electronic theses and dissertations. Paper 1177
162. Chi CC, Lin T (2013) Oxy-oil combustion characteristics of an existing furnace. Appl Energy 102:923–930
163. U.S. Department of Energy (2010) Energy sources: electric power-fossil fuel. http://www.energy.gov/energysources/electricpower.htm
164. Buhre BJP, Elliott LK, Sheng CD, Gupta RP, Wall TF (2005) Oxy-fuel combustion technology for coal-fired power generation. Prog Energy Combust Sci 31:283–307
165. Wall TF (2007) Combustion processes for carbon capture. Proc Combust Inst 31(1):31–47
166. Scheffknecht G, Al-Makhadmeh L, Schnell U, Maier J (2011) Oxy-fuel coal combustion—a review of the current state-of-the-art. Int J Greenhouse Gas Control 5:16–35
167. Chen L, Yong SZ, Ghoniem AF (2012) Oxy-fuel combustion of pulverized coal: characterization, fundamentals, stabilization and CFD modeling. Prog Energy Combust Sci 38(2):156–214
168. Borah RC, Ghosh P, Rao PG (2011) A review on devolatilization of coal in fluidized bed. Int J Energy Res 35(11):929–963
169. Mathekga HI, Oboirien BO, North BC (2016) A review of oxy-fuel combustion in fluidized bed reactors. Int J Energy Res 40(7):878–902
170. Kab H, Tappe S, Krautz HJ (2008) The combustion of dry lignite under oxy-fuel process conditions in a 0.5-MW_{th} test plant. Energy Procedia 1:423–430
171. Miklaszewski EJ, Zheng Y, Son SF (2013) Oxy-fuel combustion: laboratory experiments and pilot scale tests. Fuel 104:452–461
172. Jia L, Tan Y, McCalden D, Wu Y, He I, Symonds R, Anthony EJ (2012) Commissioning of a 0.8 MW_{th} CFBC for oxy-fuel combustion. Int J Greenhouse Gas Control 7:240–243
173. Hong J, Field R, Gazzino M, Ghoniem AF (2010) Operating pressure dependence of the pressurized oxy-fuel combustion power cycle. Energy 35(12):5391–5399
174. Chen HT, Wu W (2015) Efficiency enhancement of pressurized oxy-coal power plant with heat integration. Int J Energy Res 39(2):256–264

175. Tumsa TZ, Mun TY, Lee U, Yang W (2017) Effects of coal characteristics to performance of a highly efficient thermal power generation system based on pressurized oxy-fuel combustion. Int J Energy Res 41(1):127–138
176. Zebian H, Gazzino M, Mitsos A (2012) Multi-variable optimization of pressurized oxy-coal combustion. Energy 38:37–57
177. Donato F, Favini B, Giacomazzi E, Picchia FR, Arcidiacono N, Cecere D, Creta F (2010) Numerical modelling of pulverized coal oxy-combustion. Report RdS/2010/30
178. Hong J, Kirchen P, Ghoniem AF (2013) Laminar oxy-fuel diffusion flame supported by an oxygen-permeable-ion-transport membrane. Combust Flame 160(3):704–717
179. Stadler H, Beggel F, Habermehl M, Persigehl B, Kneer R, Modigell M, Jeschke P (2011) Oxy-fuel coal combustion by efficient integration of oxygen transport membranes. Int J Greenhouse Gas Control 5(1):7–15
180. Al-Makhadmeh L, Maier J, Al-Harahsheh M, Scheffknecht G (2013) Oxy-fuel technology: an experimental investigation into oil shale combustion under oxy-fuel conditions. Fuel 103:421–429
181. Marek E, Swiatkowski B (2014) Experimental studies of single particle combustion in air and different oxy-fuel atmospheres. Appl Therm Eng 66:35–42
182. Burkett J (2006) Reborn in the USA, modern power systems. Available from: http://www.modernpowersystems.com/story.aspstoryCode=2037552
183. Chiesa P, Consonni S (1999a) Shift reactors and physical absorption for low-CO_2 emission IGCCs. ASME J Eng Gas Turbines Power 121:295–305
184. Chiesa P, Lozza G (1999b) CO_2 emission abatement in IGCC power plants by semi-closed cycles: part A—with oxygen-blown combustion. ASME J Eng Gas Turbines Power 121:635–641
185. Chiesa P, Lozza G (1999c) CO_2 emission abatement in IGCC power plants by semi-closed cycles: part B—with air-blown combustion and CO_2 physical absorption. ASME J Eng Gas Turbines Power 121:642–648
186. Sanz W, Mayr M, Jericha H (2010) Thermodynamic and economic evaluation of an IGCC plant based on the Graz cycle for CO_2 capture. In: ASME turbo expo 2010, paper GT2010–22189, Glasgow, UK
187. Pronske K, Trowsdale L, Macadam S, Viteri F (2006) An overview of turbine and combustor development for coal-based oxy-syngas systems. In: ASME turbo expo 2006, paper GT2006–90816, Barcelona, Spain
188. Davidson RM, Santos S (2010) Oxy-fuel combustion of pulverised coal. IEA Clean Coal Centre. Report no CCC/168, ISBN 978-92-9029-488-7, June 2010
189. Liu H, Zailani R, Gibbs BM (2005) Pulverized coal combustion in air and in O_2/CO_2 mixtures with NO_x recycle. Fuel 84(16):2109–2115
190. Kiga T, Takano S, Kimura N, Omata K, Okawa M, Mori T, Kato M (1997) Characteristics of pulverized-coal combustion in the system of oxygen/recycled flue gas combustion. Proc Third Int Conf Carbon Dioxide Remov 38(supplement 0):S129
191. Molina A, Shaddix CR (2007) Ignition and devolatilization of pulverized bituminous coal particles during oxygen/carbon dioxide coal combustion. Proc Combust Inst 31(2):1905–1912
192. Zhang J, Kelly KE, Eddings EG, Wendt JOL (2011) CO_2 effects on near field aerodynamic phenomena in 40-kW, co-axial, oxy-coal, turbulent diffusion flames. IJGCC 5(special issue/suppl):S47–S57
193. Stanger R, Wall T, Spörl R et al (2015) Oxy-fuel combustion for CO_2 capture in power plants. Int J Greenhouse Gas Control 40:55–125
194. Burchhardt U, Griebe S (2013) Tests and results of Vattenfall's oxy-fuel pilot plant. In: 3rd oxy-fuel combustion conference (OCC3), Ponferrada, Spain, Sept 2013
195. Copin D (2013) The storage dimension of the oxy-combustion based integrated CCS project of Lacq and Rousse. In: 3rd oxy-fuel combustion conference (OCC3), Ponferrada, Spain, Sept 2013

196. Otero P (2013) Keeping the flame alight. An overview to CIUDEN's R&D activities—the way forward. In: 3rd oxy-fuel combustion conference (OCC3), Ponferrada, Spain, Sept 2013
197. Zheng C (2014) Some important aspects about oxy-fuel combustion technology. In: 39th clear water conference, Clearwater, Fl, USA, June 2009
198. Ekstrom C, Schwending F, Biede O, Franco F, Haupt G, Koeijer G, Papapavlou C, Rokke PE (2009) Techno-economic evaluations and benchmarking of pre-combustion CO_2 capture and oxy-fuel process development in the European ENCAP project. Energy Procedia 1(1):4233–4240
199. Colombo KE, Bolland O, Khorton VV, Stiller C (2009) Simulation of an oxygen membrane-based combined cycle power plant: part load operation with operational and material constraints. Energy Environ Sci 2:1310–1324
200. Hong J, Chaudhry G, Brisson JG, Field R, Gazzino M, Ghoniem AF (2009) Analysis of oxy-fuel combustion power cycle utilizing a pressurized coal combustor. Energy 34 (9):1332–1340
201. Pak PS, Lee YD, Ahn KY (2010) Characteristics and economic evaluation of a power plant applying oxy-fuel combustion to increase power output and decrease CO_2 emission. Energy 35:3230–3238
202. Jamal A, Pham TV, Al-Juaied M (2012) Combined hydrogen and electricity production with CO_2 capture using liquid petroleum fuels. Saudi Aramco J Technol Winter 2012:15–22
203. Zak GM, Mancini ND, Mitsos A (2013) Integration of thermal desalination methods with membrane-based oxy-combustion power cycles. Desalination 311:137–149
204. Gunasekaran S, Mancini ND, Mitsos A (2014) Optimal design and operation of membrane-based oxy-combustion power plants. Energy 70:338–354
205. Zebian H, Rossi N, Gazzino M, Cumbo D, Mitsos A (2013) Optimal design and operation of pressurized oxy-coal combustion with a direct contact separation column. Energy 49:268–278
206. Job M, Bartela L, Skorek-Osikowska A (2013) Analysis of the use of waste heat in the oxy-combustion power plant to replace the steam cycle heat regeneration. J Power Technol 93(3):133–141
207. Park SK, Kim TS, Sohn JL, Lee YD (2011) An integrated power generation system combining solid oxide fuel cell and oxy-fuel combustion for high performance and CO_2 capture. Appl Energy 88(4):1187–1196
208. Skorek-Osikowska A, Bartela L, Kotowicz J, Job M (2013) Thermodynamic and economic analysis of the different variants of a coal-fired, 460-MW power plant using oxy-combustion technology. Energy Convers Manag 76:109–120
209. IEA Green House Gas R&D Programme (2005) Oxy combustion processes for CO_2 capture from power plant. IEA greenhouse gas R&D programme, report number 2005/9
210. IEA Greenhouse Gas R&D Programme (IEAGHG) report (2012) Emissions of substances other than CO_2 from power plants with CCS. Mar 2012 (03)
211. Scholes CA, Smith KH, Kentish SE, Stevens GW (2010) CO_2 capture from pre-combustion processes—strategies for membrane gas separation. Int J Greenhouse Gas Control 4:739–755

Chapter 3
Ion Transport Membranes (ITMs) for Oxygen Separation

3.1 Introduction

The high efficiency penalty associated with using cryogenic O_2 separation units in oxy-combustion systems called for alternative methods for O_2 production. One of these methods is the use of ion transport membranes (ITMs) for O_2 separation from air. These ITMs have the capability of extracting oxygen from air at high temperatures (above 700 °C). The permeation of oxygen through the ion transport membranes depends on the membrane type, thickness, operating temperature, and the difference in oxygen partial pressure across the membrane [1]. It is expected that the utilization of membranes in gas separation processes would increase five times by the year 2020 [2]. The integration of these membranes in oxy-combustion reactors led to the design of oxygen transport reactors (OTRs), where oxygen–air separation occurs at one side while the fuel combustion occurring at the other membrane side [3, 4]. Currently, numerous studies are conducted for improving membrane performance and chemical stability under more demanding operational conditions. Most of the reported work focuses on ITMs utilization in the oxy-combustion of gaseous fuels [5, 6]. On the other hand, the availability of large amounts of low-quality fuels (heavy petroleum fuels, bituminous coal, petroleum coke, and lignite) created the need to develop new technologies that combine oxy-combustion with carbon capture utilizing ion transport membranes.

The performance of oxygen separation membranes has been investigated extensively in the literature considering different membrane materials. Xu and Thomson [7] explicitly constructed a model for oxygen permeation across LSCF ($La_{0.6}Sr_{0.4}Co_{0.2}Fe_{0.8}O_{3-\delta}$) membranes. They performed a series of experimental measurements over wide ranges of operating O_2 partial pressure and temperature. The results revealed that the permeation flux of oxygen increases while increasing the inlet gas temperature. They also reported that the bulk diffusion controls the oxygen flux at high temperatures. Rui et al. [8] investigated the characteristics of oxygen permeation across oxygen ionic ceramic membranes under finite-rate

M. A. Nemitallah et al., *Oxyfuel Combustion for Clean Energy Applications*, Green Energy and Technology, https://doi.org/10.1007/978-3-030-10588-4_3

reaction conditions. The results showed that the combustion in the permeate side of the membrane reduces the partial pressure of the oxygen in the permeation side which, consequently, increases the oxygen permeation flux. Akin and Lin [9] applied different oxidation reaction kinetics mechanisms including very fast reaction and no reaction mechanisms. The results were compared with the experimental data of oxygen permeation flux through a BYS ($Bi_{1.5}Y_{0.3}Sm_{0.2}O_{3-\delta}$) membrane for two reaction conditions of ethane and methane. The results showed higher oxygen permeation flux, by an order of magnitude, in case of ethane as compared to the case of methane due to the faster reaction rate of ethane. Habib et al. [10] studied the effect of reactivity of methane, in the permeation side of a stagnation flow ITM reactor, on the oxygen permeation flux across a LSCF-6428 ($La_{0.6}Sr_{0.4}Co_{0.2}Fe_{0.8}O_{3-\delta}$) ionic ceramic membrane. Sharp increase in oxygen permeation flux was reported in the reacting flow case as compared to the case of non-reacting flow. This can be attributed to the increase of temperature due to combustion which reduces the bulk diffusion resistance of the membrane, and as a result, oxygen permeation flux is increased.

Hong et al. [11] studied numerically the operation of a LSCF ion transport membrane reactor (ITMR) under reaction conditions while incorporating a detailed gas-phase model of methane. The results showed that the oxygen permeation flux depends on the geometry of the reactor, temperature of the membrane and feed and sweep flow rates. Ben-Mansour et al. [12] and Nemitallah et al. [13] investigated the performance of a LSCF-1991 ($La_{0.1}Sr_{0.9}Co_{0.9}Fe_{0.1}O_{3-\delta}$) ITMR under non-reacting and reacting flow conditions over wide ranges of operating conditions. They distinguished a set of parameters that have a significant effect on oxygen permeation flux across the membrane. These parameters include inlet gas temperature, inlet methane concentration, and reactor geometry. Hong et al. [14] studied the interactions between oxygen permeation and combustion of methane in the permeation side of a LSCF membrane. Enhancement of oxygen permeation flux was reported while increasing the inlet gas temperature and inlet fuel concentration. Hunt et al. [15] measured the oxygen concentration normal to the surface of a LCF ($La_{0.9}Ca_{0.1}FeO_{3-\delta}$) membrane and the total oxygen permeation flux in a stagnation flow ITMR. Also, a multi-step surface exchange model for oxygen permeation was developed. Using the same ITMR, Kirchen et al. [16] investigated experimentally the permeation of oxygen characteristics under oxy-combustion conditions. The results showed that in order to use an ITMR for a complete fuel conversion (i.e., oxy-combustion applications); the inlet fuel concentration should be kept below a certain limit to satisfy the stoichiometric conditions for combustion.

Recently, the idea of using oxygen transport reactors (OTRs) for large-scale power generation in power plants has been implemented by Mancini and Mitsos [17]. A design of a monolith structure ITM reactor was proposed to be used for power generation based on detailed numerical simulations. The study resulted in an OTR having total volume of 1000 m^3, total number of feed and permeate channels of 100,000 (50,000 per each stream), surface area of 266,700 m^2, height of 4.75 m, length of 44.44 m, and power output in the range from 300 to 500 MWe based on cycle first law efficiency. Following the same concept, Nemitallah et al. [18]

developed a multi-channel compact OTR design for gas turbine combustion application based on detailed three-dimensional (3D) numerical simulations. The resultant OTR has a length of 0.9 m, height of 3.35 m, volume of 10 m^3, surface area of 2700 m^2, total number of channels per stream of 25,000, and output power in the range from 5 to 8 MWe based on the first law efficiency.

In this chapter, the process of oxygen separation in ion transport membranes is studied in detail considering different membrane materials and comparing their performance. The application of oxy-combustion technology in OTRs is discussed considering the recent advances in membrane technology for oxygen separation under reacting conditions. The operation of OTRs under oxy-combustion of gaseous fuel is presented. Trending applications of OTR technology are discussed including OTRs for syngas production, combustion utilizing liquid fuels in OTRs, splitting H_2O to produce H_2, and utilization of CO_2.

3.2 Oxygen Separation Membranes

In recent years, many researchers considered oxy-fuel combustion to be the most promising carbon capture technology. However, because of the high energy penalty caused by oxygen production using cryogenic processes, there has been a need for developing other technologies for oxygen separation from air. In this regard, dense mixed ionic–electronic conducting ceramic (MIEC) membranes have shown good potential for use in oxy-fuel combustion technology when operating at temperatures in the range from 700 to 900 °C [19, 20]. In ITMs, oxygen with high partial pressure at membrane feed side permeates to the permeate side of low oxygen partial pressure. The oxygen flux depends on many factors including the membrane material and thickness, the operating temperature, and the O_2 partial pressure on both sides of the membrane. One main challenge in designing an ITM reactor to be used in power plants is the low oxygen flux that makes the required membrane surface area to be very large [21, 22]. Accordingly, improving oxygen permeation in ITMs is a necessary step toward the efficient design of ITM reactors. Also, the mechanical and chemical membrane stability performance under actual operating conditions needs a detailed experimental investigation. A comprehensive review of the work done on the ITM membranes developed for oxy-fuel combustion up to 2010 was reported by Habib et al. [23].

Currently, many researchers are working on the development of various material syntheses that are used for making oxygen separation membranes. Any promising material for oxygen transport membrane (OTM) has to exhibit good oxygen permeability in addition to being thermally and chemically stable. Among different O_2 separation membranes, the BSCF ($Ba_{0.5}Sr_{0.5}Co_{0.8}Fe_{0.2}O_{3-\delta}$) membranes have shown the most promising oxygen permeation characteristics and chemical stability at high temperatures [24]. The oxygen permeability versus temperature of different ceramic membranes is summarized in Fig. 3.1. It is clear from the figure that BSCF membrane has the highest permeability [23]. Generally, ceramic membranes are

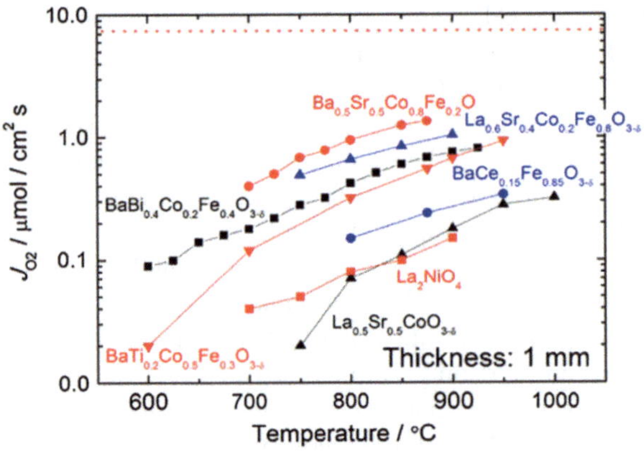

Fig. 3.1 Effect of temperature on different materials permeation flux of oxygen [23]

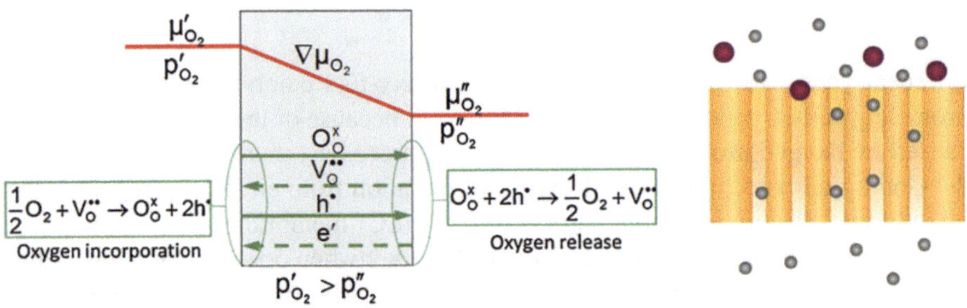

Fig. 3.2 Ion transport processes for dense ceramic (left) and porous ceramic (right) membranes [25]

classified into two main categories, namely dense and porous membranes. Ionic and molecular diffusions are the main transport mechanisms that are considered in these membranes [25]; see Fig. 3.2. In dense membranes, the mass transport occurs as a result of the diffusion of ions through the crystal structure and augmented by electrical conductivity for charge compensation. The ionic transportations of oxygen in these membranes take place through voids in the crystal structure, and accordingly, it is used for pure oxygen separation from air. Asymmetric structure membranes consisting of a very thin layer of dense membrane on a substrate of a porous structure were developed. Graded porous inter-layer(s) are sometimes used to bridging gaps between the support macropores and the dense membrane layer. Such a layer is usually made from (or infiltrated with) a catalytically active material. The structure of an asymmetric membrane with a $Ba_{0.5}Sr_{0.5}Co_{0.8}Fe_{0.2}O_{3-\delta}$ activation layer [25] is depicted in Fig. 3.3.

Li et al. [26] performed measurements for oxygen permeation, under vacuum conditions, through an asymmetric membrane built of a dense BSCF layer supported on a porous structure of the same material. This BSCF membrane has a thickness of

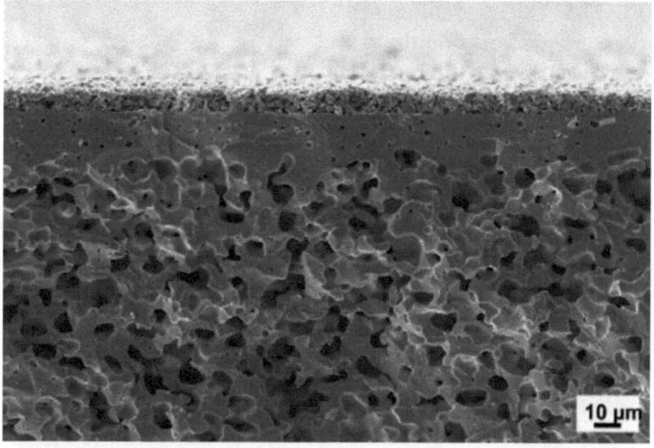

Fig. 3.3 Cross section of an asymmetric ceramic membrane assembly comprising porous support, dense ceramic membrane layer, and a porous activation layer fabricated from $Ba_{0.5}Sr_{0.5}Co_{0.8}Fe_{0.2}O_{3-\delta}$ [25]

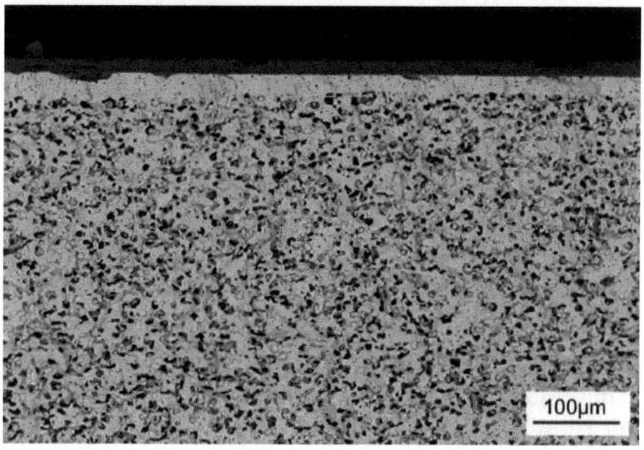

Fig. 3.4 SEM micrograph of the asymmetric ceramic membrane with 20-μm-dense membrane layer and 900-μm-porous substrate [26]

1.9 mm and was produced by tape casting on top of a 50 μm layer of the same material. The sintered membrane layer has a thickness of 20 μm, and the support was of 900 μm thickness. A micrograph of the asymmetric membrane after sintering is shown in Fig. 3.4. The SEM images were used to determine the dense and porous support layers porosities that were found to be 5 and 34%, respectively. The figure clearly shows the well attachment of the dense membrane layer to the porous substrate. Figure 3.5 depicts the effect of the gradient of P_{O2} (created by varying the oxygen content between 20 to 100% in the air feed side) on the permeating rate of oxygen through this asymmetric membrane when operating at 900 °C and a vacuum pressure of 150 mbar at the permeate side. The results indicated that a considerable increase in the flux of oxygen permeation with the increase of the oxygen concentrations in the feed side, which actually increases the permeation driving force. The maximum achieved oxygen flux (J_{O2}) was 32.5 ml min^{-1} cm^{-2} [STP] with pure oxygen in the feed side. Under similar conditions, the dense BSCF membranes produced oxygen flux of $J_{O2} < 5$ ml min^{-1} cm^{-2} [STP], for a pressure gradient, ln ($P_{O2,\ feed}/P_{O2,permeate}$), of 4.

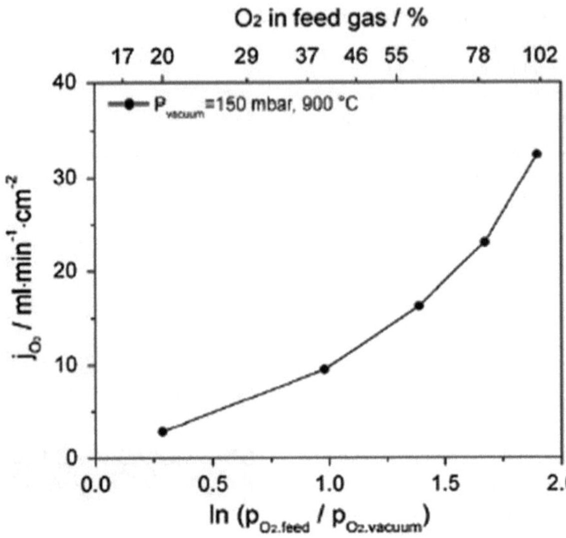

Fig. 3.5 Effect of different pressure gradients on the rate of oxygen permeation across asymmetric BSCF ceramic membrane at 900 °C (oxygen concentration in feed gas: 0.2–1.0; vacuum pressure at permeate side: 150 mbar) [26]

In an earlier work, Baumann et al. [27] synthesized BSCF membranes by tape casting and co-firing. These membranes are composed of two layers: one was the top gastight layer about 70 μm thick, and the other was the porous substrate about 830 μm thick on which this layer was deposited. A very thin layer of BSCF (about 17 μm thick) was also deposited on the top gastight layer to act as an activation layer for oxygen permeation, thus measuring its effect on the overall oxygen permeability. They found that this activation layer increased the surface exchange rate and thus increased the permeability. A flux of oxygen permeation of 12.2 ml/cm^2/min was achieved at 1000 °C when the air was used as the feed and argon as the sweep, whereas a flux of 67.7 ml/cm^2/min was achieved at 1000 °C when pure oxygen is the feed and argon is the sweep gas. Another way to enhance oxygen permeation is doping or substitution with different elements. Haworth et al. [28] substituted some yttrium in place of iron in BSCF forming the perovskite-type oxide $Ba_{0.5}Sr_{0.5}Co_{0.8}Fe_{0.2-x}Y_xO_{3-\delta}$ and optimized the concentration of yttrium by varying the value of x from 0 to 0.2. It was observed that there was an improvement in the oxygen permeability for the values $x = 0$–0.15. The best result for oxygen permeability was observed for $x = 0.025$, and a flux of 2.05 ml/cm^2/min was obtained for a BSCFY membrane of 1.2 mm thickness at a temperature of 900 °C showing an enhancement of more than 1.5 times as compared to the pure BSCF membrane which gave a flux of about 0.79 ml/cm^2/min. There were two main reasons attributed to this increase in the oxygen permeability with yttrium substitution: One was the increase in the concentration of oxygen vacancies with yttrium substitution, and the other was the increase in the mobility of oxygen ions due to the expansion of the crystal lattice.

Menzler et al. [29] tried preparing the asymmetric BSCF membranes with the tape casting technology. They reported that the porosity could be controlled by using various types of pore-forming agents. Using different types of starch (rice, corn, and potato starches) resulted in a change in porosity reaching 32, 41, and 35% for average pore sizes of 1.4, 1.7, and 4 μm, respectively, as shown in Fig. 3.6. In addition to screen printing of the dense membrane layer, another technology, namely green-in-green tape casting, is being explored. In this regard, three asymmetric membranes, prepared using this technology, are displayed in Fig. 3.6. A comparison between the oxygen flux permeating through the 1-mm-thick bulk membrane and supported asymmetric thin membrane is shown in Fig. 3.7 for BSCF and LSCF membranes. The development of hollow fiber (capillary) membranes was carried out by a number of researches. Buysse et al. [30] investigated the stability and performance of gastight, macrovoid-free, and sulfur-free BSCF capillary membranes fabricated through phase-inversion spinning and sintering. They chose a sulfur-free phase-inversion polymer to get a pure BSCF crystal phase. In order to achieve a gastight membrane free of macrovoids, special care was given to the polymer solution and ceramic spinning suspension. Comparing the performance of

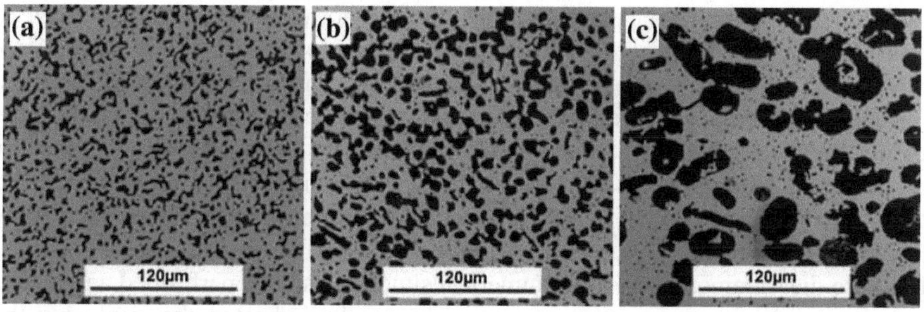

Fig. 3.6 SEM micrographs of BSCF membrane supports considering different starch types used as pore-forming agent; **a** rice, **b** corn, and **c** potato (polished cross sections) [29]

Fig. 3.7 Effect of temperature on oxygen permeation flux considering BSCF- and LSCF-based ceramic membranes [29]

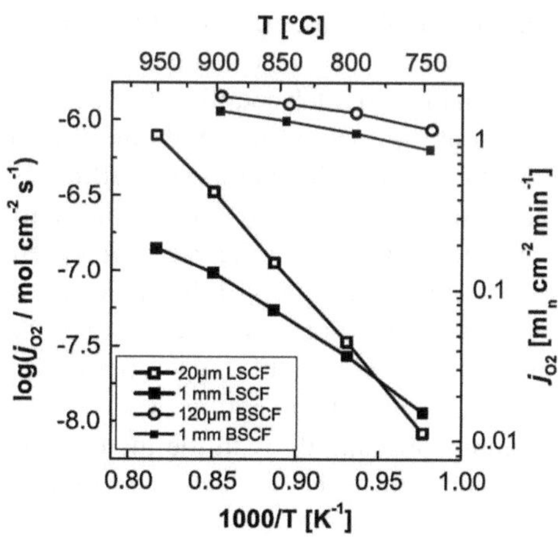

sulfur-free and sulfur-contaminated BSCF capillaries of same dimensions, they found a significant effect of sulfur contamination on both oxygen permeation rate and activation energy for oxygen transport.

3.3 Gaseous Oxy-fuel Combustion in OTRs

High-temperature oxygen–air separation is usually achieved using ceramic membranes due to its high selectivity and thermochemical stability. The utilization of these membranes has significantly increased because of the low cost of O_2 production in comparison with cryogenic techniques. The integration of these membranes in oxy-combustion reactors led to the design of oxygen transport reactors (OTR) where oxygen–air separation occurs at one membrane side while fuel combustion occurring at the other side. Experimental measurements of O_2 permeation through perovskite membranes have shown relatively high flux [31]. These membranes are capable of oxygen–air separation at 700–900 °C and can be formed in tubular or planar shapes, thus enabling the formation of compact oxygen transport reactors. Habib et al. [23] reported a comprehensive review of the work done, up to 2011, on oxy-combustion in conventional and ion transport membranes.

One of the operating parameters affecting oxygen permeation across ITMs is the thermochemical reactions that normally occur near the membrane surface. In their recent work, Kirchen et al. [16] developed an ion transport membrane reactor to study permeation of oxygen and oxy-fuel combustion. A planar stagnation flow reactor of a finite gap with optical and probe access to the reaction zone was used for these measurements. As well, numerical simulations were performed and results are compared with these measurements. The results indicated that replacing the inert sweep gas by a reactive gas tends to an increase in the flux of oxygen which can be attributed to the reduction of O_2 partial pressure near the membrane surface. It was also shown that increasing the sweep CH_4 fraction results in an increase the oxygen flux up to a certain percentage above which the oxygen flux remains unchanged. The interaction between oxygen permeation and fuel oxidation in the sweep side of ITM was also studied by Hong et al. [32]. The study was based on a mathematical model coupling the rate of O_2 permeation dependence on the conditions at the surface of membrane and the detailed chemistry and transport in the neighborhood of the membrane surface assuming that there is no catalytic effect of the surface of the membrane on the hydrocarbon or syngas reactions. In order to obtain the spatial and time variation of temperature and species concentration in the flow domain, the details of the chemistry of the gas-phase and species transportation were modeled and were incorporated in the computational model. The results showed that the rates of oxygen permeation and methane conversion depend on the sweep gas inlet temperature. The results showed an increase of the oxygen permeation substantially with the increase of the sweep gas inlet temperature and fuel concentration. This was achieved by enhancing oxidation reactions and products transportation as well as heat generation closer to the membrane. Consequently, the temperature of the gas in the vicinity of the membrane increases and the oxygen concentration decreases. It was indicated that considerable conversion of the fuel and higher oxygen flux could be achieved by higher rates of reactions at higher gas inlet temperatures and higher fuel concentration with no catalytic activities of the membrane surface. After reaching a given upper limit of the fuel contents, the oxidation kinetic rates are reduced due to excessive heat loss to the membrane, thus limiting the rate of permeating oxygen. Furthermore, it was concluded that the channel height and the sweep gas flow rate have slight effect on oxygen permeation and fuel conversion rates due to the presence of the reaction zone close to the membrane surface.

A computational model for simulating oxygen permeation through a high electronic to ionic conductivity ratio ion-conducting membrane was developed by Xu and Thomson [7]. For modeling, the resistances related to surface kinetics on each membrane side and bulk diffusion were presented and validated with the experimental results obtained by varying membrane temperature and partial pressure for LSCF-6428 membrane. The results revealed the strong dependence of oxygen permeation on partial pressure on both of membrane sides, membrane thickness, and temperature. The model was found to be applicable to all perovskite membranes that have higher electronic to ionic conductivity ratio. The model parameters like bulk diffusion coefficient (D_v) and rate constants for surface kinetics (k_r and k_f) were found by fitting the experimental data. The surface exchange resistances are found to be small as compared to the resistance for bulk diffusion. Another computational model for simulating the performance of a specially designed ion transport membrane (ITM) reactor was developed by Hong et al. [33]. They focused on the influence of surface chemistry, temperature, feed, and sweep flow rates on fuel conversion and oxygen transport. The ITM reactor was considered to have a symmetric configuration and the geometry was reduced to one half of the reactor size for reducing computational time. Methane was used as a fuel and it reacts in oxygen environment permeating through the ITM. The detailed chemistry was reduced to 36 species and 217 reactions for oxy-fuel combustion conditions. A mathematical model was developed for oxygen permeation in terms of oxygen partial pressure on feed and sweep sides with the help of experimental results and proposed models available in the literature. The proposed model in this study was proved to be more accurate over a broad operating range. It was found that the oxygen permeation strongly depends on ITM temperature, flow rates, and fuel conversion on the surface. The fuel oxidation reduces partial pressure of oxygen in the sweep side, and hence, the permeation is enhanced. Amato et al. [34] computationally and experimentally investigated the effect of CO_2 dilution of oxy-fuel combustion of methane on CO and O_2 emissions. They concluded that burning the fuel in oxygen diluted by CO_2 leads to higher CO emission due to the higher CO_2 levels and the slower combustion of the intermediate CO formed. The features of oxy-combustion of methane in a cylindrical oxygen-transport reactor (OTR) were numerically analyzed by Ben-Mansour et al. [15]. They assumed the reactor walls to be constructed from dense, non-porous, mixed-conducting ceramic membranes. These membranes permeate only oxygen from air surrounding the reactor into the reaction zone (combustion chamber). Inside this chamber, a reaction takes place between the permeated oxygen and the sweeping gas comprising CO_2 and CH_4 producing H_2O and CO_2. Their model was based on the conservation equations of mass, momentum, energy, and species in the flow domain. The local values of the membrane surface temperature and oxygen partial pressure resulted in a variation of the local oxygen permeation rate over the surface of the membrane. The effects of CH_4 and CO_2 content of the mixture composition as well as the mass flow rate on the flow and reaction processes were investigated. The study revealed that combining separation with reaction in OTR results in a significant increase in the rate of oxygen permeation by a factor of approximately 2.5 in comparison with O_2 separation-only unit. Moreover, it was found that most of the combustion heat transfers to the airside while a part of it heats the O_2-permeating flux. At high mixture mass flow rate, the OTR operation using a rich mixture resulted in low conversion of methane. The authors recommended improving the combustion process via dividing the OTR into a series of units and adding the fuel at stages along the reactor network. Zanganeh et al. [35] used simulated process data to provide a comparison between different options for CO_2 capture in the case of refinery oxy-fuel combustion. They conducted simulations and cost analysis for oxy-combustion combined with CO_2 capturing of two-refinery fuel gases considering four different modes of combustion. These modes include the base case of using air as oxidizer in addition to three oxy-combustion cases that contain wet and dry recycle as well pure oxygen. The results indicated that oxy-combustion is a feasible technique to separate CO_2 from refinery fuel gases in addition to the production of CO_2-rich stream from which CO_2 can be separated using relatively straightforward physical processes. The process neither requires any further treatment nor produces any form of wastes that could be produced via other CO_2 capturing techniques. Luo et al. [36] applied a glycine–nitrate-based fast single-spot combustion synthesizing procedure to manufacture a new oxygen-transporting dual-phase CO_2-stable membrane of the composition 40 wt% $Mn_{1.5}Co_{1.5}O_{4-\delta}$ with 60 wt% $Ce_{0.9}Pr_{0.1}O_{2-\delta}$ (40MCO–60CPO). Characterization of the membrane indicated a stable oxygen permeation flux for more than 60 h while pure CO_2 was the sweeping gas. This stability advocates the use of 40MCO–60CPO material as a good candidate for separating air–oxygen for oxy-fuel processes.

3.4 Trending Applications of OTR Technology

3.4.1 OTRs for Syngas Production

3.4.1.1 Characteristics of Syngas OTM System

Syngas, consisting mainly of H_2 and CO, is extensively used in synthesizing of methanol, higher alcohols, aldehydes and in Fischer–Tropsch processes. It is also

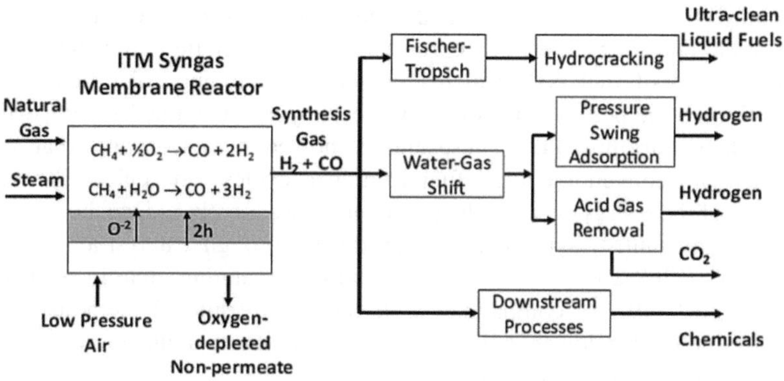

Fig. 3.8 Operation principal of the ITM syngas reactor and processes that can utilize the produced syngas

utilized in the ferrous and non-ferrous industries as a reduction agent. The ITM oxygen separation characteristics led to the design of compact ceramic membrane chemical reactors for production of syngas. Integrating oxygen separation with thermal reforming within the reactor eliminates the need for the energy-expensive cryogenic air separation unit. This makes the process of producing syngas much easier in remote areas such as oil platforms or where the construction cost is as high as that in the north and south poles [37]. The reactor can be also utilized as part of a chemical process to produce hydrogen from methane/natural gas. Figure 3.8 shows the concept of operation of ITM syngas reactor as well as other processes that can utilize the produced syngas [38]. Syngas and hydrogen have high potential for use in cars and small-size hydrogen energy applications. In addition, syngas can be used in internal combustion engines as fuel additive or as a fuel for fuel cells. Conventionally, syngas is obtained using endothermic reaction of steam and natural gas on nickel-based catalysts [39–41]. Mokheimer et al. [42] presented an experimentally validated CFD model for steam methane reforming in a nickel-based catalyst-assisted reformer. Their results revealed that the methane conversion increases with the increase of the steam contents in the feed mixture. They also reported that the thermodynamic limit of the methane conversion increases with the decrease of the reformer pressure. The drawback of this process is its high cost and considerable emissions of nitrogen oxides during the reformer heating. To overcome this, many researchers investigated the use of low-temperature heating sources such as solar or waste energy to avail the heat required for the fuel reforming process for production of syngas or hydrogen. In this regard, Simakov et al. [43] reviewed the chemistry and system designs for solar thermal catalyst-assisted natural gas reformers. On the other hand, Sheu et al. [44] reviewed the solar methane reforming including dry and wet redox and catalyst-assisted reforming. Said et al. [45, 46] developed and presented a CFD model to assess and optimize the performance of a catalyst-assisted membrane-based reformer to produce hydrogen via steam methane reforming at low temperature. The range of operating temperature studied was suitable to make the reformer as the receiver of a parabolic trough solar collector. Sheu et al. reviewed the integration of solar energy to conventional fossil fuel power plants in general including solar fuel reforming option [47] and the performance of such hybrid cycles integrated with dry and wet redox methane reforming [48]. These studies showed that integrating solar methane reforming with convention fossil fuel combined cycles can achieve 30% annual solar share of the energy input to the plant and consequently reduce the annual CO_2 emission significantly. Syngas is also produced through catalyst-assisted partial oxidation of methane (natural gas) process which is gaining great interest because of its low cost [49]. In comparison with other conventional methods, the ITM syngas production has lower capital cost and higher thermal efficiency added to the more efficient carbon dioxide capture [50].

To optimally utilize the syngas produced by solar-assisted reformers, many of the recent research focused on studying the fundamental characteristics of syngas combustion for stable and clean combustions under different operating conditions including the composition of the syngas and the oxidizer. In this regard, Taamallah el. [51] conducted a comprehensive review on the effect of fuel composition, with focus on hydrogen-enriched fuels, on the combustion stability and emissions. They presented and discussed the recent advances in combustion technologies, combustion fundamentals and the pertinent experimental and numerical analysis. Alzahrani et al. [52] explored the accuracy of a selected set of kinetics and combustion mechanisms for syngas combustion. Mokheimer et al. [53] numerically investigated the stability and emission of hydrogen-enriched methane under very lean air combustion in a diffusion flam. They developed a combustion mechanism that can closely fit with the experimental results by Sanusi et al. [54]. From their numerical analysis, Mokheimer et al. [53] could determine the optimum equivalence ratio at which the CO emission can be eliminated and NO_x emission can be limited below 5 ppm. This optimal equivalence ratio was 0.45 for a hydrogen-enriched methane fuel mixture with 30% H_2.

3.4.1.2 Syngas Catalytic OTRs

Many different materials and membrane geometries have been investigated for ceramic membrane reactors. Membrane processes using a feed of natural gas at high pressure, with low air pressure, achieve the best economics. Using a feed of natural gas at high pressure allows the production of high-pressure syngas product to match feed requirements for downstream processes without additional compression. This arrangement also avoids the high capital and operating expenses that would result when using high-pressure compressed air [38]. The use of microchannel (MC) systems for partial oxidation of methane (POM) has many advantages compared with the conventional catalyst-assisted reactors. The first is the high surface to volume ratio, which results in a considerable increase in heat and mass transfer, thus allowing higher rates of heat transfer for either heating or cooling the reaction mixture as well as suppressing hot/cold spots formation [55]. The second is providing a narrow distribution of the residence time of the reacting mixture within the catalytic layer leading to an increase in the catalyst selectivity and efficiency [56]. The main challenge is in the manufacturing of MC reactors starting from preparing the MC plates and depositing the catalysts on top of it. Different methods for the manufacturing of such channels are available; however, MC reactors designed for use in POM processes operate at high temperatures (700 °C), imposing constraints on the MC plate's material selection. Accordingly, either silicon or various ferrous alloys including stainless steel, Nichrome, Nicrofer, and Fecralloy were used in manufacturing the MC plates because of their high heat conductance and thermal stability [57].

The performance of the catalyst layer in MC reactors was the subject of several investigations. For example, the use of Rh/Al_2O_3 deposited on microchannel Nicrofer plates resulted in Fe and Cr oxide layers formation that weakened the catalyst activities. On the contrary, a Fecralloy reactor displayed a stable performance at 700 °C added to 57% conversion of methane and 84% CO selectivity [58]. A similar performance (methane conversion of 58% and selectivity toward CO of 81% when operating at 780 °C) was achieved in using a microreactor with the channel comprised of 50-μm Fecralloy foil plates with a Rh/ZrO_2 catalyst coating [59]. The main cause of high-temperature stable thermal performance of Fecralloy is segregating the alumina particles on the alloy surface while forming a dense layer of oxide. The good adhesion of supported catalysts to the thermally treated surfaces is another advantage of Fecralloy. Other metals commonly used as catalysts in POM are the transition metals (Ni, Co, and Fe) and noble metals (Ru, Rh, Pd, Pt, and Ir) [60]. Unfortunately, noble metals are very expensive while catalysts comprising Ni, Co, and Fe may suffer from cocking [61]. Using catalytic bimetals (such as $Ni–Al_2O_3$ with Pt additives, Rh modified via adding rare earth metal oxides) offered significant improvements in the catalytic performance combined with increase in the selectivity and activity of POM at small durations of contact at temperatures more than 700 °C [62]. In addition, using Ce–Zr mixed oxides in Ni-containing catalysts prevents the formation of the coke due to the oxides high oxygen mobility [63].

The direct conversion of methane to syngas was studied by Tsai et al. [64] using a disk-shaped $La_{0.2}Ba_{0.8}CO_{0.2}–Fe_{0.8}O_{3-\delta}$ membrane reactor operating at 1123 K. The study revealed that packing a 5% Ni/Al_2O_3 catalyst directly on the membrane reactor side resulted in a significant improvement in O_2 permeation and CH_4 conversion (fivefold increase in O_2 permeation and a fourfold increase in CH_4 conversion) in comparison with that of the blank reactor. The drawbacks of the disk-shaped membrane are the low ratio of volume of the reaction zone to the total volume [64, 65]. In this configuration, a large amount of methane may pass through the reactor without participating in the reaction, which results in a low conversion of methane. The use of tubular membranes with small diameters may provide a partial solution to this problem [66]. Balachandran et al. [67] were the first and perhaps the only group that used a reactor utilizing tubular dense ceramic membranes to study methane conversion to syngas. Using tubular non-perovskite $SrCO_{0.5}FeO_x$ membrane reactor operating at about 1173 K assisted by Rh-based reforming catalyst, they reported more than 98% methane conversion with 90% CO selectivity for partial oxidation of methane to syngas. The recently developed Ni/g–Al_2O_3 catalyst of improved performance makes it attractive to study methane conversion to syngas utilizing a tubular dense ceramic membrane reactor packed with Ni/g–Al_2O_3 catalyst. The effects of many operating parameters on the oxy-combustion of syngas in the OTR were reported and analyzed by Habib et al. [68] who reported their CFD simulations for oxy-combustion of syngas in a tube OTR heated by hot air in an annular space surrounding the tube reactor. In their study, they investigated the oxygen–air separation from one side of the LSCF-6428 membrane to the other side where fuel is burned by oxygen diluted by a sweeping stream of recycled CO_2.

3.4.2 Combustion Utilizing Liquid Fuels in OTRs

Based on the above discussion, it is clear that there is a need for application of an emission reduction technique while using liquid fuels. The oxy-combustion of soot-free liquid fuels (light fuels) may be a good solution toward emission reduction, and the produced CO_2 can easily be captured after H_2O condensation. This technique could be applied with some difficulties in ITMRs. As per the open literature, most of the conducted works on ITMRs are done considering gaseous fuels, mainly methane. Actually, ITMs are still under material development aiming at the development of new materials that can be used to separate oxygen in enough amounts for complete and stable combustion of the fuel. The application of such kind of reactors, ITMRs, is limited with the use of light gaseous fuels due to the limitation of separated amount of oxygen. The case becomes more complicated when liquid fuels are to be used in ITMRs. However, the concept of operation of ITMRs is depending on recirculation of a portion of the exhaust gases to heat up the incoming fuel and sweep the separated oxygen through the reactor. Part of the heat carried by exhaust gases can be utilized in evaporating liquid fuel before it reaches the membrane surface. Ion transport membranes are designed to separate oxygen in a gas-phase medium in both sides of the membrane, and the use of liquid fuel will badly affect the oxygen separation. However, the use of liquid fuels in ITMRs is limited to only light liquid fuels due to limitations in oxygen permeation and sensitivity of the membrane material to soot formation.

Recently, Nemitallah and Habib [69] performed a numerical investigation of the performance of a button-cell ITMR while using methanol, as a light liquid fuel. Table 3.1 provides a summary of the applied numerical models. In this study, air is fed to the membrane feed side through the inner tube and a portion of oxygen is separated through the ITM while the flow is in contact with the membrane surface, and the remaining stream of oxygen-depleted air exits the ITM reactor through the annular gap between the two concentric tubes. In the ITM reactor permeate side, fuel is being injected through a plain orifice atomizer through an orifice with a diameter of 0.5 mm at a temperature of 300 K. Portion of hot recycled CO_2 is fed to the permeate side through the inner tube and around the fuel injector orifice to help

Table 3.1 Summary of the applied numerical models to simulate oxy-combustion of methanol inside a button-cell ITM reactor [69]

Parameter/model	Description
Membrane reactor	Button cell [69]
Membrane material	La_2NiO_4 (LNO) [70]
Oxygen permeation model	ABn model [17, 71]
Droplet collision and breakup models	– Droplet collision model [72] – Droplet distortion and breakup [73–75]
CFD modeling	– The discrete phase model was solved using the Euler–Lagrange approach – A source/sink term, S_i, is added to mass and species transport equations to account for oxygen permeation – Mass, momentum, energy conservation, and species transport equations are solved in the 2D axisymmetric domain – Fluent 12.1 software is used along with user-defined functions (UDFs) to solve for the S_i term
Radiation model	Discrete ordinates (DO) radiation model [69]
Reaction kinetic model	Finite-rate single-step reaction kinetic model of methanol: $CH_3OH + 1.5\ O_2 \rightarrow CO_2 + 2\ H_2O$

heating and evaporating the injected liquid methanol. After its evaporation, fuel is burned with permeated oxygen across the membrane near the membrane surface in the permeate side. The hot flue gases are allowed to leave the reactor through the annulus in the permeate side. Portion of the heat contained in the exhaust gas is allowed to pass across the inner tube surface to help in heating up and evaporating the liquid fuel. The influences of fuel concentration (combustion vs. separation only), gas-phase temperature (combustion), and sweep flux rate (combustion) on liquid fuel evaporation and combustion inside the ITM reactor are investigated [69]. Validations of both the reaction kinetics and the oxygen permeation models are performed. To validate the reaction kinetic model, the present model results and the experimentally obtained data by Lacas et al. [76] are compared in terms of temperature distributions under pure oxygen combustion (100% O_2). Also, the oxygen permeation mode is validated using the experimental data by Xu and Thomson [7] in terms of oxygen permeation flux across a LSCF membrane under variable temperature and sweep flux rates. The comparisons of results proved the validity of the applied models; more details can be obtained from [69].

Some of the results of this study are presented in Figs. 3.9 and 3.10. Figure 3.9 shows the effects of inlet fuel concentration on oxygen permeation flux under both reacting and non-reacting flow conditions. Due to the consumption of oxygen in the combustion process in the permeate side of the membrane under the reacting flow conditions, the oxygen partial pressure driving force across the membrane is increased, and as a result, oxygen permeation flux is enhanced as compared to the obtained values under non-reacting flow conditions. In addition, the membrane temperature is increased due to combustion resulting in more oxygen permeation

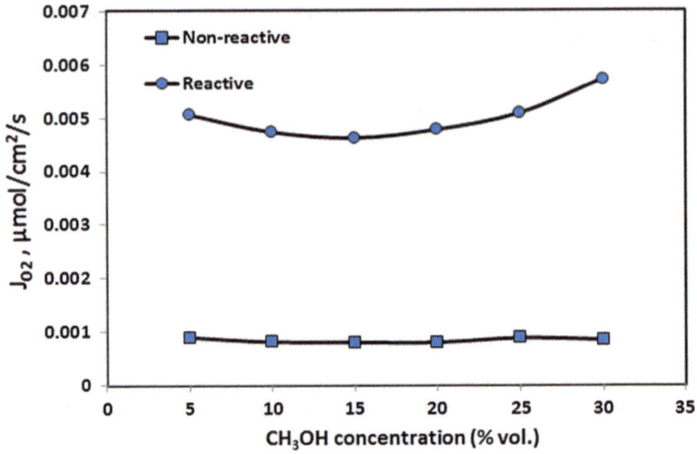

Fig. 3.9 Effects of inlet fuel concentration on oxygen permeation flux considering reacting and non-reacting flow conditions [69]

flux as shown in Fig. 3.9. For the considered range of fuel concentrations, oxygen flux for the reacting flow is increased in average by about five times the value of non-reacting flow case. Figure 3.10 presents the plots of axial distributions of fuel evaporation rate under reacting and non-reacting flow conditions over a range of inlet fuel concentrations. As a matter of fact, methanol has a low boiling temperature (around 338 K), and it has been introduced to the atomizer in all cases at 300 K. So, slight amount of heat is required to start its evaporation. For both reacting and non-reacting flows, CO_2 is supplied at 1173 K which is high enough to guarantee fuel evaporation after short distance from the inlet section as shown in Fig. 3.10. For this reason, very slight differences are obtained between the values of fuel evaporation rates for the cases of reacting and non-reacting flows. Also, the mixtures with lower fuel concentration resulted in faster evaporation of the fuel.

3.4.3 Membranes for Splitting H_2O to Produce H_2

Hydrogen energy is one of the clean, environment-friendly, and efficient energy sources that could be generated from renewable sources. These characteristics make hydrogen a valuable source of energy for industry as well as power production. It is well known that many countries are mainly depending on fossil fuels for energy generation; consequently, energy production results in massive amounts of greenhouse gas (GHG) emissions. Therefore, hydrogen energy technology is highly recommended for clean energy production in the coming decades. Recently, hydrogen has been used in the integrated gasification combined cycle (IGCC), which encourages the use of hydrogen enrichment in many power generation applications with low CO_2 emissions.

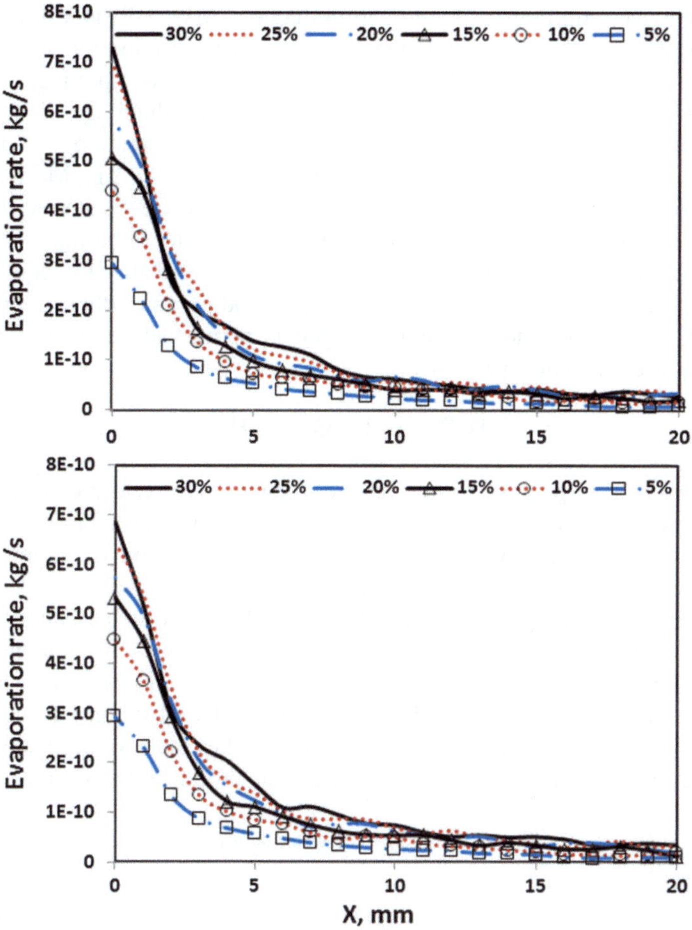

Fig. 3.10 Effects of inlet fuel concentration on fuel evaporation rate considering non-reacting (upper plots) and reacting (lower plots) flow conditions [69]

Nowadays, the most conventional way to produce hydrogen is steam reforming. This technique does not consume high energy, but emits massive amounts of CO_2 and CO, which, in turn, makes hydrogen no longer a clean energy source. Better techniques have been introduced in the area of water splitting, including (1) photocatalytic water splitting, (2) inorganic membranes, and (3) ion transport membranes. Photocatalytic water splitting is a technique to produce cleaner hydrogen, which can decompose water into oxygen and hydrogen by utilizing sunlight with the aid of photocatalysts [77–80]. Inorganic membranes were proven to have superior performance for hydrogen production and purification because of the increasing demand for pure hydrogen for petrochemical applications [81–84]. The dense ceramic membranes have attracted the attention for gas separation applications because of their capability to transport or separate oxygen [85–88]. If the membrane water splitting for hydrogen production (MWSHP) is a direct result of oxygen permeation through membranes, the oxygen permeation rate is an important parameter in determining hydrogen production via water splitting [89]. Other crucial factors include (1) oxygen ion and electron conductivity, (2) gradient of oxygen partial pressure across the membrane, (3) temperature, (4) membrane thickness, and (5) surface oxygen exchange kinetics [87, 90, 91]. More details are found in the review done by Li et al. [92] on mixed ionic–electronic conducting MIEC membranes for hydrogen production through water splitting.

The use of ion transport membranes (ITMs) indicated that the equilibrium of water-splitting reaction can be driven to the product side (thus improving the hydrogen permeation rate), as more oxygen is permeated across the membrane according to the reaction [93, 94]:

$$H_2O_{(g)} \leftrightarrow 0.5\,O_{2(g)} + H_{2(g)} \tag{1}$$

The dissociation process of this reaction generates very low concentrations of oxygen and hydrogen. Therefore, new membranes have been introduced that can provide high oxygen permeation fluxes. These membranes have different materials compositions as per the summary in Table 3.2.

Production of hydrogen from the widely and freely available water on earth is considered one of the major challenges because the process is unstable. Franca et al. [105] studied the MIEC perovskite membrane $La_{0.6}Sr_{0.4}Co_{0.2}Fe_{0.8}O_{3-\delta}$ for oxygen permeation and hydrogen production by membrane-based steam reforming, where syngas production is coupled with water splitting. The results of their study showed good stability, and they reported that hydrogen production occurred due to the

Table 3.2 Summary of different membranes for water splitting

Membranes	Type	References
GeO_2–Gd–Ni	Mixed oxygen ion and electron conducting (MIEC)	Balachandran et al. [87]
CeO_2–Gd–NiO		Balachandran et al. [88] Li et al. [92]
$BaCo_xFe_yZr_{1-x-y}O_{3-\delta}$	Perovskite-type mixed metal oxides	Jiang et al. [95] Caro et al. [96] Jiang et al. [97], [98]
$La_{0.3}Sr_{0.7}FeO_{3-\delta}$		Evdou et al. [99]
$La_{0.7}Sr_{0.3}FeO_{3-\delta}$		Nalbandian et al. [100]
$La_{0.7}Sr_{0.3}Cu_{0.2}Fe_{0.8}O_{3-\delta}$		Park et al. [82] Balachandran et al. [101]
$SrFeCo_{0.5}O_x$	Non-perovskite type	Ikeguchi et al. [102, 103] Ma and Balachandran [104]

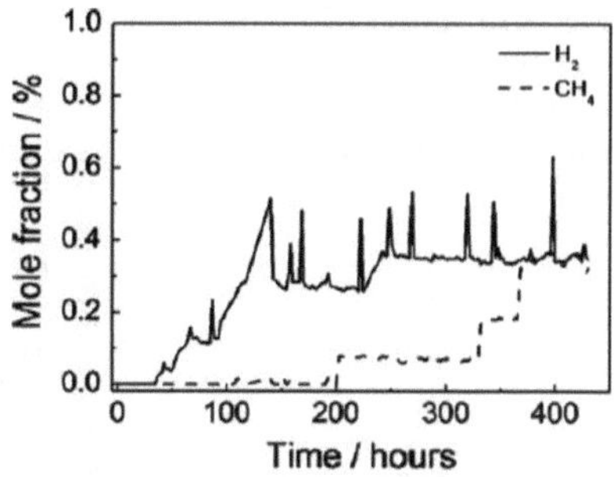

Fig. 3.11 Outlet composition from membrane-based steam reforming at 900 °C on the shell side (water splitting)

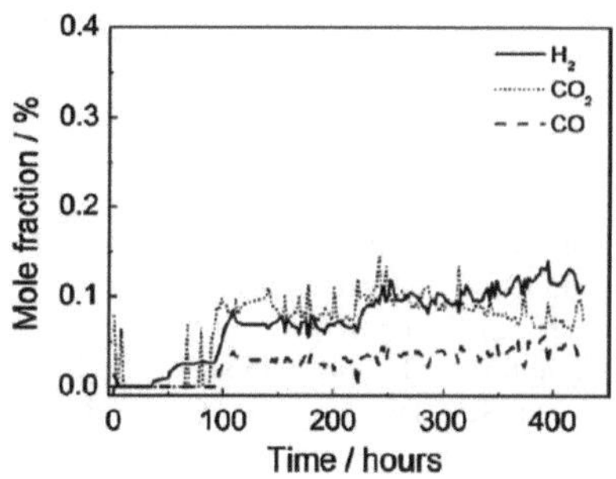

Fig. 3.12 Outlet composition from membrane-based steam reforming at 900 °C on lumen side (methane oxidation side)

oxygen flux and not the surface reaction. Their results are summarized in Figs. 3.11 and 3.12. Figure 3.11 shows the outlet composition of hydrogen and methane on the water-splitting side, while Fig. 3.12 presents the mole fractions of hydrogen, carbon dioxide, and carbon monoxide on the lumen or methane oxidation side.

Balachandran et al. [188] studied the use of a MIEC membrane to produce hydrogen by water dissociation. Hydrogen production was investigated over ranges of parameters including temperature, membrane thickness, water pressure, and oxygen chemical potential. The study revealed that the rate of hydrogen production due to water splitting increases with moisture concentration and oxygen chemical potential. Also, the production rate increases with decreasing the membrane thickness as shown in Fig. 3.13. In another experiment conducted by Lee et al. [106], the hydrogen production rate was also reported to vary inversely with thickness as shown in Fig. 3.14.

Balachandran et al. [87] investigated further the hydrogen production by water dissociation using mixed-conduction dense ceramic membranes. They reported several ways to produce higher amounts of hydrogen via water splitting. One way is through decreasing the membrane thickness. They reported that hydrogen production is increased by producing finer microstructure to enhance the surface kinetics. Hydrogen production by 1-mm-thick SFC2 membranes was higher compared to CGO/Ni ones. Naito and Arashi [93] investigated hydrogen production from direct water splitting at elevated temperature using ZrO_2–Y_2O_3 and ZrO_2–TiO_2–Y_2O_3 membranes. Water was vaporized and dissociated at an elevated

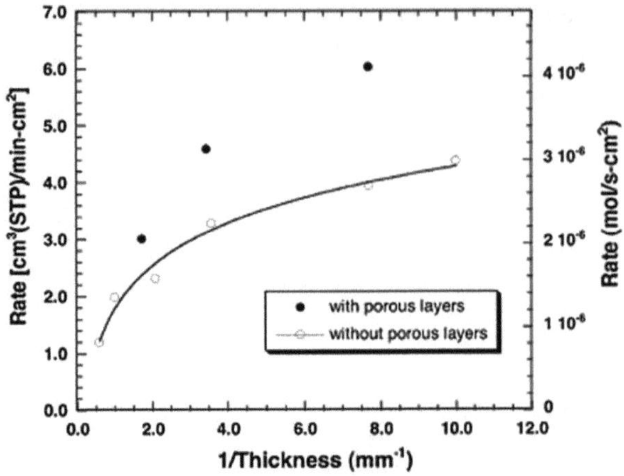

Fig. 3.13 Effect of membrane thickness on hydrogen production rate with and without porous layers [88]

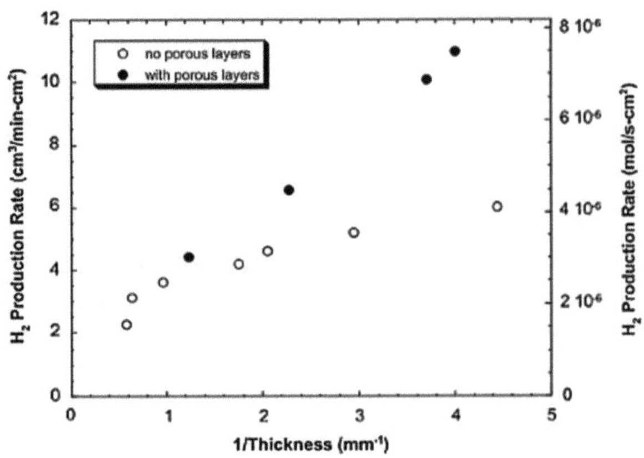

Fig. 3.14 Hydrogen production rate versus inverse of thickness for SFC membrane with and without porous SFC layers at 900 °C [106]

temperature of 2000 K, and oxygen was permeated simultaneously through the membrane by the partial pressure gradient of oxygen. Thus, water dissociation was increased which enhanced hydrogen production. Figs. 3.15 and 3.16 present the hydrogen production rate in mol/s/cm^2 with respect to the oxygen partial pressure gradient at elevated temperature. The higher the temperature the higher the produced hydrogen flux for the considered two membranes.

The effect of oxygen partial pressure gradient across the water-splitting membrane has been investigated by Park et al. [82]. They studied the performance of LSCF-7328 asymmetric membrane at a temperature of 900 °C. They reported that as the oxygen partial pressure gradient increases on the hydrogen production side, the partial pressure gradients of H_2O and hydrogen increase simultaneously. Figure 3.17 represents the results of hydrogen production rate with respect to the partial pressure of H_2O on the hydrogen production side.

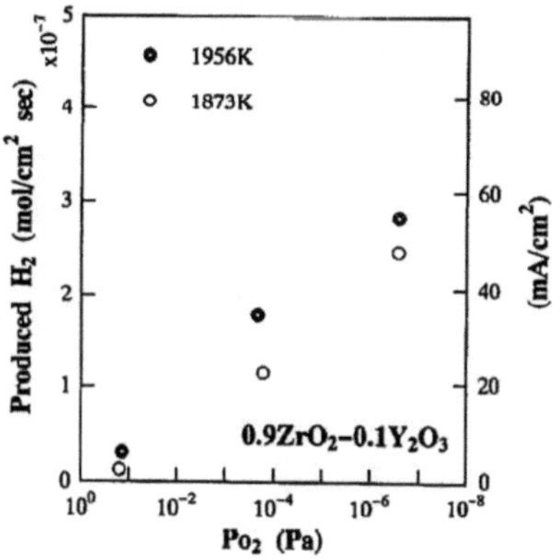

Fig. 3.15 Hydrogen production rate and oxygen partial pressure on the outer side of ZrO_2–Y_2O_3 membrane [93]

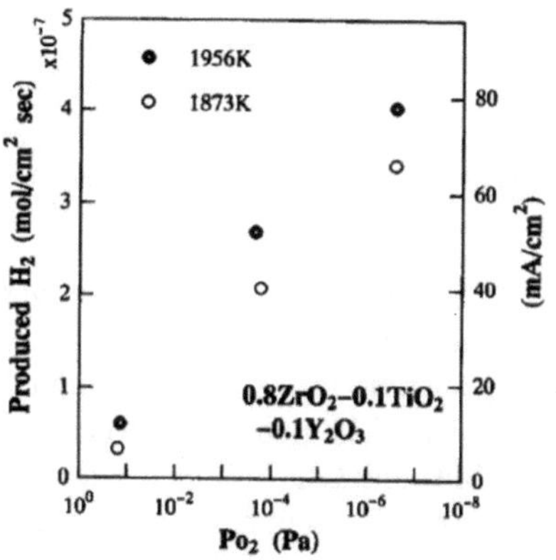

Fig. 3.16 Hydrogen production rate and oxygen partial pressure on the outer side of ZrO_2–TiO_2–Y_2O_3 membrane [93]

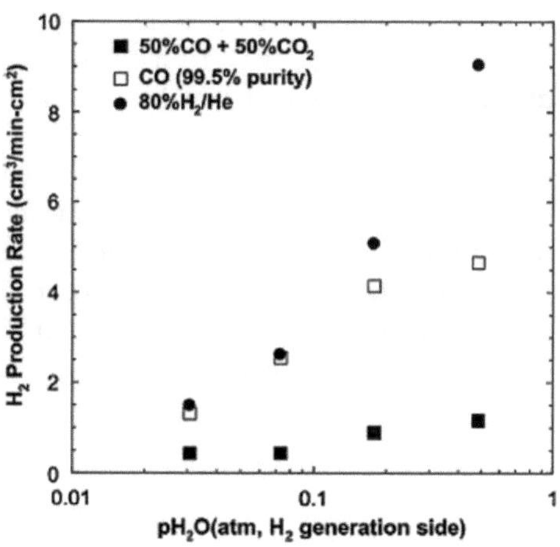

Fig. 3.17 Hydrogen production rates of LSCF-7328 membrane in terms of partial pressure of H_2O in the hydrogen production side at 900 °C [82]

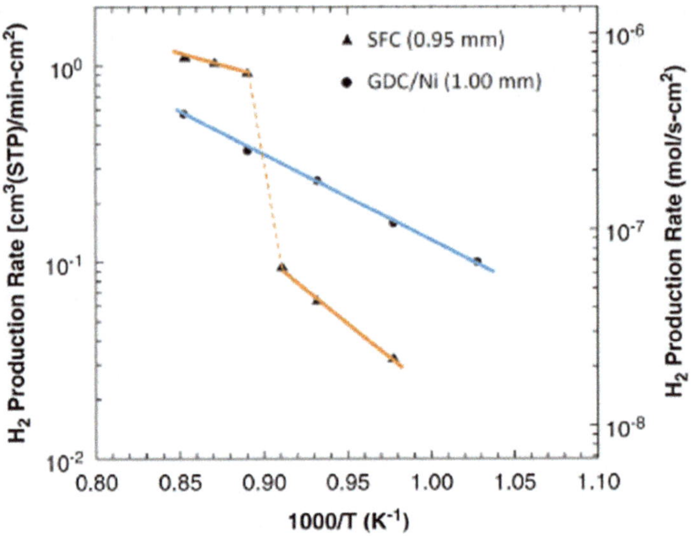

Fig. 3.18 Hydrogen production rate for SFC and GDC/Ni membranes [108]

The effect of temperature is very important due to its impact on increasing the permeated flux. Balachandran et al. [107] studied the hydrogen production by water dissociation using mixed-conducting dense ceramic membranes as presented in Fig. 3.18. The hydrogen production rate is plotted versus the inverse of temperature for SFC and GDC/Ni membranes in the temperature range of 700–900 °C. Note the sharp reduction in hydrogen production rate at ~1100 K for SFC membrane. It was thus recommended that it is more feasible to use the GDC/Ni membrane at low temperatures because it does not show a phase transition within the considered temperature range. Meng et al. [108] studied the bifunctional performances of $BaCe_{0.95}Tb_{0.05}O_{3-\delta}$ (BCTb) membranes for power generation and hydrogen production. They reported that below 700 °C, BCTb has dominant proton conductivity, while above this temperature the electronic conductivity becomes more significant. They investigated a microtubular solid-oxide fuel cell (SOFC), and the study revealed that hydrogen permeation recorded the highest flux of about 0.53 mL/cm^2/min over the studied range of temperature. They recorded the hydrogen permeation flux with respect to the helium flow rate in mL/min over a range of temperatures, from 700 to 850 °C, as shown in Fig. 3.19. Also, Table 3.3 provides a comparison of hydrogen production rates considering different membranes.

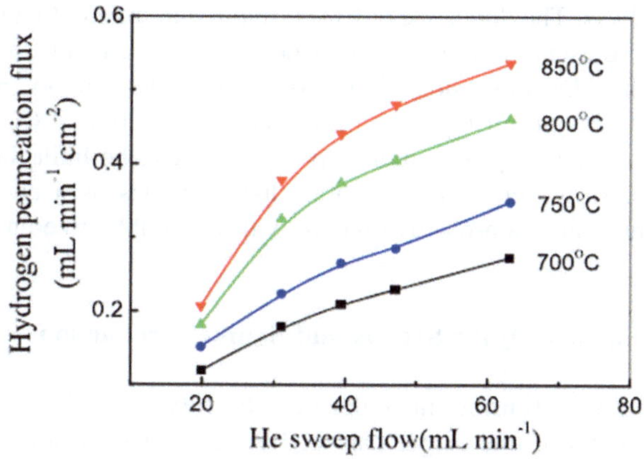

Fig. 3.19 Hydrogen permeation fluxes through BCTb-coated NIO–BCTb hollow fiber membrane as a function of flow rate of sweep helium gas at different temperatures ranged from 700 to 850 °C [108]

Table 3.3 Comparison of hydrogen production rates considering different membranes

Membrane material	Membrane configuration	Temperature range (K)	References	H_2 production rate
LSCF-6428	Microtubular	1173	Franca et al. [105]	
CGO/NI	0.13-mm-thick cermet with coarse microstructure	973–1173	Balachandran et al. [88]	6.0 $cm^3/min/cm^2$
CGO/Ni	0.09-mm-thick cermet with porous surface layers and 1-mm-thick SFC2	773–1173	Balachandran et al. [107]	10.0 $cm^3/min/cm^2$
0.9ZrO–0.1Y_2O_3 and 0.8ZrO_2–0.1TiO_2–0.1Y_2O_3	MIEC	1873–1956	Naito and Arashi [93]	4.0×10^{-7} mol/s/cm^2

3.4.4 Membranes for CO_2 Utilization

One of the main problems in carbon capture technologies is the storage of the huge amount of CO_2 either in deep geological formations or in deep oceans [109, 110]. The utilization of captured CO_2 in various industries offers a better alternative and is currently becoming popular especially because of the concerns associated with dumping it underground or in oceanic beds. Two approaches have been adopted for CO_2 utilization, namely the direct utilization and the conversion of CO_2 to chemical and energy products. The direct use of CO_2 is applied in different industries, such as soft drinks, food preservation, fire extinguishers, water treatment, and packing. The second approach is the conversion of CO_2 to chemicals and energy products, which was found to be very promising, as it may result in reduction in CO_2 capture cost. Furthermore, a closed loop of carbon capture cycle can be built during the combustion process. However, CO_2 has some disadvantages as a chemical reactant because of its inert and non-reactive nature with low Gibbs free energy.

3.4.4.1 Splitting of CO_2 for Syngas and Synfuel Production

Direct utilization of CO_2 using microalgae can be very promising knowing the fact that cultivating 1 ton of microalgae can fix 1.8 tons of CO_2 from the environment [111]. Indirect utilization of CO_2 include using it as a source of carbon for the synthesis of various valuable chemicals and fuels via CO_2 hydrogenation, CO_2 cycloaddition to epoxides, and CO_2 carbonylation of amines or alcohols, as discussed by Dai et al. [112] and Razali et al. [113]. Existing chemical industries utilize CO_2 conversion to produce urea and organic carbonates. The other way of recycling CO_2 is to convert it into synthetic fuels, such as methanol, dimethyl carbonate (DMC), and dimethyl ether (DME) [113–117]. The development of these energy products via renewable energy resources will not only reduce the burden on fossil fuels but also mitigate the threats of global warming.

CO_2 splitting has been of keen interest to the scientists for decades, as a tool to reduce the impact of damage caused by CO_2 emissions into the environment. The approach to reduce CO_2 emissions can be exploited to produce CO as a raw material for various chemical industries or to synthesize syngas. A detailed review of the production of syngas from CO_2 and H_2O splitting and its utilization has been presented by Nguyen and Blum [118]. An important application of syngas production is its conversion into synthetic liquid fuels through the Fischer–Tropsch (F–T) process. The produced synfuel is a promising alternative fuel for the transportation industry, as it can be safely and efficiently used without altering the motor engine technology. The F–T process, if combined with the hot temperature H_2O/CO_2 co-electrolysis process, can provide an effective way to harness and store clean energy for the transportation sector. An economic assessment of the combined process through modeling and sensitivity analysis is presented by Fu et al. [119]. The most crucial economic constraint was assessed to be the electricity price as the combined cycle is energy intensive.

The renewable energy source that can be effectively employed for syngas production is concentrated solar power (CSP). Agrafiotis et al. [120] provided a thorough survey of the development, evolution, and status of CSP-aided syngas production technology. They asserted that although the redox-pair-based thermochemical cycles, aided with CSP for hydrogen or syngas production, are technically competitive with conventional technologies like solar-panel-powered electrolysis, further research is required to achieve this in practice. The focus should be on the improvement of heat recuperation systems and the development of right functional materials at competitive costs.

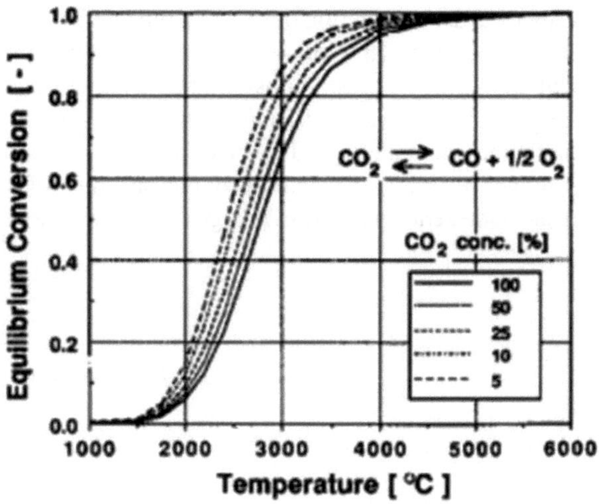

Fig. 3.20 Equilibrium conversion as a function of temperature [121]

Itoh et al. [121] performed an experimental investigation of CO_2 splitting enhanced with an yttria-stabilized zirconia (YSZ) membrane. Thermal decomposition of CO_2 was characterized as a highly endothermic reaction and can be expressed by the following equation:

$$CO_2 \rightleftharpoons CO + \frac{1}{2}O_2 \tag{2}$$

However, the reaction requires elevated temperature, and the equilibrium concentration favors the reactant side. Figure 3.20 shows the conversion of CO_2 as a function of temperature.

The use of YSZ membrane reactor system was found to enhance the direct thermal decomposition of CO_2. An experimental study by Graves et al. [122], on the production of syngas by co-electrolysis of CO_2 and H_2O, showed a favorable initial performance slightly lower than that of H_2O electrolysis but significantly higher than that of CO_2 electrolysis. They found that Ni/YSZ electrode degradation was dominant at low current density operation of the solid-oxide cells, whereas at higher current density, the degradation of lanthanum strontium manganite (LSM) electrode and serial resistance showed the major contribution to the loss of performance. Similar observations were reported by Zhan et al. [123] in their study on Ni–YSZ electrode-based solid-oxide electrolysis cells to produce syngas. Nigara and Cales [124] showed that the thermal decomposition of CO_2 can be significantly enhanced using stabilized zirconia membrane. They found a net increase in the thermodynamic equilibrium conversion from 1.2 to 21.5% at 1954 K.

The second-order rate coefficient for the thermal decomposition of CO_2 was investigated experimentally by Oehlschlaeger et al. [125]. The calculated rate coefficients were found to be close to previous findings giving an uncertainty of ±20% in the experimental study. Furthermore, they claimed that the incubation period prior to the thermal decomposition of CO_2 in a reflected shock wave experiment was observed for the first time. They presented a master equation for the calculation of both the incubation time and decomposition rate coefficients and suggested a model incorporating a collision energy transfer parameter and energy transfer probability density function to adequately predict the experimental incubation times.

Galvez et al. [126] examined the splitting of CO_2 via two-step cycle-based metal-oxide redox reaction. The governing reactions can be expressed as:

$$M_xO_y \rightleftharpoons xM + \frac{y}{2}O_2 \tag{3}$$

$$xM + \frac{y}{2}CO_2 \rightleftharpoons M_xO_y + \frac{y}{2}C \tag{4a}$$

$$xM + yCO_2 \rightleftharpoons M_xO_y + yCO \tag{4b}$$

The first step is the endothermic thermal dissociation of metal oxide into metal and oxygen, while the second step is exothermic, where the reduced metal reacts with carbon dioxide to regenerate the metal oxide and form carbon or carbon monoxide. Concentrated solar beam was used for the high-temperature endothermic reaction. Solar-chemical conversion efficiencies of 39% were reported for Zn/ZnO cycle and 29% for FeO/Fe_3O_4. Figure 3.21 presents the reaction extents for various

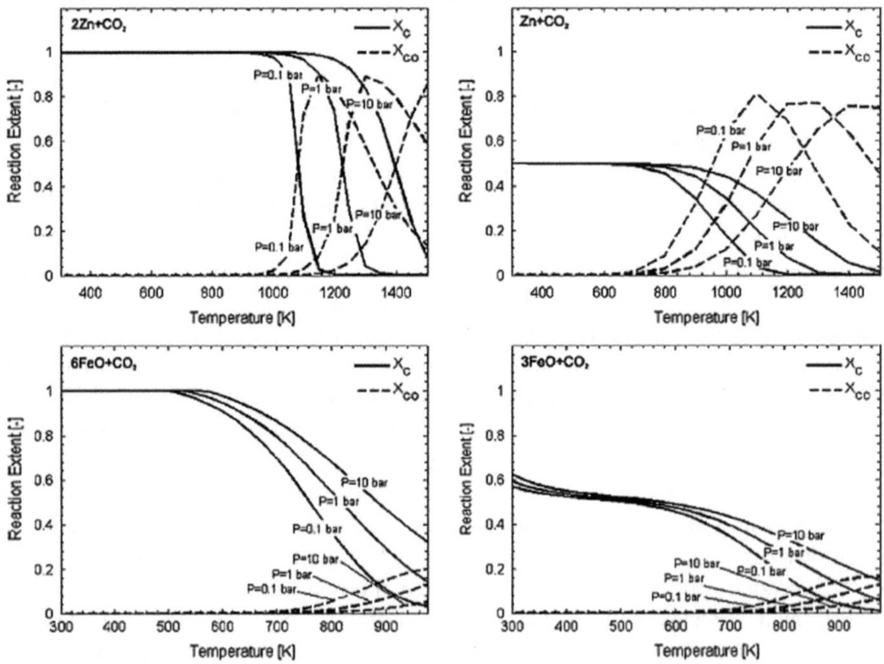

Fig. 3.21 Reaction extent variation as a function of temperature at different operating pressure for 2Zn + CO_2, Zn + CO_2, 6FeO + CO_2, and 3FeO + CO_2 reaction systems [126]

reaction systems at pressures of 0.1, 1, and 10 bar as functions of temperature. With increasing pressure, the reaction extent reduces for CO and increases for carbon. The formation of carbon is favored at higher pressures as per the Le Chatelier's principle. Further insight into the kinetic analysis of the two-step cycle-based H_2O/CO_2 splitting reaction to form syngas was performed by Stamatiou et al. [127]. They discussed detailed reaction mechanism to serve the guidelines for the design of a reactor to produce syngas via H_2O/CO_2 splitting.

Venstrom and Davidson [128] performed a thermodynamic analysis of the second step of Zn/ZnO cycle. They stated that the heterogeneous oxidation of zinc vapor has two basic advantages over the reaction of solid or liquid zinc; i.e., the reaction is rapid and characterized with complete zinc conversion. They reported that, without heat recuperation, the splitting efficiency of water and carbon dioxide raised from 6 to 27 and 31%, respectively. With heat recuperation, the maximum theoretical cycle efficiency was found to be 38 and 41% for water and carbon dioxide splitting, respectively, where 95% of the energy required to form zinc vapor is supplied from the exothermic oxidation reaction. Investigations by Chen et al. [129] suggested that electrocatalytic splitting of water and CO_2 can be efficiently achieved by using simple Cu(II)/Cu(0) electrodes in a two-compartment electrochemical cell in a neutral aqueous solution. The main advantages of using this kind of electrochemical cell are the simple nature of catalysts, solution condition, and cell configuration.

Experimental investigations of the production of syngas by H_2O/CO_2 splitting via ceria-based redox reaction cycle were carried out by Furler et al. [130]. They performed ten consecutive cycles of H_2O/CO_2 splitting in a span of 8 h, powered by solar concentrator using porous ceria maintained at 1800 K inside the solar cavity receiver. The results indicated the feasibility of syngas production via ceria-based redox reaction cycle in a repetitive and controlled volume as required in a practical solar fuel application. However, the prolonged deposition of ceria vapor on the parabolic concentrator affected the radiative power input, resulting in lower temperature level, and consequently lower production yield. In another study, Furler et al. [131] performed an experimental investigation of the thermochemical splitting of CO_2 in a solar cavity receiver using a reticulated porous ceramic (RPC). They found an average solar-to-fuel energy conversion efficiency of 1.73% with a peak value of 3.53%, which is four times higher than the previously reported value. Lorentzou et al. [132] studied the catalytic properties of pure ceria and zirconia-doped ceria for the solar thermochemical splitting of CO_2 and H_2O. The reactions involved thermal reduction of ceria-based compound and then the oxidation of ceria with CO_2/H_2O to form CO/H_2. They found that the doping of ceria with zirconia enhances the reduction yield from 5% for pure non-stoichiometric ceria to about 20% for the doped one, thus increasing the CO/H_2 production significantly. However, unlike pure zirconia, a drop in the reduction yield was observed with the doped ceria due to lower thermal stability. Thermal resistance of the doped ceria basically depends on the synthesis route, and they asserted that the doped ceria synthesized by Pechini method can improve the production capacity without inhibiting the thermal cyclic capability.

Smestad and Steinfeld [133] presented a detailed review of the photochemical and thermochemical splitting of H_2O/CO_2 using metal-oxide catalysts. They reported that TiO_2 is the most stable and useful metal oxide for photochemical splitting of H_2O/CO_2. They examined the kinetics to compare the photochemical reaction with AgCl to the thermochemical one with Ag_2O. A maximum theoretical solar-to-fuel energy conversion efficiency of the photochemical reaction has been reported to be 26.8%, while can exceed 30% for the thermochemical conversion system if most of the sensible heat is recuperated during the reaction cycle.

A detailed review of the CO_2 splitting characteristics via Boudouard reaction is presented by Lahijani et al. [134]. Reduction of CO_2 in the presence of solid carbon (char) to form CO was thoroughly investigated by the authors in terms of the parameters governing the reactivity of char, its structural features, and operational parameters. After an extensive literature survey on thermochemical splitting of carbon dioxide, Rayne [135] concluded that since solid phases of CO_2 and CO require pressures in GPa for solidification, non-traditional phases of CO_2/CO require large energy inputs, and hence, this is not a viable strategy for carbon dioxide splitting.

3.4.4.2 Conversion of CO_2 into Methanol

The conversion of CO_2 to other compounds that can store energy chemically is a viable technique for CO_2 utilization. This will contribute to the replacement of fossil fuels as well as to reducing the effect of greenhouse gases and global warming. Among these energy products are the conversion of CO_2 to methanol [136, 137]. To produce methanol, CO_2 could be combined with hydrogen (separated or electrolyzed from water), compressed, then reacted to produce methanol and water. Estimations of the possible yield range from 30 MMt (million metric tons) of methanol produced per year to over 300 MMt of CO_2 per year [138], with the amount of CO_2 per tones of methanol ranging from 3.1 to 14 tones. The potential market of CO_2 conversion to methanol will be significantly increasing if methanol consumption increases and if methanol is able to replace methane for energy production.

3.4.4.3 Use of CO_2 for Enhancing Oil Recovery

In recent years, the application of cyclic CO_2 injection to enhance the recovery of light oil has been examined [139]. Recent studies have been performed in this field and suggested that the principles beyond the oil recovery mechanisms are oil swelling, oil viscosity reduction, and gas relative permeability [140–142]. The cyclic injection technique is composed of three steps: (1) injection phase, where gas is injected into the active well; (2) soaking phase, where the well is shut in to allow the fluid to dissipate into the formation; and (3) production phase, where the well is operated for production. Gamadi et al. [143] studied experimentally the cyclic CO_2

injection to improve shale oil recovery. They presented the potential beyond applying the technology and reported several benefits associated with the cyclic CO_2 injection for effective oil recovery. Their study revealed that oil recovery has been improved from 33 to 85%, and they indicated that cyclic CO_2 injection is a promising method to improve shale oil recovery.

3.4.4.4 Conversion of CO_2 into Plastics

Due to the abundance of carbon dioxide, which is also inexpensive and non-toxic, it is considered an attractive raw material for incorporation into important industrial processes. The increase in fossil fuel cost, coupled with the need for cheap plastics, is forcing the industry to reduce the production cost of plastics by using CO_2 as a cheap bio-renewable resource that can potentially solve several problems related to plastics production. The catalytic bonding of CO_2 and epoxies to create carbonates or polycarbonate has proven to be very promising techniques in the implementation of CO_2 as a major component in a wide variety of plastics products.

Armstrong et al. [144] reported that the industrial progress has allowed us to envisage a resolution in which CO_2 becomes an increasingly valuable resource for producing several products. Zhang et al. [145] surveyed the sustainable chemistry and reported that certain types of organic molecules can contribute to the capture of CO_2 from air–fuel combustion and its conversion to new plastic materials. Satthawong et al. [146] reported a different point of view regarding CO_2 conversion into plastics. They claimed that, if CO_2 could be captured cheaply enough, converting it to more chemicals will require more energy, which might come from fossil fuel plants. The conversion process could require tremendous cost and emit much more CO_2 than the captured and consumed amount. Aresta et al. [147] discussed the capture and utilization of CO_2 for the sake of reducing greenhouse gas emissions.

Existing CO_2 utilization technologies, based on the available literature, are summarized, quantified and reported in Table 3.4. These technologies make use of the captured CO_2 from power plants and gas turbines, as main emitters of CO_2. Some of these technologies have immense potential while others have limited potential for commercialization.

Table 3.4 Summary of CO_2 utilization technologies [138]

Technology	References	Description
Algae cultivation	Benemann [148] Wilson et al. [149] Wang et al. [150]	CO_2 can be added to the growth process of algae, significantly increasing its productivity
Baking soda	Huang and Tan [110] Song [151]	Another type of mineralization, CO_2 can be contacted with sodium-rich brine, producing sodium bicarbonate, or baking soda
Bauxite residue	Dilmore et al. [152] Power et al. [153]	CO_2 can be injected into bauxite residue slurry to partially neutralize the slurry
Beverage carbonation	Descoins et al. [154] Oelkers et al. [155]	High-purity CO_2 can be used to carbonate beverages
Chemical synthesis	North et al. [156] Ma et al. [157]	CO_2 can be used as a feedstock in the production of a variety of chemicals, including acetic acid, alcohols, and sugars
Coffee decaffeination	Chen et al. [158] Franca [159]	CO_2 can be used in its supercritical state as the solvent for decaffeinating coffee
Concrete curing	Zhan et al. [160], [161]	At a precast concrete production facility, on-site waste CO_2 from flue gas can be permanently stored as limestone within the concrete
Electronics	Zhang and Han [162]	CO_2 is used, mostly as a cleaning fluid, in a small number of printed circuit board manufacturing applications
Enhanced coal bed methane (ECBM) recovery	Zuo-tang et al. [163] Ozdemir [164]	CO_2 can be injected into a coal seam to displace methane, which is recovered at the surface
Enhanced geothermal systems (EGS)	Pruess [165] Olasolo et al. [166]	Supercritical CO_2 could be used either as the circulating heat exchange fluid or as the working fluid in a supercritical CO_2 geothermal power cycle
Enhanced oil recovery	Ferguson et al. [167]	CO_2, if injected into a depleted oil reservoir, decreases oil viscosity and allows the oil to flow to a production well more easily. The CO_2 can be permanently stored in the depleted oil reservoir when production concludes
Desalination	McGinnis et al. [168] Al-Hallaj et al. [169] El-Naas et al. [170]	CO_2, mixed with H_2O brine at high pressure and low temperature, forms a hydrate of CO_2 surrounded by H_2O molecules. The hydrate is removed and rinsed, and then goes through multiple stages to remove dissolved solids in the brine, resulting in an exhaust stream of potable water
Fire suppression technology	Guo and Fu [171] Tilman et al. [172] Lenihan et al. [173]	CO_2 can be applied to a fire to reduce the oxygen level low enough to stop combustion. It is also used in fire extinguishers and fire protection systems
Food processing, preservation, and packaging	Ahmed and Alam [174] Han [175] Vaclavik and Christian [176]	CO_2 can be used during food processing for things such as cooling or spoilage prevention, and during packaging as an inert atmosphere to extend the shelf life of food products. Many industries, such as food packaging and wine making, use CO_2 as an inert gas
Formic acid	Schaub and Paciello [177] Kortlever et al. [178]	CO_2 can be electro-reduced to produce formic acid and O_2
Horticulture	Prior et al. [179] Marbel et al. [180]	Plant growth in a greenhouse occurs at an optimal level of CO_2 concentration. CO_2 can be captured from on-site processes or can be provided by off-site industries
Injection to conventional methanol synthesis	Huff and Sandford [181] Zhang et al. [182]	CO_2 can be injected upstream of the methanol reformer during conventional methanol synthesis, increasing the yield of the system

(continued)

Table 3.4 (continued)

Technology	References	Description
Methanol	Wang et al. [183] Kothandaraman et al. [184]	CO_2 can be combined with H_2, compressed, and reacted to produce methanol and water
Mineral carbonation	Huijgen et al. [185] Mayoral et al. [186]	CO_2 can be used to produce a replacement aggregate product for the construction industry by contacting a medium-concentrate CO_2 with mineral-loaded alkaline brine. The CO_2 from the gas will precipitate out as a limestone or dolomite equivalent
Pharmaceutical processes	Huisman and Gray [187] Kellaway [188]	In the pharmaceutical industry, CO_2 is used in many of the technologies listed, including inerting and supercritical fluid extraction
Polymer processing	Nalawade et al. [189] Cooper [190]	CO_2 can be used as a feedstock in a variety of polymer processing routes
Power generation	Pipitone and Bolland [191] Lombardi [192] Beer [193]	Supercritical CO_2 could be used as the working fluid in various power cycles, regardless of the heat source
Pulp and paper processing	Oral et al. [194] Gavrilescu [195]	CO_2 can be used to decrease acidity during pulp washing operations
Refrigerant gas	Kauf [196], Colasson and Haberschill [197], Agrawal and Roberts [198]	CO_2 can be used as a refrigerant gas replacement for more toxic gases
Supercritical CO_2 as a solvent	Nalawade et al. [189] Henon et al. [199]	Supercritical CO_2 can be used as a solvent for high-pressure extraction and to separate targeted compounds. It is also used in industries for dry cleaning for certain processes due to its low critical temperature
Water treatment	Lee et al. [200]	CO_2 can be used to remineralize water after desalination process using the reverse osmosis (RO), or for the sake of reducing acidity
Welding	Sun et al. [201]	CO_2 is used to prevent oxidation of the welding metal

3.5 Conclusions

In this chapter, the use of ion transport membranes (ITMs) for oxygen separation is studied for different applications. Performance of ITMs in terms of oxygen separation and stability is examined for different membrane materials. The permeation rate of oxygen has been improved for all membranes under partial oxidation or oxy-combustion of hydrocarbon fuel in the permeate side. The increased temperature level due to combustion reduced the resistance of the membrane material for oxygen permeation. Trending applications of ITMs are discussed including OTRs for syngas production, combustion utilizing liquid fuels in OTRs, membranes for splitting H_2O to produce H_2, and membranes for CO_2 utilization. The study revealed that there is an exciting potential for application of ITMs in different fields. The use of OTRs for substitution of a conventional gas turbine combustor or fire-tube boiler furnace is discussed in Chap. 6 of this book with complete design details and limitations.

References

1. Balachandran U, Kleefisch MS, Kobylinski TP, Morissette SL, Pei S (1997) Oxygen ion-conducting dense ceramic membranes. Assigned to Amoco Co., US Patent 5,639,437
2. Bernardo P, Drioli E, Golemme G (2009) Membrane gas separation: a review of state of the art. Ind Chem Eng 48(1):4638–4663
3. Farooqui AE, Badr HM, Habib MA, Ben-Mansour R (2014) Numerical investigation of combustion characteristics in an oxygen transport reactor. Int J Energy Res 38(5):638–651
4. Habib MA, Ahmed P, Ben-Mansour R, Badr HM, Kirchen P, Ghoniem AF (2013) Modeling of a combined ion transport and porous membrane reactor for oxy-combustion. J Membr Sci 446:230–243
5. Ben-Mansour R, Habib MA, Badr HM, Nemitallah MA (2012) Characteristics of oxy-fuel combustion in an oxygen transport reactor. Energy Fuels 26(7):4599–4606
6. Nemitallah MA, Habib MA, Mezghani K (2015) Experimental and numerical study of oxygen separation and oxy-combustion characteristics inside a button-cell LNO-ITM reactor. Energy 84:600–611
7. Xu SJ, Thomson WJ (1999) Oxygen permeation rates through ion-conducting perovskite membranes. Chem Eng Sci 54:3839–3850
8. Ruia Z, Lia Y, Lin YS (2009) Analysis of oxygen permeation through dense ceramic membranes with chemical reactions of finite rate. Chem Eng Sci 64:172–179
9. Akin FT, Lin YS (2004) Oxygen permeation through oxygen ionic or mixed-conducting ceramic membranes with chemical reactions. J Membr Sci 231:133–146
10. Habib MA, Ben Mansour R, Nemitallah MA (2013) Modeling of oxygen permeation through a LSCF ion transport membrane. Comput Fluids 76:1–10
11. Hong J, Kirchen P, Ghoniem AF (2012) Numerical simulation of ion transport membrane reactors: Oxygen permeation and transport and fuel conversion. J Membr Sci 407:71–85
12. Ben-Mansour R, Nemitallah MA, Habib MA (2013) Numerical investigation of oxygen permeation and methane oxy-combustion in a stagnation flow ion transport membrane reactor. Energy 54:322–332
13. Nemitallah MA, Habib MA, BenMansour R (2013) Investigations of oxy-fuel combustion and oxygen permeation in an ITM reactor using a two-step oxy-combustion reaction kinetics model. J Membr Sci 432:1–12
14. Hong J, Kirchen P, Ghoniem AF (2013) Interactions between oxygen permeation and homogeneous-phase fuel conversion on the sweep side of an ion transport membrane. J Membr Sci 428:309–322
15. Hunt A, Dimitrakopoulos G, Kirchen P, Ghoniem AF (2014) Measuring the oxygen profile and permeation flux across an ion transport ($La_{0.9}Ca_{0.1}FeO_{3-\delta}$) membrane and the development and validation of a multi-step surface exchange model. J Membr Sci 468:62–72
16. Kirchen P, Apo DJ, Hunt A, Ghoniem AF (2013) A novel ion transport membrane reactor for fundamental investigations of oxygen permeation and oxy-combustion under reactive flow conditions. Proc Combust Inst 34:3463–3470
17. Mancini ND, Mitsos A (2011) Ion transport membrane reactors for oxy-combustion; part II: analysis and comparison of alternatives. Energy 36:4721–4739

18. Nemitallah MA, Habib MA, Ben-mansour R, Ghoniem AF (2014) Design of an ion transport membrane reactor for gas turbine combustion application. J Membr Sci 450:60–71
19. Mezghani K, Hamza A, Habib MA, Lee D, Shao-Horn Y (2015) Effect of microstructure and thickness on oxygen permeation of $La_2NiO_{4+\delta}$ membranes. Ceram Int 42(1):666–672
20. Habib MA, Ahmed P, Ben-Mansour R, Mezghani K, Alam Z, Shao-Horn Y, Ghoniem AF (2015) Experimental and numerical investigation of $la_2NiO_{4+\delta}$ membranes for oxygen separation: geometry optimization and model validation. J Energy Res Technol 137 (3):03110
21. Wang L, Imashuku S, Grimaud A, Lee D, Mezghani K, Habib MA, Shao-Horn Y (2013) Enhancing oxygen permeation of electronically short-circuited oxygen-ion conductors by decorating with mixed ionic-electronic conducting oxides. ECS Electrochem Lett 2(11):77–81
22. Imashuku S, Wang L, Mezghani K, Habib MA, Shao-Horn Y (2013) Oxygen permeation from oxygen ion-conducting membranes coated with porous metals or mixed ionic and electronic conducting oxides. J Electrochem Soc 160(11):148–153
23. Habib MA, Badr HM, Ahmed SF, Ben-Mansour R, Mazghani K, Imashuku GJ, Shao-Horn Y, Mancini N, Mitsos A, Kirchen P, Ghoneim AF (2011) A review of recent developments in carbon capture utilizing oxy-fuel combustion in conventional and ion transport membrane systems. Int J Energy Res 35(9):741–764
24. Salehi M, Pfaff EM, Junior RM, Bergmann CP, Diethelm S, Neururer C, Graule T, Grobety B, Clemens FJ (2013) $Ba_{0.5}Sr_{0.5}Co_{0.8}Fe_{0.2}O_{3-\delta}$ (BSCF) feedstock development and optimization for thermoplastic forming of thin planar and tubular oxygen separation membranes. J Membr Sci 443:237–245
25. Baumann S, Meulenberg WA, Buchkremer HP (2013) Manufacturing strategies for asymmetric ceramic membranes for efficient separation of oxygen from air. J Eur Ceram Soc 33(7):1251–1261
26. Li X, Kerstiens T, Markus T (2013) Oxygen permeability and phase stability of $Ba_{0.5}Sr_{0.5}Co_{0.8}Fe_{0.2}O_{3-\delta}$ perovskite at intermediate temperatures. J Membr Sci 438:83–89
27. Baumann S, Serra JM, Lobera MP, Escolástico S, Schulze-Küppers F, Meulenberg W (2011) Ultrahigh oxygen permeation flux through supported $Ba_{0.5}Sr_{0.5}Co_{0.8}Fe_{0.2}O_{3-\delta}$ membranes. J Membr Sc 377(1):198–205
28. Haworth P, Smart S, Glasscock J, Diniz da Costa JC (2011) Yttrium doped BSCF membranes for oxygen separation. Sep Purif Technol 81(1):88–93
29. Menzler NH, Han F, Van Gestel T, Schafbauer W, Schulze-Küppers F, Baumann S, Uhlenbruck S et al (2013) Application of thin-film manufacturing technologies to solid oxide fuel cells and gas separation membranes. Int J Appl Ceram Technol 10(3):421–427
30. Buysse C, Kovalevsky A, Snijkers F, Buekenhoudt A, Mullens S, Luyten J, Kretzschmar J, Lenaerts S (2011) Development, performance and stability of sulfur-free, macrovoid-free BSCF capillaries for high temperature oxygen separation from air. J Membr Sci 372 (1–2):239–248
31. Liu H, Pang Z, Tan X, Shao Z, Sunarso J, Ding R, Liu S (2009) Enhanced oxygen permeation through perovskite hollow fiber membranes by methane activation. Ceram Int 35 (4):1435–1439
32. Hong J, Kirchen P, Ghoniem AF (2013) Interactions between oxygen permeation and homogeneous-phase fuel conversion on the sweep side of an ion transport membrane. J Memb Sci 428:309–322
33. Hong J, Kirchen P, Ghoniem AF (2012) Numerical simulation of ion transport membrane reactors: Oxygen permeation and transport and fuel conversion. J Memb Sci 407:71–85
34. Amato A, Hudak R, Noble DR, Scarborough D, Peter AD, Seitzman JM, Lieuwen TC (2010) Methane oxy-combustion for low CO_2 cycles: measurements and modeling of CO and O_2 emissions. In: ASME turbo expo 2010: power for land, sea, and air, pp 213–222
35. Zanganeh KE, Shafeen A, Thambimuthu K (2005) A comparative study of refinery fuel gas oxy-fuel combustion options for CO_2 capture using simulated process data. In: Greenhouse gas control technologies, pp 1117–1123

36. Luo H, Jiang H, Klande T, Liang F, Cao Z, Wang H, Caro J (2012) Rapid glycine-nitrate combustion synthesis of the CO_2-stable dual phase membrane $40Mn_{1.5}Co_{1.5}O_{4-\delta}$–$60Ce_{0.9}Pr_{0.1}O_{2-\delta}$ for CO_2 capture via an oxy-fuel process. J Membr Sci 423:450–458
37. Dyer PN, Richards RE, Russek SL, Taylor DM (2000) Ion transport membrane technology for oxygen separation and syngas production. Solid State Ionics 134:21–33
38. Miller CF, Chen J, Carolan MF, Foster EP (2014) Advances in ion transport membrane technology for syngas production. Catal Today 228:152–157
39. Izquierdo U, Barrio VL, Cambra JF, Requies J, Güemez MB, Arias PL, Kolb G, Zapf R, Gutiérrez AM, Arraibi JR (2012) Hydrogen production from methane and natural gas steam reforming in conventional and microreactor reaction systems. Int J Hydrogen Energy 37:7026–7033
40. Liu Z, Chu B, Zhai X, Jin Y, Cheng Y (2012) Total methanation of syngas to synthetic natural gas over Ni catalyst in a micro-channel reactor. Fuel 95:599–605
41. Rostrup-Nielsen JR (1993) Production of synthesis gas. Catal Today 18:305–324
42. Mokheimer EM, Hussain MI, Ahmed S, Habib MA, Al-Qutub AA (2015) On the modeling of steam methane reforming. J Energy Res Technol 137:012001
43. Simakov DS, Wright MM, Ahmed S, Mokheimer EM, Román-Leshkov Y (2015) Solar thermal catalytic reforming of natural gas: a review on chemistry, catalysis and system design. Catal Sci Technol 5:1991–2016
44. Sheu EJ, Mokheimer EM, Ghoniem AF (2015) A review of solar methane reforming systems. Int J Hydrogen Energy 40:12929–12955
45. Said SAM, Simakov DS, Mokheimer EM, Habib MA, Ahmed S, Waseeuddin M et al (2015) Computational fluid dynamics study of hydrogen generation by low temperature methane reforming in a membrane reactor. Int J Hydrogen Energy 40(8):3158–3169
46. Said SAM, Simakov DSA, Waseeuddin M, Román-Leshkov Y (2016) Solar molten salt heated membrane reformer for natural gas upgrading and hydrogen generation: a CFD model. Sol Energy 124:163–176
47. Sheu EJ, Mitsos A, Eter AA, Mokheimer EM, Habib MA, Al-Qutub A (2012) A review of hybrid solar–fossil fuel power generation systems and performance metrics. J Sol Energy Eng 134(4):041006
48. Sheu EJ, Mokheimer EM, Ghoniem AF (2015) Dry redox reforming hybrid power cycle: performance analysis and comparison to steam redox reforming. Int J Hydrogen Energy 40:2939–2949
49. Delsman ER (2005) Microstructured reactors for a portable hydrogen production unit. Ph.D thesis, Technische Universiteit Eindhoven, Eindhoven
50. Christopher CM, Bennett DL, Carolan MF, Miller CF, Steppan JJ, Waldron WE (2006) ITM syngas: ceramic membrane technology for lower cost conversion of natural gas. In: AIChE Spring National Meeting 25, 2006
51. Taamallah S, Vogiatzaki K, Alzahrani F, Mokheimer EM, Habib MA, Ghoniem AF (2015) Fuel flexibility, stability and emissions in premixed hydrogen-rich gas turbine combustion: technology, fundamentals, and numerical simulations. Appl Energy 154:1020–1047
52. Alzahrani FM, Sanusi YS, Vogiatzaki K, Ghoniem AF, Habib MA, Mokheimer EM (2015) Evaluation of the accuracy of selected syngas chemical mechanisms. J Energy Res Technol 137:042201
53. Mokheimer EM, Sanusi YS, Habib MA (2016) Numerical study of hydrogen-enriched methane–air combustion under ultra-lean conditions. Int J Energy Res 40(6):743–762
54. Sanusi YS, Habib MA, Mokheimer EM (2015) Experimental study on the effect of hydrogen enrichment of methane on the stability and emission of nonpremixed swirl stabilized combustor. J Energy Res Technol 137:032203
55. Schubert K, Brandner J, Fichtner M, Linder G, Schygulla U, Wenka A (2001) Microstructure devices for applications in thermal and chemical process engineering. Microscale Thermophys Eng 5:17–39
56. Ehrfeld W, Hessel V, Löwe H (2000) Microreactors: new technology for modern chemistry. Wiley-VCH, Weinheim, p 2000

57. Vorontsov VA, Gribovskiy AG, Makarshin LI, Andreev DV, Ylianitsky VY, Parmon VN (2014) Influence of a reaction mixture streamline on partial oxidation of methane in an asymmetric microchannel reactor. Int J Hydrogen Energy 39:325–330
58. Enger BC, Walmsley J, Bjørgum E, Lødeng R, Pfeifer P, Schubert K, Holmen A, Venvik HJ (2008) Performance and SEM characterization of Rh impregnated microchannel reactors in the catalytic partial oxidation of methane and propane. Chem Eng J 144:489–501
59. Schneider A, Mantzaras J, Jansohn P (2006) Experimental and numerical investigation of the catalytic partial oxidation of CH_4/O_2 mixtures diluted with H_2O and CO_2 in a short contact time reactor. Chem Eng Sci 61:4634–4649
60. Sadykov V, Bobrova L, Pavlova S, Simagina V, Makarshin L, Parmon V, Ross JRH, Veen ACV (2012) Syngas generation from hydrocarbons and oxygenates with structured, catalysts. Nova Science Publishers, New York
61. Enger BC, Lødeng R, Holmen A (2008) A review of catalytic partial oxidation of methane to synthesis gas with emphasis on reaction mechanisms over transition metal catalysts. Appl Catal A 346:1–27
62. Vernikovskaya NV, Bobrova LN, Pinaeva LG, Sadykov VA, Zolotarskii IA, Sobyanin VA, Buyakou I, Kalinin V, Zhdanok S (2007) Transient behavior of the methane partial oxidation in a short contact time reactor: modeling on the base of catalyst detailed chemistry. Chem Eng J 134:180–189
63. Silva FA, Resende KA, Da Silva AM, De Souza KR, Mattos LV, Montes M, Souza-Aguiar EF, Noronha FB, Hori CE (2012) Syngas production by partial oxidation of methane over $Pt/CeZrO_2/Al_2O_3$ catalysts. Catal Today 180(1):111–116
64. Tsai CY, Dixon AG, Moser WR, Ma YH (1997) Dense perovskite membrane reactors for the partial oxidation of methane to syngas. AIChE J 43:2741
65. Li S, Jin W, Huang P, Xu N, Shi J, Payzant MZEA, Ma YH, Hu MZC (1999) Perovskite-related ZrO_2-doped $SrCO_{0.4}Fe_{0.6}O_{3-\delta}$ membrane for oxygen permeation. AIChE J 45:276
66. Yang C, Xu N, Shi J (1998) Experimental and modeling study on a packed-bed membrane reactor for partial oxidation of methane to formaldehyde. Ind Eng Chem Res 37:2601
67. Balachandran U, Dusk JT, Maiya PS, Ma B, Mieville RL, Kleefisch MS, Udovich CA (1997) Ceramic membrane reactor for converting methane to syngas. Catal Today 36:265
68. Habib MA, Salaudeen SA, Nemitallah MA, Ben-Mansour R, Mokheimer EM (2016) Numerical investigation of syngas oxy-combustion inside a LSCF-6428 oxygen transport membrane reactor. Energy 96:654–665
69. Nemtallah MA, Habib MA (2017) Numerical investigation of liquid methanol evaporation and oxy-combustion in a button-cell ITM reactor. Appl Therm Eng 112:378–391
70. Smith JB, Norby T (2006) On the steady state oxygen permeation through $La_2NiO_{4+\delta}$ membranes. J Electrochem Soc 153:233–238
71. Mancini ND, Mitsos A (2011) Ion transport membrane reactors for oxy-combustion-part I: intermediate fidelity modeling. Energy 36:4701–4720
72. O'Rourke PJ (1981) Collective drop effects on vaporizing liquid sprays. Ph.D thesis, Princeton University, Princeton, New Jersey
73. Taylor GI (1963) The shape and acceleration of a drop in a high speed air stream. Technical report in the scientific papers of G.I
74. O'Rourke PJ, Amsden AA (1987) The TAB method for numerical calculation of spray droplet breakup. SAE technical paper, SAE:872089
75. Ranz WE, Marshall WR (1952) Evaporation from drops, part II. Chem Eng Prog 48: 173–180
76. Lacas F, Leroux B, Darabiha N (2005) Experimental study of air dilution in oxy-liquid fuel flames. Proc Combust Inst 30:2037–2045
77. Yu S-C, Huang C-W, Liao C-H, Wu JCS, Chang S-T, Chen K-H (2011) A novel membrane reactor for separating hydrogen and oxygen in photocatalytic water splitting. J Memb Sci 382:291–299. https://doi.org/10.1016/j.memsci.2011.08.022

78. Hisatomi T, Kubota J, Domen K (2014) Recent advances in semiconductors for photocatalytic and photoelectrochemical water splitting. Chem Soc Rev 43:7520–7535. https://doi.org/10.1039/C3CS60378D
79. Maeda K (2011) Photocatalytic water splitting using semiconductor particles: history and recent developments. J Photochem Photobiol C Photochem Rev 12:237–268. https://doi.org/10.1016/j.jphotochemrev.2011.07.001
80. Kato H, Kudo A (2003) Photocatalytic water splitting into H_2 and O_2 over various tantalate photocatalysts. Catal Today 78:561–569. https://doi.org/10.1016/S0920-5861(02)00355-3
81. Park CY, Lee TH, Dorris SE, Lu Y, Balachandran U (2011) Oxygen permeation and coal-gas-assisted hydrogen production using oxygen transport membranes. Int J Hydrogen Energy 36:9345–9354. https://doi.org/10.1016/j.ijhydene.2011.04.090
82. Park CY, Lee TH, Dorris SE, Balachandran U (2010) Hydrogen production from fossil and renewable sources using an oxygen transport membrane. Int J Hydrogen Energy 35:4103–4110. https://doi.org/10.1016/j.ijhydene.2010.02.025
83. Mundschau MV, Xie X, Evenson CR IV, Sammells AF (2006) Dense inorganic membranes for production of hydrogen from methane and coal with carbon dioxide sequestration. Catal Today 118:12–23. https://doi.org/10.1016/j.cattod.2006.01.042
84. Rahimpour MR, Arab Aboosadi Z, Jahanmiri AH (2012) Synthesis gas production in a novel hydrogen and oxygen perm-selective membranes tri-reformer for methanol production. J Nat Gas Sci Eng 9:149–159. https://doi.org/10.1016/j.jngse.2012.06.007
85. Song SJ, Moon JH, Lee TH, Dorris SE, Balachandran U (2008) Thickness dependence of hydrogen permeability for Ni–$BaCe_{0.8}Y_{0.2}O_{3-\delta}$. Solid State Ionics 179:1854–1857. https://doi.org/10.1016/j.ssi.2008.05.012
86. Jeon SY, Choi MB, Park CN, Wachsman ED, Song SJ (2011) High sulfur tolerance dual-functional cermet hydrogen separation membranes. J Memb Sci 382:323–327. https://doi.org/10.1016/j.memsci.2011.08.024
87. Balachandran U, Lee TH, Dorris SE (2007) Hydrogen production by water dissociation using mixed conducting dense ceramic membranes. Int J Hydrogen Energy 32:451–456. https://doi.org/10.1016/j.ijhydene.2006.05.010
88. Balachandran U, Lee TH, Wang S, Dorris SE (2004) Use of mixed conducting membranes to produce hydrogen by water dissociation. Int J Hydrogen Energy 29:291–296. https://doi.org/10.1016/S0360-3199(03)00134-4
89. Jeon SY, Im HN, Singh B, Hwang JH, Song SJ (2013) A thermodynamically stable $La_2NiO_4/Gd_{0.1}Ce_{0.9}O_{1.95}$ bilayer oxygen transport membrane in membrane-assisted water splitting for hydrogen production. Ceram Int 39:3893–3899. https://doi.org/10.1016/j.ceramint.2012.10.233
90. Xu SJ, Thomson WJ (1997) Perovskite-type oxide membranes for the oxidative coupling of methane. AIChE J 43:2731–2740. https://doi.org/10.1002/aic.690431319
91. Kharton VV, Yaremchenko AA, Kovalevsky AV, Viskup AP, Naumovich EN, Kerko PF (1999) Perovskite-type oxides for high-temperature oxygen separation membranes. J Memb Sci 163:307–317. https://doi.org/10.1016/S0376-7388(99)00172-6
92. Li W, Zhu X, Cao Z, Wang W, Yang W (2015) Mixed ionic-electronic conducting (MIEC) membranes for hydrogen production from water splitting. Int J Hydrogen Energy 40:3452–3461. https://doi.org/10.1016/j.ijhydene.2014.10.080
93. Naito H, Arashi H (1995) Hydrogen production from direct water splitting at high temperatures using a ZrO_2–TiO_2–Y2O_3 membrane. Solid State Ionics 79:366–370. https://doi.org/10.1016/0167-2738(95)00089-O
94. Lede J, Lapicque F, Villermaux J (1983) Production of hydrogen by direct thermal decomposition of water. Int J Hydrogen Energy 8:675–679. https://doi.org/10.1016/0360-3199(83)90175-1
95. Jiang H, Cao Z, Schirrmeister S, Schiestel T, Caro J (2010) A coupling strategy to produce hydrogen and ethylene in a membrane reactor. Angew Chemie Int Ed 49:5656–5660. https://doi.org/10.1002/anie.201000664

96. Jiang H, Liang F, Czuprat O, Efimov K, Feldhoff A, Schirrmeister S et al (2010) Hydrogen production by water dissociation in surface-modified BaCo(x)Fe(y)Zr(1-x-y)O(3-delta) hollow-fiber membrane reactor with improved oxygen permeation. Chemistry 16:7898–7903. https://doi.org/10.1002/chem.200902494
97. Jiang H, Wang H, Liang F, Werth S, Schirrmeister S, Schiestel T et al (2010) Improved water dissociation and nitrous oxide decomposition by in situ oxygen removal in perovskite catalytic membrane reactor. Catal Today 156:187–190. https://doi.org/10.1016/j.cattod.2010.02.027
98. Jiang H, Wang H, Werth S, Schiestel T, Caro J (2008) Simultaneous production of hydrogen and synthesis gas by combining water splitting with partial oxidation of methane in a hollow-fiber membrane reactor. Angew Chem Int Ed 47:9341–9344. https://doi.org/10.1002/anie.200803899
99. Evdou A, Nalbandian L, Zaspalis VT (2008) Perovskite membrane reactor for continuous and isothermal redox hydrogen production from the dissociation of water. J Memb Sci 325:704–711. https://doi.org/10.1016/j.memsci.2008.08.042
100. Nalbandian L, Evdou A, Zaspalis V (2009) $La_{1-x}Sr_xMO_3$ (M = Mn, Fe) perovskites as materials for thermochemical hydrogen production in conventional and membrane reactors. Int J Hydrogen Energy 34:7162–7172. https://doi.org/10.1016/j.ijhydene.2009.06.076
101. Balachandran U, Dorris SE, Lu Y, Emerson JE, Park CY, Lee TH, et al (2010) Development of dense ceramic membranes for hydrogen separation. Adv Membr Technol Appl 3:1–41. https://doi.org/10.1016/s0167-2991(01)80347-5
102. Ikeguchi M, Ishii K, Sekine Y, Kikuchi E, Matsukata M (2005) Improving oxygen permeability in $SrFeCo_{0.5}O_x$ asymmetric membranes by modifying support-layer porous structure. Mater Lett 59:1356–1360. https://doi.org/10.1016/j.matlet.2004.12.042
103. Ikeguchi M, Yoshino Y, Kanie K, Nomura M, Kikuchi E, Matsukata M (2003) Effects of preparation method on oxygen permeation properties of $SrFeCo_{0.5}Ox$ membrane. Sep Purif Technol 32:313–318. https://doi.org/10.1016/s1383-5866(03)00048-0
104. Ma B, Balachandran U (1998) Phase stability of $SrFeCo_{0.5}O_x$ in reducing environments. Mater Res Bull 33:223–236. https://doi.org/10.1016/s0025-5408(97)00214-6
105. Franca RV, Thursfield A, Metcalfe IS (2012) $La_{0.6}Sr_{0.4}Co_{0.2}Fe_{0.8}O_3$ microtubular membranes for hydrogen production from water splitting. J Memb Sci 389:173–181. https://doi.org/10.1016/j.memsci.2011.10.027
106. Lee TH, Park CY, Dorris SEBU (2008) Hydrogen production from steam using oxygen transport membranes. ECS Trans 13:379–384
107. Balachandran, Lee TH, Dorris SE (2007) Hydrogen production by water dissociation using mixed conducting dense ceramic membranes. Int J Hydrogen Energy 32:451–456. https://doi.org/10.1016/j.ijhydene.2006.05.010
108. Meng X, Shang Y, Meng B, Yang N, Tan X, Sunarso J, et al (2016) Bi-functional performances of $BaCe_{0.95}Tb_{0.05}O_{3-\delta}$-based hollow fiber membranes for power generation and hydrogen permeation. JECS J Eur Ceram Soc. https://doi.org/10.1016/j.jeurceramsoc.2016.06.041
109. Albo J, Luis P, Irabien A (2010) Carbon dioxide capture from flue gases using a cross-flow membrane contactor and the ionic liquid 1-ethyl-3-methylimidazolium ethylsulfate. Ind Eng Chem Res 49:11045–11051. https://doi.org/10.1021/ie1014266
110. Huang CH, Tan CS (2014) A review: CO_2 utilization. Aerosol Air Qual Res 14:480–499. https://doi.org/10.4209/aaqr.2013.10.0326
111. Fernández F, González-López C (2012) Conversion of CO_2 into biomass by microalgae: how realistic a contribution may it be to significant CO_2 removal? Appl Microbiol
112. Dai W, Luo S, Yin S, Au C (2009) The direct transformation of carbon dioxide to organic carbonates over heterogeneous catalysts. Appl Catal Gen
113. Razali N, Lee K, Bhatia S, Mohamed A (2012) Heterogeneous catalysts for production of chemicals using carbon dioxide as raw material: a review. Renew Sustain
114. Ganesh I (2011) Conversion of carbon dioxide to methanol using solar energy. Curr Sci

115. Olah G (2013) Towards oil independence through renewable methanol chemistry. Angew Chemie Int Ed
116. Omae I (2012) Recent developments in carbon dioxide utilization for the production of organic chemicals. Coord Chem Rev
117. Rihko-Struckmann, L, Peschel A (2010) Assessment of methanol synthesis utilizing exhaust CO_2 for chemical storage of electrical energy. Ind Eng Chem Res
118. Nguyen VN, Blum L (2015) Syngas and synfuels from H_2O and CO_2: current status. Chem Ing Tech 87(4):354–375
119. Fu Q, Mabilat C, Zahid M, Brisse A, Gautier L (2010) Syngas production via high-temperature steam/CO2 co-electrolysis: an economic assessment. Energy Environ Sci 3(10):1382
120. Agrafiotis C, Roeb M, Sattler C (2015) A review on solar thermal syngas production via redox pair-based water/carbon dioxide splitting thermochemical cycles. Renew Sustain Energy Rev 42:254–285
121. Itoh N, Sanchez MA, Xu W-C, Haraya K, Hongo M (1993) Application of a membrane reactor system to thermal decomposition of CO_2. J Memb Sci 77(2):245–253
122. Graves C, Ebbesen SD, Mogensen M (2011) Co-electrolysis of CO_2 and H_2O in solid oxide cells: performance and durability. Solid State Ionics 192(1):398–403
123. Zhan Z, Kobsiriphat W, Wilson JR, Pillai M, Kim I, Barnet SA (2009) Syngas production by coelectrolysis of CO_2/H_2O: the basis for a renewable energy cycle. Energy Fuels 23 (6):3089–3096
124. Nigara Y, Cales B (1986) Production of carbon monoxide by direct thermal splitting of carbon dioxide at high temperature. Bull Chem Soc Jpn 59(6):1997–2002
125. Oehlschlaeger MA, Davidson DF, Jeffries JB, Hanson RK (2005) Carbon dioxide thermal decomposition: observation of incubation. Zeitschrift für Phys Chemie 219(5):555–567
126. Galvez ME, Loutzenhiser PG, Hischier I, Steinfeld A (2008) CO_2 splitting via two-step solar thermochemical cycles with Zn/ZnO and FeO/Fe_3O_4 redox reactions: thermodynamic analysis. Energy Fuels 22:3544–3550
127. Stamatiou A, Loutzenhiser PG, Steinfeld A (2010) Solar syngas production from H_2O and CO_2 via two-step thermochemical cycles based on Zn/ZnO and FeO/Fe_3O_4 redox reactions: kinetic analysis. Energy Fuels 24(4):2716–2722
128. Venstrom LJ, Davidson JH (2011) Splitting water and carbon dioxide via the heterogeneous oxidation of zinc vapor: thermodynamic considerations. J Sol Energy Eng 133(1):011017
129. Chen Z, Kang P, Zhang M-T, Stoner BR, Meyer TJ (2013) Cu(ii)/Cu(0) electrocatalyzed CO_2 and H_2O splitting. Energy Environ Sci 6(3):813
130. Furler P, Scheffe JR, Steinfeld A (2012) Syngas production by simultaneous splitting of H_2O and CO_2 via ceria redox reactions in a high-temperature solar reactor. Energy Environ Sci 5 (3):6098
131. Furler P, Scheffe J, Gorbar M, Moes L, Vogt U, Steinfeld A (2012) Solar thermochemical CO_2 splitting utilizing a reticulated porous ceria redox system. Energy Fuels 26(11): 7051–7059
132. Lorentzou S, Karagiannakis G, Pagkoura C, Zygogianni A (2011) CO_2 and H_2O splitting for thermochemical production of solar fuels using nonstoichiometric ceria and ceria/zirconia solid solutions. Energy Fuels 25:4836–4845
133. Smestad GP, Steinfeld A (2012) Review: photochemical and thermochemical production of solar fuels from H_2O and CO_2 using metal oxide catalysts. Ind Eng Chem Res 51: 11828–11840
134. Lahijani P, Zainal ZA, Mohammadi M, Mohamed AR (2015) Conversion of the greenhouse gas CO_2 to the fuel gas CO via the Boudouard reaction: a review. Renew Sustain Energy Rev 41:615–632
135. Rayne S (2008) Thermal carbon dioxide splitting: a summary of the peer-reviewed scientific literature. Nat Preced 1(250):1–17
136. Ganesh I (2011) Conversion of carbon dioxide to methanol using solar energy. Curr Sci 101:731–733. https://doi.org/10.4236/msa.2011.210190

137. Ganesh I (2014) Conversion of carbon dioxide into methanol—a potential liquid fuel: fundamental challenges and opportunities (a review). Renew Sustain Energy Rev 31:221–257. https://doi.org/10.1016/j.rser.2013.11.045
138. Thomas D, Thomas B, Jesse Goellner RM (2013) Novel CO_2 utilization concepts. https://doi.org/10.1016/j.enconman.2007.12.029
139. Thomas GA, Mcclure TGM (1991) Feasibility of cyclic CO_2 injection for light-oil recovery. 179–184
140. Torabi F, Qazvini Firouz A, Kavousi A, Asghari K (2012) Comparative evaluation of immiscible, near miscible and miscible CO_2 huff-n-puff to enhance oil recovery from a single matrix-fracture system (experimental and simulation studies). Fuel 93:443–453. https://doi.org/10.1016/j.fuel.2011.08.037
141. Li JH, Bao R, Qin B, Jiang T (2013) Numerical simulation of foamy oil stability using natural gas huff and puff for ultra-deep heavy oil reservoir, vol 318. https://doi.org/10.4028/www.scientific.net/AMM.318.405
142. Li G, Li X (2011) Numerical simulation for gas production from hydrate accumulated in Shenhu area, South China sea, using huff and puff method. Huagong Xuebao/CIESC J 62:458–468
143. Gamadi TD, Sheng JJ, Soliman MY, Menouar H, Watson MC, Emadibaladehi H (2014)An experimental study of cyclic CO_2 injection to improve shale oil recovery. In: SPE improved oil recovery symposium, pp 1–9. https://doi.org/10.2118/169142-ms
144. Armstrong K, Styring P (2015) Assessing the potential of utilization and storage strategies for post-combustion CO_2 emissions reduction. Front Energy Res 3:1–9. https://doi.org/10.3389/fenrg.2015.00008
145. Zhang Y, Chan JYG (2010) Sustainable chemistry: imidazolium salts in biomass conversion and CO_2 fixation. Energy Environ Sci 3:408–417. https://doi.org/10.1039/B914206A
146. Satthawong R, Koizumi N, Song C, Prasassarakich P (2015) Light olefin synthesis from CO_2 hydrogenation over K-promoted Fe–Co bimetallic catalysts. Catal Today 251:34–40. https://doi.org/10.1016/j.cattod.2015.01.011
147. Aresta M, Dibenedetto A, Angelini A (2013) The changing paradigm in CO_2 utilization. J CO_2 Util 3–4:65–73. https://doi.org/10.1016/j.jcou.2013.08.001
148. Benemann JR (1997) CO2 mitigation with microalgae systems. Energy Convers Manag 38: S475–S479. https://doi.org/10.1016/S0196-8904(96)00313-5
149. Wilson MHH, Groppo J, Placido A, Graham S, Morton SA, Santillan-Jimenez E, et al (2014) CO_2 recycling using microalgae for the production of fuels. Appl Petrochem Res 4 (1):41–43. https://doi.org/10.1007/s13203-014-0052-3
150. Wang B, Li Y, Wu N, Lan CQ (2008) CO_2 bio-mitigation using microalgae. Appl Microbiol Biotechnol 79:707–718. https://doi.org/10.1007/s00253-008-1518-y
151. Song C (2002) CO_2 Conversion and utilization: an overview. CO_2 Convers Util 809:1–2. https://doi.org/10.1021/bk-2002-0809.ch001
152. Dilmore R, Lu P, Allen D, Soong Y, Hedges S, Fu JK et al (2008) Sequestration of CO_2 in mixtures of bauxite residue and saline wastewater. Energy Fuels 22:343–353. https://doi.org/10.1021/ef7003943
153. Power G, Gräfe M, Klauber C (2011) Bauxite residue issues: I. Current management, disposal and storage practices. Hydrometallurgy 108:33–45. https://doi.org/10.1016/j.hydromet.2011.02.006
154. Descoins C, Mathlouthi M, Le Moual M, Hennequin J (2006) Carbonation monitoring of beverage in a laboratory scale unit with on-line measurement of dissolved CO_2. Food Chem 95:541–553. https://doi.org/10.1016/j.foodchem.2004.11.031
155. Oelkers EH, Gislason SR, Matter J (2008) Mineral carbonation of CO_2. Elements 4:333–337. https://doi.org/10.2113/gselements.4.5.333
156. North M, Pasquale R, Young C (2010) Synthesis of cyclic carbonates from epoxides and CO_2. Green Chem 12:1514. https://doi.org/10.1039/c0gc00065e
157. Ma J, Sun N, Zhang X, Zhao N, Xiao F, Wei W et al (2009) A short review of catalysis for CO_2 conversion. Catal Today 148:221–231. https://doi.org/10.1016/j.cattod.2009.08.015

158. Chen Y, Brown PH, Hu K, Black RM, Prior RL, Ou B, et al (2011) Supercritical CO_2 decaffeination of unroasted coffee beans produces melanoidins with distinct NF-κB inhibitory activity. J Food Sci 76. https://doi.org/10.1111/j.1750-3841.2011.02304.x
159. Franca AS (2016) Coffee: decaffeination. Encycl Food Health 232–236. http://dx.doi.org/10.1016/B978-0-12-384947-2.00183-5
160. Zhan BJ, Xuan DX, Poon CS, Shi CJ (2016) Effect of curing parameters on CO_2 curing of concrete blocks containing recycled aggregates. Cem Concr Compos 71:122–130. https://doi.org/10.1016/j.cemconcomp.2016.05.002
161. Zhan B, Poon C, Shi C (2013) CO_2 curing for improving the properties of concrete blocks containing recycled aggregates. Cem Concr Compos 42:1–8. https://doi.org/10.1016/j.cemconcomp.2013.04.013
162. Zhang X, Han B (2007) Cleaning using CO_2-based solvents. Clean Soil, Air, Water 35:223–229. https://doi.org/10.1002/clen.200700007
163. Zuo-tang W, Guo-xiong W, Rudolph V, Diniz da Costa JC, Pei-ming H, Lin X (2009) Simulation of CO_2-geosequestration enhanced coal bed methane recovery with a deformation-flow coupled model. Proc Earth Planet Sci 1:81–89. https://doi.org/10.1016/j.proeps.2009.09.015
164. Ozdemir E (2009) Modeling of coal bed methane (CBM) production and CO_2 sequestration in coal seams. Int J Coal Geol 77:145–152. https://doi.org/10.1016/j.coal.2008.09.003
165. Pruess K (2006) Enhanced geothermal systems (EGS) using CO_2 as working fluid-a novel approach for generating renewable energy with simultaneous sequestration of carbon. Geothermics 35:351–367. https://doi.org/10.1016/j.geothermics.2006.08.002
166. Olasolo P, Juárez MC, Morales MP, Damico S, Liarte IA (2016) Enhanced geothermal systems (EGS): a review. Renew Sustain Energy Rev 56:133–144. https://doi.org/10.1016/j.rser.2015.11.031
167. Ferguson RC, Nichols C, Van Leeuwen T, Kuuskraa VA (2009) Storing CO2 with enhanced oil recovery. Energy Proc 1:1989–1996. https://doi.org/10.1016/j.egypro.2009.01.259
168. McGinnis RL, Hancock NT, Nowosielski-Slepowron MS, McGurgan GD (2013) Pilot demonstration of the NH_3/CO_2 forward osmosis desalination process on high salinity brines. Desalination 312:67–74. https://doi.org/10.1016/j.desal.2012.11.032
169. Al-Hallaj S, Parekh S, Farid MM, Selman JR (2006) Solar desalination with humidification-dehumidification cycle: review of economics. Desalination 195:169–186. https://doi.org/10.1016/j.desal.2005.09.033
170. El-Naas MH, Al-Marzouqi AH, Chaalal O (2010) A combined approach for the management of desalination reject brine and capture of CO_2. Desalination 251:70–74. https://doi.org/10.1016/j.desal.2009.09.141
171. Guo TN, Fu ZM (2007) The fire situation and progress in fire safety science and technology in China. Fire Saf J 42:171–182. https://doi.org/10.1016/j.firesaf.2006.10.005
172. Tilman D, Reich P, Phillips H, Menton M, Patel A, Vos E et al (2000) Fire suppression and ecosystem carbon storage. Ecology 81:2680–2685
173. Lenihan JM, Bachelet D, Neilson RP, Drapek R (2008) Simulated response of conterminous United States ecosystems to climate change at different levels of fire suppression, CO_2 emission rate, and growth response to CO_2. Glob Planet Change 64:16–25. https://doi.org/10.1016/j.gloplacha.2008.01.006
174. Ahmed J, Alam T (2012) An overview of food packaging: material selection and the future of packaging. In: Handbook of food process design, p 1237–1283. https://doi.org/10.1002/9781444398274.ch41
175. Han JH (2005) Innovations in food packaging. In: Innovations in food ackaging, p 517. https://doi.org/10.1016/b978-012311632-1/50046-2
176. Vaclavik V, Christian E (2008) Food preservation and processing. Essent Food Sci 425–446. https://doi.org/10.1007/978-0-387-69940-0_17
177. Schaub T, Paciello RA (2011) A process for the synthesis of formic acid by CO_2 hydrogenation: thermodynamic aspects and the role of CO. Angew Chemie Int Ed 50:7278–7282. https://doi.org/10.1002/anie.201101292

178. Kortlever R, Balemans C, Kwon Y, Koper MTM (2015) Electrochemical CO_2 reduction to formic acid on a Pd-based formic acid oxidation catalyst. Catal Today 244:58–62. https://doi.org/10.1016/j.cattod.2014.08.001
179. Prior SA, Brett Runion G, Christopher Marble S, Rogers HH, Gilliam CH, Allen Torbert H (2011) A review of elevated atmospheric CO_2 effects on plant growth and water relations: implications for horticulture. HortScience 46:158–162
180. Christopher Marble S, Prior SA, Brett Runion G, Allen Torbert H, Gilliam CH, Fain GB (2011) The importance of determining carbon sequestration and greenhouse gas mitigation potential in ornamental horticulture. HortScience 46:240–244
181. Huff CA, Sanford MS (2011) Cascade catalysis for the homogeneous hydrogenation of CO_2 to $_M$ethanol. J Am Chem Soc 133:18122–18125. https://doi.org/10.1021/ja208760j
182. Zhang C, Jun KW, Kwak G, Lee YJ, Park HG (2016) Efficient utilization of carbon dioxide in a gas-to-methanol process composed of CO_2/steam-mixed reforming and methanol synthesis. J CO_2 Util 16:1–7. https://doi.org/10.1016/j.jcou.2016.05.005
183. Wang WH, Himeda Y, Muckerman JT, Manbeck GF, Fujita E (2015) CO_2 hydrogenation to formate and methanol as an alternative to photo- and electrochemical CO_2 reduction. Chem Rev 115:12936–12973. https://doi.org/10.1021/acs.chemrev.5b00197
184. Kothandaraman J, Goeppert A, Czaun M, Olah GA, Prakash GKS (2016) Conversion of CO_2 from air into methanol using a polyamine and a homogeneous ruthenium catalyst. J Am Chem Soc 138:778–781. https://doi.org/10.1021/jacs.5b12354
185. Huijgen WJJ, Comans RNJ, Witkamp GJ (2007) Cost evaluation of CO_2 sequestration by aqueous mineral carbonation. Energy Convers Manag 48:1923–1935. https://doi.org/10.1016/j.enconman.2007.01.035
186. Mayoral MC, Andrés JM, Gimeno MP (2013) Optimization of mineral carbonation process for CO_2 sequestration by lime-rich coal ashes. Fuel 106:448–454. https://doi.org/10.1016/j.fuel.2012.11.042
187. Huisman GW, Gray D (2002) Towards novel processes for the fine-chemical and pharmaceutical industries. Curr Opin Biotechnol 13:352–358. https://doi.org/10.1016/S0958-1669(02)00335-X
188. Kellaway I (2001) Transport processes in pharmaceutical systems, vol 228. https://doi.org/10.1016/s0378-5173(01)00823-7
189. Nalawade SP, Picchioni F, Janssen LPBM (2006) Supercritical carbon dioxide as a green solvent for processing polymer melts: processing aspects and applications. Prog Polym Sci 31:19–43. https://doi.org/10.1016/j.progpolymsci.2005.08.002
190. Cooper AI (2000) Polymer synthesis and processing using supercritical carbon dioxide. J Mater Chem 10:207–234. https://doi.org/10.1039/a906486i
191. Pipitone G, Bolland O (2009) Power generation with CO_2 capture: technology for CO_2 purification. Int J Greenhouse Gas Control 3:528–534. https://doi.org/10.1016/j.ijggc.2009.03.001
192. Lombardi L (2003) Life cycle assessment comparison of technical solutions for CO_2 emissions reduction in power generation. Energy Convers Manag 44:93–108. https://doi.org/10.1016/S0196-8904(02)00049-3
193. Beér JM (2007) High efficiency electric power generation: the environmental role. Prog Energy Combust Sci 33:107–134. https://doi.org/10.1016/j.pecs.2006.08.002
194. Oral J, Sikula J, Puchyr R, Hajny Z, Stehlik P, Bebar L (2005) Processing of waste from pulp and paper plant. J Clean Prod 13:509–515. https://doi.org/10.1016/j.jclepro.2003.09.005
195. Gavrilescu D (2008) Energy from biomass in pulp and paper mills. Environ Eng Manag J 7:537–546
196. Kauf F (1999) Determination of the optimum high pressure for transcritical CO_2-refrigeration cycles. Int J Therm Sci 38:325–330. https://doi.org/10.1016/S1290-0729(99)80098-2

197. Colasson S, Haberschill P (2010) Effect of refrigerant charge on global performances of a transcritical CO_2 heat pump. Sustain. In: Refrigerant heat pump technology conference Stock, Sweden, pp 1–7. https://doi.org/10.3969/j.issn.0258-2724.2010.05.005
198. Agrawal R Roberts M (2000) Dual mixed refrigerant cycle for gas liquefaction
199. Hénon FE, Camaiti M, Burke AL, Carbonell RG, DeSimone JM, Piacenti F (1999) Supercritical CO_2 as a solvent for polymeric stone protective materials. J Supercrit Fluids 15:173–179. https://doi.org/10.1016/s0896-8446(99)00005-4
200. Lee SY, Seo S, Broda JC, Pal S, Santoro RJ (2000) An experimental estimation of mean reaction rate and flame structure during combustion instability in a lean premixed gas turbine combustor. Proc Combust Inst 28:775–782. https://doi.org/10.1016/S0082-0784(00)80280-5
201. Sun J, Liu X, Tong Y, Deng D (2014) A comparative study on welding temperature fields, residual stress distributions and deformations induced by laser beam welding and CO_2 gas arc welding. Mater Des 63:519–530. https://doi.org/10.1016/j.matdes.2014.06.057

Chapter 4
Novel Approaches for Clean Combustion in Gas Turbines

4.1 Introduction

Nowadays, power demand is growing globally, and access to reliable, affordable energy is a critical issue. The International Energy Agency (IEA) reported that, by 2020, the global economy will grow by about 3.5% annually and the total population will rise by about one billion. Based on such forecasts, the IEA expects the total energy demand to be raised more than 30% by 2040 [1]. Additional 6700 GW of power is expected to be added in the next 25 years considering newly installed and retired gas power plants. Gas turbines for power production are characterized by high cycle efficiency, more than 50% in modern 250 MW-class combined cycles, and low NO_x in natural gas-fueled turbines under premixed combustion conditions [2]. Gas turbines also account for all commercial aero-propulsion power production systems fueled by kerosene. This growing market for gas turbines encouraged the investments in this sector for clean power production. The pressure of strict regulations on emissions has pushed the manufacturers of gas turbine combustors to invent environmental combustors while maintaining high combustion efficiency and wider operability limits. NO_x emissions are considered as the most serious kinds of pollutants. The production of NO_x emissions within the combustor is mainly function of the combustion temperature and the residence time of the combustion products in the elevated temperature zones. NO_x can be produced within the combustor through various mechanisms [3–5]; however, the dominant mechanism within the zones of elevated temperatures is the thermal NO_x (Zeldovich) mechanism [6]. Lowering the combustion temperature even by few degrees can result in notable reduction in NO_x emissions [7]. For same given oxygen and nitrogen fractions, the produced amount of thermal NO_x in few seconds at a temperature of the flame of 1800 K is in the same order of the produced amount of NO_x in milliseconds when the flame temperature is raised to 2100 K [6].

In the last three decades, intensive research and development work has been conducted to advance combustion technologies for clean power production

M. A. Nemitallah et al., *Oxyfuel Combustion for Clean Energy Applications*, Green Energy and Technology, https://doi.org/10.1007/978-3-030-10588-4_4

especially for gas turbine applications. In the 1990s, the US Department of Energy (DOE) started a collaborative program including gas turbine manufacturers, universities, and laboratories to develop advanced turbine systems with an improved cycle efficiency of up to 60% [8, 9]. In 2000, another program called Vision 21 was initiated by the US DOE to develop fuel-flexible electrical generation facilities with the main objective of reducing NO_x emissions while capturing CO_2 [10]. Different techniques have been proposed by researchers to reduce emissions out of gas turbines toward clean energy production for the control of global warming. Those techniques include either modification in the existing combustion systems or inventing novel combustion burners to control the emissions. Some of those clean burning techniques have already been implemented in industrial gas turbines for electrical power production, and the others are still under research and development. One of the most common clean burning techniques is the dry low NO_x technology, which has been implemented in many industrial gas turbine units by different manufacturers. Those gas turbine units include Siemens KG2-3G, Ansaldo Energia AE94.3A, and General Electrics (GE) GE LM2500. However, research is still going trying to improve the performance of the gas turbine while adapting DLN technology to further reduce the emissions. The subject of the present chapter is to discuss in detail the status of the research and development of novel techniques for clean power production in gas turbine. Those techniques include flame-type variability, burner design, fuel flexibility, and oxidizer flexibility.

In fact, different flame types have different combustion and emission characteristics and changing flame mode can result in significant changes in the produced emissions. Non-premixed flames have been used in gas turbines for power generation thanks to their strong stability behavior over wide ranges of loading conditions [11–13]. Non-premixed flames result in stoichiometric combustion zones within the combustor, and consequently, elevated temperature spots are created within the combustor, which raises the level of NO_x emissions [14]. Converting the combustion mode from non-premixed to premixed prevents the creation of stoichiometric combustion zones within the combustor as the reactants are premixed upstream of the combustor. This results in reduction in combustion temperature, and accordingly, NO_x emissions are reduced [15]. However, premixing the reactants upstream of the combustor results in fluctuations in the flow field that interact with the pressure field. This interaction results in various kinds of combustion instabilities, which adversely affect the engine operation [16–18]. Colorless distributed combustion (CDC) is another technique to control emissions through the control of flame type within the combustor. The idea here is to create a uniform mixture within the combustor through fast mixing between incoming fresh air, fuel, and burned gases to eliminate the possibility of the creation of stoichiometric combustion zone to control flame temperature and, accordingly, control NO_x emissions. CDC can result in improved performance of the gas turbine combustor in terms of uniform thermal field, low NO_x and CO emissions, and improved combustion stability at reduced noise level. Also, low-swirl injector (LSI) combustion is an effective technique to control of the flame and to control the emissions. The low-swirl combustion technique exploits the same wavelike principal of premixed turbulent flames. The free burning of the premixed flame in a divergent flow region under reduced swirling flow results in a suspended flame wave "standing wave" without the need for flame anchoring technique. The flame is suspended in a location where the flame speed is the same as the flow velocity. This results in more stable flame and leads to possibilities for flashback, and blowout is reduced.

The design of the burner is also an effective technique to control emissions out of gas turbine combustors. Different burner designs with different combustion and emissions characteristics are discussed in the present chapter. Those designs include swirl-stabilized burners, dry low NO_x (DLE) burners, dry low emission (DLE) burners, burners for catalytic combustion, perforated plate burners, environmental burners including environmental vortex (EV), sequential environmental vortex (SEV), and advanced environmental vortex (AEV), and micromixer burners. The designs of these burners are introduced in this chapter with their performance in terms of flame stabilization and emissions. Fuel flexibility approach may be an option to control instabilities associated with premixed mode of combustion and improving the overall combustion and emissions characteristics. It is concluded that there is a rising interest in switching the operation of combined-cycle power plants from natural gas to hydrogen-enriched methane or to syngas to increase the flame stability limits and control the emissions of NO_x as well. The progress of the effects of hydrogen enrichment on premixed combustion and emissions characteristics is also discussed in the present chapter. Also, oxidizer flexibility can play a significant role in controlling gas turbine emissions. Air combustion using hydrocarbon fuels is the main sources for both NO_x and CO_2 emissions. Due to the existence of nitrogen in air, NO_x emissions will be created in the combustion zone and the rate of production will increase depending on the combustion temperature. There have been many approaches in reducing the CO_2 emissions via carbon capture and sequestration (CCS) technologies. Among the current techniques, oxy-fuel combustion offers a brighter future for carbon dioxide emission-free society as the fuel is burned using O_2 to produce H_2O and CO_2. The CO_2 is separated after H_2O condensation. This technique of oxy-fuel burning can also eliminate NO_x emissions due to the absence of nitrogen within the combustor. The application of the oxy-fuel combustion technique in gas turbines is discussed in this chapter considering the associated operability issues and their effects on the performance of the gas turbine combustor. Finally, the feasibility of the different technologies and trends for applications for controlling gas turbine emissions is introduced at the end of the current chapter including market products utilizing these technologies.

4.2 Adaptation of Gas Turbines to Regulations of Pollutant Emissions

In the 1970s, when emission controls were first introduced, the pollutant of primary concern to regulators shifted to be concerned with NO_x. For the relatively low levels of NO_x emission reduction initially targeted, the injection of steam into the flame zone produced the targeted reduction in NO_x emissions with an insignificant effect on the performance. Also, the emissions of other pollutants including CO and volatile organic compounds (VOCs) did not increase in significant amounts. During the 1980s, more strict policies were imposed requiring a significant reduction in NO_x emissions. Based on that, further attempts were made to utilize steam injection to ensure achieving the targeted emission values. As a result, cycle performance and part lives are affected badly, and the emission rates for other pollutants including CO and VOC increased significantly. At this stage, the research was forced to develop other techniques for controlling emissions, which led to the development of the lean premixed (LPM) combustion technique [19].

4.2.1 Emission Regulatory Overview

4.2.1.1 Clean Air Act (CAA)

Congress enacted the Clean Air Act (CAA) in 1970 to consider growing concerns about the quality of atmospheric air [20]. Based on this act, National Ambient Air Quality Standards (NAAQS) were established to control the emissions for criteria pollutants. Areas of the country that exceed the NAAQS are considered non-attainment for that pollutant. In non-attainment areas, the US Regulatory Structure imposes more stringent air pollution control programs. The CAA is a federal law covering the entire country. But, states and local governments are permitted to create and implement more stringent air pollution strategies than those required by the CAA.

4.2.1.2 New Source Performance Standards (NSPS)

In the early 1970s, the emissions control requirements for nitrogen oxides (NO_x) were first applied to control the emissions of gas turbines by the Los Angeles County Air Pollution Control District (LAAPCD) and the San Diego Air Pollution Control District (SDAPCD). As a solution to match these regulations, water was injected into the combustor in the flame zone in order to cool down the combustion temperature. When half as much water as fuel was injected into the flame zone, NO_x emissions were reduced by about 40%. Under such operating conditions, the emission level achieved was approximately 75 ppmvd (parts per million by volume, dry) on oil. The US EPA used these data and other data to develop new source performance standards that went into effect in September 1979. Turbines with heat input over 10 million Btu/h, generating less than 30 MW electrical output, and supplying less than one-third of their electrical output to an electric utility, are required to meet a NO_x emission standard of 150 ppm, corrected for efficiency. The details of this NSPS are presented in 40CRF60 Subpart GG. The emergency turbines are excused from this standard, as are special types of turbines. Electric utility turbines having heat input above 100 million Btu/h should meet a NO_x emissions standard of 75 ppm, corrected for efficiency. Today, most of the working turbines can achieve NO_x emissions of 42–25 ppm or less without any means for post-combustion control. Based on that, the existing NSPS is not typically a controlling regulation for gas turbines emissions [19].

In July 2004, EPA modified the NSPS for gas turbines in a direct final ruling. The significant modification is that the new LPM turbines that commence construction after July 8, 2004, are required to use a NO_x Continuous Emissions Monitoring System (CEMS). Also, the owners can monitor continuously the engine parameters that give indication when the turbine is working out of the lean premixed mode of combustion. On February 18, 2005, EPA suggested standards of performance for new stationary gas turbines in 40CFR60 Subpart KKKK. The new standards reflect the changes in the turbine design and, accordingly, NO_x emissions control technologies and are intended to make the emission limits up to date with the performance of present gas turbines.

4.2.1.3 New Source Review

In addition to the AAQS, the CAA developed an air permitting program called New Source Review (NSR). The NSR is divided into two main programs: Prevention of Significant Deterioration (PSD) and Non-Attainment NSR. Each of these programs applies to "major sources" and "major modifications." However, in Non-Attainment NSR at much lower thresholds, a source can be considered "major."

4.2.1.4 Best Available Control Technology (BACT)

If a source triggers PSD review, so the owner should define the appropriate level of emissions controls for the pollutants that exceed the specific limits. In attainment areas, the standard for evaluation is the BACT. The BACT determination puts achievable emissions limitations considering environmental, energy, and economic effects of applying the emission control technology required to meet that limitation. Now, the EPA asks for a "top-down" BACT analysis to be followed for all PSD Permit applications [21]. Based on that, BACT is a "living" standard, which is more stringent over time unless new, low-cost, and more efficient emission control technologies are developed. For a certain application, BACT can only be defined in the form of current demonstrated technology. Generic BACT requirements do not exist as BACT determinations are site-specific. However, for gas turbines with power greater than 25 MW, there have been recent BACT determinations for NO_x as low as 5–2 ppm.

4.2.1.5 Lowest Achievable Emission Rate (LAER)

LAER determination is used in major sources/modifications in non-attainment areas. For class or source category, LAER is defined as the strictest emission control system achieved in practice. The main difference between BACT and LAER is that LAER does not take care of economic effects when evaluating emission control technologies of pollutants.

4.3 Types of Flame

The mode of the flame and how fuel and oxidizer are introduced to the combustor has direct effect on combustion temperature and emissions within the combustor. In this section, different flame modes are introduced toward the applications of clean power production in gas turbines.

4.3.1 Non-premixed/Premixed Flames

In diffusion flames, a reaction sheet is created forming the flame border as shown in Fig. 4.1. This figure represents a coaxial swirling flow jet, the most common flow configurations which support diffusion flames. The diffusion flame starts at laminar mode, but it shifts to turbulent diffusion flame when a further increase of flow rate occurs. This could form a flicker at the flame top. The flame length is also increased due to the increase in turbulence level. The fuel concentration is highest on the centerline as the fuel is introduced from the centerline pipe and slightly decreases to reach zero at the flame reaction sheet. The concentration of oxidizer is highest at the wall and tends to zero at the flame reaction sheet.

This diffusion (non-premixed) kind of flames creates stoichiometric combustion zones on the thin flame sheet resulting in elevated temperature spots within the combustor. This results in the dissociation of reacting species to produce emissions, mainly NO_x. To solve this issue, flame mode can be turned to the premixed combustion mode. In lean premixed approach, fuel is mixed with air upstream within the mixing length. Many studies revealed that the application of the technique of premixed flames is a promising solution that can bring down the concentration of NO_x by a single-digit value in ppm [23–25]. However, the conversion of the flame from the diffusion type (non-premixed) to the premixed type results in

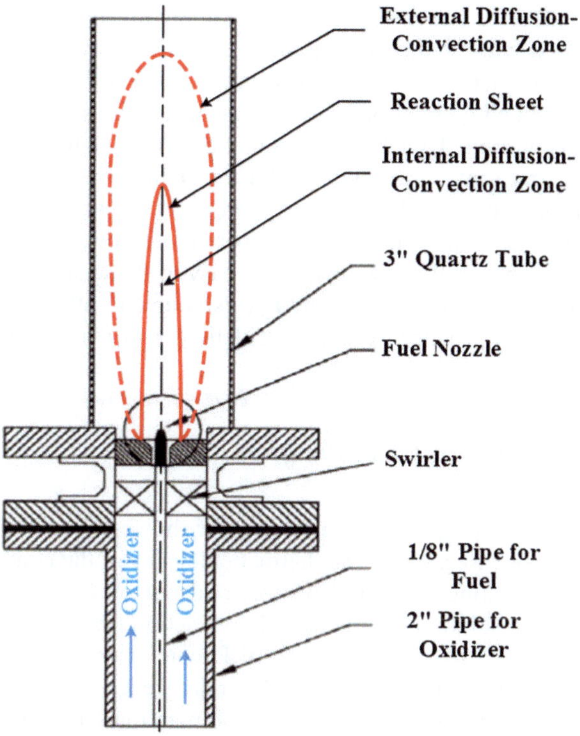

Fig. 4.1 Reaction zones of coaxial burner holding jet diffusion flame [22]

increased levels of flame instabilities which brings its own challenges to burner designers and manufacturer. The variations in flame temperature for both conventional diffusion and DLE premixed flames are presented in Fig. 4.2 (left). The maximum temperature corresponds to the conventional diffusion flame, while

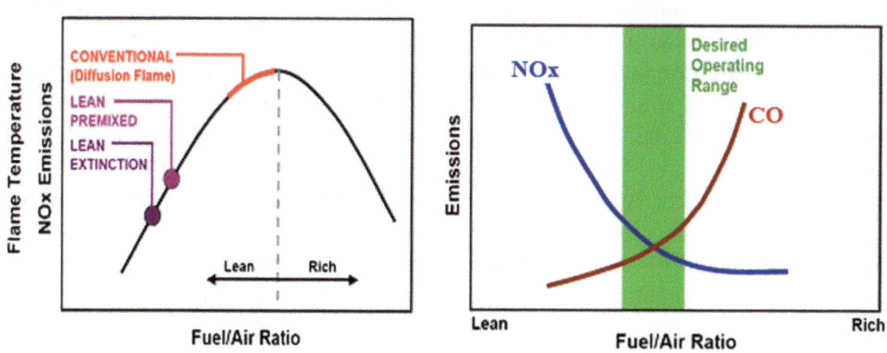

Fig. 4.2 Effect of fuel-to-air ratio on flame temperature (left) and CO and NO_x emissions (right) [26]

the minimum temperature corresponds to the lean premixed flame. The highest flue gas temperature occurs at the stoichiometric condition, and consequently, NO_x emissions have a higher value at this operating condition. Decreasing temperature leads to reduction in NO_x emissions and increase in CO emission while operating at lean extinction condition. To achieve ultra-low NO_x emissions (<10 ppm) and low CO emission at the same time, the combustor should be operated near to lean extinction limit or at the decided operating range, as shown in Fig. 4.2 (right). At the lean extinction condition, the flame temperature is low, and the flame is very weak and unstable, which then becomes very easy to flameouts. These combustion instabilities occur when the heat release rate is coupled with the acoustics of the combustion system to produce pressure oscillations and fluctuations.

4.3.2 MILD/Flameless Combustion

The moderate or intense low oxygen dilution or the so-called MILD combustion approach is developed to enhance the thermal efficiency of the combustion process, by means of recirculating the waste heat carried out by flue gases. MILD combustion is one of the solutions to maintain low reaction temperature; consequently, the NO_x formation will be reduced [27]. MILD combustion has been defined as a process in which the temperature of the inlet reactants is higher than the self-ignition temperature of the mixture, and the maximum increase in temperature during the combustion process is less than the self-ignition temperature of the mixture due to the high dilution levels. In other words, MILD combustion occurs when we do have a highly diluted, highly preheated reacting mixture [28, 29]. Such operating conditions result in a combustion process characterized by distributed reaction zone, uniformly distributed temperature combustor, no visible flame, reduced noise level, and highly reduced levels of soot, NO_x, and CO emissions [30, 31]. Such combustion conditions can be obtained through several techniques depending on the typologies of the processes and systems involved. For instance, premixed mixture can be ignited by hot exhaust gases through internal or external exhaust gas recirculation (EGR) [32, 33].

MILD combustion or flameless combustion occurs when a combustible mixture of air and fuel is diluted/mixed with flue gases injected into the combustion chamber away from the air stream to be mixed with the exhaust/flue gases [34, 35], as shown in Fig. 4.3. Meanwhile, air is injected at a high velocity into the combustion chamber directly to be mixed with the exhaust gases. Then ignition occurs, and flameless combustion is formed with no visible flame. In this kind of combustion, the flame front is formed though not being concentrated. The temperature gradient is much smoother in case of flameless combustion. This can be attributed to the gradual burning of fuel under exhaust gas recirculation. Flameless combustion is also known as flameless oxidation (FLOX), and there is a considerable work done under this name [36–40]. The results are promising for complete control of NO emissions from a variety of combustion sources. The adaptation of FLOX in

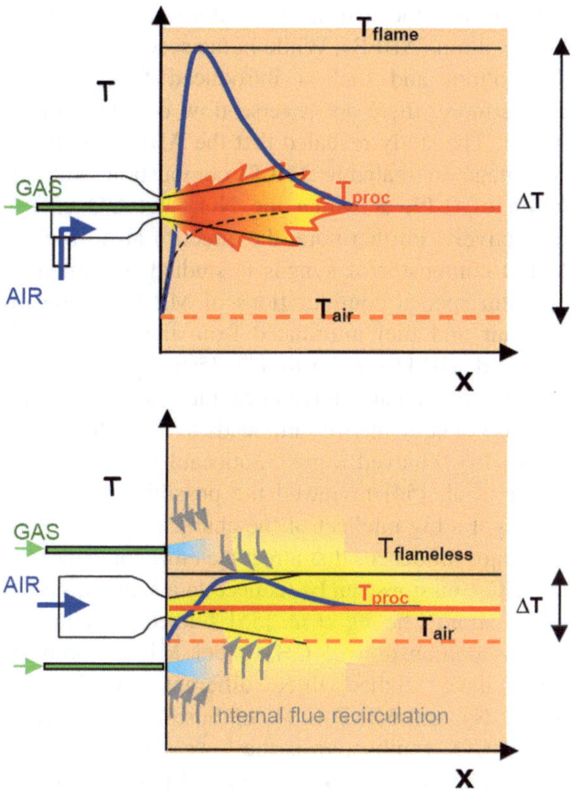

Fig. 4.3 Comparison between conventional flame with no air preheating (top) and flameless combustion with air preheating (bottom) [41]

industrial furnaces has proven reliability and acceptance [36]. There are many benefits of the flameless combustion as compared to conventional flame combustion. With flameless combustion, there is no flame front, no visible flame, neither noise nor roar are present, and CO and NO_x are emitted at low levels. There are alternative names that describe the same phenomena of MILD combustion like high-temperature air combustion (HTAC), dilute combustion, and fuel direct injection (FDI). The principle of such approaches is diluting the combustible mixture with the recirculated combustion products to slow down the reactions. The aim is to reduce the flame speed to reduce the NO_x formation. This can be achieved by reducing the reaction intensity and distributes the heat release over a large space.

MILD combustion has recently been under investigation on a jet-in-hot co-flow setup [42–44], laboratory-scale furnace [45–47], and semi-industrial furnaces [48, 49]. Numerically, eddy dissipation concept model is often applied for modeling MILD combustion in chambers and furnaces [50, 51]. A recent study by Huang et al. [52] has revealed that when air and fuel are introduced in the parallel direction, the exhaust emissions are reduced. They have introduced four different configurations for the combinations of air and fuel jets based on the effect of the flow field on MILD combustion. The first configuration occurs when both air and fuel are introduced from the bottom of the combustor and labeled as Air Bottom Fuel Bottom (ABFB). While in the second configuration, the air is introduced from the bottom and fuel is introduced from the top and is defined as ABFT. Furthermore, there are reverse flow configurations as well labeled as ATFT and ATFB. The study revealed that the ABFB configuration has recorded a significant advantage in realizing MILD combustion with operational lean at equivalence ration of 0.49, at which the NO_x and CO emissions recorded 4 and 39 ppm, respectively. Furthermore, the effect of air-to-fuel injection nozzle arrangement on MILD combustor of syngas is studied by Huang et al. [53]. They used a set of different special configurations of MILD combustion. The configuration that has both air and fuel introduced from the opposite side of the combustor exit has achieved MILD combustion at leaner condition than others. They reported that above the critical equivalence ratio at which mild combustion occurs, further increase in equivalence ratio leads to the reduction of maximum flame temperature. They also achieved a great/noticeable reduction in both NO_x and CO emissions. Khidr et al. [54] reviewed the possibilities toward lower gas turbine emissions. They got a big intellect of the characteristics of flameless combustion systems and the major influences that affect the emission concentrations. They reported that NO_x and CO emissions can be reduced to a single digit, lower than 10 ppm, using MILD combustion. Cheong et al. [55] studied the emissions of NO and CO from counterflow combustion of CH_4 under MILD combustion and oxy-fuel combustion. They have studied three different injection conditions including oxy-fuel, MILD-N_2, and MILD-CO_2 combustion. It was observed that the NO emission of MILD CO_2 combustion is the lowest for all cases as shown in Fig. 4.4. Ye et al. [56] studied experimentally the MILD combustion of pre-evaporated liquid fuels. They studied the influence of fuel type and equivalence ratio on the combustion stability and emissions. They reported, as per Fig. 4.5, reduction in NO_x emissions when the fuel is introduced with nitrogen instead of air.

MILD combustion systems have been successfully adapted in some industrial applications; however, broad application of such technology in the industry is resisted by the lack of fundamental insight into this combustion regime [34]. This

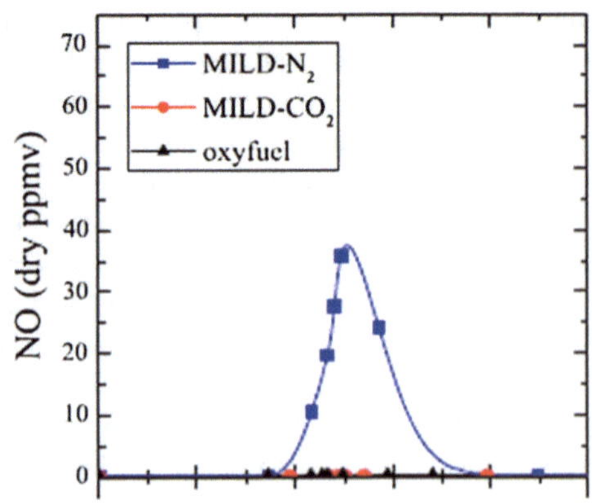

Fig. 4.4 Comparison of NO emissions in three cases: MILD-N_2, MILD-CO_2, and oxy-fuel combustion [55]

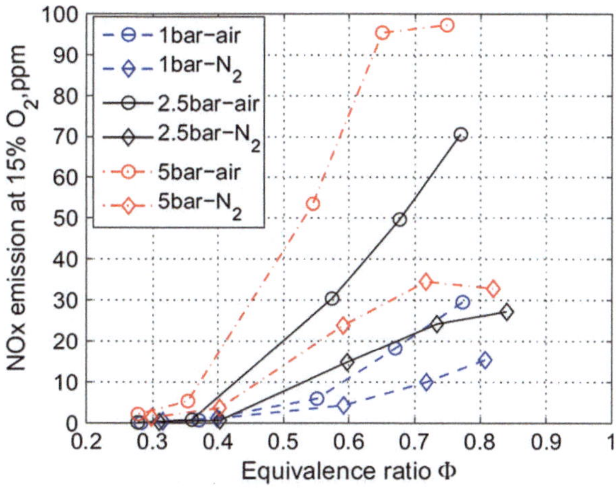

Fig. 4.5 NO_x emissions when fuel is introduced with N_2 as compared to air [56]

combustion regime requires perfect mixing between the incoming fresh gases and the recirculated hot exhaust gases [57–59]. Also, the severe inlet conditions in MILD combustion may result in the occurrence of dynamic phenomenologies [60, 61]. Such instabilities depend mainly on the working parameters including degree of dilution, residence time, and mixture composition, which need to be investigated in order to widen the operability limits for wider range of applications of MILD combustion.

4.3.3 Colorless Distributed Combustion (CDC)

The future power demand necessitates the development of fuel-flexible gas turbine combustors with ultra-low emissions. Gas turbine combustors based on distributed combustion technology showed promising performance enhancement in the stability, ultra-low CO and NO_x emissions, noise reduction, thermal field uniformity at higher combustion intensities, higher efficiency as well as control of combustion instabilities. Distributed combustion can be attained by swirling the flow to ensure required mixing of the fuel, injected air as well as the hot reactive gases inside the combustion zone before igniting the mixture. The distributed combustion flames have no visual signatures and hence are called colorless flames because of insignificant observable emissions in comparison with the other conventional flames. This review of colorless distributed combustion (CDC) will concentrate on high combustion intensities as applied to stationary gas turbines. The colorless distributed combustion reported in previous studies demonstrated substantial improvement in ultra-low CO and NO_x emissions, pattern factor, and low noise emission levels [62–64]. Also, CDC has been studied at high energy release intensities under several operational conditions [65], aiming at technology development for aviation applications [66]. Uniformity of mixture preparation inside the combustor to prevent the occurrence of hot spots and thin reaction zone in the flame

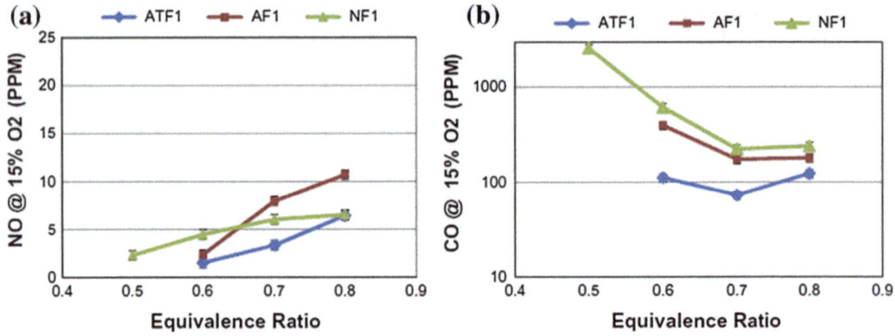

Fig. 4.6 NO and CO emissions for non-premixed combustion mode [65]

by controlling the mixing of fuel stream and hot reactive gases as well as the air stream has been a crucial requirement in distributed combustion. Fast mixing of fuel/oxidizer species and high recirculation of the reactive species lead to the auto-ignition of the mixture under distributed reaction condition. This characteristic prevents the formation of hot-spot regions as well as thin reaction zones in the flame, thereby mitigating the formation and emission of thermal NO_x from Zeldovich thermal mechanism [67, 68]. Prompt NO_x (Fenimore [69]) and fuel NO_x are the other routes of NO_x formation, apart from the thermal (Zeldovich [67, 68]) mechanism. Thermal NO_x can be alleviated via hot-spot zones elimination in the combustor arising from inadequate mixture preparation before ignition. The thermal NO_x constituted most of gas turbine combustors NO_x emissions.

The flame-holding devices or low speed required for flame stabilization in the conventional gas turbine combustors are not required in the case of the CDC. The fresh mixture is mixed with the hot reactive gases to raise the mixture temperature thereby causing spontaneous auto-ignition in the whole combustor as compared to conventional flames characterized with a thin flame front with high-temperature gradients. Also, ultra-low CO and NO_x emissions have been achieved using swirl combustion with tangential air entry for both non-premixed and premixed systems considering different combustor configurations, namely ATF1, AF1, and NF1 [65]. NO emissions lower than 2 ppm were achieved for a methane combustion at equivalence ration of 0.7 and energy release intensity of 36 MW/m^3-atm as shown in Fig. 4.6 [65].

4.3.4 Low-Swirl Injector (LSI) Combustion

To achieve low emission values, the operating conditions of the DLN combustors must be near the lean stability limits where instability, flame blowout, and noise can significantly affect the reliability and performance of the engines. There have been several efforts to lessen the effects of these potential problems based on passive controls (e.g., air and/or fuel staging) [70] and active controls (e.g., auto-feedback loop) [71, 72]. The costly cleanup of exhaust gases and catalytically supported combustion are among other alternative techniques that are also being employed. Certainly, these techniques will lead to intricate combustion devices necessitating firmly controlled sensors and actuators along with many supplementary components. For combustion systems operating on gaseous fuels apart from natural gas, there is further aggravation of combustion instabilities due to the air–fuel ratios as well as fuel content variabilities. This necessitates the optimization and/or reengineering of the injectors and combustors to accommodate the modifications in the combustion characteristics. A nascent premixed combustion technology for fuel-flexible gas turbine combustors based on unique low-swirl combustion phenomenon proffers a promising solution to fuel flexibility concerns. This technology was developed using a model research burner at the Lawrence Berkeley National Laboratory for the purpose of fundamental studies (15 kW); its operating principles have been studied and reported [73–75]. This combustion method has been patented, and it was based on the aerodynamics of propagating turbulent premixed flames. It is a robust, simple, and readily compatible technology to achieve strict emissions target without significant alteration of the current system configuration, or effect on costs, efficiency, and turndown. Low-swirl combustion technology has been applied to industrial process burners. Since late 2003, low-swirl burners (LSBs) having power range of 150 kW–7.5 MW with ultra-low emissions of 4–7 ppm CO and NO_x (both at 3% O_2 for furnaces) with turndown of 10:1 have been in existence [70]. The existing knowledge acquired from the laboratory studies which provided critical data for systems scaling and resolving issues related to systems integration has been the key to the commercialization pathways [76, 77]. Low-swirl combustors utilizing natural gas are also in the development stage. A prototype LSI for 5–7 MW engine rig is tested to be economically viable solution that permits DLN turbines to achieve emission targets below 5 ppm (at 15% O_2 for gas turbines) for both CO and NO_x [78, 79]; see Fig. 4.7. The low-swirl combustion concept is compatible for burning both hydrogen-enriched fuels and other hydrocarbon fuels. This can be achieved by adjusting the flow field to accommodate the changes in turbulent flame speeds of the premixed flames. This method entails the basic knowledge of the combustion characteristics for each fuel, e.g., flame temperature, flame speeds, as well as better understanding of how the flame is coupled with and responds to the turbulent flow fields. This method has been proved to be valid based on the developed industrial LSBs operating on flexible fuels. Prototypes have been tested with ethylene, propane, natural gas diluted with exhaust gas (up to 30%) [77], as well as large hydrogen-content refinery gases (up to 50% H_2) [80].

Premixed flames consume the reacting species in the form of self-sustaining reacting waves that propagate at flame speeds controlled by the composition of the mixture, the thermodynamic condition, as well as the turbulent intensities. While the diffusion flames from non-premixed mixture do not propagate (i.e., move through the reactants medium), combustion only occurs at the mixing zone formed between the oxidizer and the fuel streams. The low-swirl combustion approach utilizes the wavelike behavior of premixed turbulent flames and is only applicable

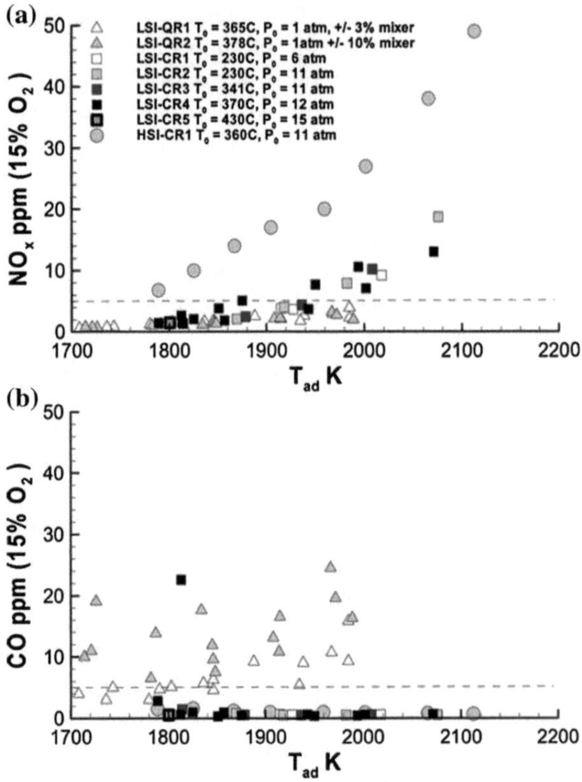

Fig. 4.7 NO_x and CO emissions for low-swirl mode (LSI) and high-swirl mode (HSI) at high-pressure conditions [79]

to premixed combustors. The idea is to consider a fast-moving turbulent premixed flame as a "standing wave" which remains still without any need to anchor the flame as shown in Fig. 4.8. This can be achieved by allowing the premixed flame to burn freely in divergent flow regions produced by a low swirling flow.

An important condition is to ensure low-swirl rates enough to constrain the vortex breakdown—a precursor to the formation of strong recirculation and flow reversal [82]. Under such a condition, the swirl motion induces mean radial aerodynamic strain by radially expanding the flow as it exits the combustor. The radial expansion of the flow causes the mean axial velocity to decrease linearly. This velocity "downramp" is the main characteristic that warrants a robust configuration for free propagation of premixed turbulent flames and stabilizes at a position where the flame speed becomes equal and opposite to the local flow speed. There will be no flame flashback since the flame propagation cannot be faster than the exit velocity. The flow divergence mitigates flame blow-off by providing a broader region for the natural settlement of the flame. More importantly, slight flow transients or mixture inhomogeneity can only cause shifting of the flame position thereby minimizing the tendencies of flameout. This becomes a self-adjusting technique for the flame stability with respect to transients as well as some sudden changes.

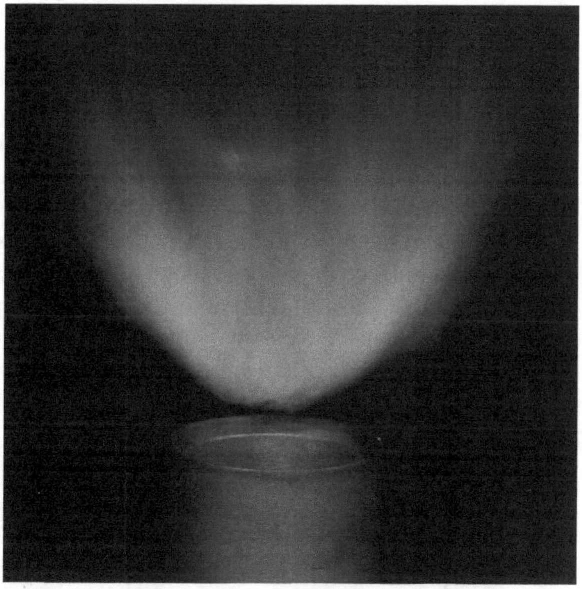

Fig. 4.8 Flame generated by a low-swirl burner [81]

4.4 Burner Design

4.4.1 Swirl-Stabilized Burners

4.4.1.1 Turbulent Reacting Flow

Turbulent premixed combustion is the operation mode of several engineering equipment, e.g., stationary gas turbines and spark ignition engines [83–85]. To make the modern power generation and propulsion devices more compact, mixing is strongly enhanced by operating these devices under highly turbulent conditions. The turbulent flow conditions arise from both high-speed turbulent inlet flow (free stream turbulence) and the turbulence generated by the flame stabilization scheme in a combustor [86]. In this equipment, the combustion processes are associated with large values of turbulence intensities, i.e., ratio of the root-mean-square (RMS) to the mean of the reactants velocity being close to 50% [87]. Several experimental setups associated with Bunsen-type, bluff body-stabilized, opposed jets, and swirl-stabilized flames have been developed in the past to investigate the characteristics of turbulent premixed flames at relatively large values of turbulence intensities [88, 89].

To investigate the effect of turbulence on premixed flames, experimental studies have been performed on different flame geometries which are classified in the "Envelope" category (Bunsen-type flames), "Oblique" category (V-shaped flames), "Unattached" category (low-swirl or counterflow flames), and propagating flame kernels [89]. Effects of turbulence in modifying the flame front structure have been demonstrated and summarized [70, 90]. The canonical flame configuration is a bluff body-stabilized inverted conical flame anchored at its apex on a disk-shaped bluff body. Several studies have focused on the characteristics of planar V-shaped flames subject to various levels of free stream turbulence [91–96]. The experimental results by [91, 92] showed that for low turbulence intensity (4–6%), the flame front is weakly wrinkled and can be somewhat represented by two straight flame fronts. For moderate turbulence intensity (about 10%), the flame front topology has been shown to depend on the distance from the flame holder. Flame segments near the bluff body featured weak wrinkles, like the low turbulence intensity conditions, while the flame segments in the downstream region showed stronger corrugations. For higher turbulence intensities (about 17%), strong levels of wrinkling along with the formation of cusps, mushroom-shaped structures, localized quenching, and pockets of reactants have been reported using laser tomography technique by Kheirkhah and Gülder [92]; see Fig. 4.9. The flame brush thickness is an important

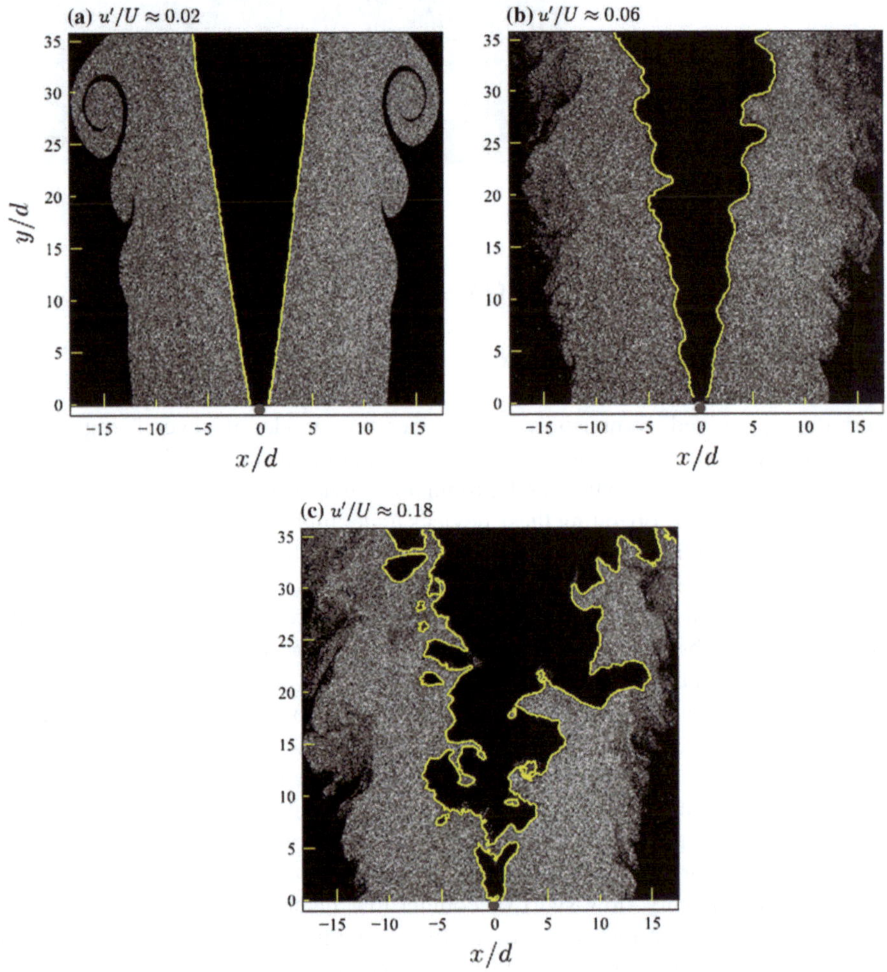

Fig. 4.9 Images of the flame front topology with different turbulence intensities [92]

parameter which represents the approximate distance over which a flame front exists. Measurements by Kheirkhah and Gülder [91, 92] and Namazian et al. [93] have shown that the brush thickness is strongly dependent on the vertical distance away from the flame holder, mean and root-mean-square of the streamwise velocity at the burner exit, integral length scale of the flow field, and the equivalence ratio.

Most of the measurements done for V-shaped flames in the past have employed laser tomography techniques [91, 92, 94, 95]. This technique uses liquid oil droplets which evaporate near the flame front, causing notable change in light intensities revealing a tentative flame front boundary. However, detailed measurements of the reaction zone in turbulent flames are necessary for the validation of combustion models. A useful diagnostic technique to achieve this objective is the simultaneous planner laser-induced fluorescence (PLIF) imaging of OH and CH_2O which was first demonstrated in laminar flames by Paul and Najm [97]. Application of this diagnostic technique to turbulent flames was demonstrated by Böckle et al. [98]. In their work, simultaneous PLIF imaging of OH and CH_2O technique was used to obtain the reaction zones marked by the overlapping regions of PLIF of OH and CH_2O profiles and was validated with simultaneous Rayleigh scattering temperature measurements in a turbulent Bunsen-type flame. Application of this technique to study flame structures near blow-off events for bluff body-stabilized flames has been demonstrated by Kariuki et al. [99, 100]. The advantage of this technique to study highly turbulent premixed flames has been demonstrated in the recent works of Skiba et al. [101] and Zhou et al. [102, 103] for piloted jet flame configuration. Broadening of the preheat region represented by CH_2O PLIF signal along with the observation of shredded flamelets has been reported.

4.4.1.2 Swirling Flow

Swirl stabilization is one of the most effective techniques for flame stabilization especially for gas turbine combustion applications. This technique can be applied for both non-premixed (diffusion) and premixed flames effectively. Swirling the flow results in the creation of outer and/or inner recirculation zones. The creation of the recirculation zones enables low velocity regions, and this helps to stabilize the combustion process. The hot products interact with the incoming reactants within the recirculation zones to form a more reactive mixture. However, the strength and the number of the recirculation zones depend on the design of the fuel nozzle, fuel type, oxidizer composition, and how the oxidizer is introduced to the combustor. This should directly affect the stability and emission characteristics of different flames.

Recirculation zones are essential within the combustor for flame stabilization, improved fuel–air mixing, and better performance of the combustion system [104–108]. Swirling flow is a traditional technique that is adapted to generate intense flow recirculation within the combustor to promote flame stability over wide ranges of operating conditions. Free-standing inner recirculation zone with proven aerodynamic flame holding characteristics can be regulated to create a free standing central recirculation zone with proven aerodynamic flame anchoring characteristics.

Wide ranges of flame configurations have been investigated experimentally [106, 107, 109] and numerically [110–112] to configure the nature of turbulent reacting flows and flame stabilization and emission reduction techniques. One of the most promising new combustion technologies for emission control is the fully/partial lean premixed combustion technique [113, 114]. This technique has been adapted by gas turbine and internal combustion engine manufacturers for control of NO_x and soot emissions [109, 110]. However, the characteristics of lean premixed combustion need to be explored more, in terms of distribution and fluctuation of local equivalence ratio, flame stabilization technique (swirl, back step, or bluff body), and turbulence–chemistry interaction, in order to get full advantages of this technique [105, 109, 110, 114].

Flame stabilization mechanism and geometry of the combustor are the most common parameters affecting gas turbine operation. The design of the gas turbine combustor determines the strength of the created recirculation zones which directly affect flame stability [115]. The effects of hydrogen enrichment of propane on the structure and size of the recirculation zones within a step combustor under premixed combustion conditions were investigated by Hong et al. [116]. The results showed that at lower operating equivalence ratio, two recirculation zones, primary and secondary zones, were created. The size of the secondary zone was reduced while increasing the equivalence ratio until it disappears at a certain equivalence ratio. Hydrogen enrichment resulted in improved combustion temperature and movement of the flame core toward the combustor step. Flame stability over a bluff body was studied by Li and Gutmark [117] with and without the use of the center body. The results showed reduced levels of oscillations and improved stability characteristics of the flame when the center body is recessed. Flame–vortex interaction was studied by Altay et al. [118] in a step combustor under premixed combustion conditions considering different operating conditions and fuel composition. The results showed that stable long flames are obtained at lower equivalence ratios, while unstable flames are obtained at higher equivalence ratios. Combustion instabilities were investigated by Speth and Ghoniem [119] in a swirl-stabilized gas turbine combustor burning syngas under premixed combustion conditions. The results proved the significant effects of combustor geometry and fuel composition on flame stabilization behavior [120, 121]. The effect of fuel composition on laminar flame speed is presented in Fig. 4.10 with higher flame speed at lower CO concentrations [121].

Recently, a new flame stabilization technique called trapped vortex combustor has been introduced for better mixing between air, fuel, and hot recycled exhaust stream [122]. Currently, this technique is under the development to be adapted in conventional and novel combustion systems [123]. In this technique, fuel and air are injected into a cavity and a stable vortex structure is created to stabilize the generated flame within the cavity. Combustion may continue downstream with additional fuel injection in a second cavity in axisymmetric combustion systems [122] or in the main combustion chamber in annular cavity combustion systems [123]. These geometries adapt well with diffusion and premixed flames and result in more stable flames.

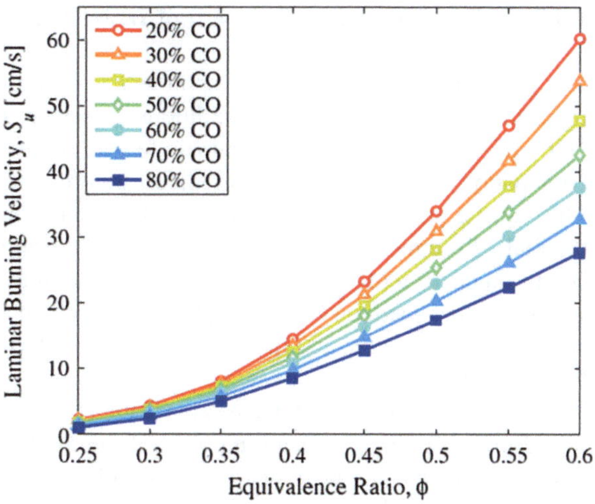

Fig. 4.10 Laminar burning velocities of syngas flames with respect to the equivalence ratio and fuel composition [121]

As a next generation, reduced emissions low fuel consumption combustors are expected to adapt multiple reaction zone technique under partially premixed operating conditions [124–126]. There are three basic means for the reduction of gas turbine emissions including injection of steam as diluent into the flame zone inside conventional combustion systems adapting diffusion flames, post-treatment of exhaust stream for catalytic removal of CO and NO_x emissions, and modification of the combustion system to adapt lean premixed combustion technique to limit the creation of combustion pollutants within the combustor. The use of steam injection to control emissions of NO_x requires frequent inspections and results in reduced hardware life of the combustor. However, steam injection technique is still applied, and the dry combustion technique became the most preferred combustion technique for the control of NO_x emissions. DLN (dry low NO_x) was the first acronym to be used; however, the DLE (dry low emission) acronym became more common with the requirement to reduce NO_x emissions without increasing both CO and unburned hydrocarbons.

4.4.2 DLN/DLE Burners

It is well known that the conventional diffusion flame has a greater stability while operating within a range of equivalence ratios. Meanwhile, NO_x emission concentration is much higher as compared to premix DLN combustion [127]. Basically, in DLN systems, stable combustion needs precise fuel and air flow control in the combustion chamber at all operating conditions [128]. Factors such as fuel flexibility, higher and lower calorific values (HHV and LHV), ambient temperature, and operating conditions can disturb the combustion stability. Furthermore, in DLN systems, the interaction of turbulent chemistry and flow is substantial [129].

4.4.2.1 Construction and Design of DLN/DLE Burners

The principle beyond the DLE approach is to burn a substantial portion of the injected fuel at lean premixed conditions to prevent high concentrations of thermal NO_x emissions. Furthermore, the main feature of this system is the premixing of the combustible mixture (fuel plus oxidizer) before entering the combustion chamber. This can be attributed to decrease the flame temperature and consequently reduce thermal NO_x emission. This condition results in bringing the full-capacity operating condition down with a much lower flame temperature and near to the lean operating limits as presented in Fig. 4.11 [129]. Under such condition, it is very hard to control CO emissions, and the engine offloads will increase the probability of flame blowout and engine shutdown.

The design of DLE systems has more than two injection points of fuel. The main injection fuel point is consuming 97% of the total injected fuel flow rate, and it is injected with the air flow downstream and before swirler at the entrance to the premixing pipe. The pilot flame is created by the pilot fuel which is approximately 3%, and it is injected into the combustion chamber without premixing. Figure 4.12 compares the DLE combustor with the conventional one. To stabilize the flame for both cases, a swirler is mounted in DLE to create the required recirculation zones. The size of DLE fuel injector is bigger due to the premixing of fuel and air before entering the combustion chamber.

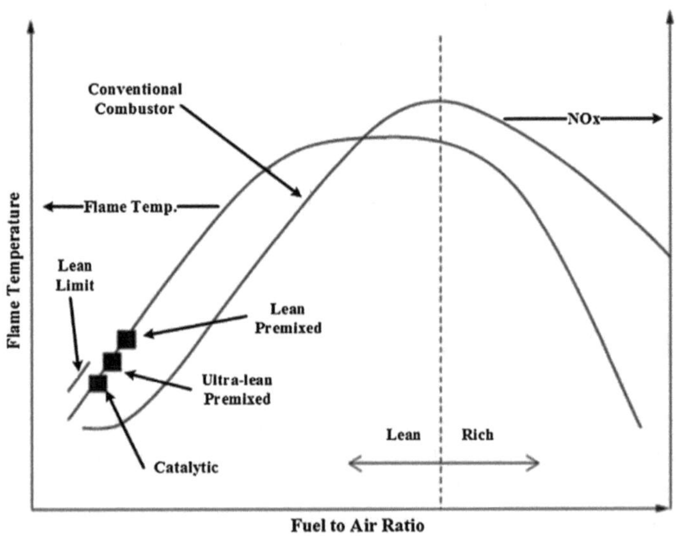

Fig. 4.11 Distributions of flame temperature and NO_x emissions for change of fuel–air ratio from lean to rich conditions [129]

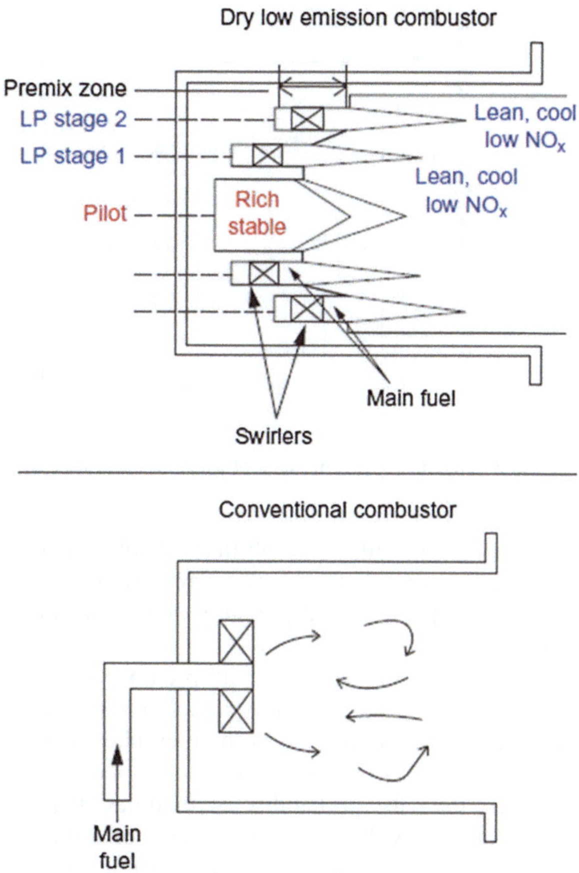

Fig. 4.12 Schematic diagram showing the main differences between DLE and conventional combustors [129]

4.4.2.2 Emissions and Combustion Characteristics of DLN/DLE Burners

The formation of NO_x emissions and the operability window are presented in Fig. 4.13 for a premixed DLN burner. The flame temperature is brought down near to the lean premixed limit than in diffusion or conventional combustors; some precaution should be taken immediately when the operating load is decreased to avoid flame extinction [130]. In case of no action was taken quickly, blow-off occurs because of decreasing the mixture strength; it becomes too lean to hold a flame. A little proportion of the fuel is meant to be burned in rich condition to provide a pilot zone, while the remaining is burned in lean condition.

Often, gas turbines experience operating issues with the DLN/DLE systems; the common issues are combustion instability, flashback, and auto-ignition. These problems may lead to power losses; because if any failure is sensed by the control

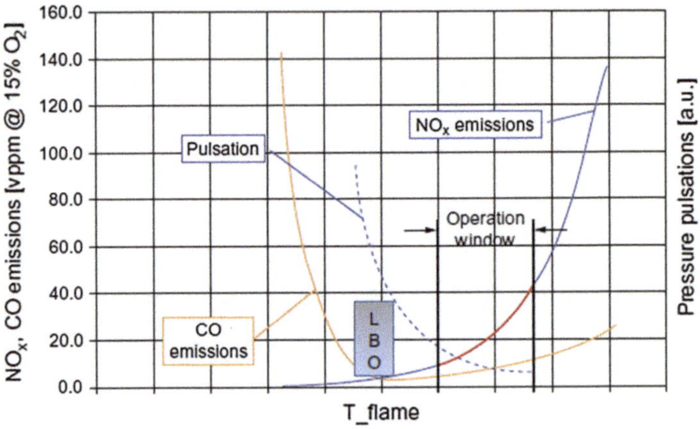

Fig. 4.13 NO_x and CO emissions of a lean premix burner [129]

unit, the engine will switch off immediately. Simply, the definition of auto-ignition is the self-ignition of fuel/oxidizer mixture. For a fuel/oxidizer mixture, at a certain elevated temperature and pressure, there is a finite time before self-ignition takes place.

DLN/DLE systems have multi-premix modules at the tip of the combustor to premix fuel/air mixture uniformly. For the sake of avoiding auto-ignition phenomena, the residence time of the fuel tube should be lower than the auto-ignition delay time [131]. If the auto-ignition phenomenon exists in the multi-premix modules, then the probability of modules damage will increase. Consequently, these damages will need either repair or replacement of damaged parts. Some technicians are exposed to frequent engine switch-off due to the auto-ignition issues. The engine suppliers to handle such issue have not been well covered. Fuel auto-ignition delay times exist; however, the literature reviews revealed that there is fuel variability. There are several reasons for auto-ignition, and they could be summarized as: (1) long auto-ignition delay time, (2) fuel flexibility, and (3) fuel residence time miscalculated.

Flashback occurs in the premix ducts when the local flame speed is faster than the velocity of the incoming combustible mixture [132]. Flashback usually occurs during unexpected or sudden transient conditions, for example, when the compressor surge occurs. As a result of compressor surge, a sudden change of air velocity consequently leads to flashback. Unfortunately, when the flame reaches the exit of the premixing pipe, the pressure drop will cause a reduction in the mixture velocity and leads also to flashback. Some cooling techniques can provide some protection during the flashback. The combustor should be coupled with fuel control valves of a fast-acting type to minimize the effect of a flashback.

As a matter of fact, DLN burners can achieve very low NO_x and CO emissions while adapting stable flame over wide ranges of operating conditions [133]. For instance, the DLN-2.6 gas turbine combustor has achieved emission rates of lower than 9 ppm NO_x (in reference to 15% O_2) for the GE MS7001FA industrial gas turbine using natural gas as a fuel [134]. The emissions can even be reduced more when coupling DLN technique with exhaust gas recirculation. General Electric, through its global research center, performed experimental tests to evaluate the performance of a DLN F-class gas turbine adapting exhaust gas recirculation for post-combustion capture of CO_2 [135]. They reported that combining DLN burners with exhaust gas recirculation system can further reduce NO_x emissions. Adaptation of DLN burners to handle hydrogen, from renewable energy sources, represents a future alternative technology for low emission power generation in gas turbines. Ayed et al. [136] studied the increased energy density for the DLN micromix hydrogen combustion principle. The results showed potential for energy density increase and reduction of combustor complexity and production cost. A program called "High Hydrogen Turbine Program" was initiated between GE and the US Department of Energy. The progress of this program is presented in the study by Lacy et al. [137] along with the advancements toward the main goal of low NO_x emissions while capturing CO_2.

4.4.3 Catalytic Combustion

Catalytic combustion is a combustion process of a mixture consisting of fuel and oxygen reacted on the surface of a catalyst, resulting in oxidation of the fuel. This combustion process occurs with no flame and at very low temperatures than those associated with conventional/diffusion combustion. Further reducing the operating temperature of the combustion chamber, catalytic combustion also produces low emissions of NO_x as compared to conventional combustion. This combustion technique has the potential to decrease the NO_x emissions to minimum values, approximately a single-digit number. Furthermore, this technique will achieve an ultra-low CO emissions and unburned hydrocarbons (UHCs). Nevertheless, such technique reduces the probability of blow-off and combustion instabilities. Furthermore, there is no need for expensive cleaning systems and it has minimum losses of efficiency as compared to conventional gas turbines [138–141].

Recently, catalytic combustion is commonly used to eliminate NO_x and CO pollutants from exhaust gases and it has an enormous potential in power generation applications. The principle beyond catalytic combustion is that the fuel will react on the surface of catalyst. This catalyst can hold/stabilize the combustion of lean combustible mixtures with very low adiabatic flame temperature <1500 °C. Hence, the combustion chamber temperature will be brought down below 1500 °C and NO_x emissions will be reduced [142]. Studies revealed that reduction in NO_x emissions in catalytic combustors is really promising as compared to low combustion temperature. Apparently, the reaction on the surface of catalytic produces zero NO_x.

Full-load testing is performed in catalytic combustor developed by General Electric (GE) for the gas turbine type MS9001E. The MS9001E combustor operates at full capacity, and the exhaust temperature was about 1190 °C. The main

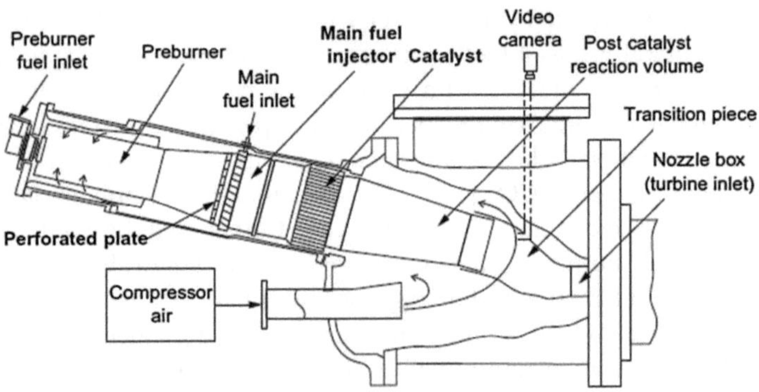

Fig. 4.14 Schematic diagram showing the main components of the catalytic burner [129]

components of the GE catalytic burner are shown in Fig. 4.14. There are three major components in the GE catalytic combustion system: the pre-burner, fuel injector, and the catalytic reactor. The catalytic combustor has exciting potential in the application of gas turbines in new combined-cycle power plants, as the NO_x emissions will be reduced below 2 ppm.

4.4.4 Perforated Plate Burners

Flame stabilization has been the most central criteria in the design of gas combustors. Flashback, quenching, and blow-off are among the stabilization problems associated with premixed flames. The stability phenomenon for premixed flames depends on physical processes, such as reactants mixing with unburnt gases, combustor wall cooling, flame curvature and stretching. Several researchers have made enormous contributions by studying the dynamic and steady-state characteristics of premixed flames with focus in the flame–wall interactions, flame acoustics, stability of steady flames, liftoff, blow-off, flashback, etc. [143–147]. Perforated plate-stabilized burners adapting premixed kind of flames are used extensively in small household and large industrial burners. For perforated plate/matrix holes as well as micromixer burners, the flame instability can be attributed to the effects of hole/jet diameter and spacing, heat transfer to the face plates. Kedia and Ghoniem [147] investigate the stability as well as mechanisms leading to the blow-off of a lean premixed methane–air flames anchored on a heat-conducting perforated plate combustor using 2D numerical modeling. They concluded that the blow-off criterion was attributed to the combined effect of flame stretching and heat transfer to the burner plate. The thermo-acoustic instability analyses conducted by Timmermans et al. [148] using hot-wire anemometry and particle image velocimetry near the wake of a perforated plate revealed low-frequency instabilities in laminar to turbulent transition regions of the flow. Rodrigues and Fernandes [149] examined the stability of propane and methane flames in a matrix perforated plate burner under the effects of holes spacing and diameters, as well as the number of holes on the plate with respect to the velocity gradient (U/D). It was concluded that the flame stability was neither dependent on the number of holes nor on the holes diameters but depends on the spacing between the holes—the smaller the hole spacing, the more stable the flame. Similar conclusions were made by Veetil et al. [150] when they conducted 3D numerical simulations to study the steady-state characteristics of laminar premixed syngas–air flames using perforated plate combustor. They found higher flame liftoff with strong recirculation at wider adjacent hole spacing. Jithin et al. [151] also found the flame to stabilize further downstream of the perforated plate as the spacing between the adjacent holes increases during laminar premixed propane–air combustion.

The combustion efficiency and the mechanism of heat transfer in a perforated plate burner determine the overall thermal performance of the burner. Hindasageri et al. [152] conducted experimental and numerical investigation on premixed methane–air flames heat flux distribution from a multi-hole burner at low Reynolds numbers using different geometric holes arrangements. They reported that even though the distribution of the heat flux was less influenced by the holes arrangement, the larger spacing between adjacent holes resulted in more heat flux distribution at a constant mass flow rate. This signifies more effective heat transfer. However, the lean blow-off stability criteria limit the wideness of the hole pitches.

The effect of the spacing between burner and plate on thermal characteristics of staggered and inline configurations of the perforated plate burner has also been recently studied by Kuntikana and Prabhu [153]. They observed improvement in the thermal efficiency of the burner with more even heat flux distribution when the Reynolds number of the mixture increased, and the maximum heat transfer was attained at stoichiometric mixture due to the highest flame temperatures. The increase in mixture equivalence ratio to the fuel-rich conditions instigated drop in the thermal efficiency, but the distribution of the heat flux was nearly unaffected. The effect of ambient air entrainment at increased plate burner spacing decreases the

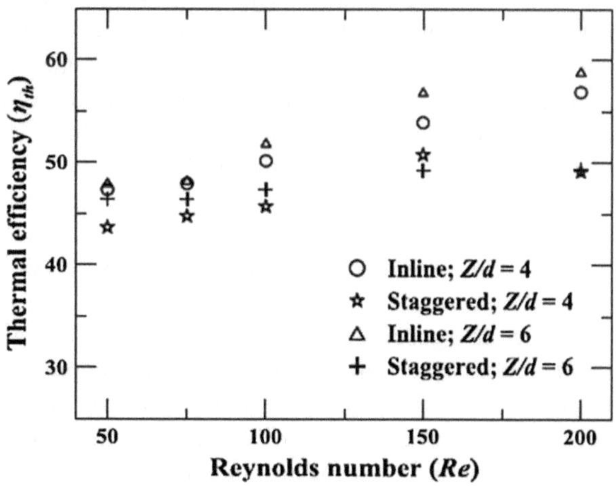

Fig. 4.15 Thermal efficiency of multi-port burners with respect to Reynolds number at equivalence ratio of unity [153]

thermal efficiency of the burner [153], as shown in Fig. 4.15. Lee and Hwang [154] experimentally examined the effect of flow distribution on flame stability and the emission characteristics on a cylindrical porous baffle plate burner. They reported the increase in thermal efficiency and temperature by increasing the size of the main hole and decrease in flame blowout by introducing the flame holder. Application of distribution mesh burners and the sizing of the main and retention holes diameters significantly affect the thermal performance of multi-holes premixed burners. Moghaddam et al. [155] also reported improvement in thermal efficiency due to higher temperature as the result of increase in main holes diameter, while increase in thermal efficiency due to the increase in the diameters of the retention holes was insignificant. The size of the retention holes should be considered to decrease the total flow velocity and bring the flame closer to the burner surface, although a diameter change did not considerably improve temperature and thermal efficiency.

4.4.5 *Environmental EV/SEV/AEV Burners*

Different techniques for reducing NO_x emissions have been tried since the 1970s up to the present date starting with the wet low NO_x technique [156]. The injection of water or steam into the combustion zone was the traditional practice to control the emissions of NO_x. As an advantage, the power output of the gas turbine is improved with the increase of the amount of injected water. Unfortunately, water or steam with the required demineralized properties is not available for many applications, which exerted more pressure on gas turbine manufacturers to develop dry techniques to control NO_x emissions. Due to the dependence of NO formation on the combustion temperature, NO_x emissions can be reduced if the mixing process is separated from the combustion process, and the combustion process is performed under very lean conditions, i.e., lean premixed combustion. The lean premix combustion technique has been proven as an effective technique for controlling NO_x emission significantly. Though being in use, the lean premixed combustion system is complex and prone to deterioration. The first generation of premixed combustion burners had two basic disadvantages. The first drawback is that the system does not have an effective method for controlling auto-ignitions. The second drawback is the use of elongated premix inlet tube for air settlement and generation of homogeneous mixture of air and fuel.

4.4.5.1 Concept of Operation of EV/SEV/AEV Burners

The second generation of premix burners was the environmental vortex (EV) burner, which left the basic idea of many present designs premix and diffusion burners. The concept of operation of the EV burner is derived from the basic concepts of vortical flows [157]. The burner uses the vortex breakdown of an intense swirling flow to control the inner recirculation zone to obtain a stable flame

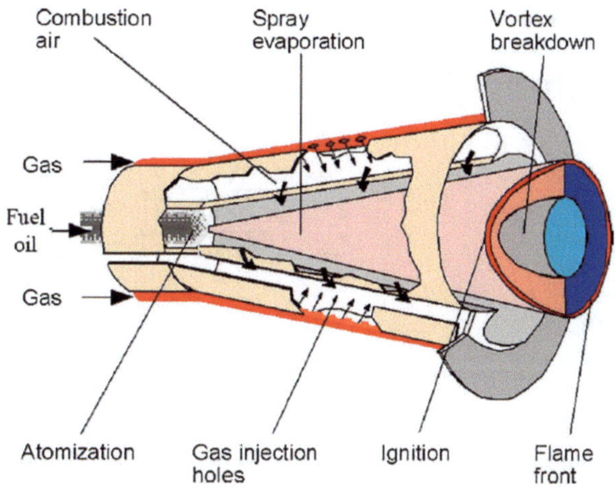

Fig. 4.16 A schematic diagram showing an EV burner [158]

without using swirler or center bodies as shown in Fig. 4.16. The main features of the flow inside the EV burner are: (1) The flow is accelerated before the vortex breakdown to prevent flame flashback, (2) a strong inner recirculation zone is created after vortex breakdown to stabilize the generated flame, and (3) the generated flame in the free flow is anchored very well due to the aerodynamic fixation of the vortex breakdown.

As a novel technique to improve turbine efficiency while operating under low inlet temperature, the sequential environmental vortex (SEV) burner has been developed. The SEV combustion system is based on the use of the EV burner in a first annular combustion chamber followed by the use of SEV burner in a second combustor, as shown in Fig. 4.17. Combining the concepts of DLN and SEV burners in one unit results in high power density gas turbine [159]. The compressor at pressure ratio greater than 30:1 delivers twice the pressure ratio of a traditional-compressor. About 50% of the total fuel to the gas turbine is burned in the first

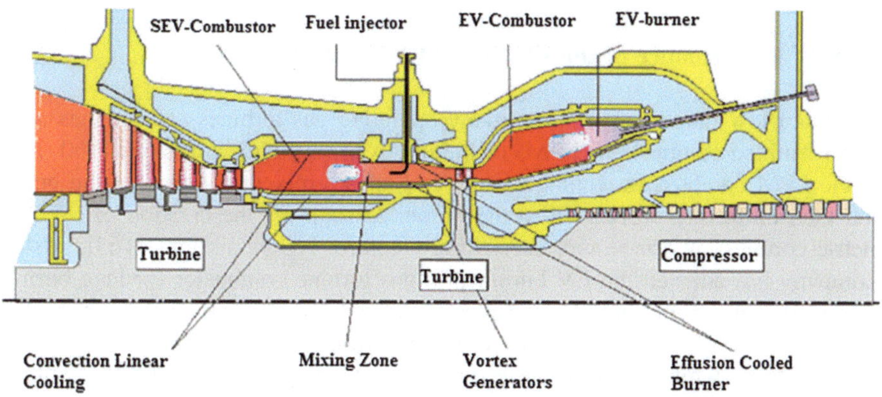

Fig. 4.17 A schematic diagram showing the components of the SEV burner [159]

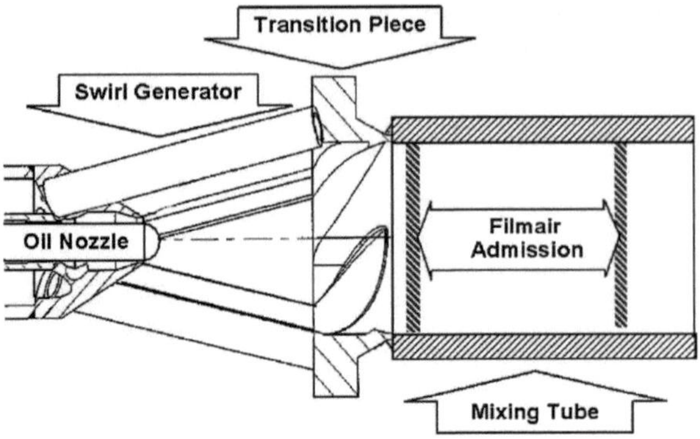

Fig. 4.18 A schematic diagram showing the main features of the AEV burner [156]

combustion stage, EV burner, to preheat the compressed air. Then, the combustion products are expanded in a high-pressure turbine resulting in pressure drop by a factor of 2.0. Additional cooling air is used, and the remaining fuel is burned, in the second combustion stage. The combustion gases are heated to the limit temperature at gas turbine inlet to be expanded in a low-pressure turbine of four stages.

The Advanced Environmental Vortex (AEV) burner was developed for the use in combustion systems burning liquid fuels. The burner utilizes four inlet slots, instead of two, to ensure strong inlet radial flow and protect the swirler body against liquid fuel impingement. The use of mixing tube enables longer fuel evaporation time and better gas phase mixing as shown in Fig. 4.18. In the conical swirl generator, a jet flow with different small body vortex cores is generated. At the swirler exit, the centerline axial velocity is of two times the value of the bulk velocity. This high-velocity jet flow represents a barrier against flame flashback as this velocity is maintained through the entire length of the mixing tube. The use of additional air (in the order of 10% of the total air) enhances the axial flow velocity even close to the walls of the mixing tube to suppress flashback along the mixing tube [156].

4.4.5.2 Performance of the EV/SEV/AEV Burners

To obtain low flame temperature, many operating gas turbines and engines adapt lean premixed combustion (LPM) technique to control NO_x emissions [160–162]. In the LPM combustion technique, fuel and air are mixed upstream of the burner and then uniformly introduced to the burner to suppress the creation of stoichiometric combustion zones and, accordingly, control NO_x emissions [163]. Alstom Company has adapted the EV burner in a gas turbine combustor for lean burn of premixed fuel–air mixture [164]. Several studies have been conducted on the laboratory scale on EV burners trying to understand the physics behind the improved flame stability while adapting the EV burner technology. Cho et al. [165] studied numerically the NO_x reduction mechanism in an EV burner by evaluating the unmixedness parameter with the mixing characteristics and NO_x emissions.

The results showed a strong correlation between the emission of NO_x and the unmixedness parameter, which proves that NO_x emissions can be reduced through enhanced mixing of air and fuel. Biagioli and Guthe [166] performed large eddy simulations (LES) of the ALSTOM EV double-cone burner to study the effects of pressure and unmixedness of air and fuel on the emissions of NO_x. They reported, based on the results, a negative NO_x pressure exponent at fully premixed operation across a rather wide range of operating equivalence ratio, but a positive exponent at unmixedness levels of industrial burners. Biagioli [167] studied experimentally and numerically the stabilization mechanism of premixed flames in the ALSTOM EV double-cone burner. The results showed that the flame propagates with negative speed in the axial direction indicating that the flame is completely stabilized in the inner recirculation zone through the symmetry axis of the burner. Biagioli et al. [168] measured experimentally the combustion dynamics associated with premixed combustion in a prototype version of ALSTOM EV burner. They linked, based on the measurements, between the unsteadiness of heat release and the generated combustion dynamics. This was confirmed through the observed positive link between flame displacement tendency and the recorded pressure pulsations.

4.4.6 Micromixer Burners

4.4.6.1 Micromixer Design

A micromixer, as the name implies, is based on the principle of mixing fuel and oxidizer on a miniature scale. Multiple straight tubes of millimeter-scale diameter are arranged in a parallel array between front and back faceplates, which forms a plenum surrounding the tubes between the faceplates. The incoming oxidizer flow is divided among those tubes, and the fuel is injected from the surrounding plenum through numerous sub-millimeter-scale side holes in the tube walls. Fuel is, thus, injected in a jet-in-cross-flow fashion into the oxidizer stream. The axial location of fuel holes is chosen to create a premixing region inside each tube, which guarantees a fully developed, fully premixed jet exiting the tube. To minimize the premixing length, two or more fuel holes are evenly spaced around the perimeter of each tube at the same axial location. Flow conditioners are utilized within the fuel plenum to ensure uniform distribution of fuel among all injection holes. From a combustion perspective, the faceplate geometry (jet spacing and diameter and number of jets) is tailored to optimize emissions and control combustor operability. The pressure drop of oxidizer flow across micromixer and the operational equivalence ratio are both chosen carefully to prevent flame flashback.

One of the key aspects behind the success of micromixer technology is its innate flexibility to accommodate staging, scalability, and fuel dilution and flexibility. York et al. [169] performed a full-can durability test of micromixer nozzles at the combustor inlet and firing conditions of General Electric's F-class gas turbine conditions at full engine load with air as oxidizer and different fuel blends of hydrogen, natural gas, and nitrogen. During 100 h of accumulated firing, NO_x emissions were recorded in single-digit ppm using a blend of hydrogen and nitrogen as fuel with more than 90% hydrogen by volume. Funke et al. [170] conducted an experimental/numerical study of the impact of momentum flux ratio on flame anchoring and NO_x emissions in micromixer nozzles using hydrogen as fuel and air as oxidizer. They showed that the anchoring mechanism of the flamelets has a strong impact on NO_x formation. Dodo et al. [171] tested micromixer air combustion of IGCC-syngas fuel simulants containing hydrogen, methane, and nitrogen with a hydrogen content of 40–65%. Three fuel blends were considered with 0, 30, and 50% carbon capture rates. Stable combustion was observed for all blends with single-digit ppm NO_x.

4.4.6.2 Lean Direct Injection (LDI) Hydrogen Micromix Combustion

The use of hydrogen as the primary fuel for power production in gas turbines is gaining increasing attention, because hydrogen combustion does not generate CO, CO_2, or unburned hydrocarbons [172]. Hydrogen fuel can be created by a variety of renewable energies and then stored and sent to a gas turbine for continuous or peak load power production. Certain chemical processes also produce H2-rich fuels, such as coke-oven gas, that can be burned in gas turbines for power generation. Similarly, H_2-rich syngas can be created by reforming hydrocarbon fuels or through the gasification of coal or solid wastes. Coal-derived syngas consists primarily of H_2 and CO with smaller amounts of N_2, CO_2, and hydrocarbons. Syngas fuel has been successfully used for several decades in gas turbines of integrated gasification combined-cycle (IGCC) plants. The strict environmental legislations to govern greenhouse gas emissions in recent years have created significant interest in using IGCCs for pre-combustion capture of CO_2. If the water-gas shift reaction is used to convert CO in syngas to CO_2, the latter can be captured resulting in a very H_2-rich syngas fuel. This high hydrogen content makes syngas more reactive than natural gas, so typical syngas combustors implement diffusion-type burners to eliminate the risk of flashback. If air is used for combustion, nitrogen oxide (NO_x) emissions become the only potential concern because of the high-temperature stoichiometric zones associating diffusion flames. NO_x abatement is commonly done using large amounts of diluents, such as N_2 or steam.

Flashback, and consequently hardware damage, is generally considered as the most formidable challenge with hydrogen combustion due to extreme flame speeds. Several technologies for safe low NO_x hydrogen combustion have been studied. Swirl-based dry low NO_x (DLN) premixers have been modified from natural gas to high H_2 combustion [173]. The major drawback is that unsteady and complex secondary flows, which may benefit flame stability in natural gas combustors, can create multiple paths for flashback with syngas operation [174]. Axial injection of high H_2 fuel from moveable lances downstream of the swirl vanes was thus examined to reduce the tendency for flashback [175] during atmospheric testing.

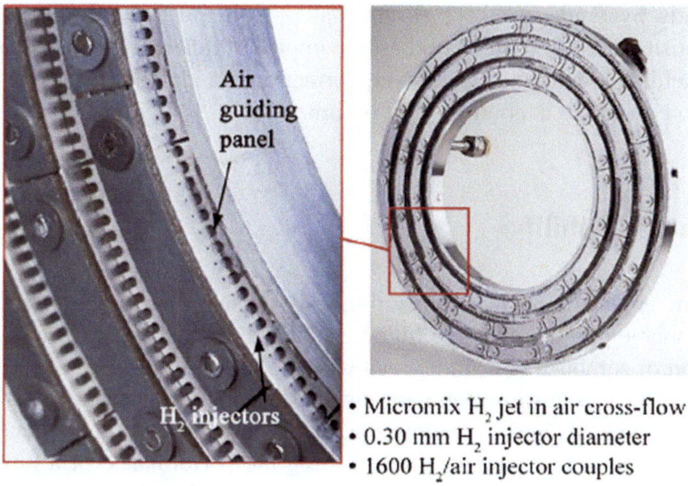

Fig. 4.19 A schematic diagram showing the main components of a micromixer combustion burner [181, 182]

Lean direct injection (LDI) is another approach to eliminate the risk of flashback in hydrogen combustion [176, 177]. LDI designs are based on injecting fuel from numerous (hundreds or thousands) locations into excess air at the combustion zone to achieve rapid mixing. Several studies developed and tested small-scale distributed premixer arrays, referred to as micromixers, to safely achieve low emissions with hydrogen fuel. These micromixers mimic the LDI approach but include a small premixing zone and may approach the operability characteristics of LDI if designed carefully. The studied configurations include jet-in-cross-flow mixing [178], co-flow mixing [179], and swirl-based mixing in small cups with radial and axial inflow [180]. Figure 4.19 shows a micromix combustion chamber with about 1600 miniature injectors, which was designed and successfully tested for a small-size Auxiliary Power Unit (APUGTCP36-300) [181, 182]. The GTCP36-300 requires about 1.6 MW thermal energy converted to shaft power generating electrical and pneumatic power up to 335 kW. The combustion section consists of an annular reverse flow combustion chamber in which the micromix combustor is to be integrated.

Within previous studies, the influence of combustion modeling and burner design parameters on flow field, temperature distribution, and flame structure has been studied for a low energy density burner configuration having a fuel injector diameter of 0.3 mm [183, 184]. The study discussed by Ayed et al. [184] has shown potential to reduce NO_x emission of the burner by controlling lateral cool air flows around the flame, which are established at given geometric parameters. The findings of the study by Ayed et al. [184] have been applied to design a high energy density burner with an injector diameter of 1 mm (increasing the heat rate per injector by more than 11 times). This burner has been analyzed numerically and experimentally in the study by Funke et al. [185] and has proven low NO_x ability at increased energy density. Based on these studies, the impact of different geometric parameters on flow field, flame structure, and NO_x formation are identified, and the micromix combustion principle is continuously optimized.

4.5 Fuel Flexibility

Gas turbine engines for power generation use diffusion flame combustors thanks to their reasonable performance and higher stability characteristics. Currently, the use of this kind of combustors is under review as the diffusion flame combustors are the main source of combustion pollutant emissions with an unacceptable higher concentration of NO_x. Robust, low emission, and highly efficient power generation are the most important key parameters measuring the performance of a power generation system. Many approaches have been introduced and developed to achieve these goals including catalytic combustion and lean premixed combustion (LPM). The catalytic combustion was found to be highly expensive with low durability and safety as well. While in lean premixed combustion, air and fuel are introduced upstream of the burner to make sure that they are homogenously mixed. In gas turbines, the combustion chambers are operated under excess air dilution to control the flame temperature; consequently, thermal NO_x emissions are significantly reduced and could be eliminated. However, premixed combustion flames are subjected to two main kinds of combustion instabilities including static and dynamic instabilities. Static instabilities are related to blow-off and flashback phenomena, while the dynamic instabilities are characterized mainly due to thermo-acoustics [186, 187]. Flame anchoring mechanism and combustor geometry are the main factors affecting combustion stability in premixed combustors. Fuel flexibility approach is considered as an effective technique to control combustion instabilities within the combustor. Hydrogen addition to hydrocarbon fuels is a solution to reduce NO_x emissions and improve stability limits in lean premixed gas turbine combustors. The recent technologies affecting the design and development of fuel-flexible combustor to avoid flashback, auto-ignition, and combustion dynamics associated with premixed have been discussed [188]. The authors reviewed the fuel variability and stability in premixed combustion systems and the effect of fuel composition on premixed flame stability and emissions. The progress of the effects of using the lean premixed combustion characteristics has been reviewed. It is concluded that there is a rising interest in switching the operation of combined-cycle power plants from diffusion to premixed combustion; this is due to the decrease in NO_x emission.

4.5.1 *Effects of Fuel Flexibility on Gas Turbine Operation*

Any change in fuel composition affects the flame characteristics, in terms of adiabatic flame temperature and laminar/turbulent burning velocity, due to the associated changes in chemical and thermo-physical properties of different fuels [189]. In addition to the changes of flame properties, there will be some changes in the combustor operations such as blowout, flashback, and dynamic instabilities characteristics. Different fuels may have different adiabatic flame temperatures at the same operating load and equivalence ratio. The adiabatic flame temperature within the combustor is of particular interest to the manufacturers of gas turbine combustors as it determines the combustor efficiency and the required cooling load for safe operation within the limit that liner material can withstand [119]. Laminar burning velocity is also one of the most important characteristics of the flame that determines flames' propagation in a premixed reactant mixture and combustor operability limits [190, 191]. Figure 4.20 presents the typical laminar flame speed for common fuels as function of the operating equivalence ratio. Among the considered fuels, hydrogen resulted in the highest laminar flame speed [192]. This may be attributed to the higher molecular diffusivity and the higher chemical reactivity of H_2 compared to other fuels. The higher burning velocity characteristic of hydrogen supports its use as flame-enhancing fuel at low operating loads or in premixed combustion near lean blow-out limits to widen the operability range of the gas turbine combustor [193].

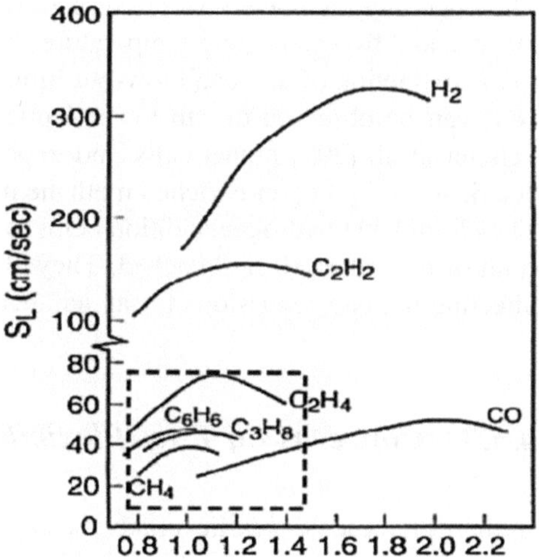

Fig. 4.20 Laminar burning velocity for different common fuels at different operating equivalence ratios [193]

4.5.2 H_2-Enriched Premixed Combustion

Enrichment of hydrocarbons with hydrogen is a solution to reduce NO_x emissions in gas turbines. The addition of small amount of hydrogen to hydrocarbon fuel can enhance lean flame stability limits [194]. Due to higher flame temperature, the addition of hydrogen increases the production of NO_x emissions [195–197]. Kim et al. [198] experimentally studied the flame characteristics of H_2-enriched air–methane swirling premixed flames. The results show that the NO concentration at stronger swirl strengths increases as the hydrogen amount increases. This can be attributed to the higher combustion temperature as a result of reducing the amount of recirculated gases. It is also shown that the flame temperature decreases as the swirl strength decreases which reduces the NO concentration. Kim et al. [199] investigated the combustion characteristics such as NO_x and CO emissions under lean premixed gas turbine combustor at pressure up to 3.5 bar with air preheat temperature of 450 K. The results show that NO_x emissions can be reduced by providing uniform fuel–air gas mixture to the combustor. Reducing the residence time of the hot gases in the combustor helps to reduce the NO_x emission. Aladawy et al. [200] numerically investigated the influence of turbulence on NO_x emissions in a lean premixed combustor burning air–methane mixture at an inlet temperature of 495 K, pressure of 5 atm, and at equivalence ratios from 0.54 to 0.85. The results show that as the temperature increases in the flame region, the NO_x emission increases with respect to the decrease of turbulence mixing time scale. They concluded that as the turbulent mixing time decreases the flame length flame residence time decreases. Griebel et al. [201] experimentally studied the lean blow-out limits and NO_x emissions of lean premixed, hydrogen-enriched methane/air flames at high pressure. The operating parameters were pressure of 14 bar, bulk velocity range from 32 to 80 m/s, and two preheating temperatures 673 and 773 K. The results show that due to the extinction of the lean blow-out limits for hydrogen addition, lower minimum NO_x can be observed due to lower flame temperature at lower equivalence ratio. Schefer et al. [202] numerically and experimentally studied the combustion characteristics of hydrogen-enriched methane in a lean premixed swirl-stabilized burner. They found that hydrogen addition helps to reduce CO emissions as the lean stability limit of natural gas is approached. They also found that hydrogen addition was not affecting the NO_x emissions for an adiabatic flame temperature.

4.5.3 Concerns on Fuel Flexibility

In the last decades, significant advances in combustion technologies, especially for gas turbines, have been achieved. The adaptation of lean premixed combustion techniques resulted in significant reduction in NO_x emissions up to the level of single digit. Although this achievement is notable, the developed burners are of tight control in terms of operating conditions and fuel type. Operation of the premix burners outside the specified operating ranges results in excess emissions and even failure in the hardware system due to flame flashback and combustion dynamics. Most of the newly developed burners for reduced emissions are designed to work on gaseous fuels within the specified operating range. This makes the process of fuel variability or operation in unique combined (biomass gasification and highly humidified) cycles very challenging.

4.5.3.1 Concerns on Hydrogen

Hydrogen combustion is one of the promising approaches toward zero-emission power cycles. The development of hydrogen burner for gas turbines with similar performance of natural gas burners is a very challenging process due to the high combustion temperature of hydrogen. The requirements of stable flame, high combustion efficiency, and low NO_x emissions can only be achieved at rapid mixing of fuel with air. This is hardly can be achieved in case of hydrogen combustion due to its highly reactive nature.

4.5.3.2 Concerns on Medium Heating Value Fuels

Oxygen-blown gasification of coal or heavy residuals is the common sources of medium heating value fuels for gas turbine applications called syngas fuels. This syngas consists mainly of H_2 and CO in comparison with natural gas (mainly methane); the syngas fuel requires smaller volume of air for stoichiometric combustion resulting in higher adiabatic flame temperature. Adding to this, hydrogen has higher burning velocity and very short ignition delay as compared to natural gas, which makes the control of flame flashback a very challenging process. The standard premixing technique for burning hydrogen-rich fuels may not be applicable.

4.5.3.3 Concerns on Low Heating Value Fuels

Air-blown gasification of coal or heavy residuals is the common sources of low heating value fuels for gas turbine applications. Those low energy fuels contain ammonia in most of the cases making the process of NO_x reduction very problematic. Also, those fuels are characterized by low burning velocity and low adiabatic flame temperature as compared to natural gas or fuels of medium heating value. This can be considered as an advantage based on the concept of formation of thermal NO_x at an elevated temperature. However, because of the great volume of the flow into the combustor, the aerodynamics of the combustor are influenced significantly, which necessitates redesigning of the combustor. The oxidation of CO is also a concern in this case.

4.6 Oxidizer Flexibility

Oxidizer flexibility is considered as one of the effective techniques toward the reduction of emissions for clean combustion applications. Air combustion results in the emission of massive amounts of NO_x due to the presence of nitrogen with high concentration in normal air. Oxygen-enriched air combustion or pure oxy-combustion is considered as effective techniques toward the reduction of NO_x emissions through the reduction of the concentration of nitrogen in the oxidizer mixture. The combination between both of the two techniques, LPM combustion and oxidizer flexibility, can result in complete control of NO_x emissions. In this section, LPM air combustion and oxygen-enriched air and pure oxygen combustion are discussed with the associated NO_x emissions and flame stability.

4.6.1 Oxy-fuel Combustion

Carbon dioxide is considered as one of the primary greenhouse gas emissions due to combustion of fossil fuels. Legislation and public awareness led to strict policies for control of the emissions of greenhouse gases. Also, the reduction of oil prices slows down the rate of application of renewable clean sources of energy [203, 204]. With these challenges, treatment technologies should be applied to control the emissions of greenhouse gases especially CO_2. Oxy-fuel combustion technology is one of the most promising carbon capture technologies with the flexibility of application in existing power plants with slight modifications and at reasonable cost [205–208]. The most dangerous pollutants because of burning natural gas are the nitrogen oxides, which are formed due to combustion of air in the presence of nitrogen. There are various methods for controlling or eliminating NO_x emissions including pre-treatment, process modifications, post-treatment, and combustion modification. Other pollutants that should be highly taken into consideration are CO_2 and CO emissions. Another method to completely eliminate NO_x emissions is burning the fuel in pure oxygen environment (oxy-combustion) [209, 210]. However, the combustion and radiative heat transfer characteristics are different in cases of oxy-fuel and air–fuel combustion [211–215]. This can be attributed to the differences in the physical properties of CO_2 and N_2 [208, 216–219]. The replacement of N_2 by CO_2 in the oxidizer mixture impacts the flame in terms of changes in volumetric heat capacity, mixture density, and transport properties in terms of mass diffusivity, thermal conductivity, and dynamic viscosity.

In the oxy-combustion technology, N_2 is separated from air using air separation unit, the remaining O_2 is used as oxidizer, and the combustion process results in flue gas mixture consisting mainly of CO_2 and H_2O. When the water vapor is condensed and left, CO_2 facilitates its capture and sequestration to totally eliminate its release into the atmosphere. However, burning the fuel with almost pure oxygen as an oxidizer will result in high flame temperature [12, 220]. To prevent the

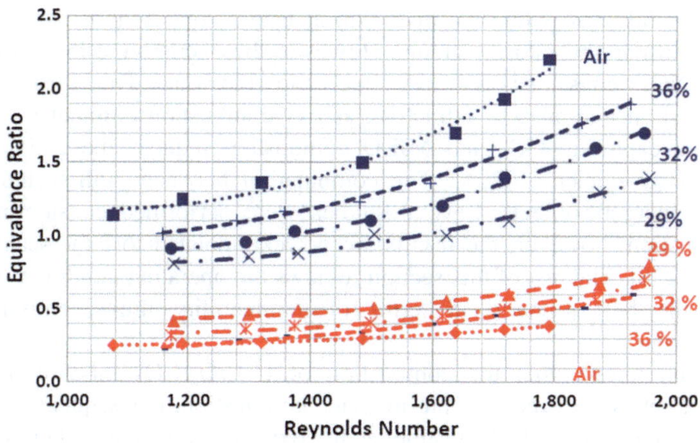

Fig. 4.21 Flammability limits in oxy-fuel combustion of different O_2 concentrations and comparison with air–fuel combustion [223]

excessive temperatures, a fraction of CO_2 from the flue/exhaust gases is circulated and mixed with the oxidizer [221]. Oxy-combustion technology reduces the exhaust gas flow rate, and accordingly, the heat losses and the size of the flue gas treatment unit are reduced [222]. On the other hand, the elevated CO_2 concentrations in the oxidizer cause decrease in chemical kinetics and, accordingly, decrease in laminar flame speed. Recently, Rashwan et al. [223] illustrated the influence of CO_2 addition, in terms of thermal conductivity, kinematic viscosity, and mass diffusivity, on flame stability as per Fig. 4.21. Actually, carbon dioxide has higher density than N_2, which affects gas volume, flame shape, and pressure drop. The density of air is 0.43 kg/m^3 at 800 K, while the densities of CO_2/O_2 mixture are 0.62, 0.61, and 0.60 kg/m^3 at oxygen fractions of 29, 32, and 36%, respectively. The higher density of CO_2 leads to higher heat capacity of CO_2/O_2 mixtures with respect to air (at 800 K, the volumetric heat capacity of air is 480 J/m^3/K, while those of CO_2/O_2 mixtures at oxygen fractions of 29, 32, and 36% are 480, 705, 700, and 690 J/m^3/K, respectively). The high volumetric heat capacity of CO_2/O_2 mixtures with respect to air directly reduces temperature level, flame speed, and flame stability. In addition to the changes in densities and in volumetric heat capacities, flame speed is also affected by gas transport properties. The thermal conductivity and the dynamic viscosity of N_2 and CO_2 gases at different temperatures and the mass diffusivity of O_2 in both N_2 and CO_2 are available in [203].

Oxy-fuel combustion technology has different degrees of freedom that confined the operational space from the air–fuel combustion. Such oxy-fuel combustion flames are more likely to be operated near to stoichiometric to reduce both the fuel and O_2 consumption with a controlled ratio of O_2/CO_2 to control the combustion temperature. It is well known that there are three combustion issues associated with oxy-fuel combustion that must be addressed for $CH_4/O_2/CO_2$ combustion systems. The operation at higher equivalence ratios or at rich mixtures is associated with higher CO emissions, while the operation at leaner conditions is associated with higher O_2 flow in the exhaust stream. However, the operation at low temperatures for low NO_x emissions is most likely to blowout. The idea of MILD oxy-combustion may also be attractive, as oxy-combustion eliminates NO_x emissions while capturing CO_2 and MILD combustion reduces the emissions also, improving the efficiency [224]. The combination between both techniques can result in synergetic effects. Several studies were performed considering MILD oxy-combustion; however, the results are not extrapolated to large-scale applications [225–228]. The rest of this section reviews the operating condition which can maintain the best scenario in terms of flammability limits, emissions, and cost-effectiveness.

It is well known that the combustion of $CH_4/O_2/CO_2$ mixture has slower chemical kinetics than those of air–methane combustion. Consequently, the flame stability becomes more challengeable, as the oxy-fuel combustion flames are more prone to blowout. Ditaranto and Hals [229] reported that the process of oxy-fuel burning requires oxygen fraction of at least 30% to perform in a stable manner as compared to air combustion. This 9% increase in oxygen fraction as compared to air is compensating for the lower heat capacity of nitrogen. Nemitallah and Habib [205] studied experimentally and numerically the characteristics of oxynon-premixed flames in a gas turbine coaxial combustor. The results showed that the stability of oxy-flames is adversely affected by the drop of oxygen fraction below 25%. Amato et al. [230] studied the flame extinction mechanisms of premixed flames. The results showed that the use of CO_2 as a diluent adversely affects the flammability limits. This may be attributed to the slower rates of chemical reactions. Rashwan et al. [203] studied the effect of partially premixing on oxy-fuel combustion, and they compared the results with the cases of air–fuel combustion. They studied three oxygen fractions, namely 29, 32, and 36%. The results showed wider stability limits under air combustion operation as compared to those of oxy-combustion operation.

Ramadan et al. [231] investigated the effects of oxidizer flexibility on flame stability and recorded the stability maps by comparing the effect of using different oxidizer while burning natural gas in a confined space and using a bluff body stabilizer with different blockage ratios. The results showed that that oxygen-enriched air flames have wider stability ranges of the flame than those in cases of air flames and oxy-flame. Jerzak and Kuźnia [232] studied the flammability limits of the combustion of natural gas in different oxidizing environments, namely air, oxygen-enriched air, and oxy-mixture of O_2 and CO_2. The tests were performed considering two swirl numbers, namely 0.69 and 1.35. The results showed that oxy-flames have the lowest flammability limits at both swirl numbers. The most favorable stable flame operation was obtained for oxygen-enriched air (25% O_2) at swirl number of 1.35. They reported that the addition of 25% O_2 to the air combustion adversely affects the flashback limits and enhances blow-off limits. Shi et al. [233] studied experimentally the oxy-fuel combustion of methane over a range of oxygen fractions. The influence of changing oxygen fraction from 0.21 to 0.86 is investigated and reported. The study confirmed some well-known observations

Table 4.1 Comparison of different clean burning technologies for gas turbine applications

Different novel technologies	Major remarks and differences	NO_x and CO reduction	Advantages	Disadvantages	References
Non-premixed flames	Fuel and oxidizer are introduced separately to the combustor	• This technique produces a lot of NO_x emissions and classified as an inefficient technique	• Wider stability limits • High turndown ratio	High concentrations of NO_x and CO emissions	[22, 205, 226]
Lean premixed flames	Fuel is mixed with air or oxidizer upstream within the mixing length	• NO_x emissions reduced as this technique is associated with low combustion temperature • To achieve ultra-low NO_x and CO emissions, the combustor should be operated near to the lean extinction limits	• Ultra-low NO_x and CO emissions	• Tiny stability limits • Low turndown ratio	[23–26]
MILD/flameless combustion	Developed to enhance thermal efficiency by recirculating the waste heat carried out by the flue gases	• This technique maintains very low reaction temperature; in turn, the NO_x formation will be reduced	• No flame front • No visible flame • No noise • No roar • Ultra-low NO_x and CO • High thermal efficiency	The flame front is not concentrated	[27–36]

(continued)

Table 4.1 (continued)

Different novel technologies	Major remarks and differences	NO_x and CO reduction	Advantages	Disadvantages	References
Colorless distributed combustion (CDC)	This technique can be attained by swirling the flow to ensure required mixing of the fuel, air, and the hot reactive gases inside the combustor	• Ultra-low CO and NO_x noise reduction	• High stable combustion • Higher combustion intensity • Higher efficiency • Low noise level	The flame front is not concentrated	[54–56, 62, 63]
Low-swirl injector (LSI) combustion	Stringent ultra-low emission, operating DLN combustor near to the lean stability limits	• Stringent ultra-low emission	Ultra-low emission	Operating near to the lean blow-off limits will inversely affect the reliability and performance of the engines	[67–69]
Swirl-stabilized burners	Creating recirculation zones will make the power generation equipment more compact; mixing is strongly enhanced with turbulent flow fields	• Swirl combustion reduces the NO_x emissions as it is enhancing the mixing process	Low NO_x emissions	• The pressure drop occurs during the swirling process	[162, 234–236]
DLN/DLE burners	The principle beyond the DLE approach is to burn a substantial portion of the injected fuel at lean premixed conditions	• Decreases the flame temperature and to reduce the thermal NO_x	Low NO_x emissions	• This condition results in bringing the full-capacity operating condition down • Controlling CO emissions can be very hard to achieve, and engine load will increase the probability of flame blowout	[183, 237, 238]

(continued)

Table 4.1 (continued)

Different novel technologies	Major remarks and differences	NO_x and CO reduction	Advantages	Disadvantages	References
Catalytic combustion	Combustion process of a mixture consisted of fuel and oxygen reacted on the surface of a catalyst, resulting in oxidation of the fuel	• Decreases NO_x emissions to minimum values	• Achieve ultra-low NO_x • Reduce the probability of blow-off • Reduce combustion instabilities	Combustion on the surface of the reactor has low stability and low durability	[239–241]
Perforated plate burners	The flame instability can be attributed to the effects of hole/jet diameter and spacing, heat transfer to the face plates	• Leads to low NO_x emissions	Flame stability affect by the holes spacing. The smaller the spacing, the more stable the flame	Hard to be manufactured	[118, 144, 242–245]
Environmental EV/SEV/AEV burners	The wet low NO_x technique; the injection of water or steam into the combustion zone was the traditional practice to control the emissions of NO_x	• Wet low NO_x emissions	The power output of the gas turbine is improved with the increase of the amount of injected water	Water or steam with the required demineralized properties is not available for many applications, which exerted more pressure on gas turbine manufacturers to develop dry techniques to control NO_x emissions	[246, 247]
Micromixer burners	The flame instability can be attributed to the effects of hole/jet diameter and spacing, heat transfer to the face plates	• Strong impact on NO_x formation	Its innate flexibility to accommodate staging, scalability, and fuel dilution and flexibility	The pressure drop of oxidizer flow across micromixer and the operational equivalence ratio are both chosen carefully to prevent flame flashback	[248–250]

regarding the oxy-fuel flames behaviors. They reported that the flame stabilization range is very small and confined by flashback at higher values of oxygen fraction and by blow-off at lower values of oxygen fraction. The results showed an increase in temperature and laminar flame speed while increasing the oxygen fraction. For stable flame operation, the lowest obtained oxygen fraction is around 0.50, while for the upper stable limit, the highest oxygen fraction is about 0.86.

Based on the above discussion of different technologies for clean combustion in gas turbines, Table 4.1 provides a summary comparison among them in terms of emission reduction, advantages and disadvantages. The translation of such technologies into market products, feasibility, and performance of different gas turbines are discussed in Sect. 4.9 in the current chapter.

4.7 Other Routes for NO_x Formation and Treatment

Other than the thermal NO_x generation mechanism proposed by Zeldovich [251], significant amount of NO_x is as well generated via prompt NO_x and fuel-bound NO_x. Fenimore [69] was the first to identify the second mechanism leading to NO_x formation called prompt NO_x. Prompt NO_x is characterized with fuels having high energy densities like natural gas, which forms free radicals in the fuel-rich flame regions in the combustion zone [252]. Not only prevalent with high-density fuels, significant NO_x formations are also attributed to some combustion conditions, like in low-temperature combustion, short residence times, and fuel-rich environments created by gas turbines, surface burners, and some multistage combustion systems [253]. Prompt NO_x formation is proportional to the number of carbon atoms present per unit volume. The actual formation of the prompt NO_x encompasses intricate chain of reactions with many intermediate species. The most common hydrocarbon fuel fragments that contribute in prompt NO_x formation include CH, CH_2, C, C_2H, in descending order of influence [254]. The amount of HCN formed is a function of the hydrocarbon radicals. The studies conducted by Schefer et al. [255] indicated that most of the NO_x formation at the base of flame is prompt NO_x produced from the CH. Fuel-bound NO_x formation is a complex phenomenon due to the varying structure of the nitrogen chemical bonding to the parent hydrocarbon molecules. According to Glarborg et al. [256], most of the fuel-bound nitrogen is converted to NH_3 and HCN, which react with other combustion radicals to form NO_x. The rate of fuel-bound NO_x formation can be higher than that of thermal NO_x since the diatomic nitrogen bonds are stronger than the carbon–nitrogen bonds and hence more likely to be easily broken [257].

The two main categories of NO_x emissions control techniques in clean combustion technology are the combustion controls and the treatment of the flue gas. The techniques involve the reduction of flame temperature; reduction in the concentration of oxygen in the primary flame zone; or reburning NO_x back to nitrogen and oxygen. The control processes include low excess air combustion—which reduces the concentration of NO_x in the flame region [258]; recirculation of flue gas

—which reduces the concentration of oxygen in the primary flame zone and the recycled gas acts as a heat sink, and thus decreases the flame peak temperature [259]; water or steam injection—dissociation of steam into oxygen and hydrogen increases the concentrations of the reducing agents, which drastically lowers the rate of NO_x formation [260]; reduce firing rate—which ensures larger heat loss to the combustor walls in a fixed size combustor thereby lowering the flame temperature; reburn by creating a secondary flame zone via adding hydrocarbon fuel outside of the primary combustion zone, which decrease NO_x formation tendencies by reaction with hydrocarbon radicals with overfired air in the secondary zone; employing ultra-low NO_x burners—lower NO_x levels of up to 10–15 ppm (dry basis) was demonstrated for gas-fired industrial furnaces and boilers [261]. The flue gas treatment practices include the selective catalytic and non-catalytic reduction methods [262], the catalytic absorption technique [263], and the low-temperature oxidation with absorption method [264].

4.8 Parallel Development of Combustor Liner Materials

Driven by the need for improvement in emissions control as well as elevated firing temperatures characterized by the current progress in clean combustion for gas turbines, substantial development efforts are in progress to come up with combustor materials that can satisfy the intended goals. In gas turbine engines, the boundaries of combustor liner experience the highest temperature as the combustion occurs through the boundaries [265, 266]. As the result of the higher local combustor temperature, the liner undergoes fatigue and creep failures, pressure loading, surface oxidation, erosion, sulfidation, etc. [267, 268]. Significant improvement has been made in developing combustor liner materials with high-temperature creep rupture strength without forfeiting its corrosion and oxidation resistance. Historically, combustor liner materials were made of nickel-based superalloys sheets. Between the periods of 1960 and 1980, a high creep strength material, Hastelloy X, was predominantly used. Subsequently, Nimonic 263, having higher creep strength, was introduced. The cobalt-based superalloy, Haynes 188, has been recently developed as a result of the increase in the need for elevated firing temperatures in recent models of gas turbines. Those recent gas turbines demonstrated good oxidation and sulfidation resistance and are metallurgically stable with good ductility at higher temperatures and enhanced creep rupture strength [269–272].

Other titanium-based alloys offer formable, thermo-mechanical fatigue resistance, oxidation, and creep resistance materials, subject to thermal barrier coatings (TBCs) for application in higher thermal loading conditions [273]. These coatings act as insulating layer with low thermal properties to reduce the base metal temperature as well as providing some corrosion and oxidation resistance. Recently, several protective coatings are employed in gas turbine combustor liners development to increase their thermal performance and durability and reduce surface [274–277]. Mishra [278] affirms that application of Yttria-stabilized zirconia thermal barrier coating to gas turbine combustor liner was found to be very effective in improving the thermal fatigue resistance. Ceramic matrix composite (CMC) materials are also considered as combustor liner materials. They have proven to be thermally stable, enhanced fatigue durability, and better thermal conductivities. Brewer [279] succeeded in testing the capability of CMC combustor liner in a test rig combustor with the aim of deploying same on a commercial jet engine. In a similar fashion, Liu et al. [280] reported significant progress in the development of a SiC/SiC ceramic matrix composite material for aeronautic combustor liner applications. Incorporating ceramic liners in an industrial gas turbine combustor has the potential of lowering NOx emissions to below 10 ppmv and CO emissions to about 25 ppmv as reported by [281].

4.9 Feasibility of Different Combustion Technologies and Future Challenges

In order to reduce the combustion temperature and control NO_x emissions, original equipment manufacturers (OEMs) developed processes that depend on air as a diluent through premixing both air and fuel upstream of the combustion chamber. This process is called lean premixed combustion. The lean premixed combustion technology is referred to by a variety of trade names in gas turbine manufacturing field including Rolls-Royce's dry low emission (DLE) process, Siemens-Westinghouse's and General Electric's dry low NO_x (DLN) processes, and solar turbines' SoLo NO_x process. In cases of burning natural gas, most of the commercially available gas turbines achieve low NO_x emissions within the range of 15–25 ppm by volume depending on the manufacturer, application, and turbine model. Some OEMs have achieved single-digit NO_x emissions.

GE's dry low NO_x emission system was selected to show the operation and evolution of a lean premixed staged combustion system in response to continuous changes and efficiency improvements. The GE dry low NO_x (DLN-1) combustor is a two-stage premixed combustor designed for burning natural gas and able to operate using liquid fuels. The combustor consists of four main components: fuel injection system, venture, liner, and cap/center body assembly. These components form two stages within the combustion chamber. In the premixed combustion mode, fuel and air are mixed in the first stage and a lean uniform unburned air–fuel mixture is delivered to the next stage. The DLN-1 system has four different operating modes including:

1. Primary: Fuel is fed to the primary nozzles only. Flame is only in the primary stage. This operating mode is used to ignite and operate the combustor over low to mid-loads, up to a limit of a preselected reference combustion temperature.
2. Lean–lean: Fuel is fed to primary and secondary nozzles. The flame exists in both the primary and secondary zones. This operating mode is used for mid-loads between two preselected reference combustion temperatures.

3. Secondary: Fuel is fed only to the secondary nozzle. The flame exists only in the secondary zone. This operating mode is a transition period between the modes of lean–lean and premix. This mode is required to extinguish the flame in the primary zone, before reintroducing the fuel to what acts as the primary premixing zone.
4. Premix: Fuel is fed to both the primary and secondary nozzles. Flame exists only in the secondary stage. This operating mode is approached near the design point of the reference combustion temperature. The optimum emission rates are obtained in the premix combustion mode.

At low loads, less than 20% of the base load, emissions of NO_x and CO from the DLN-1 are similar to those obtained from standard combustion systems utilizing diffusion flames.

Other OEMs produce similar systems with the notable exception being Alstom. The sequential combustion DLN technology was initially developed by ABB for the gas turbine models, GT24 and GT26, before being used by Alstom. Combustion process occurs in the primary DLN combustion chamber (EV^{tm}) followed by the addition of fuel in the second (SEV^{tm}) combustor placed after the first row of turbine blades. In 1997, this DLN technology was commercialized and the principal of thermodynamic reheat was applied. As a matter of fact, the SEV^{tm} combustor does not participate in NO_x production, and as a result, the sequential combustion results in low emissions of NO_x [282].

The OEMs are continuously improving the performance of gas turbines while meeting the strict requirements for controlling emissions by emission regulators. The gas turbines became available as F-technology in the late 1980s with their higher combustion temperatures; the OEMs were directed to improve their DLN combustion systems in order to keep emission rates within the required limits, abut 25 ppmvd. Based on the studies performed by GE, it was concluded that the use of air in the combustor should be limited to the amount required to be mixed with fuel for combustion. The DLN-2 design was implemented based on repackaging of the DLN-1 premixing technology and eliminating the venture and center body assemblies that needed cooling air. The DLN-2 combustor is a dual-mode single-stage combustion system that can handle gaseous and liquid fuels for combustion. For gaseous fuel operation, the combustor works at low loads (<50% load) on a diffusion mode and on a premixed mode at higher loads. Oil operation on the DLN-2 combustion system is in the diffusion operating mode across the entire load range, with injection of diluent to control the emissions of NO_x. The DLN-2 combustion system has a single firing zone created by the cap face and the liner. About 90% of the gaseous fuel is injected through the premixer radial gas injection spokes under low emissions operating mode, and the fuel is mixed with the combustion air in the tubes surrounding the five nozzles. The premixer tubes are integrated into the cap assembly. In this system, air and fuel are mixed and the flows leave the five tubes at high velocity toward the flame zone, where lean combustion occurs resulting in low NO_x emissions. Two mechanisms are utilized for stabilizing the flame including the vortex breakdown created by the swirling flow leaving the premixers and the sudden liner expansion.

Due to continuous pressure applied by emission regulators, the OEMs developed combustion systems with low emission rates below 9 ppm and this was achieved in the early 1990s. During this period, the DLN-2.6 combustion system was developed by GE for the Frame 7FA machine. This machine permitted about 6% of additional air to be passed through the premixers in the combustion chamber. The modification in air splits was done through reductions in the cooling air flows through the cap and the liner, requiring increased cooling effectiveness. The most important feature of the DLN-2.6 combustion system was the addition of a sixth burner within the center of the other five DLN-2 burners. Fueling the nozzle at the center separately from the other outer nozzles enables the modulation of the fuel-to-air ratio with respect to the outer nozzles. Another important feature of the DLN-2.6 combustion system was the complete elimination of the diffusion combustion mode, which needed additional loading and unloading systems. The H system^{tm} combustor by GE called DLN-2.5 utilizes a simplified scheme for combustion mode staging in order to obtain low emission rates over the whole premixed load range. The most important feature associated with this variant is that there are only three modes of combustion including piloted premix, diffusion, and full premix [283]. The required modifications to reduce emission rates from the DLN combustor by Siemens-Westinghouse from 15 to 9 ppm are predominately the utilization of a premixed pilot and support housing design changes [284].

All the above clean combustion technologies are considered as mature technologies that are ready for application in industrial gas turbines. For the sake of comparison, Table 4.2 summarizes the most recent achieved emission values for the different emission control technologies considered in the present study. Though low emission levels can be achieved using different technologies, a technology may provide better performance under certain loading conditions or while using another kind of fuel. As a matter of fact, gas turbine operation is very sensitive to many parameters, including loading condition, fuel type, oxidizer, burner design, emission reduction level, and flame type and stability. So, all advantages cannot be combined in one combustion system. In other words, efficiency usually interferes to a certain limit with the emission reduction level, and the performance of a turbine using a certain fuel is different when another fuel is used. Based on this review, the main future challenge for the OEMs is the development of low emission fuel-flexible gas turbines. This requires the design of versatile combustors that can handle different fuels in different flame modes under different loading conditions to achieve high performance at low emission levels under all operating conditions. OEMs are continuously improving the combustion technology to meet the continuously lowered emission rates requirements by the regulators. Research and development efforts are continuously advancing technology and provide significant contributions to the design and manufacturing techniques for further enhancing gas turbine performance while controlling the emission rates and the costs.

Table 4.2 Achieved emission values for different emission control technologies

Reference	Application	Achieved emission value
Advanced vortex combustion [26]	DLE premixed combustor Ultra-low NO_x emissions	NO_x emissions <10 ppm
Huang et al. [52]	MILD combustion with different fuel and air injection configurations	NO_x emissions <4 ppm CO emissions <39 ppm
Khidr et al. [54]	MILD combustion	NO_x and CO emissions <10 ppm
Cheong et al. [55]	MILD-N_2 combustion MILD-CO_2 combustion	NO and CO emissions <5 ppm
Ye et al. [56]	MILD combustion of pervaporated liquid fuels	NO_x emissions <4 ppm
Khalil and Gupta [63, 65]	Non-premixed and premixed systems Ultra-low CO and NO_x emissions	NO emissions <2 ppm
Cheng et al. [76] Littlejohn et al. [77]	Low-swirl burners (LSBs)	NO_x and CO emissions 4–7 ppm
Johnson et al. [78] Nazeer et al. [79]	Prototype LSI of DLN turbine	NO_x and CO <5 ppm
Forzatti [142]	Catalytic combustion	NO_x emissions <2 ppm

4.10 Conclusions

In this chapter, different novel combustion techniques for clean combustion in gas turbines have been presented. These novel combustion techniques include flame variability, burner design, and fuel and oxidizer flexibility approaches. This chapter started with the flame variability approach. The combustion and emission characteristics of different flame types including non-premixed/premixed, moderate or intense low oxygen dilution (MILD) flameless combustion, colorless distributed combustion (CDC), and low-swirl injector (LSI) combustion flames have been studied, and their limitations for application have been discussed. The conversion of the flame from the diffusion (non-premixed) to the premixed combustion model controls the combustion temperature and emissions; however, flame stability is adversely affected. MILD and CDC showed reduced emission levels; however, the performance of the gas turbine needs to be improved to obtain wider operability limits for MILD combustion and to improve the mixing for CDC. This is followed by discussing the burner design approach. Novel burner designs for clean burning in gas turbines have been studied in detail considering swirl stabilized, dry low NO_x (DLN) and dry low emission (DLE), catalytic combustion, perforated plate, environmental vortex (EV), sequential environmental vortex (SEV), advanced environmental vortex (AEV), and lean direct injection (LDI) micromixer burners. Among the different burner designs, the EV burners have proven the lowest level of

NO_x emissions at wider flame stability range. Then, fuel flexibility approaches have been investigated considering mainly hydrogen-enriched combustion, and the associated concerns about fuel variability technique have been discussed. Hydrogen-enriched combustion has been proven as an effective technique toward the control of flame stability and emissions; however, some technical concerns on the use of hydrogen need to be addressed in research especially mixing of H_2 with air. Also, oxidizer flexibility approach has been studied considering lean premixed (LPM) air and oxy-fuel combustion and both techniques have been evaluated and compared in terms of performance and emissions. Oxy-combustion technique eliminated NO_x emissions while capturing CO_2; however, the oxygen separation cost is still an issue rewarding the application of this technology. At the end of this chapter, the feasibility of the different clean combustion techniques in gas turbines has been discussed and the available market products utilizing such novel technologies are presented. Research and development efforts are required for advancing different technologies for further enhancing gas turbine performance while controlling the emissions at reasonable cost.

References

1. International Energy Agency (2017) World energy outlook
2. Correa SM (1993) A review of NO_x formation under gas-turbine combustion conditions. Combust Sci Technol 87(1–6):329–362
3. Miller JA, Bowman CT (1989) Mechanism and modeling of nitrogen chemistry in combustion. Prog Energy Combust Sci 15:287–338
4. Hayhurst AN, Lawrence AD (1992) Emissions of nitrous oxide from combustion sources. Prog Energy Combust Sci 18:529–552
5. Gascoin N, Yang Q, Chetehouna K (2017) Thermal effects of CO_2 on the NO_x formation behavior in the CH_4 diffusion combustion system. Appl Therm Eng 110:144–149
6. Kang Y, Wei S, Zhang P, Lu X, Wang Q, Gou X, Huang X, Peng S, Yang D, Ji X (2017) Detailed multi-dimensional study on NO_x formation and destruction mechanisms in dimethyl ether/air diffusion flame under the moderate or intense low-oxygen dilution (MILD) condition. Energy 119:1195–1211
7. Zeldovich J (1946) The oxidation of nitrogen in combustion and explosions. Eur Phys J A 21:577–628
8. Layne AW (2000) Developing the next generation gas turbine systems—a national partnership. ASME Paper No. 2000-GT-176
9. Diakunchak IS, Gaul GR, McQuiggan G, Southall LR (2002) Siemens Westinghouse advanced turbine systems program final summary. ASME Paper No. GT-2002-30654
10. Der VK (1999) In: 13th U.S.–Korean joint workshop on energy and environment, Sep 1999
11. Kashir B, Tabejamaat S, Jalalatian N (2015) A numerical study on combustion characteristics of blended methane-hydrogen bluff-body stabilized swirl diffusion flames. Int J Hydrogen Energy 40:6243–6258
12. Habib MA, Nemitallah MA, Ahmed P, Sharqawy MH, Badr HM, Muhammad I, Yaqub M (2015) Experimental analysis of oxygen-methane combustion inside a gas turbine reactor under various operating conditions. Energy 86:105–114
13. Yu B, Lee S, Lee CE (2015) Study of NO_x emission characteristics in CH_4/air non-premixed flames with exhaust gas recirculation. Energy 91:119–127

14. Gao X, Duan F, Lim SC, Yip MS (2013) NO_x formation in hydrogen–methane turbulent diffusion flame under the moderate or intense low-oxygen dilution conditions. Energy 59:559–569
15. Li YH, Chen GB, Lin YC, Chao YC (2015) Effects of flue gas recirculation on the premixed oxy-methane flames in atmospheric condition. Energy 89:845–857
16. Altay HM, Hudgins DE, Speth RL, Annaswamy AM, Ghoniem AF (2010) Mitigation of thermoacoustic instability utilizing steady air injection near the flame anchoring zone. Combust Flame 157:686–700
17. Ghoniem AF, Park S, Wachsman A, Annaswamy A, Wee D, Altay HM (2005) Mechanism of combustion dynamics in a backward-facing step stabilized premixed flame. Proc Combust Inst 30:1783–1790
18. Lee K, Kim H, Park P, Yang S, Ko Y (2013) CO_2 radiation heat loss effects on NO_x emissions and combustion instabilities in lean premixed flames. Fuel 106:682–689
19. Bender WR (2007) Lean premixed combustion. Technology & Management Services, Inc., Gaithersburg, MD 20879
20. Hydari NH, Yousuf AA, Ellis HM (2002) Comparison of the most recent BACT/LAER determinations for combustion turbines by state air pollution control agencies
21. Calcagni J. U.S. Environmental Protection Agency, Memorandum dated June 13, 1989. Transmittal of background statement on “top down” BACT
22. Rashwan SS, Nemitallah MA, Habib MA (2016) Review on premixed combustion technology: stability, emission control, applications, and numerical case study. Energy Fuels 30:9981–10014
23. Habib MA, Rashwan SS, Nemitallah MA, Abdelhafez AA (2017) Stability maps of non-premixed methane flames in different oxidizing environments of a gas turbine model combustor. Appl Energy 189:177–186
24. Yanzhao A, Jaasim M, Vallinayagam R, Vedharaj S, Im HG, Johansson B (2018) Numerical simulation of combustion and soot under partially premixed combustion of low-octane gasoline. Fuel 211:420–431
25. Yilmaz H, Cam S, Yilmaz I (2017) Effect of micro combustor geometry on combustion and emission behavior of premixed hydrogen/air flames. Energy 135:585–597
26. Advanced vortex combustion. http://www.ramgen.com/tech_vortex_conventional.html
27. Huang X, Tummers MJ, Roekaerts DJEM (2017) Experimental and numerical study of MILD combustion in a lab-scale furnace. Energy Procedia 120:395–402
28. De Joannon M, Sorrentino G, Cavaliere A (2012) MILD combustion in diffusion-controlled regimes of hot diluted fuel. Combust Flame 159:1832–1839
29. Cavaliere A, De Joannon M (2004) Mild combustion. Prog Energy Combust Sci 30(4): 329–366
30. Chen S, Liu H, Zheng C (2012) Methane combustion in MILD oxyfuel regime: influences of dilution atmosphere in co-flow configuration. Energy 121:159–175
31. Duwig C, Li B, Li ZS, Aldén M (2012) High resolution imaging of flameless and distributed turbulent combustion. Combust Flame 159(1):306–316
32. Mastorakos E, Taylor A, Whitelaw J (1995) Extinction of turbulent counterflow flames with reactants diluted by hot products. Combust Flame 102(1–2):101–114
33. De Joannon M, Matarazzo A, Sabia P, Cavaliere PA (2007) Mild combustion in homogeneous charge diffusion ignition (HCDI) regime. Proc Combust Inst 31(2):3409–3416
34. Galletti C, Parente A, Tognotti L (2007) Numerical and experimental investigation of a mild combustion burner. Combust Flame 151:649–664
35. Coelho PJ, Peters N (2001) Numerical simulation of a mild combustion burner. Combust Flame 124:503–518
36. Wiinning JA, Wiinning JG (1997) Flameless oxidation to reduce thermal no-formation. Prog Energy Combust Sci 23:81–94
37. Luhmann H, Carrasco Maldonado F, Spörl R, Scheffknecht G (2017) Flameless oxidation of liquid fuel oil in a reverse-flow cooled combustion chamber. Energy Procedia 120:222–229

38. Goh KHH, Geipel P, Hampp F, Lindstedt RP (2013) Regime transition from premixed to flameless oxidation in turbulent JP-10 flames. Proc Combust Inst 34:3311–3318
39. Veríssimo AS, Rocha AMA, Costa M (2013) Experimental study on the influence of the thermal input on the reaction zone under flameless oxidation conditions. Fuel Process Technol 106:423–428
40. Weidmann M, Verbaere V, Boutin G, Honoré D, Grathwohl S, Goddard G, Gobin C, Dieter H, Kneer R, Scheffknecht G (2015) Detailed investigation of flameless oxidation of pulverized coal at pilot-scale (230 kWth). Appl Therm Eng 74:96–101
41. Milani A, Wünning J (2002) What is flameless combustion? In: IFRF online combustion handbook, pp 1–171, ISSN 1607-9116
42. Oldenhof E, Tummers MJ, Van Veen EH, Roekaerts DJEM (2009) Ignition kernel statistics of Delft jet-in-hot coflow flames. Eur Combust Meet
43. Oldenhof E, Tummers MJ, Van Veen EH, Roekaerts DJEM (2010) Ignition kernel formation and lift-off behaviour of jet-in-hot-coflow flames. Combust Flame 157:1167–1178
44. Dally BB, Karpetis AN, Barlow RS (2002) Structure of turbulent non-premixed jet flames in a diluted hot coflow. Proc Combust Inst 29:1147–1154
45. Szegö GG, Dally BB, Nathan GJ (2008) Scaling of NO_x emissions from a laboratory-scale mild combustion furnace. Combust Flame 154:281–295
46. Veríssimo AS, Rocha AMA, Costa M (2011) Operational, combustion, and emission characteristics of a small-scale combustor. Energy Fuels 25:2469–2480
47. Lupant D, Pesenti B, Lybaert P (2010) Influence of probe sampling on reacting species measurement in diluted combustion. Exp Therm Fluid Sci 34:516–522
48. Mancini M, Weber R, Bollettini U (2002) Predicting NO_x emissions of a burner operated in flameless oxidation mode. Proc Combust Inst 29:1155–1163
49. Cho ES, Shin D, Lu J, de Jong W, Roekaerts DJEM (2013) Configuration effects of natural gas fired multi-pair regenerative burners in a flameless oxidation furnace on efficiency and emissions. Appl Energy 107:25–32
50. Li P, Mi J, Dally BB, Craig RA, Wang F (2011) Premixed moderate or intense low-oxygen dilution (MILD) combustion from a single jet burner in a laboratory-scale furnace. Energy Fuels 25:2782–2793
51. Graça M, Duarte A, Coelho PJ, Costa M (2013) Numerical simulation of a reversed flow small-scale combustor. Fuel Process Technol 107:126–137
52. Huang M, Zhang Z, Shao W, Xiong Y, Liu Y, Lei F, Xiao Y (2014) Coal-derived syngas MILD combustion in parallel jet forward flow combustor. Appl Therm Eng 71:161–168
53. Huang M, Xiao Y, Zhang Z, Shao W, Xiong Y, Liu Y, Liu Z, Lei F (2015) Effect of air/fuel nozzle arrangement on the MILD combustion of syngas. Appl Therm Eng 87:200–208
54. Khidr KI, Eldrainy YA, EL-Kassaby MM (2017) Towards lower gas turbine emissions: flameless distributed combustion. Renew Sustain Energy Rev 67:1237–1266
55. Cheong KP, Li P, Wang F, Mi J (2017) Emissions of NO and CO from counterflow combustion of CH_4 under MILD and oxyfuel conditions. Energy 124:652–664
56. Ye J, Medwell PR, Varea E, Kruse S, Dally BB, Pitsch HG (2015) An experimental study on MILD combustion of prevaporised liquid fuels. Appl Energy 151:93–101
57. Sorrentino G, Goktolga U, De Jannon M, Van Oijen J, Cavaliere A, De Goey P (2017) An experimental and numerical study of MILD combustion in a cyclonic burner. Energy Procedia 120:649–656
58. Doan NA, Swaminathan N, Minamoto Y (2018) DNS of MILD combustion with mixture fraction variations. Combust Flame 189:173–189
59. Ye J, Medwell PR, Dally BB, Evans MJ (2016) The transition of ethanol flames from conventional to MILD combustion. Combust Flame 171:173–184
60. Sabia P, De Joannon M, Fierro S, Tregrossi A, Cavaliere A (2007) Hydrogen-enriched methane mild combustion in a well stirred reactor. Exp Thermal Fluid Sci 31:469–475
61. De Joannona M, Cavaliereb A, Faravellic T, Ranzic E, Sabiab P, Tregrossi A (2005) Analysis of process parameters for steady operations in methane mild combustion technology. Proc Combust Inst 30:2605–2612

62. Arghode VK, Gupta AK (2010) Effect of flow field for colorless distributed combustion (CDC) for gas turbine combustion. J Appl Energy 78:1631–1640
63. Khalil AEE, Gupta AK (2011) Swirling distributed combustion for clean energy conversion in gas turbine applications. J Appl Energy 88:3685–3693
64. Arghode VK, Gupta AK, Bryden KM (2012) High intensity colorless distributed combustion for ultra low emissions and enhanced performance. J Appl Energy 92:822–830
65. Khalil AEE, Gupta AK (2011) Distributed swirl combustion for gas turbine application. J Appl Energy 88:4898–4907
66. Arghode VK, Gupta AK (2012) Fuel dilution and liquid fuel operational effects on ultra-high thermal intensity distributed combustor. J Appl Energy 95:132–138
67. Tsuji H, Gupta AK, Hasegawa T, Katsuki M, Kishimoto K, Morita M (2003) High temperature air combustion: from energy conservation to pollution reduction. CRC Press
68. Al-Halbouni A, Giese A, Tali E, Leicher J, Albus R, Görner K (2017) Combustor concept for industrial gas turbines with single digit NO_x and CO emission values. Energy Procedia 120:134–139
69. Fenimore CP (1971) In: 13th symposium (international) on combustion. The Combustion Institute, p 373
70. Richards GA, Straub DL, Robey EH (2003) Passive control of combustion dynamics in stationary gas turbines. J Propul Power 195:795–810
71. Mongia HC, Held TJ, Hsiao GC, Pandalai RP (2003) Challenges and progress in controlling dynamics in gas turbine combustors. J Propul Power 195:822–829
72. Muruganandam TM, Nair S, Scarborough D, Neumeier Y, Jagoda J, Lieuwen T, Seitzman J, Zinn B (2005) Active control of lean blowout for turbine engine combustors. J Propul Power 215:807–814
73. Cheng RK (1995) Velocity and scalar characteristics of premixed turbulent flames stabilized by weak swirl. Combust Flame 101:1–14
74. Shepherd IG, Cheng RK (2001) Measurements of the turbulent burning velocity and the structure of premixed flames on a low-swirl burner. Combust Flame 1273:2066–2075
75. Cheng RK, Shepherd IG, Bedat B, Talbot L (2002) Premixed turbulent flame structures in moderate and intense isotropic turbulence. Combust Sci Technol 1741:29–59
76. Cheng RK, Yegian DT, Miyasato MM, Samuelsen GS, Pellizzari R, Loftus P, Benson C (2000) Scaling and development of low-swirl burners for low-emission furnaces and boilers. Proc Combust Inst 28:1305–1313
77. Littlejohn D, Majeski MJ, Tonse S, Castaldini C, Cheng RK (2002) Laboratory investigation of an ultralow NO_x premixed combustion concept for industrial boilers. Proc Combust Inst 29:1115–1121
78. Johnson MR, Littlejohn D, Nazeer WA, Smith KO, Cheng RK (2005) A comparison of the flowfields and emissions of high-swirl injectors and low-swirl injectors for lean premixed gas turbines. Proc Combust Inst 30:2867–2874
79. Nazeer WA, Smith KO, Sheppard P, Cheng RK, Littlejohn D (2006) Full Scale testing of a low swirl fuel injector concept for ultra-low NO_x gas turbine combustion systems. In: ASME turbo expo, Paper No. GT2006-90150
80. Littlejohn D, Cheng RK (2006) Fuel effects on a low-swirl injector for lean premixed gas turbines. Proc Combust Inst 31:3155–3162
81. Cheng RK, Littlejohn D, Nazeer WA, Smith K (2008) Laboratory studies of the flow field characteristics of low-swirl injectors for adaptation to fuel-flexible turbines. J Eng Gas Turb Power 130:021501-1
82. Beér JM, Chigier NA (1972) Combustion aerodynamics. New York
83. Peters N (2000) Turbulent combustion, 1st edn. Cambridge University Press
84. Glassman I, Yetter RA (2008) Combustion, 4th edn. Elsevier Inc
85. Kuznetsov RV, Sabel'nikov VA, Libby PA (1990) Turbulence and combustion, 1st edn. Hemisphere Publishing Corporation

86. Chowdhury BR, Cetegen BM (2017) Experimental study of the effects of free stream turbulence on characteristics and flame structure of bluff-body stabilized conical lean premixed flames. Combust Flame 178:311–328
87. Koutmos P, McGuirk JJ (1989) Isothermal flow in a gas turbine combustor—a benchmark experimental study. Exp Fluids 7:344–354
88. Clavin P (1985) Dynamic behavior of premixed flame fronts in laminar and turbulent flows. Prog Energy Combust Sci 11:1–59
89. Driscoll JF (2008) Turbulent premixed combustion: flamelet structure and its effect on turbulent burning velocities. Prog Energy Combust Sci 34:91–134
90. Lipatnikov AN, Chomiak J (2002) Turbulent flame speed and thickness: phenomenology, evaluation, and application in multi-dimensional simulations. Prog Energy Combust Sci 28:1–74
91. Kheirkhah S, Gülder ÖL (2013) Turbulent premixed combustion in V-shaped flames: characteristics of flame front. Phys Fluids 25(055107):1–23
92. Kheirkhah S, Gülder ÖL (2014) Topology and brush thickness of turbulent premixed V-shaped flames. Flow Turbul Combust 93:439–459
93. Namazian M, Shepherd IG, Talbot L (1986) Characterization of the density fluctuations in turbulent V-shaped premixed flames. Flame 64:229–308
94. Shepherd IG (1996) Flame surface density and burning rate in premixed turbulent flames. In: Symposium (international) on combustion, vol 26. Elsevier, pp 373–379
95. Sattler SS, Knaus DA, Gouldin FC (2002) Determination of three-dimensional flamelet orientation distributions in turbulent V-flames from two-dimensional image data. Proc Combust Inst 29:1785–1792
96. Robin V, Mura A, Champion M, Degardin O, Renou B, Boukhalfa M (2008) Experimental and numerical analysis of stratified turbulent V-shaped flames. Combust Flame 153:288–315
97. Paul PH, Najm HN (1998) Planar laser-induced fluorescence imaging of flame heat release rate. In: Symposium (international) on combustion, vol 27. pp 43–50
98. Böckle S, Kazenwadel J, Shin DI, Schulz C, Wolfrum J (2000) Simultaneous single-shot laser-based imaging of formaldehyde, OH, and temperature in turbulent flames. Proc Combust Inst 28:279–286
99. Kariuki J, Dowlut A, Balachandran R, Mastorakos E (2016) Heat release imaging in turbulent premixed ethylene–air flames near blow-off. Flow Turbul Combust 96:1039–1051
100. Kariuki J, Dowlut A, Yuan R, Balachandran R, Mastorakos E (2015) Heat release imaging in turbulent premixed methane–air flames close to blow-off. Proc Combust Inst 35:1443–1450
101. Skiba AW, Wabel TM, Temme J, Driscoll JF (2016) Experimental assessment of premixed flames subjected to extreme turbulence. In: 54th AIAA aerospace sciences meeting, p 1454
102. Zhou B, Brackman C, Li Z, Alden M, Bai X (2015) Simultaneous multi-species and temperature visualization of premixed flames in the distributed reaction zone regime. Proc Combust Inst 35:1409–1416
103. Zhou B, Brackmann C, Li Q, Wang Z, Petersson P, Li Z, Aldén M, Bai XS (2015) Distributed reactions in highly turbulent premixed methane/air flames: part I. Flame structure characterization. Combust Flame 162:2937–2953
104. Cavaliere A, De Joannon M (2004) Mild combustion. Prog Energy Combust Sci 30:329–366
105. Masri AR, Barlow RS, Fiechtner GJ, Fletcher DF (1998) Instantaneous and mean compositional structure of bluff-body stabilized nonpremixed flames. Combust Flame 114:119–148
106. Masri AR, Kalt PAM, Barlow RS (2004) The compositional structure of swirl-stabilised turbulent nonpremixed flames. Combust Flame 137:1–37
107. Olivani A, Solero G, Cozzi F, Coghe A (2007) Near field flow structure of isothermal swirling flows and reacting non-premixed swirling flames. Exp Thermal Fluid Sci 31:427–436
108. Vanoverberghe K (2004) Flow, turbulence and combustion of premixed swirling jet flames. PhD thesis, University of Leuven

109. Meier W, Duan XR, Weigand P (2005) Reaction zone structures and mixing characteristics of partially premixed swirling CH_4/air flames in a gas turbine model combustor. Proc Combust Inst 30:835–842
110. El-Asrag H, Menon S (2007) Large eddy simulation of bluff-body stabilized swirling non-premixed flames. Proc Combust Inst 31:1747–1754
111. Sengissen AX, Van Kampen JF, Huls RA, Stoffels GGM, Kok JBW, Poinsot TJ (2007) LES and experimental studies of cold and reacting flow in a swirled partially premixed burner with and without fuel modulation. Combust Flame 150:40–53
112. Sadiki A, Maltsev A, Wegner B, Flemming F, Kempf A, Janicka J (2006) Unsteady methods (URANS and LES) for simulation of combustion systems. Int J Therm Sci 45:760–773
113. Derek Dunn-Rankin D (ed) (2008) Lean combustion technology and control. Academic Press, Elsevier
114. Gore JP, Zhan NJ (1996) NO_x emission and major species concentrations in partially premixed laminar methane/air co-flow jet flames. Combust Flame 105:414–427
115. Yadav NP, Kushari A (2009) Visualization of recirculation in a low aspect ratio dump combustor. J Flow Vis Image Process 16:127–136
116. Hong S, Shanbhogue SJ, Ghoniem AF (2015) Impact of fuel composition on the recirculation zone structure and its role in lean premixed flame anchoring. Proc Combust Inst 35:1493–1500
117. Li G, Gutmark EJ (2005) Effect of exhaust nozzle geometry on combustor flow field and combustion characteristics. Proc Combust Inst 30:2893–2901
118. Altay HM, Speth RL, Hudgins DE, Ghoniem AF (2009) Flame–vortex interaction driven combustion dynamics in a backward-facing step combustor. Combust Flame 156:1111–1125
119. Speth RL, Ghoniem AF (2009) Using a strained flame model to collapse dynamic mode data in a swirl-stabilized syngas combustor. Proc Combust Inst 32:2993–3000
120. Speth RL, Hong S, Shanbhogue SJ, Ghoniem AF (2011) Mode selection in flame-vortex driven combustion instabilities. In: 49th AIAA aerospace sciences meeting, p 236
121. Speth RL (2010) Fundamental studies in hydrogen-rich combustion: instability mechanisms and dynamic mode selection. PhD thesis, Massachusetts Institute of Technology, USA
122. Hsu KY, Goss LP, Roquemore WM (1998) Characteristics of a trapped-vortex combustor. J Propul Power 14:57–65
123. Patrignani L, Losurdo M, Bruno C (2010) Numerical studies of the integration of a trapped vortex combustor into traditional combustion chambers. Int Flame Res Found Combust J 4:201003
124. Kim W, Im ZS, Do ZH, Mungal MG (2010) Flame liftoff height dependence on geometrically modified bluffbodies in a vitiated flow. Exp Fluids 49:27–41
125. Nogenmyr KJ, Kiefer J, Li ZS, Bai XS, Alden M (2010) Numerical computations and optical diagnostics of unsteady partially premixed methane/air flames. Combust Flame 157:915–924
126. Schneider E, Maltsev A, Sadiki A, Janicka J (2008) Study on the potential of BML-approach and G-equation concept-based models for predicting swirling partially premixed combustion systems: URANS computations. Combust Flame 152:548–572
127. Asgari B, Amani E (2017) A multi-objective CFD optimization of liquid fuel spray injection in dry-low-emission gas-turbine combustors. Appl Energy 203:696–710
128. Dunn-Rankin D (ed) (2011) Lean combustion: technology and control. Academic Press
129. Boyce MP (2012) Gas turbine engineering handbook, 4th edn. Butterworth-Heinemann, Oxford, UK
130. Tang Q, Liu H, Li M, Yao M (2017) Optical study of spray-wall impingement impact on early-injection gasoline partially premixed combustion at low engine load. Appl Energy 185:708–719
131. Asai T, Dodo S, Karishuku M, Yagi N, Akiyama Y, Hayashi A (2015) Performance of multiple-injection dry low-NO_x combustors on hydrogen-rich syngas fuel in an IGCC pilot plant. J Eng Gas Turbines Power 137:091504

132. Ebi D, Clemens NT (2016) Experimental investigation of upstream flame propagation during boundary layer flashback of swirl flames. Combust Flame 168:39–52
133. Nord LO, Andersen HG (2004) A study of parameters affecting the combustion stability and emissions behavior of ALSTOM heavy-duty gas turbines. ASME Turbo Expo 1:93–100
134. Vandervort CL (2000) Dry, 9 ppm NO_x/CO combustion system for "F" class industrial gas turbines. ASME Turbo Expo 2:1–7
135. ElKady AM, Evulet A, Brand A, Ursin TP, Lynghjem A (2008) Exhaust gas recirculation in DLN F-class gas turbines for post-combustion CO_2 capture. ASME Turbo Expo 3:847–854
136. Ayed AH, Kusterer K, Funke HW, Keinz J, Kazari M, Kitajima J, Horikawa A, Okada K, Bohn D (2014) Numerical study on increased energy density for the DLN micromix hydrogen combustion principle. ASME Turbo Expo 4A:1–12
137. Lacy B, Ziminsky W, Lipinski J, Varatharajan B, Yilmaz E, Brumberg J (2008) Low emissions combustion system development for the ge energy high hydrogen turbine program. ASME Turbo Expo 3:617–624
138. Pfefferle LD, Pfefferle WC (1987) Catalysis in combustion. Catal Rev Sci Eng 29:219–267
139. Hayes RE, Kolaczkowski ST (1997) Introduction to catalytic combustion. Gordon and Breach Science Publishers, Reading, UK
140. Dalla Betta RA (1997) Catalytic combustion gas turbine systems: the preferred technology for low emissions electric power production and co-generation. Catal Today 35:129–135
141. Forzatti P, Groppi G (1999) Catalytic combustion for the production of energy. Catal Today 54:165–180
142. Forzatti P (2003) Status and perspectives of catalytic combustion for gas turbines. Catal Today 83:3–18
143. Ducruix S, Schuller T, Durox D, Candel S (2003) Combustion dynamics and instabilities: elementary coupling and driving mechanisms. J Propul Power 19:722–734
144. Altay HM, Kedia KS, Speth RL, Ghoniem AF (2010) Two-dimensional simulations of steady perforated-plate stabilized premixed flames. Combust Theor Model 14:125–154
145. Mallens R, De Goey L (1998) Flash-back of laminar premixed methane/air flames on slitand tube burners. Combust Sci Technol 136:41–54
146. Lammers F, De Goey L (2003) A numerical study of flash back of laminar premixed flames in ceramic-foam surface burners. Combust Flame 133:47–61
147. Kedia KS, Ghoniem AF (2012) Mechanisms of stabilization and blowoff of a premixed flame downstream of a heat-conducting perforated plate. Combust Flame 159:1055–1069
148. Timmermans J, Vanierschot M, Van den Bulck E (2011) Flow instabilities in the near wake of a perforated plate. In: 9th national congress on theoretical and applied mechanics, Brussels
149. Rodrigues JM, Fernandes EC (2014) Stability analysis and flow characterization of multi-perforated plate premixed burners. In: The 17th international symposium on applications of laser techniques to fluid mechanics, Lisbon, Portugal
150. Veetil JE, Rajith C, Velamati RK (2016) Numerical simulations of steady perforated-plate stabilized syngas air pre-mixed flames. Int J Hydrogen Energy 41:13747–13757
151. Jithin E, Kishore VR, Varghese RJ (2014) Three-dimensional simulations of steady perforated-plate stabilized propane–air premixed flames. Energy Fuels 28:5415–5425
152. Hindasageri V, Kuntikana P, Vedula RP, Prabhu SV (2015) An experimental and numerical investigation of heat transfer distribution of perforated plate burner flames impinging on a flat plate. Int J Therm Sci 94:156–169
153. Kuntikana P, Prabhu S (2017) Thermal investigations on methane-air premixed flame jets of multi-port burners. Energy 123:218–228
154. Lee PH, Hwang SS (2013) Formation of lean premixed surface flame using porous baffle plate and flame holder. J Therm Sci Technol 8:178–189
155. Moghaddam MHS, Moghaddam MS, Khorramdel M (2017) Numerical study of geometric parameters effecting temperature and thermal efficiency in a premix multi-hole flat flame burner. Energy 125:654–662

156. DÃ¶bbeling K, Hellat J, Koch H (2007) 25 years of BBC/ABB/Alstom lean premix combustion technologies. J Eng Gas Turbines Power 129:2–12
157. Sattelmayer T, Felchlin MP, Haumann J, Hellat J, Styner D (1992) Second-generation low-emission combustors for ABB gas turbines: burner development and tests at atmospheric pressure. ASME J Eng Gas Turbines Power 114:118–125
158. Sattelmayer T, Polifke W, Winkler D, Döbbeling K (1996) NO_x-abatement potential of lean-premixed GT-combustors. ASME J Eng Gas Turbines Power 120:48–59
159. Güthe F, Hellat J, Flohr P (2008) The reheat concept: the proven pathway to ultralow emissions and high efficiency and flexibility. J Eng Gas Turbines Power 131:021503
160. Peter F, Martin Z, Rudolf L, Stefano B, Christian M (2007) Development and design of Alstom's staged fuel gas injection EV burner for NO_x reduction. In: ASME turbo expo, GT2007–27730
161. Mallampalli HP, Fletcher TH, Chen JY (1998) Evaluation of CH_4/NO_x reduced mechanisms used for modeling lean premixed turbulent combustion of natural gas. J Eng Gas Turbines Power 120:703–712
162. Huang Y, Yang V (2009) Dynamics and stability of lean-premixed swirl-stabilized combustion. Prog Energy Combust Sci 35:293–364
163. Sohn CH, Cho HC (2005) A CFD study on thermo-acoustic instability of methane/air flames in gas turbine combustor. J Mech Sci Technol 19:1811–1812
164. Lefebvre AH (1998) Gas turbine combustion, 2nd edn. Taylor & Francis, pp 350–352
165. Cho CH, Baek GM, Sohn CH, Cho JH, Kim HS (2013) A numerical approach to reduction of NO_x emission from swirl premix burner in a gas turbine combustor. Appl Therm Eng 59:454–463
166. Biagioli F, Güthe F (2007) Effect of pressure and fuel–air unmixedness on NO_x emissions from industrial gas turbine burners. Combust Flame 151:274–288
167. Biagioli F (2006) Stabilization mechanism of turbulent premixed flames in strongly swirled flows. Combust Theor Model 10:389–412
168. Biagioli F, Güthe F, Schuermans B (2008) Combustion dynamics linked to flame behaviour in a partially premixed swirled industrial burner. Exp Thermal Fluid Sci 32:1344–1353
169. York WD, Ziminsky WS, Yilmaz E (2013) Development and testing of a low NO_x hydrogen combustion system for heavy-duty gas turbines. J Eng Gas Turbines Power 135:022001
170. Funke HHW, Börner S, Keinz J, Kusterer K, Kroniger D, Kitajima J, Kazari M, Horikawa A (2012) Numerical and experimental characterization of low NO_x micromix combustion principle for industrial hydrogen gas turbine applications. ASME Turbo Expo 2:1069–1079
171. Dodo S, Asai T, Koizumi H, Takahashi H, Yoshida S, Inoue H (2011) Combustion characteristics of a multiple-injection combustor for dry low-NO_x combustion of hydrogen-rich fuels under medium pressure. Proc ASME Turbo Expo 2:467–476
172. De Robbio R (2017) Innovative combustion analysis of a micro-gas turbine burner supplied with hydrogen-natural gas mixtures. Energy Procedia 126:858–866
173. Cheng R (2008) Adaption of the low-swirl burner technology for syngas and H_2 gas turbines. In: Proceedings of the future of gas turbine technology: fourth international conference, Paper No. IGTC08_P45, Brussels, Belgium
174. Lieuwen T, McDonnell V, Santavicca D, Sattelmayer T (2008) Burner development and operability issues associated with steady flowing syngas fired combustors. Combust Sci Technol 180:1169–1192
175. Brunetti I, Riccio G, Rossi N, Cappelletti A, Bonelli L, Marini A, Paganini E, Martelli F (2011) Experimental and numerical characterization of lean hydrogen combustion in a premix burner prototype. In: Proceedings of ASME turbo expo, Paper No. GT2011-45623, Vancouver, BC, Canada
176. Weiland N, Sidwell T, Strakey P (2011) Testing of a hydrogen dilute diffusion array injector at gas turbine conditions. In: Proceedings of ASME turbo expo, Paper No. GT2011-46596, Vancouver, BC, Canada

177. Marek C, Smith T, Kundu K (2005) Low emission hydrogen combustors for gas turbines using lean direct injection. In: Proceedings of the 41st joint propulsion conference, Tucson, AZ, Paper No. AIAA-2005-3776
178. Funke H, Boerner S, Krebs W, Wolf E (2011) Experimental characterization of low NO_x micromix prototype combustors for industrial gas turbine applications. In: Proceedings of ASME turbo expo, Paper No. GT2011-45305, Vancouver BC, Canada
179. Asai T, Dodo S, Koizumi H, Takahashi H, Yoshida S, Inoue H (2011) Effects of multiple-injection-burner configurations on combustion characteristics for dry low-NO_x combustion of hydrogen-rich fuels. In: Proceedings of ASME turbo expo, Paper No. GT2011-45295, Vancouver BC, Canada
180. Lee H, Hernandez S, McDonnell V, Steinthorsson E, Mansour A, Hollon B (2009) Development of flashback resistant low-emission micro-mixing fuel injector for 100% hydrogen and syngas fuels. In: Proceedings of ASME turbo expo, Paper No. GT2009-59502, Orlando, FL
181. Funke HHW, Recker E, Börner S, Bosschaerts W (2009) LES of jets in cross-flow and application to the micromix hydrogen combustion. In: Proceedings of the 19th international symposium on air breathing engine, ISABE, 1309, Montreal, Canada
182. Ayed AH, Kusterer K, Funke HHW, Keinz J, Bohn D (2017) CFD based exploration of the dry-low-NO_x hydrogen micromix combustion technology at increased energy densities. Propul Power Res 6:15–24
183. Ayed AH, Kusterer K, Funke HHW, Keinz J, Striegan C, Bohn D (2015) Experimental and numerical investigations of the dry-low-NO_x hydrogen micromix combustion chamber of an industrial gas turbine. Propul Power Res 4:123–131
184. Ayed AH, Kusterer K, Funke HHW, Striegan C, Bohn D (2015) Improvement study for the dry-low-NO_x hydrogen micromix combustion technology. Propul Power Res 4:132–140
185. Funke HHW, Keinz J, Kusterer K, Ayed AH, Kazari M, Kitajima J, Horikawa A, Okada K (2015) Experimental and numerical study on optimizing the DLN micromix hydrogen combustion principle for industrial gas turbine applications. In: ASME turbo expo, GT2015-42043, Montréal, Canada
186. Taamallah S, Shanbhogue SJ, Ghoniem AF (2016) Turbulent flame stabilization modes in premixed swirl combustion: physical mechanism and Karlovitz number-based criterion. Combust Flame 166:19–33
187. Lipatnikov AN (2017) Stratified turbulent flames: recent advances in understanding the influence of mixture inhomogeneities on premixed combustion and modeling challenges. Prog Energy Combust Sci 62:87–132
188. Ditaranto M, Anantharaman R, Weydahl T (2013) Performance and NO_x emissions of refinery fired heaters retrofitted to hydrogen combustion. Energy Procedia 37:7214–7220
189. Bouvet N, Halter F, Chauveau C, Yoon Y (2013) On the effective Lewis number formulations for lean hydrogen/hydrocarbon/air mixtures. Int J Hydrogen Energy 38:5949–5960
190. Lieuwen T, McDonell V, Petersen E, Santavicca D (2008) Fuel flexibility influences on premixed combustor blowout, flashback, autoignition, and stability. J Eng Gas Turbines Power 130:11506
191. Taamallah S, Vogiatzaki K, Alzahrani FM, Mokheimer EMA, Habib MA, Ghoniem AF (2015) Fuel flexibility, stability and emissions in premixed hydrogen-rich gas turbine combustion: technology, fundamentals, and numerical simulations. Appl Energy 154:1020–1047
192. Glassman I (2008) Combustion, 4th edn. https://doi.org/10.1017/cbo9781107415324.004
193. Sankaran R, Im HG (2006) Effects of hydrogen addition on the Markstein length and flammability limit of stretched methane/air premixed flames. Combust Sci Technol 178:1585–1611
194. Shanbhogue SJ, Sanusi YS, Taamallah S, Habibb MA, Mokheimer EMA, Ghoniem AF (2016) Flame macrostructures, combustion instability and extinction strain scaling in swirl-stabilized premixed CH_4/H_2 combustion. Combust Flame 163:494–507

195. Hawkes ER, Chen JH (2004) Direct numerical simulation of hydrogen-enriched lean premixed methane–air flames. Combust Flame 138:242–258
196. Choundhuri AR, Gollahalli SR (2000) Combustion characteristics of hydrogen–hydrocarbon hybrid fuels. Hydrogen Energy 25:451–462
197. Choundhuri AR, Gollahalli SR (2003) Characteristics of hydrogen–hydrocarbon composite fuel turbulent jet flame. Hydrogen Energy 28:445–454
198. Kim S, Arghode K, Gupta K (2009) Flame characteristics of hydrogen-enriched methane–air premixed swirling flames. Int J Hydrogen Energy 34:1063–1073
199. Kim S, Arghode K, Gupta K (2009) Combustion characteristics of a lean premixed LPG–air combustor. Int J Hydrogen Energy 34:1045–1053
200. Aladawy S, Lee J, Abdelnabi B (2017) Effect of turbulence on NO_x emission in a lean perfectly-premixed combustor. Fuel 208:160–167
201. Griebel P, Boschek E, Jansohn P (2007) Lean blowout limits and NO_x emissions of turbulent, lean premixed, hydrogen-enriched methane/air flames at high pressure. J Eng Gas Turbines Power 129:404–410
202. Schefer RW, Wicksall DM, Agrawal AK (2002) Combustion of hydrogen-enriched methane in a lean premixed swirl-stabilized burner. Proc Combust Inst 29:843–851
203. Rashwan SS, Ibrahim AH, Abou-Arab TW, Nemitallah MA, Habib MA (2016) Experimental investigation of partially premixed methane–air and methane–oxygen flames stabilized over a perforated-plate burner. Appl Energy 169:126–137
204. Rashwan SS, Ibrahim AH, Abou-Arab TW, Nemitallah MA, Habib MA (2017) Experimental study of atmospheric partially premixed oxy-combustion flames anchored over a perforated plate burner. Energy 122:159–167
205. Nemitallah MA, Habib MA (2013) Experimental and numerical investigations of an atmospheric diffusion oxy-combustion flame in a gas turbine model combustor. Appl Energy 111:401–415
206. Nemitallah MA, Habib MA, Badr HM (2017) Design of a multi-can carbon-free gas turbine combustor utilizing multiple shell-and-tube OTRs for ZEPP applications. J Nat Gas Sci Eng 46:172–187
207. Guo J, Hu F, Jiang X, Huang X, Li P, Liu Z, Zheng C (2017) Experimental and numerical investigations on heat transfer characteristics of a 35 MW oxy-fuel combustion boiler. Energy Procedia 114:481–489
208. Li B, Shi B, Zhao X, Ma K, Xie D, Zhao D, Li J (2018) Oxy-fuel combustion of methane in a swirl tubular flame burner under various oxygen contents: operation limits and combustion instability. Exp Thermal Fluid Sci 90:115–124
209. Habib MA, Salaudeen SA, Nemitallah MA, Ben-Mansour R, Mokheimer EMA (2016) Numerical investigation of syngas oxy-combustion inside a LSCF-6428 oxygen transport membrane reactor. Energy 96:654–665
210. Scheffknecht G, Al-Makhadmeh L, Schnell U, Maier J (2011) Oxy-fuel coal combustion—a review of the current state-of-the-art. Int J Greenh Gas Control 5:16–35
211. Khalil AEE, Gupta AK (2017) The role of CO_2 on oxy-colorless distributed combustion. Appl Energy 188:466–474
212. Álvarez JFG, De Grado JG (2016) Study of a modern industrial low pressure turbine for electricity production employed in oxy-combustion cycles with CO_2 capture purposes. Energy 107:734–747
213. Wang Z, Liu M, Cheng X, He Y, Hu Y, Ma C (2017) Experimental study on oxy-fuel combustion of heavy oil. Int J Hydrogen Energy 42:20306–20315
214. Taamallah S, Chakroun NW, Watanabe H, Shanbhogue SJ, Ghoniem AF (2017) On the characteristic flow and flame times for scaling oxy and air flame stabilization modes in premixed swirl combustion. Proc Combust Inst 36:3799–3807
215. Jimenez S, Gonzalo-Tirado C (2017) Properties and relevance of the volatile flame of an isolated coal particle in conventional and oxy-fuel combustion conditions. Combust Flame 176:94–103

216. Kedia KS, Ghoniem AF (2015) The blow-off mechanism of a bluff-body stabilized laminar premixed flame. Combust Flame 162:1304–1315
217. Tuttle SG, Chaudhuri S, Kostka S, Kopp-Vaughan KM, Jensen TR, Cetegen BM, Renfro MW (2012) Time-resolved blowoff transition measurements for two-dimensional bluff body-stabilized flames in vitiated flow. Combust Flame 159:291–305
218. Plee SL, Mellor AM (1979) Characteristic time correlation for lean blowoff of bluff-body-stabilized flames. Combust Flame 35:61–80
219. Dawson JR, Gordon RL, Kariuki J, Mastorakos E, Masri AR, Juddoo M (2011) Visualization of blow-off events in bluff-body stabilized turbulent premixed flames. Proc Combust Inst 33:1559–1566
220. Granados DA, Chejne F, Mejía JM (2015) Oxy-fuel combustion as an alternative for increasing lime production in rotary kilns. Appl Energy 158:107–117
221. Aneke M, Wang M (2015) Process analysis of pressurized oxy-coal power cycle for carbon capture application integrated with liquid air power generation and binary cycle engines. Appl Energy 154:556–566
222. Fu Q, Kansha Y, Song C, Liu Y, Ishizuka M, Tsutsumi A (2014) A cryogenic air separation process based on self-heat recuperation for oxy-combustion plants. Appl Energy 162:1114–1121
223. Rashwan SS, Ibrahim AH, Abou-Arab TW (2015) Experimental investigation of oxy-fuel combustion of CNG flames stabilized over a perforated-plate burner. In: 18th international flame research foundation, Friesing, Munich, pp 1–11
224. Adamczyk WP, Bialecki RA, Ditaranto M, Gladysz P, Haugen NEL, Katelbach-Wozniak A, Klimanek A, Sladek S, Szlek A, Wecel G (2017) CFD modeling and thermodynamic analysis of a concept of a MILD-OXY combustion large scale pulverized coal boiler. Energy 140:1305–1315
225. Chen L, Yong SZ, Ghoniem AF (2012) Oxy-fuel combustion of pulverized coal: characterization, fundamentals, stabilization and CFD modeling. Prog Energy Combust Sci 38:156–214
226. Li P, Mi J, Dally B, Wang F, Wang L, Liu Z et al (2011) Progress and recent trend in MILD combustion. Sci China Technol Sci 54:255–269
227. Stadler H, Ristic D, Foerster M, Schuster A, Kneer R, Scheffknecht G (2009) NO_x-emissions from flameless coal combustion in air, Ar/O_2 and CO_2/O_2. Proc Combust Inst 32:3131–3138
228. Perrone D, Amelio M (2016) Numerical investigation of oxy-mild combustion of pulverized coal in a pilot furnace. Energy Procedia 101:1191–1198
229. Ditaranto M, Hals J (2006) Combustion instabilities in sudden expansion oxy–fuel flames. Combust Flame 146:493–512
230. Amato A, Hudak B, D'Souza P, D'Carlo P, Noble D, Scarborough D, Seitzman J, Lieuwen T (2011) Measurements and analysis of CO and O_2 emissions in $CH_4/CO_2/O_2$ flames. Proc Combust Inst 33:3399–3405
231. Ramadan IA, Ibrahim AH, Abou-Arab TW, Rashwan SS, Nemitallah MA, Habib MA (2016) Effects of oxidizer flexibility and bluff-body blockage ratio on flammability limits of diffusion flames. Appl Energy 178:19–28
232. Jerzak W, Kuźnia M (2016) Experimental study of impact of swirl number as well as oxygen and carbon dioxide content in natural gas combustion air on flame flashback and blow-off. J Nat Gas Sci Eng 29:46–54
233. Shi B, Zhua Z, Wanga N, Lub P, Ishizukac S (2015) An experimental study on oxy-fuel combustion of methane under various oxygen mole fractions. In: 8th international symposium on coal combustion, Beijing, China
234. Huang Y, Yang V (2005) Effect of swirl on combustion dynamics in a lean-premixed swirl-stabilized combustor. Proc Combust Inst 30:1775–1782
235. Tunçer O, Kaynaroglu B, Karakaya MC, Kahraman S, Çetiner-Yildirim O, Baytas C (2014) Preliminary investigation of a swirl stabilized premixed combustor. Fuel 115:870–874
236. Strakey P, Sidwell T, Ontko J (2007) Investigation of the effects of hydrogen addition on lean extinction in a swirl stabilized combustor. Proc Combust Inst 31:3173–3180

237. Davis LB, Black SH (2000) GE power systems, GER-3568G
238. Evulet AT, ELKady AM, Branda AR, Chinn D (2009) On the performance and operability of GE's dry low NO_x combustors utilizing exhaust gas recirculation for postcombustion carbon capture. Energy Procedia 1:3809–3816
239. Trimm D (1983) Catalytic combustion. Appl Catal 7:249–282
240. Mantzaras J (2008) Catalytic combustion of syngas. Combust Sci Technol 180:1137–1168
241. Li WB, Wang JX, Gong H (2010) Catalytic combustion of VOCs on non-noble metal catalysts. Catal Today 148:81–87
242. Noiray N, Durox D, Schuller T, Candel S (2007) Passive control of combustion instabilities involving premixed flames anchored on perforated plates. Proc Combust Inst 31:1283–1290
243. Noiray N, Durox D, Schuller T, Candel S (2009) Dynamic phase converter for passive control of combustion instabilities. Proc Combust Inst 32:3163–3170
244. Altay HM, Speth RL, Hudgins DE, Ghoniem AF (2009) The impact of equivalence ratio oscillations on combustion dynamics in a backward-facing step combustor. Combust Flame 156:2106–2116
245. Wei H, Gao D, Zhou L, Feng D, Chen R (2017) Different combustion modes caused by flame-shock interactions in a confined chamber with a perforated plate. Combust Flame 178:277–285
246. Wood S, Harris AT (2008) Porous burners for lean-burn applications. Prog Energy Combust Sci 34:667–684
247. Rørtveit GJ, Zepter K, Skreiberg Q, Fossum M, Hustad JE (2002) A comparison of low-NO_x burners for combustion of methane and hydrogen mixtures. Proc Combust Inst 29:1123–1129
248. Bottausci F, Cardonne C, Meinhart C, Mezić I (2007) An ultrashort mixing length micromixer: the shear superposition micromixer. Lab Chip 7:396–398
249. Bhagat AS, Peterson EK, Papautsky I (2007) A passive planar micromixer with obstructions for mixing at low Reynolds numbers. J Micromech Microeng 17:1017–1024
250. Tofteberg T, Skolimowski M, Andreassen E, Geschke O (2010) A novel passive micromixer: lamination in a planar channel system. Microfluid Nanofluidics 8:209–215
251. Zeldovich YA (1947) Oxidation of nitrogen in combustion. Academy of Sciences of USSR, Institute of Chemical Physics, Moscow-Leningrad
252. Liss B, Wilson BR, Wilson BW (2017) Mobil tensor holdings LLC and mobile tensor Ip holdings LLC. U.S. patent 20170218284A1
253. Barnes FJ, Bromly JH, Edwards TJ, Madngezewsky R (1988) NO_x emissions from radiant gas burners. J Inst Energy 155:184–188
254. Lieuwen TC, Yang V (2013) Gas turbine emissions. Cambridge Aerospace Series, University Press, Cambridge
255. Schefer RW, Namazian M, Kelly J (1991) In: Combustion research facility news, vol 3, number 4. Sandia
256. Glarborg P, Jensen AD, Johnsson JE (2003) Fuel nitrogen conversion in solid fuel fired systems. Prog Energy Combust Sci 29:89–113
257. Schnelle KB Jr, Dunn RF, Ternes ME (2015) Air pollution control technology handbook. CRC Press, pp 271–283
258. Booth RC, Kosvic TC, Parikh NJ (1997) Presented at EPRI-DOE-EPA combined utility air pollutant control symposium, Washington, DC, 25–29 Aug
259. Wood SC (1994) Select the right IMOx control technology. Chem Eng Progr 90:32
260. John Zink Company (1993) Burner design parameters for flue gas NO_x control. Technical Paper 4010B
261. Bortz SJ et al (1997) Ultra-low NO_x rapid mix burner demonstration at Con Edison's 59th Street Station. Presented at EPRI-DOE-EPA combined utility air pollutant control symposium, Washington, DC, 25–29 Aug
262. Chen W et al (1997) Deviation and application of a global SNCR model in maximizing NO_x reduction. Presented at EPRI-DOE-EPA combined utility air pollutant control symposium, Washington, DC, 25–29 Aug

263. Reyes BE, Cutshaw TR (1999) SCONOx™ catalytic absorption system. Western Energy, p 13
264. Ferrell R (2000) Controlling NO (x) emissions: a cooler alternative. Pollut Eng 32:50
265. Saravanamuttoo HIH, Rogers GFC, Cohen H (2001) Gas turbine theory. Pearson Education
266. Giampaolo T (2003) The gas turbine hand book: principles and practices, 2nd edn. Fairmont Press Inc, Lilburn
267. Lefebvre AH (1998) Gas turbine combustion. Taylor & Francis, London
268. Han JC, Dutta S, Ekkad S (2012) Gas turbine heat transfer and cooling technology. CRC Press
269. Schilke PW, Foster AD, Pepe JJ (2004) Advanced gas turbine materials and coatings. General Electric Company
270. Hicks B (1987) High-temperature sheet materials for gas turbine applications. Mater Sci Technol 3:772–781
271. Zonfrillo G, Giovannetti I, Manetti M (2008) Material selection for high temperature applications. Meccanica 43:125–131
272. Halford GR, Saltsman JF, Verrilli MJ, Arya VINODK (1992) In: Advances in fatigue lifetime predictive techniques. ASTM International
273. Kumar GM, Rose JBR (2015) Numerical comparative study on convective heat transfer coefficient in a combustor liner of gas turbine with coating. Int J Mech Eng Res. ISSN 0973-4562
274. Ragupathy R, Mishra RK, Misal RD (2011) Life analysis of TBC on an aero engine combustor based on in-service failures data. J Aerosp Sci Technol 63:158–164
275. Ragupathy R, Panigrahi SK, Mishra RK (2013) Effects of interface roughness on the life estimation of a thermal barrier coating layer. Int J Surf Sci Eng 7:269–284
276. Padture NP, Gell M, Jordan EH (2002) Thermal barrier coatings for gas-turbine engine applications. Science 296:280–284
277. Rajendran R (2012) Gas turbine coatings—an overview. Eng Fail Anal 26:355–369
278. Mishra RK (2017) Life enhancement of gas turbine combustor liner through thermal barrier coating. J Fail Anal Prev 17:914–918
279. Brewer D (1999) HSR/EPM combustor materials development program. Mater Sci Eng A 261:284–291
280. Liu Y, Sun X, Sethi V, Nalianda D, Li YG, Wang L (2017) Review of modern low emissions combustion technologies for aero gas turbine engines. Prog Aerosp Sci 94:12–45
281. Miriyala N, Fahme A, van Roode M (2001) In: ASME turbo expo 2001: power for land, sea, and air. American Society of Mechanical Engineers, New Orleans, Louisiana, June 2001, p V004T02A009
282. Alstom power, sequential combustion. http://www.power.alstom.com/home/equipment_systems/turbine/gas_turbines/GT24_andGT26__188MW_and_281MW.html
283. Matta RK, Mercer GD, Tuthill RS (2000) Power systems for the 21st century-"H" gas turbine combined-cycles. GE Power Systems, Schenectady, NY
284. Bland R, Ryan W, Abou-Jaoude K, Bandaru R, Harris A, Rising B (2004) Siemens W501F gas turbine: ultra low NO_x combustion system development. Siemens Westinghouse Power Corporation, POWER-GEN International

Chapter 5
Modeling of Combustion in Gas Turbines

5.1 Introduction

In order to develop high efficiency, low emission, and compact combustors for different applications (such as large-scale gas turbine systems, automotive and aero-engines) capable of stable operation over a wide range of operating conditions, fundamental mechanisms controlling combustion behavior in such systems must be elucidated. Understanding turbulent combustion mechanisms in such configurations has been challenging and remains an active area of research in the combustion community [1]. Numerical modeling and simulations as well as experimental techniques are being increasingly used, therefore, to predict the performance of such systems, for developing a fundamental understanding of combustion dynamics and flame stability. It is also important for the assessment of the combustion performance in terms of power density, efficiency, and emissions, which is critical to their large-scale implementation in practical applications. While detailed measurements of various gaseous species, temperature, and velocity profiles inside the combustor can be performed using advanced laser diagnostics with relative ease, repeated measurement of all these variables is an expensive task, especially when incorporating geometric changes to modify the combustor performance. Computational fluid dynamics (CFD) methods have great potential in predicting the performance of these combustors, and the approach is relatively easy to implement, flexible, and cost-effective as compared to a detailed experimental investigation.

Combustion in gas turbines can be classified to either non-premixed (diffusion) combustion or premixed combustion. Non-premixed flames have been used in gas turbines for power generation thanks to their strong stability behavior over wide ranges of loading conditions [2–4]. These flames are characterized by stoichiometric combustion zones within the combustor and, consequently, elevated temperature spots which tend to raise the level of NOx emissions [5]. Converting the combustion mode from non-premixed to premixed prevents the creation of stoichiometric combustion zones within the combustor as the reactants are premixed

M. A. Nemitallah et al., *Oxyfuel Combustion for Clean Energy Applications*, Green Energy and Technology, https://doi.org/10.1007/978-3-030-10588-4_5

upstream of the combustor. This results in reduction in combustion temperature, and accordingly, NOx emissions are reduced [6]. Lean premixed combustion is a promising approach for many industrial applications (such as large-scale gas turbine systems, automotive and aero-engines) primarily because of its benefits such as lower pollutant emissions and more efficient combustion when compared to non-premixed systems. A drawback of premixed combustion, however, is that the flame is more prone to thermo-acoustic instability and/or flashback. Combustion instability, therefore, has been one of the most critical phenomena encountered during the development of combustor systems. Specifically, flames at one limit, if not anchored properly, may blow off leading to what is referred to as lean flammability limit or "static" instability. Recirculating flow in the wake of a bluff body, behind sudden expansion or downstream a swirler, is often used to expand the stability range. However, as the fuel concentration is raised, high amplitude pressure and flow oscillations, i.e., dynamic instability (thermo-acoustic or combustion instability), are often observed. This can cause flame extinction, flame flashback, structural vibration, significant noise, or even structural damage as shown in Fig. 5.1. Such complicated combustion phenomena forced the researchers to develop CFD codes that are able to solve combustion within gas turbine systems seeking better understanding of the process. In this chapter, the mechanisms considered for solving combustion in different systems under non-premixed and premixed modes are presented in detail having the main focus made on modeling of premixed combustion.

Large eddy simulation (LES) with appropriate turbulent combustion models and reaction mechanisms is considered as one of the more promising CFD approaches, balancing computational complexity and predictive accuracy. While direct numerical simulation (DNS) resolves all the turbulent scales, it is computationally expensive and impractical for high Reynolds number large-scale applications. On the other hand, the Reynolds-averaged Navier–Stokes (RANS) equations model the

Fig. 5.1 Burner assembly: (left) damaged by combustion instability and (right) new burner assembly [1]

influence of turbulence on the mean flow and hence cannot capture the details of flow unsteadiness. In LES, rather than averaging the effect of turbulence, the equations are filtered. The larger turbulent eddies are explicitly resolved and computed (as with DNS), and the smallest ones are modeled (as with RANS modeling). This results in better predictions as compared to RANS technique because the effect of turbulence is represented more accurately since the large eddies (often dependent on the geometry) are explicitly computed. The approach is computationally manageable, as the smaller scales (having more universal features) are modeled. Further, the approach is especially applicable to combustion instability studies, since such flows typically exhibit large coherent structures.

Over the past decade, 3D numerical simulations using the LES framework have been performed for both laboratory-scale research combustors and industry test rigs, to study the dynamic behavior of gas turbine combustion [1]. The effects of swirl and equivalence ratio on flame dynamics in a swirl combustor have been investigated by Stone and Menon [7], using a flamelet model (G-equation) to capture the unsteady vortex–flame and acoustic–flame interactions, and it was observed that the fluctuating pressure amplitudes are attenuated significantly for large values of the swirl number. In another study [8], the role of large-scale vortices on the combustion oscillations in a coaxial combustor with and without swirl has been examined, and a passive method of using coaxial flows to decrease the influence of large-scale vortices on the flame has been proposed. Recent work in LES of flame dynamics has also involved the study of equivalence ratio fluctuations at the combustion chamber inlet and its effect on self-excited combustion instabilities [9] for a swirl combustor. LES and experimental results were compared in terms of mean and rms (root-mean-square) fields of temperature, species, velocities, and mixture fraction PDFs corresponding to stable and unstable regimes. In a related study [10], researchers considered a premixed swirl burner for both non-reacting and reacting cases and noted a strong processing vortex core in the non-reacting flows, which disappears when combustion occurs. In another study [11], the influence of inlet flow conditions on the combustion dynamics in a lean premixed swirl-stabilized combustor has been investigated. The investigation focused on the flame bifurcation phenomena and stability boundary as a function of burner operating conditions. Studies have also been performed for a backward step combustor and the response of the combustor to acoustic wave excitations and to equivalence ratio modulations, and the results were compared to experimental measurements [12]. Additionally, work has also been performed to evaluate the performance of finite rate chemistry LES combustion models relative to the flamelet-based approaches, and it has been observed that the former approach resulted in more accurate predictions when compared with experimental measurements for a swirl-stabilized premixed flame in a laboratory gas turbine combustor [13]. The studies above have generated a significant amount of information about the combustion dynamics and flow evolution in specific geometries of concern under well-defined operating conditions. However, substantial work is still required to be able to extract phenomenological information contributing to a deeper understanding and subsequent modeling of the processes of concern, which are going to be discussed in detail in this chapter.

This chapter also includes extensive discussions on LES modeling with a focus on gas turbine combustion applications. Detailed descriptions of the models solving radiation effects and comparison between such models are presented. Species transport models are also discussed in detail to solve species concentration within the combustor. Different modified combustion mechanisms to solve reaction kinetics under oxy-fuel combustion conditions are introduced. Two detailed numerical case studies for modeling combustion in different systems are presented. In the first case study, detailed numerical modeling and the obtained results are given considering H_2-enriched methane oxy-combustion in a model gas turbine combustor. The second numerical case study considers modeling of premixed combustion in a backward-facing step combustor.

5.2 General Conservation Equations

The mathematical modeling of combustion is based on the numerical solution of the conservation equations for mass, momentum, and energy, and transport equations for scalar variables. The equations, which are elliptic and three-dimensional, can be solved to provide predictions of the flow patterns as well as thermal and emission characteristics of reacting medium inside a combustion system. These conservation equations may be expressed in the following general form [14]:

$$\frac{\partial}{\partial x_j}\left(\bar{\rho}\overline{U}_j\Phi+\bar{\rho}\overline{u_j\phi}\right)=\frac{\partial}{\partial x_j}\left[\Gamma_\phi\frac{\partial\Phi}{\partial x_j}\right]+\bar{\rho}S_\Phi \tag{5.1}$$

where Φ and ϕ are the average and fluctuating values of the dependent variable, u_j is the velocity component along the coordinate direction x_j, $\bar{\rho}$ is the fluid density, Γ_Φ is the diffusion coefficient, and S_Φ is the source term. Equation (5.1) stands for the mass conservation equation when $\Phi = 1$; the momentum conservation equation when Φ is a velocity component; the energy equation when Φ is the stagnation enthalpy; or the transport equation of a scalar when Φ is a scalar variable such as mixture fraction.

The steady-state mass, momentum, energy, and species conservation equations for Newtonian fluids can also be expressed as follows:

$$\nabla\cdot(\rho U)=0 \tag{5.2}$$

$$\nabla\cdot(\rho UU)=-\nabla P+\mu\nabla^2 U \tag{5.3}$$

$$(\rho C_P)U \cdot \nabla T = \nabla \cdot (k\nabla T) \tag{5.4}$$

$$\nabla \cdot (\rho UY_i) - \nabla \cdot (\rho D_i \nabla Y_i) = 0 \tag{5.5}$$

where u is the velocity vector, ρ is the fluid density, p is the pressure, μ is the dynamic viscosity, k is the thermal conductivity, D_i is the diffusion coefficient, and Y_i is the scalar species mass fraction. The CFD commercial software utilizing the finite-volume method, e.g., Fluent package [15], is used for representing and evaluating partial differential equations in the form of algebraic equations. The CFD calculations normally provide detailed results including velocity, temperature, species concentrations, and heat flux that are not easily obtained through experimental measurements. The discretization of the governing equations can be conducted using a segregated solver (each equation is solved separately). In order to couple the calculations of pressure and velocity, a Semi-Implicit Method for Pressure-Linked Equation (SIMPLE) scheme can be applied. Detailed descriptions of the different models, model selection, and model setup are provided in Chap. 6 for specific applications in gas turbine and boiler combustion.

5.3 Modeling of Turbulent Reacting Flows

In most combustion devices of technical importance, the flow is strongly turbulent. Flames in such devices are affected by the turbulent eddies, and the reaction zone is significantly influenced by the flow structure. At the same time, the flame front in such applications has a significant impact on the flow field resulting in considerable deviations in comparison with the corresponding isothermal flow conditions. The presence of a reaction zone can influence the flow turbulence in different ways. Heat release can result in a reduction of the turbulence intensity compared to the cold flow configuration, due to an increase in temperature-dependent kinematic viscosity, $v(T)$, and a corresponding lowering of the Reynolds number. Specifically, the increased molecular viscosity tends to homogenize the flow, reducing the local magnitude of the velocity gradient, and thus the energy transfer between the different turbulent scales. In contrast, the presence of a flame can also induce additional turbulence due to its fluctuating surface. At the flame interface, where the unburned and burnt gases mix, changes in the local density and baroclinic effects can influence flow dynamics, the intermittency may result in production of turbulent kinetic energy, and the flow may exhibit enhanced turbulence in the presence of combustion. The propagation of the flame front in a turbulent flow field typically results in a deformation of the reaction zone, which is influenced by the presence of the local flow structures. The turbulent eddies have a significant impact on the flame front: The small eddies are likely to enter the flame front, up to the reaction zone, and modify the intrinsic structure of the flame front, while the larger coherent vortex structures generally augment the wrinkling of the flame. The effect of the turbulent eddies on the flame front, in terms of the stretching and wrinkling of the interface, also depends on the velocity ratio u_t/S_L, where u_t is the turbulent flow velocity and S_L is the laminar flame speed.

5.3.1 *Modeling Non-premixed Turbulent Combustion*

The combustion of gaseous fuels involves chemical reactions between the fuel gas and the oxidizer (often air). In technical applications, flames are distinguished as: (i) premixed flame, where fuel and oxidizer are homogeneously mixed before combustion; (ii) non-premixed flame, where mixing and combustion occur simultaneously; and (iii) partially premixed combustion, where both premixed and non-premixed regions appear. The distinction between premixed and diffusion flame configurations is typically imposed by the geometry of the burner. If fuel and oxidizer are injected separately and burning takes place by mixing, a diffusion flame is formed. The process of mixing primarily controls the flame position, i.e., the spatial distribution of heat release, provided the fuel is supplied to the reaction zone at a rate sufficient to sustain the flame. On the contrary, if the fuel and oxidant are homogeneously mixed prior to ignition, a perfectly premixed flame is formed. In this scenario, the chemical aspect is dominant, in general. The diffusion flames have a distinct advantage over premixed flames, since they are safer to operate. The fuel is unable to burn before mixing with the oxidant, which prevents the flame from propagating upstream (flashback phenomenon) and damaging the system. Additionally, the system is also easier to design since there is no need to develop a section for premixing. However, a drawback of diffusion flames is that the burning efficiency is controlled and eventually reduced by the mixing of the species. The speed of chemical reactions can slow down when mixing processes do not bring fast enough the reactants into the reaction zone.

5.3.1.1 Overview of Non-premixed Combustion

In most combustion processes, the fuel and oxidizer are separated before entering the reaction zone in which they mix and burn. The flames of the combustion reactions in such cases are called "non-premixed flames" or, traditionally, "diffusion flames" because the transport of fuel and oxidizer into the reaction zone occurs primarily by diffusion. Many combustors operate in the non-premixed burning mode, often for safety reasons. Since the fuel and oxidizer are not premixed, the risk of sudden combustion (explosion) is eliminated. Chemical reactions between fuel and oxidizer occur only at the molecular level, so "mixing" between fuel and oxidizer must take place before combustion. In non-premixed combustion, the fuel and oxidizer are transported independently to the reaction zone, by convection and diffusion, where mixing of the fuel and oxidizer occurs prior to their reaction. Often, the chemical reactions are fast; hence, the burning rate is limited by the transport and mixing process rather than by the chemical kinetics. Consequently, greater flame stability can be maintained. The flame surface is where vaporized fuel and oxygen mix, forming a stoichiometric mixture. At the flame surface, combustion leads to high temperatures that sustain the flame. Non-premixed flames have been used in gas turbines for power generation thanks to their strong stability behavior over wide ranges of loading conditions [2–4]. However, the elevated level of combustion temperature on the flame surface results in dissociation of nitrogen atoms to produce a set of pollutant emissions called as NO*x* [5]. Mainly for this reason, non-premixed burning mixture is considered at a state close to extinction, and premixed combustion mode is to replace it in different combustion applications. Converting the combustion mode from non-premixed to premixed prevents the creation of stoichiometric combustion zones within the combustor as the reactants are premixed upstream of the combustor. This results in reduction in combustion temperature, and accordingly, NO*x* emissions are reduced [6].

5.3.1.2 Modeling Turbulent Non-premixed Combustion

The governing equations indicate that the flow encountered in a diffusion flame can be controlled by turbulence, which enhances mixing, rate of energy dissipation, and heat transfer rate. Several models by which such flow could be modeled are available in commercial CFD software, e.g., ANSYS Fluent. Out of these models, a standard k-ε turbulence model is considered to be effective for modeling turbulent reacting flow under high swirling flow conditions and formation of strong recirculation zones [16]. The Reynolds stresses and turbulent scalar fluxes are related to the gradients of the mean velocities and scalar variable, respectively, via exchange coefficients as follows:

$$-\rho\overline{u_j u_j} = \mu_t\left(\frac{\partial \overline{U}_i}{\partial x_j} + \frac{\partial \overline{U}_j}{\partial x_i}\right) - \frac{2}{3}\rho k \delta_{ij} \tag{5.6}$$

$$-\rho\overline{u_j \phi} = \Gamma_\Phi \frac{\partial \Phi}{\partial x_j} \tag{5.7}$$

where μ_t is the turbulent viscosity and Γ_Φ is equal to μ_t/σ_Φ. The turbulent viscosity is modeled as:

$$\mu_t = c_\mu \rho k^2/\varepsilon \tag{5.8}$$

where c_μ and σ_Φ are constants. The turbulent viscosity is obtained from the solution of the transport equations for k and ε [17]. The eddy dissipation model that describes turbulence–chemistry interaction in non-premixed combustion can be utilized to provide the production rate of species. The conservation equations of the kinetic energy of turbulence and the rate of dissipation of the kinetic energy of turbulence are [18]:

$$\frac{\partial}{\partial x_j}\left(\rho \overline{U}_j k\right) = \frac{\partial}{\partial x_j}\left(\frac{\mu_{\text{eff}}}{\sigma_k}\frac{\partial k}{\partial x_i}\right) + G_k - \rho\varepsilon \tag{5.9}$$

$$\frac{\partial}{\partial x_j}\left(\rho \overline{U}_j \varepsilon\right) = \frac{\partial}{\partial x_i}\left(\frac{\mu_{\text{eff}}}{\sigma_\varepsilon}\frac{\partial \varepsilon}{\partial x_i}\right) + C_1 G_k \frac{\varepsilon}{k} - C_2^* \rho \frac{\varepsilon^2}{k} \tag{5.10}$$

where G_k represents the generation of turbulent kinetic energy due to the mean velocity gradients and is given by:

$$G_k = -\rho \overline{u_i u_j} \frac{\partial \overline{U}_j}{\partial x_i} \tag{5.11}$$

The quantities σ_k and σ_ε are the effective Prandtl numbers for k and ε, respectively, and C_2^* is given as:

$$\mathrm{C}_2^* = \mathrm{C}_2 + \mathrm{C}_3 \tag{5.12}$$

where C_3 is a function of the term k/ε and, therefore, the model is responsive to the effects of rapid strain and streamline curvature and is suitable for modeling turbulent non-premixed flames. The model constants C_1 and C_2 have the values; C_1 = 1.42 and C_2 = 1.68. The wall functions establish the link between the field variables at the near-wall cells and the corresponding quantities at the wall.

5.3.2 *Modeling Turbulent Premixed Combustion*

5.3.2.1 Overview of Premixed Combustion

In contrast to non-premixed flame, combustion is typically more efficient in case of a premixed flame configuration. Additionally, the control of the flame speed is simpler with premixed flames than with diffusion flames. The flame velocity scales with the thermal diffusion and does not depend directly on mixing of species; it can therefore be altered by modifying the temperature of the incoming unburned mixture. Moreover, the flame temperature is directly controlled by the stoichiometry of the mixture (as opposed to diffusion flames, where the mixing-dependent flame temperature is not easily controlled). Thus, the production of NOx can be regulated, since it largely depends on the flame temperature, allowing ease in development of burners with lower pollutant emissions. However, premixed flames are more prone to flashback and thermo-acoustic instability (in particular for rich mixtures) and can have higher sensitivity to variations in inlet mixture stoichiometry and the flow (in particular for lean mixtures).

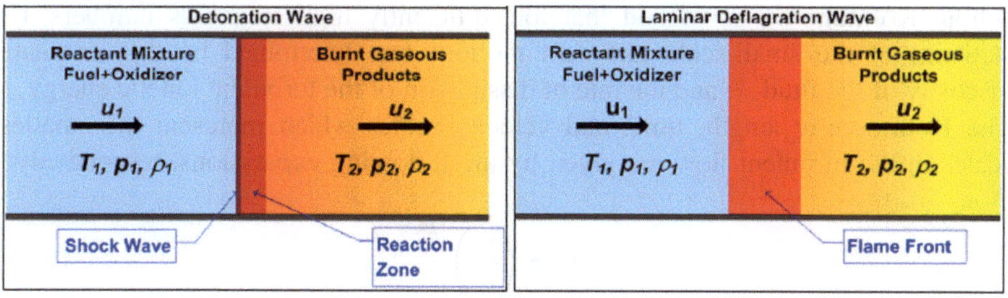

Fig. 5.2 Sketches of: (left) a detonation wave and (right) a laminar deflagration wave

5.3.2.2 Fundamentals of Premixed Flame

When fuel and oxidizer are well mixed, and the mixture ratio keeps within certain flammability limits, the mixture is able to form self-sustaining, propagating reactive waves after ignition. Thus, a premixed flame can be regarded as a wave phenomenon in that the flame propagates toward the combustible mixture while consuming it. Two modes of combustion wave propagation may be identified as per Fig. 5.2: (i) detonation, where the wave propagates at supersonic speed as a shock wave and the temperature rise across the shock exceeds the explosion limits of the mixture, and (ii) deflagration, where the wave is sustained by diffusion of species and energy from the reaction zone into the fresh gases ahead. Since diffusive processes are rather slow, the speed of the deflagration wave is in the order of 20–100 cm/s for common fuel–air combinations.

Specifically, in detonation, combustion is initiated by the advancing of a shock wave, which compresses and heats up the reactant mixture rapidly. This shock wave is then sustained by the energy released from the combustion. Contrary to this, a deflagration wave is sustained by the chemical reaction, and its traveling speed is controlled by the heat conduction and radical diffusion. A prominent difference between the detonation and the deflagration is the pressure jump across the flame. Across the detonation flame, pressure increases greatly (the pressure in burnt side is generally 10–50 times higher than that in unburnt side), while in deflagration the pressure variation is very small (almost constant pressure across the flame). The deflagration wave, which is of interest in the present combustor systems, can further be categorized as laminar and turbulent premixed flames.

5.3.2.3 Premixed Combustion Regimes

Length and Time Scales

In case of a turbulent flow field, eddies of varying sizes possess different amounts of kinetic energy. The turbulent kinetic energy contained in the large-scale eddies is continuously transferred to eddies of smaller sizes, until it is dissipated by viscous action. Kolmogorov postulated that for sufficiently high Reynolds numbers, the statistics of the small-scale turbulent motions are determined by the kinematic viscosity of the fluid, ν, and the rate of dissipation of the turbulent kinetic energy, ε. The Kolmogorov length, time, and velocity scales which represent the smallest scales in the turbulent flow are given by the following expressions, respectively:

$$\eta_k = \left(\frac{\nu^3}{\varepsilon}\right)^{1/4} \tag{5.13}$$

$$\tau_k = \sqrt{\frac{\nu}{\varepsilon}} \tag{5.14}$$

$$u'_k = (\nu\varepsilon)^{1/4} \tag{5.15}$$

Furthermore, Kolmogorov postulated that for sufficiently high Reynolds numbers, there is a range of length scales through which the energy transfer rate is uniquely determined by ε. This range of length scales defined for the region where $L_t \gg \ell \gg \eta$ (here, L_t and η refer to the integral and Kolmogorov length scales, respectively) is called the inertial sub-range. On dimensional grounds, the rate of energy transfer in this range is found to be:

$$\varepsilon \approx \frac{u'^3}{L_t} \approx \frac{k^{3/2}}{L_t} \tag{5.16}$$

where u' and k refer to the rms velocity fluctuation and turbulent kinetic energy, respectively. Another length scale of interest, which is intermediate between L_t and η_k, is the Taylor microscale, λ_T. It can be understood as the distance that a large eddy convects an eddy of size η_k during the time τ_k, and is given by:

$$\lambda_T = \left(\frac{15\nu u'^2}{\varepsilon}\right)^{1/2} \tag{5.17}$$

The fundamental time scales other than the Kolmogorov time scale include the time scale of premixed combustion, τ_c, and the turbulent mixing time scale, τ_t, given as:

$$\tau_c = \frac{\delta_L}{S_L^o}, \quad \tau_t = \frac{L_t}{u'} = \frac{k}{\varepsilon} \tag{5.18}$$

where

$$u' = \sqrt{k}, \quad L_t = \frac{u'^3}{\varepsilon} \tag{5.19}$$

where L_t, u', δ_L, k, ε, and s_L are the integral length scale, the root-mean-square (rms) velocity fluctuation, the laminar flame thickness, the turbulent kinetic energy, the rate of dissipation of the turbulent kinetic energy, and the laminar flame speed, respectively.

Dimensionless Numbers

The regimes of premixed flames have been developed to help the understanding of the possible interaction between the premixed flame front and the turbulent eddies. These regimes, commonly visualized in the form of a combustion diagram, are usually discussed in terms of velocity and length-scale ratios, quite well delimited (Fig. 5.3), and based on non-dimensional numbers defined as follows. The Reynolds number, *Re*, evaluates the convective force relative to the diffusion force. It is based on global scale, and a sole value is relevant to a geometry (which gives the reference scale, *L*) and to a duty point (which gives the reference velocity, *U*). A turbulent Reynolds number, Re_t, is similarly defined locally, using the turbulent scales, and compares the kinematic diffusion due to turbulence to the molecular diffusion:

$$Re_t = \frac{u' L_t}{\nu} \tag{5.20}$$

With the assumption $S_c = 1$, and using $\delta_L = \alpha / s_L = \nu / s_L$, this can be rewritten as:

$$Re_t = \frac{u' L_t}{S_L \delta_L} \tag{5.21}$$

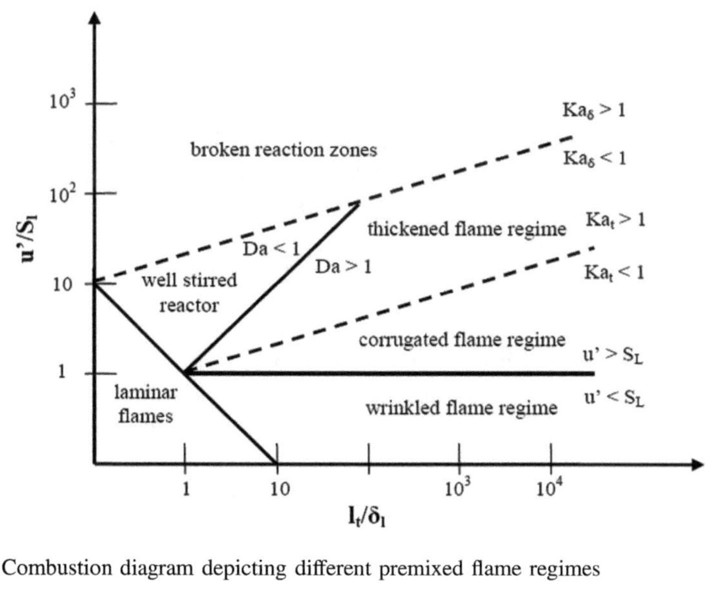

Fig. 5.3 Combustion diagram depicting different premixed flame regimes

The turbulent Reynolds number is directly expressed with turbulent variables and provides a more precise evaluation of the actual and local turbulence property, since the integral length scale L_t and velocity fluctuations u' are likely to differ at different locations of the geometry considered. The turbulent Reynolds number also measures the ratio between the largest L_t and the smallest Kolmogorov turbulent scales:

$$\frac{L_t}{\eta_k} = Re_t^{3/4} \tag{5.22}$$

The (turbulent) Damköhler number, Da, is the ratio between the characteristic time scale of the flow, based on the integral length and velocity scales, and the characteristic chemical time scale, based on the laminar heat diffusion and the laminar flame speed:

$$Da = \frac{\tau_t}{\tau_c} = \frac{L_t S_L^o}{u' \delta_L} \tag{5.23}$$

A Damköhler number smaller than unity ($Da < 1$) corresponds to slow chemistry: The turbulent time scale is smaller than the chemical time scale, and therefore turbulence is faster than combustion. Reciprocally, a Damköhler number larger than unity ($Da > 1$) indicates a fast reaction process which is most common for industrial burners.

The turbulent Karlovitz number, Ka_t, is the ratio between the characteristic chemical time scale and the characteristic time scale of the flow based on the Kolmogorov length scale:

$$Ka_t = \frac{\tau_c}{\tau_k} = \frac{\delta_L}{S_L^o}\sqrt{\frac{\varepsilon}{\nu}} = \begin{cases} \frac{\delta_L^2}{\nu}\sqrt{\frac{\varepsilon}{\nu}} = \frac{\delta_L^2}{\eta_k^2} \\ \frac{\nu}{S_L^{o2}}\sqrt{\frac{\varepsilon}{\nu}} = \frac{u_k'^2}{S_L^{o2}} \end{cases} \tag{5.24}$$

It indicates whether the smallest eddies have any influence on the flame front, and is related to the Karlovitz number, Ka, for which the flame stretch K is based on the strain rate produced by the smallest eddies:

$$Ka = K \cdot \frac{\delta_L}{S_L^o} = \frac{1}{\tau_\eta}\frac{\delta_L}{S_L^o} \tag{5.25}$$

A second turbulent Karlovitz number, Ka_δ, based on the inner layer characteristic thickness (reactive layer on the flame front), can also be defined to indicate whether the smallest eddies are small enough to enter the reactive layer, δ_r:

$$Ka_\delta = \frac{\delta_r^2}{\eta_k^2} = \delta^2 Ka_t \approx \frac{Ka}{100} \tag{5.26}$$

where $\delta = Ze^{-1}$ states the relative thickness of the reactive layer δ_r in the flame front δ_l. It is thus used to quantify the level of interaction of turbulent eddies with the flamelet layers and relates the reaction zone thickness, $\delta_r = \delta.\ \delta_L$, to the size of the Kolmogorov eddies, η^3.

Combustion Diagram and Regimes

In Fig. 5.3, the combustion diagram that highlights the different premixed combustion regimes is presented. The regions are separated by the lines of constant Reynolds, Damköhler, and Karlovitz numbers, which can be easily constructed based on the following relations between the characteristic scales:

$$\frac{u'}{S_L} = Ka^{2/3}\left(\frac{L_t}{\delta_L}\right)^{1/3} = Re_t\left(\frac{L_t}{\delta_L}\right)^{-1} = Da^{-1}\left(\frac{L_t}{\delta_L}\right) \tag{5.27}$$

The diagram is based on scaling laws applicable to homogeneous isotropic turbulence without heat release and without consideration of any non-adiabatic, non-unity Lewis number, or non-unity Schmidt number effects. Also, note that a diffusive flame thickness is employed to construct the diagram. Nevertheless, the diagram provides an order-of-magnitude estimation of the regimes characterizing the flame–turbulence interaction, with the lines $Re_t = 1$, $Da = 1$, $Ka_t = 1$, $Ka_\delta = 1$, and $u'/s_L = 1$, setting the boundaries between the various combustion regimes.

The various regimes displayed in Fig. 5.3 consist of laminar flames, well-stirred reactor, corrugated and wrinkled flamelets, thickened flame regime, and broken reaction zones. When $Re_t < 1$, the flow is essentially laminar, and the combustion lies in the laminar flame regime. As Re_t increases, different turbulent flame regimes can be identified depending on the flame–flow interactions and the resultant reaction zone configuration. The well-stirred reactor regime is principally defined for scenarios with reduced Damköhler number ($Da < 1$) and with moderate turbulence intensity ($Ka_\delta < 1$). While the turbulent time scale is smaller than the chemical time scale, the small-scale eddies are unable to disrupt the inner layer of the flame front. In this regime, turbulence effects homogenize the mixture as it undergoes combustion, with the chemical mechanism governing the reaction process; the notion of flame front is therefore irrelevant.

In the flamelet region, while the turbulent time scales are larger than the chemical time scales ($Da > 1$), the smallest scales (Kolmogorov eddies) in the flow are larger than the laminar flame thickness ($\delta_L < \eta_k$) such that the flame front is embedded in the smallest eddies and the turbulent eddies are unable to perturb the internal structure of the quasi-steady laminar flame. The flame front therefore remains thin, its local inner structure is essentially that of a laminar flame, and the interaction between flame and turbulence is purely kinematic. The flame front is considered as a continuous collection of laminar flamelets, which move and behave differently because of the local action of eddies, and the regime may be subdivided into wrinkled flamelets and corrugated flamelets. The wrinkled flame regime corresponds to the configuration where the flame is placed in a weakly turbulent flow.

If $u'/s_L < 1$, i.e., the turbulent fluctuations are rather small compared to the laminar flame propagation, turbulence cannot compete with the advancement of the flame front, and only slight wrinkling of the flame front is observed. As the turbulent fluctuations become larger than the laminar flame speed, flame topology changes begin to take place as a result of significant wrinkling of the flame front, and formation of pockets of fresh and burnt gas is also possible.

For the thin reaction zones (thickened flame regime), $\delta_L > \eta > \delta$, therefore, the Kolmogorov eddies are smaller than the flame thickness and can influence the local structure of the laminar flame front layer. They are, however, still larger than the inner layer ($Ka_\delta < 1$), suggesting that the small eddies can only penetrate into the preheated zone and the oxidation layer, but not into the inner layer. They can therefore distort the laminar structure of the flame front and also modify the flame speed, since turbulence may enhance or weaken the transport of species or energy in the preheated zone, but the chemical reactions are not influenced by turbulence. The flame front is substantially wrinkled, and the Kolmogorov eddies increase the diffusion within the flame front, so that the flame front thickness is increased leading to the formation of the thickened flamelets. In other words, for this regime, turbulence and combustion cannot be dissociated, and flame and turbulent characteristics become implicitly dependent. The thickened flamelet regime and the corrugated flame regime are of significant importance for industrial applications and can be associated with "high-intensity, small-scale" turbulence and "small-intensity, large-scale" turbulence, respectively (Fig. 5.4). For small-intensity turbulence, the Kolmogorov scale is larger than the flame thickness, and the interaction between the flame front and the turbulence field is purely kinematic; i.e., turbulence can wrinkle

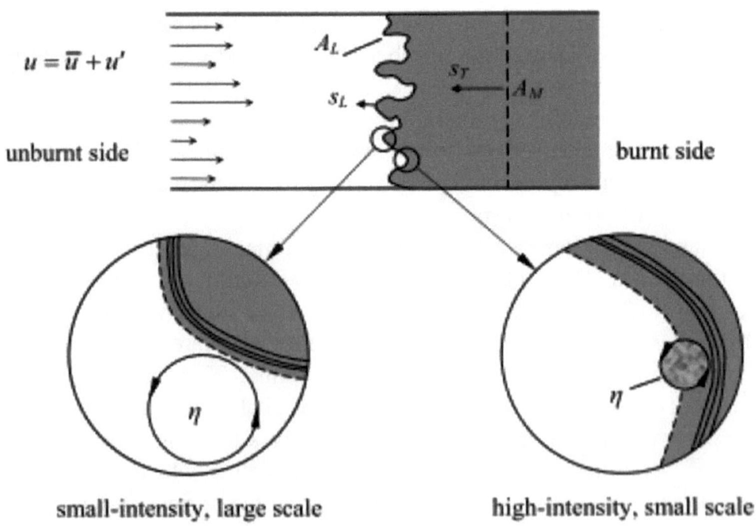

Fig. 5.4 Schematic drawing of an idealized steady premixed flame in a duct with corrugated and thickened flame front locations highlighted

the flame but cannot disturb its local structure. For high-intensity turbulence, the Kolmogorov eddy scale is smaller than the preheated zone; hence, the eddies can enter into the preheated zone and enhance the transport of radicals and heat between the reaction zone and the unburnt gas.

The broken reaction zone's combustion regime occurs when turbulence is more intensive than in the thickened flame regime, such that the Kolmogorov eddies in the flow are smaller than the inner layer thickness of the flame, $\delta > \eta$, can enter the inner layer, and influence the internal structure of the flame ($Ka_\delta > 1$). By increasing the heat and the radical loss to the preheated zone, they can perturb or suppress the chemical reactions and lead to local extinction of the flame, resulting in the broken reaction zone regime. The limit between the thickened flame regime and the broken zone regime is difficult to define. This is very much dependent on the instantaneous and local turbulence characteristics and also on the chemical properties of the mixture. In the thickened flame regime, the flamelet velocity is becoming larger than the laminar flame speed, because of the increased diffusion in the preheated zone. In the broken zone region, the effect of eddies penetrating the inner layer zone may tend to reduce the flamelet velocity and may lead to local extinction, which is classically described as quenching.

5.3.2.4 Combustion Governing Equations

As suggested earlier, the combustion process involves strong interaction between fluid mechanics, chemical reactions, and both heat and mass transfer. These phenomena are characteristically represented in conservation equations. In the following sections, the mathematical formulation of the basic conservation laws is presented. The temporal change of mass in a closed space-fixed control volume is equal to the sum of mass that is transported over the boundaries of the domain and mass that is removed or added inside the control volume per unit time. The continuity equation, in its differential form, describes the local change of the density ρ because of density fluxes through the surfaces of the volume control:

$$\frac{\partial \rho}{\partial t} + \frac{\partial (\rho u_j)}{\partial X_j} = 0 \tag{5.28}$$

where ρ is the bulk mixture density and u_i is the velocity vector component. The temporal change of momentum in a closed, space-fixed control volume is equal to the sum of volume forces f (typically the gravity effect), surface forces T (typically pressure and viscosity effects), the flux of momentum through the boundaries of the control volume, and momentum that vanishes or is produced inside the control volume per unit time. The Navier–Stokes equation, in its differential form, can be written as:

$$\frac{\partial \rho u_i}{\partial t} + \frac{\partial(\rho u_j u_i)}{\partial x_j} = -\frac{\partial p}{\partial x_i} + \frac{\partial \sigma_{ji}}{\partial x_j}\rho f_i \tag{5.29}$$

where p is the pressure field, σ is the stress tensor, and f is the external force. The Stokes formulation of the stress tensor is used to close the stress tensor in the conservation equation:

$$\sigma_{ji} = \mu\left(\frac{\partial u_i}{\partial x_j} + \frac{\partial u_j}{\partial x_i}\right) - \frac{2}{3}\mu\frac{\partial u_k}{\partial x_k}\delta_{ij} \tag{5.30}$$

where μ is the viscosity of the mixture and δ_{ij} is the Kronecker delta tensor. The velocity of species transport across the control volume boundaries, v_k, can be different from the flow velocity, u. In other words, each species is drifting or diffusing relative to the flow with the diffusion velocity: $u_k = v_k - u$. A conservation equation for the mass of each species is valid using its own velocity components $v_{k,j}$ and defining a source term w_k:

$$\frac{\partial \rho Y_k}{\partial t} + \frac{\partial(\rho Y_k v_{k,j})}{\partial x_j} = \dot{\omega}_k \tag{5.31}$$

Other than the overall mass, species mass can be destroyed or produced, since species may be destroyed or formed by chemical reactions. Therefore, a source term ω_k appears in the right-hand side of the equation. In order to write the species mass fraction conservation equation in a way similar to the previous balance equations, the left-hand side is written with the flow velocity u:

$$\frac{\partial \rho Y_k}{\partial t} + \frac{\partial(\rho Y_k u_j)}{\partial x_j} = -\frac{\partial(\rho Y_k u_{k,j})}{\partial x_j} + \dot{\omega}_k \tag{5.32}$$

The diffusion velocity depends on pressure, temperature, and concentration gradients, but usually only the concentration gradient is accounted (isotropy hypothesis). This leads to Fick's law:

$$\rho Y_k u_k = -\rho D_k \nabla Y_k \tag{5.33}$$

where D_k is the diffusion coefficient of the species k into the mixture. The species mass fraction conservation equations can eventually be written as:

$$\frac{\partial \rho Y_k}{\partial t} + \frac{\partial(\rho Y_k u_j)}{\partial x_j} = \frac{\partial}{\partial x_j}\left(\rho D_k \frac{\partial Y_k}{\partial x_j}\right) + \dot{\omega}_k \tag{5.34}$$

Note that if the mass of all species is summed up, the global mass conservation equation must be recovered. So, the sum of all species source terms and diffusion velocities should equate to zero.

The transport equation for the energy can be written as:

$$\frac{\partial \rho E}{\partial t} + \frac{\partial(\rho u_j E + u_j p)}{\partial x_j} = -\frac{\partial q_j}{\partial x_j} + \frac{\partial \sigma_{jk} u_k}{\partial x_j} \tag{5.35}$$

where q_j is the heat flux and is given in terms of the thermal conductivity, κ_t, and the temperature, T, as:

$$q_j = -\kappa_t \frac{\partial T}{\partial x_j} \tag{5.36}$$

Alternative formulations for the energy balance equation may also be written and include the enthalpy form, the temperature form, the pressure form, and the entropy form [19]. The connection between the different state variables is given by equations of state, which depend on the fluid under consideration. Often, all species and the mixture are assumed to be perfect gases, and ideal mixing is presumed as well. For a perfect gas, the ideal gas relation between pressure, temperature, and density is represented as:

$$p = \rho \frac{\Re}{M} T \tag{5.37}$$

where R is the gas constant and M is the molecular weight.

5.3.2.5 Combustion Modeling Approaches

Solving the conservation equations is challenging, particularly in case of turbulent reactive flows, which involve a wide spectrum of length and time scales. The multi-scale and multi-physics attributes of turbulent combustion necessitate the use of modeling approaches in order to simplify the mathematical description of the complex physical phenomena. Three different approaches may be identified: (i) direct numerical simulation (DNS), (ii) Reynolds-averaged Navier–Stokes (RANS), and (iii) large eddy simulation (LES). In comparison with one another, each of these methods possesses various advantages and drawbacks, and is discussed next. Figure 5.5 shows comparative results obtained by the different approaches.

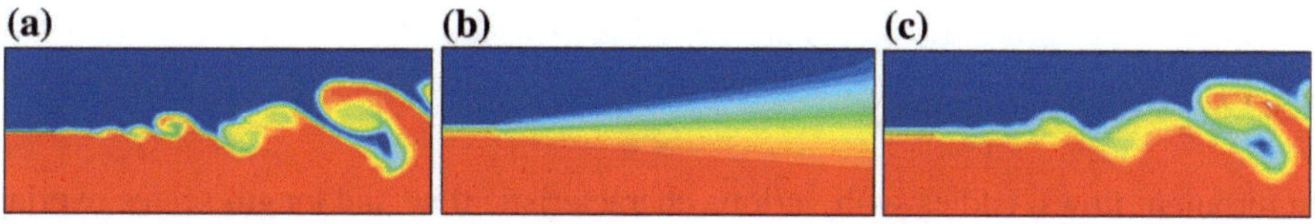

Fig. 5.5 Comparative results from **a** DNS, **b** RANS, and **c** LES

Direct Numerical Simulation (DNS)

In the DNS approach, the whole spectrum of turbulence is explicitly resolved, requiring the use of a fine mesh to enable the discretization of the smallest (Kolmogorov) eddies and the use of small time steps to capture all the dynamic flow and chemical phenomena. While modeling is avoided, the accurate resolution requirement makes DNS computationally expensive and generally impractical for engineering problems. The number of grid points required to simulate the full range of turbulence scales, even for somewhat simple flow geometries, is estimated to be proportional to $Re_t^{9/4}$. For many practical applications, the turbulence Reynolds number is very large, making computational requirements for DNS enormous. DNS can be even more unfeasible for reacting flows and/or for flows having complex geometries. It should be pointed out, however, that DNS has its place in the numerical prediction of turbulent combustion. It has been applied with considerable success to low Reynolds number flows and effectively employed as a research tool for model development.

Reynolds-Averaged Navier–Stokes (RANS) Simulations

The RANS approach introduces the notion of averages of flow variables, and time-averaged governing equations, and involves prediction the macroscopic effect of turbulence, by considering the averaged flow equations. RANS methods do not resolve any part of the turbulent fluctuations, which introduces unclosed correlation terms from the time-averaging procedure and must be modeled to include the effect of the turbulent fluctuations and the unclosed terms on the mean flow. In general, the complexity of turbulence makes it impossible for a single RANS model to represent all turbulent flows and thus some adjustment of model parameters is often required. A primary attraction of RANS approaches is being the least expensive approach of the possible methods in terms of computational effort.

Large Eddy Simulation (LES)

In case of large eddy simulation (LES) technique, rather than averaging the effect of turbulence, the equations are filtered. While the larger turbulent eddies (which have a more significant influence on the flow and are more dependent on the geometry) are explicitly resolved and computed and the effect of smallest eddies (which tends to be somewhat isotropic) is modeled using sub-filter-scale models. Thus, in contrast to RANS, there is partial resolution of the turbulent fluctuations in LES, which decreases the importance of modeling and effects of modeling uncertainty but increases the computational cost. In the next section, a more comprehensive description of the LES approach is presented.

5.3.2.6 LES Methodology

In turbulent flows, eddies with various characteristic time and length scales overlap in space, from the largest length scale (determined by the geometry of the flow)

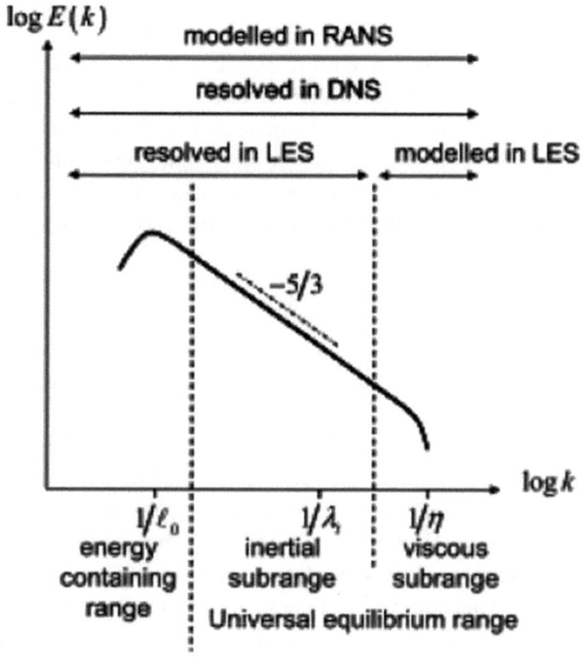

Fig. 5.6 Energy spectrum $E(k)$ as a function of wave number k

down to the smallest scales (at which the kinetic energy is dissipated into heat). In Fig. 5.6, the energy content corresponding to the eddy length scales is depicted. Under the hypothesis of homogeneous and isotropic turbulence, and assuming that the rate of production and dissipation of the turbulent kinetic energy are in balance, it has been demonstrated that turbulence follows an energy cascade, wherein energy is continuously transferred from the large eddies to the small eddies where it is dissipated into heat. The various sub-ranges may be categorized as follows:

Energy-Containing (Large-Scale) Range

The large-scale spectrum contains the small wave number eddies which account for the transfer of energy from the mean flow to turbulence. This domain of the turbulence spectrum is dominated by the mean flow characteristics (e.g., mean strain rate) and therefore depends on the geometry.

Inertial Sub-range

The inertial range or equilibrium range is the most important domain of the turbulence spectrum and comprises of the transfer of turbulent kinetic energy from the largest integral length scales (L_t) to the smallest Kolmogorov length scales (η_k).

Viscous Sub-range

The dissipation range corresponds to the domain where the turbulent kinetic energy is transformed to heat due to viscous effects. For small wave numbers, the energy per unit wave number, E, increases with a power law between κ^2 and κ^4, and the spectrum peaks at a wave number that corresponds to the integral length scale. For larger wave numbers, in the inertial sub-range, E decreases following the $\kappa^{-5/3}$ law. In this domain, the energy spectrum has the form:

$$E(\kappa) = C_k \varepsilon^{2/3} \kappa^{-5/3} \tag{5.38}$$

where C_k is the Kolmogorov constant. For wave numbers larger than the one corresponding to the Kolmogorov scale, E decreases exponentially due to viscous dissipation.

In the DNS approach, all scales down to the Kolmogorov scales are resolved, which requires fine grid resolution and renders the framework impractical for high Reynolds number flow conditions (considering currently available computational resources). However, in general, the large eddies are anisotropic (geometry dependent), while according to Kolmogorov's hypothesis the inertial sub-range and the viscous sub-range are statistically similar or universal for high Reynolds number flows; hence, they correspond to the universal equilibrium range, and local isotropic turbulence is a valid approximation at small scales. The LES approach takes advantage of the above property, i.e., existence of a universal range in turbulent flows. In LES, the large-scale turbulent motions that are dependent on the flow configuration are resolved, while the smaller scales that are universal are modeled.

The various flow quantities Φ are therefore filtered in the spectral space (components greater than a given cutoff length are suppressed) or in the physical space (weighted averaging in a given volume). The filtered quantity Φ is defined as:

$$\overline{\Phi}(x) = \int \Phi(x') g(x - x') \mathrm{d}x' \tag{5.39}$$

where g is a filter function and can take different forms, e.g., top hat and Gaussian. For non-constant density flows, as considered in this work, Favre, or mass-weighted, filtered quantity Φ is defined as:

$$\overline{\rho}\overline{\Phi}(x) = \overline{\rho\Phi} \tag{5.40}$$

Often, the grid is used as the spatial filter. In Fig. 5.7, the LES regime diagram is depicted, which highlights the non-dimensional filter width as a function of the Karlovitz number.

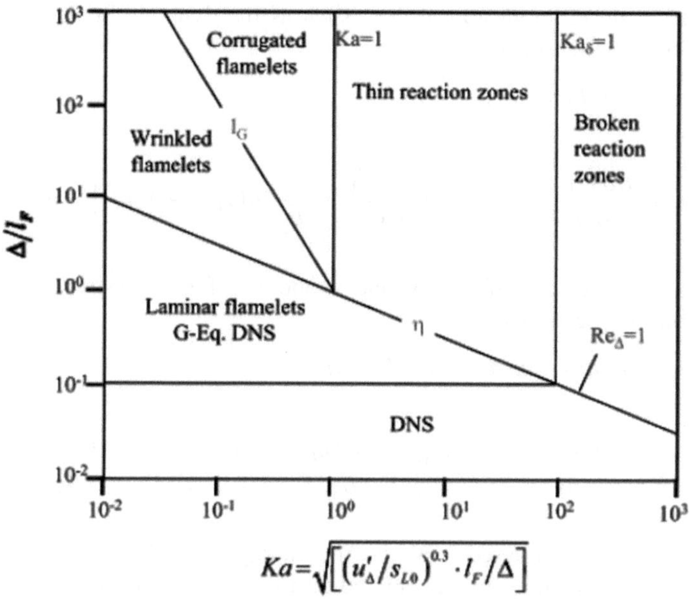

Fig. 5.7 LES regime diagram showing the non-dimensional filter width as a function of the Karlovitz number

5.3.2.7 Governing Equations for LES

The governing equations for LES, obtained by applying the Favre filtering operation to each term in the conservation equations of mass, momentum, and energy, and the species transport equations are given below.

Mass Conservation Equation

$$\frac{\partial \bar{\rho}}{\partial t} + \frac{\partial(\bar{\rho}\tilde{u}_j)}{\partial x_j} = 0 \tag{5.41}$$

where $\breve{u}$ is the filtered velocity vector.

Momentum Transport Equations

$$\frac{\partial \bar{\rho}\tilde{u}_i}{\partial t} + \frac{\partial(\bar{\rho}\tilde{u}_j\tilde{u}_i)}{\partial x_j} = -\frac{\partial \bar{p}}{\partial x_i} + \frac{\partial \tau_{u_j u_i}}{\partial x_j} \tag{5.42}$$

where σ and τ refer to the filtered viscous stress tensor and the corresponding sub-grid-scale (SGS) term, respectively. The unknown SGS correlations above can be modeled using different approaches, such as the dynamic Smagorinsky model for the SGS stress tensor term.

Species Transport Equations

$$\frac{\partial \bar{\rho}\widetilde{Y}_k}{\partial t} + \frac{\partial \bar{\rho}\widetilde{Y}_k\tilde{u}_j}{\partial x_j} = \frac{\partial}{\partial x_j}\left(\bar{\rho}D_k\frac{\partial \widetilde{Y}_k}{\partial x_j}\right) - \frac{\partial}{\partial x_j}\left[\bar{\rho}\left(Y_k u_j - \widetilde{Y}_k\tilde{u}_j\right)\right] + \bar{\omega}_k \tag{5.43}$$

where D is the molecular diffusivity, and Y_k and w_k refer to the specie mass fraction and reaction rate, respectively.

Energy and Enthalpy Conservation Equations

$$\frac{\partial \bar{\rho}\widetilde{E}}{\partial t} + \frac{\partial\left(\bar{\rho}\tilde{u}_j\widetilde{E} + \tilde{u}_j\bar{p}\right)}{\partial x_j} = \frac{\partial \bar{q}_j}{\partial x_j} + \frac{\partial \overline{\sigma_{jk}u_k}}{\partial x_j} + \frac{\gamma R}{\gamma - 1}\frac{\partial \tau_{u_j}T}{\partial x_j} + \frac{\partial \tau_{u_j u_k}u_k}{\partial x_j} \tag{5.44}$$

where $\widetilde{E}$ refers to the filtered total specific energy and q is the filtered heat flux. More details on modeling the unresolved terms can be found in [20].

5.3.2.8 LES Combustion Modeling Techniques

During turbulent combustion, chemical reactions are confined to thin reacting layers at small scales that cannot be resolved on typical LES grids. As a consequence, most of the turbulence–chemistry interactions need to be modeled. This modeling of the reaction rates presents a major challenge in turbulent premixed combustion because reaction rates are highly nonlinear functions of temperature and species mass fractions. An integral component of LES therefore is the turbulent combustion sub-grid model, which is necessary to incorporate the effect of turbulence–chemistry interactions at the under-resolved scales on the reaction rate. Various approaches have been adopted by researchers to model the filtered reaction rates [21], typically corresponding to specific flame regimes (Fig. 5.3), which may be characterized as follows.

Turbulence Mixing-Based (Finite Reaction Zone) Methods

Models based on turbulence mixing are usually a direct extension of existing RANS models, and based on the assumption that at high Damköhler number, turbulent mixing (rather than chemical reaction rate) controls the combustion process. In other words, the underlying concept is that the mean reaction rate is governed by the turbulent mixing processes, especially in scenarios where the chemistry is fast. Models falling into this group include the eddy breakup (EBU) model and the eddy dissipation concept (EDC) model. In the EBU approach, the formulation of the reactive source term is essentially estimated based on turbulence, rather than on chemical considerations. In the LES context, the reaction rate can be expressed as:

$$\dot{w} = C_{EBU}\bar{\rho}\frac{1}{\tau_t}\left(\overline{Y_p'^2}\right)^{1/2} = C_{EBU}\bar{\rho}\frac{\varepsilon}{k}\bar{c}(1-\bar{c}) \tag{5.45}$$

where τ_t is the sub-grid turbulent time scale, C_{EBU} is a model constant, and $\overline{Y_P'^2}$ is the product mass fraction variance, modeled in terms of c, the reduced temperature (reaction progress variable):

$$c = \frac{T - T_u}{T_b - T_u} \tag{5.46}$$

The drawbacks of the formulation include its dependence on the turbulent time and thus on the sub-grid turbulent velocity, and the dependence of the model constant on the flow conditions and mesh size. Furthermore, it has been suggested that the reaction rate is often overpredicted in zones with high strain rates, using the EBU approach.

Flame Front Topology Approaches

The turbulent flame in the flamelet regime can be viewed as an ensemble of locally stretched, thin laminar flames (flamelets) embedded in an otherwise non-reacting turbulent flow field. In this regime, turbulence and chemistry influence each other because the flamelet introduces heat expansion and flow acceleration across the flame front, which changes the turbulent flow field on either side of the reaction zone. On the other hand, convection of the turbulent eddies distorts the flame front, but is not able to disturb the internal structure of the flamelets. Thus, the calculation of turbulence and chemistry may be decoupled, for scenarios where the Kolmogorov length scale is larger than the flame thickness ($Ka_\delta < 1$). While this corresponds to the corrugated flamelet and wrinkled flamelet regimes, it has been shown that the applicable region for flamelet concept can be extended to $Ka < 100$, i.e., when the turbulent Karlovitz number is moderately larger than unity. For these configurations, the turbulent eddies can broaden the preheated zone and increase the heat and species diffusion; however, they are unable to penetrate the thin reaction zone because of the increased viscous dissipation by high temperature near the flame. In this range, the reaction zone remains thin, and the primary effect of increasing turbulence is to wrinkle the flame front and increase the reaction rate, without changing the thin layer structure. Thus, if the flame thickness is thin compared to the size of the turbulent vortices, the reactive layer is unaffected by turbulence and the interaction between turbulence and chemistry is purely kinematic; i.e., the flame is a thin surface that is wrinkled by the turbulent eddies, but its structure across the thickness is the same as for the laminar flame sheet.

In numerical simulations, the laminar flamelet structure is computed using flamelet equations that describe the reactive–diffusive structure in the vicinity of the flame front. The turbulent flame sheet is treated as an ensemble of quasi-laminar

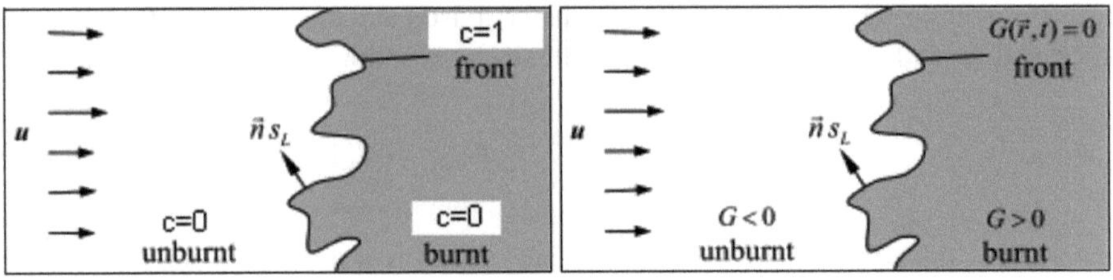

Fig. 5.8 Schematic representing the progress variable and level-set approaches

flames embedded in a turbulent flow, and the location of the flamelet in the turbulent flow field is typically denoted by a geometrical surface, which is in general an iso-c (reaction progress variable) surface or an iso-G (signed distance function) surface for premixed flames (Fig. 5.8). These models are based on the assumption that the flame sheet may be considered as propagating locally as a laminar flame, and the main effect of sub-grid turbulence is to wrinkle and stretch this flame sheet. In general, this approach ignores the internal structure of the flame and detailed chemical kinetics.

The approach represents combustion occurring at the flame front in terms of a transport equation of a reaction progress variable, c, such that $c = 0$ in the reactants and $c = 1$ in the combustion products. The progress variable, a unique parameter, evolves monotonically from the unburned region to the burned region across the flame front. Accordingly, it quantifies the progress of reaction and may be defined as a reduced temperature or reduced fuel mass fraction ($Le = 1$):

$$c = \frac{T - T_u}{T_b - T_u}, c = \frac{Y_F - F_{F,u}}{Y_{F,b} - Y_{F,u}} \tag{5.47}$$

where T, T_u, and T_b are, respectively, the local, the unburned, and burned gas temperatures. Similarly, Y_F, Y_{Fu}, and Y_{Fb} are the local, the unburned, and the burned fuel mass fractions. The transport equation for the progress variable, in the filtered form, is written as:

$$\frac{\partial \bar{\rho}\tilde{c}}{\partial t} + \frac{\partial \left(\bar{\rho}\tilde{c}_i\tilde{u}_j\right)}{\partial x_j} + \frac{\partial}{\partial x_j}\left(\bar{\rho}\widetilde{cu_j} - \bar{\rho}\tilde{c}_i\tilde{u}_j\right) = \frac{\partial}{\partial x_j}\overline{\left(\rho D\frac{\partial c}{\partial x_j}\right)} + \bar{\omega}_c \tag{5.48}$$

where the three terms on the left-hand side are unsteady effects, resolved convective flux, and unclosed transport flux, respectively. The two terms on the right-hand side denote the filtered molecular diffusion and the filtered reaction rate, respectively. The unclosed sub-grid-scale transport flux is generally modeled with a gradient-diffusion assumption:

$$\bar{\rho}\left(\widetilde{cu_j} - \bar{\rho}\tilde{c}_i\tilde{u}_j\right) = -\frac{\mu_t}{Sc_t}\nabla\tilde{c} \tag{5.49}$$

The terms on the right-hand side can be rewritten from the freely propagating one-dimensional flame analysis as:

$$\frac{\partial}{\partial x_j}\left(\overline{\rho D\frac{\partial c}{\partial x_j}}\right) + \bar{\omega}_c = \overline{\rho S_d|\nabla c|} \tag{5.50}$$

where S_d is the local displacement speed of the iso-surface c that depends on the physical–chemical characteristics of the combustible mixture and the local turbulence at the sub-grid level. The closure for the flame front displacement term, $\rho S_d\ |\nabla c|$, may be achieved in terms of the flame surface density or the flame wrinkling factor, and is discussed next.

The flame surface density, $\Sigma\nabla$, is defined as the flame surface area per unit volume, i.e., the relative quantity of flame surface within each cell, and increases with more intense wrinkling of the flame surface. The flame surface is important for combustion modeling, since the burning rate correlates with it, as the wrinkling of the flame increases the burning rate. In the FSD technique, the expression can be rewritten as:

$$\overline{\rho S_d|\nabla c|} \approx \rho_u S_L \Sigma_\Delta \tag{5.51}$$

where Σ_Δ is the flame surface density per unit volume at the sub-grid level, ρ_u is the density of the unburned gas, and S_L is the laminar flame speed. For the evaluation of Σ_Δ, there are different strategies using algebraic relations and a similarity model or a transport equation. In its simplest form, the flame surface density (FSD) model provides an algebraic expression for the amount of flame surface area per unit volume Σ at each point within the premixed turbulent flame brush. If the flamelet assumption holds, then the local flame structure remains quasi-laminar and the local propagation speed remains close to the unstrained planar laminar burning velocity, S_L. For example, an algebraic model for the LES modeling using a DNS analysis can be written as:

$$\widetilde{\Sigma}_\Delta = 4\beta\frac{\bar{c}(1-\bar{c})}{\Delta} \tag{5.52}$$

where the parameter β depends on the sub-grid-scale flame front wrinkling. The estimation of the flame surface density using transport equation(s) offers a more accurate evaluation of the reaction rate in order to resolve the progress variable. For example, one-equation and two-equation approaches have previously been developed. More details can be found in [22].

An alternative to the flame surface density is the flame surface wrinkling factor Ξ_Δ that can be interpreted as the ratio of the SGS turbulent flame surface to the flame surface projected in the propagation direction, and is expressed as:

$$\Sigma_{\Delta} = \Xi_{\Delta} |\nabla \bar{c}| \tag{5.53}$$

where Σ_M is the SGS flame surface density per unit volume and Ξ_{Δ} is the sub-grid-scale wrinkling factor. The same algebraic models as for the flame surface density can be used in this case. A model based on two transport equations, to resolve the regress variable ($b = 1 - c$) and the sub-grid-scale flame front wrinkling, respectively, has been developed. More details can be found in [23].

A method that is able to describe the temporal evolution of surfaces with arbitrary complexity in space is the level-set method. The base of this method is a scalar field, and the surface under consideration is defined as an iso-surface of this field. The well-known level-set equation has the following general form:

$$\frac{\partial \phi}{\partial t} + F(x, \nabla \phi, k)|\nabla \phi| = 0 \tag{5.54}$$

In the level-set equation, the zero level set $\varphi = 0$ corresponds to the moving front, and k is the curvature of the level-set surface. It describes the propagating front along its normal direction with a speed F, which is a function of multiple variables like x, $\nabla\varphi$, k. If the propagating front is also passively advected by an underlying flow field u, then the equation can be rewritten as:

$$\frac{\partial \phi}{\partial t} + F(x, \nabla \phi, k)|\nabla \phi| + u \cdot \nabla \phi = 0 \tag{5.55}$$

When the approach corresponds to a geometrical description of the flame front using an iso-level surface of a scalar field G (iso-contour in two dimensions), the iso-surface of G is typically fixed at $G = G_0$, and the level set of the scalar at this value represents the spatial location of the flame surface. Figure 5.8 depicts an instantaneous flame surface, which is represented by the zero level set of the quantity G, $G(\vec{r}, t) = 0$, where r denotes the spatial position and n is the unit normal vector to the front and points into the unburned zone and is defined by $\vec{n} = -\nabla G/|\nabla G|$. This zero level set divides the flow field into two zones: $G > 0$, the burnt zone, and $G < 0$, the unburned zone. The level-set equation may then be expressed as:

$$\frac{\partial G}{\partial t} + \nabla G \cdot \frac{d\vec{r}}{dt} = 0 \tag{5.56}$$

where the propagation speed of the flame front, dr/dt, results from two contributions: the flow velocity, u, and the (burning) velocity of the flame front normal to itself:

$$\frac{d\vec{r}}{dt} = u + \vec{n} S_L \tag{5.57}$$

Using the above equation, the basic propagation equation for the scalar G, the well-known G-equation, is obtained as:

$$\frac{\partial G}{\partial t} + u \cdot \nabla G = S_{\mathrm{L}} |\nabla G| \tag{5.58}$$

where S_{d} stands for the flame displacement speed. The equation describes a surface propagating in space with a velocity S_{L} (normal to itself) relative to the local unburned flow velocity, u. A turbulent flame speed relation is also derived for the averaged representation to describe the displacement of the iso-value G_0 [24]. Therefore, the adaptation of this model to different turbulent combustion regimes depends on the relevant evaluation of Sd. Further, for LES of turbulent combustion, the above equation needs to be filtered. The laminar flame speed, S_{L}, is not always equivalent to the unstretched laminar flame speed, S_{L}^0, as the effect of the local curvature and strain should be considered. The combined effects are often referred to as the flame stretch, K, defined by the fractional rate of change of a flame surface element A [25]:

$$K = \frac{1}{A}\frac{\mathrm{d}A}{\mathrm{d}t} \tag{5.59}$$

A general expression of stretch, for a thin flame sheet, can be found in [26]. If only the local curvature effects are considered in the level-set approach, the equation becomes [27]:

$$\frac{\partial G}{\partial t} + u \cdot \nabla G = S_{\mathrm{L}}(1 - L_o k) |\nabla G| \tag{5.60}$$

where L_o is the Markstein length [28] and κ is the local curvature defined as:

$$k = \nabla \cdot \vec{n} = \nabla \cdot \left(- \frac{\nabla G}{|\nabla G|} \right) \tag{5.61}$$

Different from the c-equation, the G-equation has several special properties, in that it is a Hamilton–Jacobi-type equation and is only well defined at the zero level set. Since the scalar G surrounding the flame front (zero level set) is not uniquely defined, it may be assigned arbitrarily without changing the physics of the problem and at the same time keeps the zero iso-surface of this assigned G-field coinciding with the original flame surface. By this way, the flame front becomes an iso-surface of a scalar, which is defined in the whole field; therefore, the G-equation turns into a front capturing method. To guarantee uniqueness, it has been proposed to define the surrounding G-field as a signed distance function, which can be expressed in a first-order partial differential equation: $\nabla G = 1$. Typically, therefore, the flame position is associated with $G = G_0$: The values of G are typically chosen such that $G = 0$ at the flame front, $G < 0$ in the unburned mixture, and $G > 0$ in the burned gases.

The signed distance is a well-defined physical quantity and represents the shortest distance from a local point to the zero level set but with a sign: If this point lies in the $G > 0$ side, it is positive; otherwise, it is negative. However, during the evolution of G within the whole computational domain, G will not necessarily remain as the signed distance, though the initial G-field may be initialized to satisfy the signed distance function. Therefore, it is necessary to reshape the level-set G-field to the signed distance function but without change its zero level set often. This reshaping is called signed distance re-initialization.

Once the front evolves, the coupling with the one-dimensional flame structure requires spatial information (mapping). It is therefore convenient to let the scalar G be defined as a signed distance function. At each point, within a close region of the flame front, the scalar G corresponds to the closest distance to the front. Additionally, by convention the unburned region is defined by negative values $(G < 0)$ and the burned region by positive values $(G > 0)$. Since the transport equation does not preserve the distance property, an additional equation is solved, $\nabla G = 1$ (re-initialization step). Fulfilling this requirement, the flame structure can be readily mapped onto the domain through the scalar G values. The density, temperature, and viscosity are given by a flamelet library.

To summarize, combustion modeling with the level-set method is not based on solving the species mass fraction equations. Instead, the flame is treated as a geometric entity, i.e., a surface which is convected by the flow and self-propagating normal to itself with its flame speed. For computing the flame speed, correlation equations are used. Thus, no closure of the chemical source term is required since its effect is included in the flame speed. Although the G-equation approach is an elegant description of the flame front geometry, the solution of the G-equation is numerically expensive and difficult to implement because the G-field is discontinuous due to $\nabla G = 1$. Special discretization schemes and frequent re-initialization of the scalar field are required in this case. Another difficulty is the coupling of the artificial quantity G with the mass and energy equations such that their conservation properties are preserved.

Finite-Rate Chemistry Models

In order to have an accurate prediction and analysis of combustion stability, it is essential to incorporate finite-rate chemistry effects in the computational model. In this regard, various methods have been developed to estimate the filtered reaction rates, which typically require a multiple-step (reduced) reaction mechanism. Further, unlike in the RANS, where the mean turbulent flame is solved and the computed flame is very thick, the LES approach predicts the instantaneous flame region and the flame thickness is generally smaller than the LES mesh size, especially for premixed flames. To address these concerns, the finite rate chemistry models are presented in the following sections.

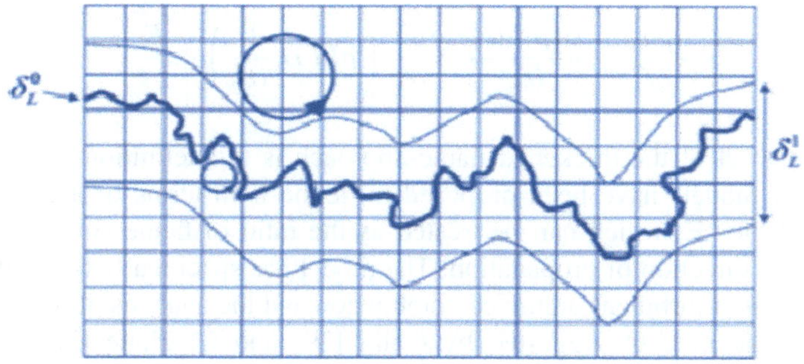

Fig. 5.9 Thickening of the reaction zone to resolve the flame front on the LES mesh

Artificially Thickened Flame Approach

The thickened flame approach involves artificially thickening the flame front so that it can be resolved on the LES grid, while maintaining the same laminar flame speed and turbulence–flame interaction, as per Fig. 5.9. From fundamental theories (dimensional analysis of a laminar premixed flame), it follows for a one-step reaction:

$$s_{\mathrm{L}}^{0} \propto \sqrt{D\bar{\omega}}, \quad \delta_{\mathrm{L}}^{0} \propto D/s_{\mathrm{L}}^{0} = \sqrt{D/\bar{\omega}} \tag{5.62}$$

where S_{L}^{0} is the laminar flame speed, δ_{L}^{0} is the laminar flame thickness, D is the diffusion coefficient, and w is the average reaction rate. Increasing the flame thickness by a factor F, while maintaining a constant flame speed, can therefore be achieved by suitably modifying the diffusivity D and the mean reaction rate B (i.e., increasing D by a factor F and reducing B by the same magnitude). Thus, the modified expressions become:

$$\overline{s_{\mathrm{L}}^{0}} \propto \sqrt{FD\frac{\bar{\omega}}{F}} = s_{\mathrm{L}}^{0}, \quad \overline{\delta_{\mathrm{L}}^{0}} \propto FD/s_{\mathrm{L}}^{0} = F\delta_{\mathrm{L}}^{0} \tag{5.63}$$

If F is sufficiently large, the thickened flame front can be resolved on the LES computational grid. Furthermore, in this method, the reaction rate is still calculated using the Arrhenius law; therefore, various phenomena (quench, ignition, etc.) may be accounted for without turning to sub-models. The thickening of the flame front, however, leads to a modified interaction between turbulence and chemistry since the Damköhler number is decreased by the factor F. The effect of turbulent vortices is to wrinkle the flame and enhance the laminar flame speed to a higher turbulent flame speed. This interaction is altered, when the flame thickness changes, since the flame becomes less sensitive to turbulence motions. To account for the wrinkling effect of the unresolved features on the (thickened) flame front, an efficiency function, E, is introduced. The filtered transport equation for the chemical species then takes the form:

$$\frac{\partial \bar{\rho} \widetilde{Y}_i}{\partial t} + \frac{\partial (\bar{\rho} \widetilde{Y}_i \tilde{u}_j)}{\partial x_j} = \frac{\partial}{\partial x_j} \left(\bar{\rho} EFD_i \frac{\partial \widetilde{Y}}{\partial x_j} \right) + \frac{E\bar{\dot{\omega}}_i}{F} \tag{5.64}$$

A key aspect of the thickened flame approach is the definition of the function E. Different models have been proposed to define E in terms of a dimensionless wrinkling factor Ξ, which can be treated as the ratio of flame surface to its projection in the direction of propagation. The model has specifically been adopted for the simulation of turbulent premixed combustion and the analysis of unsteady flame dynamics. The models suggested by Colin [29] and Charlette [30, 31] can be implemented, and a dynamic formulation should be developed in this case as discussed in the next section.

Partially Stirred Reactor (PaSR) Method

As aforementioned, premixed flames in general are much thinner than the typical LES computational cell, and assuming the entire cell as a perfect reactor is an overestimation [32, 33]. Thus, in the PaSR model, the computational cells are split into a reacting part and a non-reacting part. The reacting part is treated like a perfectly stirred reactor, in which all present species are homogeneously mixed and reacted. After reactions have taken place, the species are assumed to be mixed due to turbulence for the mixing time, and the resulting concentration gives the final concentration in the entire partially stirred cell. In other words, the PaSR model phenomenologically incorporates the sequential processes of micromixing and chemical reactions. The microscale processes responsible for the molecular mixing, as well as the dissipation of turbulent kinetic energy, concentrated in isolated regions, and occupies only a small fraction of the fluid volume. These fine flow structures in which most of the dissipation and mixing take place form topologically complex regions, whose characteristic dimensions are small compared to the LES filter width. Since most of the mixing occurs in the fine structures, the reactions also take place in these regions as the reactants are mixed at scales down to the molecular scales provided that the temperature is high enough. Therefore, each LES cell can be viewed as partially stirred reactor containing the reactive fine structures (ideally viewed as perfectly stirred reactors), exchanging mass and energy with its surroundings.

The part of the computational cell constituting the reactor is governed by the turbulent mixing time and the residence time. This part is explicitly computed. The reacting volume fraction may be estimated as the ratio between the chemical reaction time and the total conversion time in the reactor, i.e., the sum of the micromixing time and the chemical reaction time. The chemical reaction time is estimated from the laminar flame speed at the laminar flame thickness, whereas the mixing time, ranging from the sub-grid time to the Kolmogorov time, is modeled as the geometrical mean of the two. In this approach, no a priori assumption is necessary for the combustion regime:

$$k^* = \frac{\tau_c}{\tau_c + \tau_t} \tag{5.65}$$

$$\frac{c_1 - c_0}{\tau_c} = \frac{c - c_1}{\tau_t} \Rightarrow c_1 = k^* c + (1 - k^*) c_0 \tag{5.66}$$

$$RR_t = k^* \times \omega_t \tag{5.67}$$

where RR_t is the chemical source term and ω_t is the reaction rate from solving the chemical system. To summarize, the finite-rate chemistry approaches (in contrast to some of the other LES sub-grid-scale combustion modeling methods) allow the use of multiple-step (reduced) chemistry mechanisms [34], thereby enabling the reaction rates to be calculated much like a direct numerical simulation (DNS) calculation without the need of any ad hoc sub-models (to take into account phenomena such as ignition and flame–wall interactions). Use of detailed chemistry can also be important, especially when studying non-conventional combustion (such as oxy-combustion) or when specific species distributions (such as CO or OH) are required to be predicted. Another approach to account for detailed chemistry is through the use of statistical approach and is discussed next.

Statistical Techniques (PDF Approaches)

The probability density function (PDF) models are based on one-point statistical analysis: If the probability of locating a variable at a location is known, then its mean value at this location can be estimated. This idea is easily extended to take into account multiple variables. For example, the instantaneous reaction rate, a function of species mass fractions, and temperature can be written as: $\omega = \omega(Y_1, Y_2, \ldots, T)$. If the joint probability $p(Y_1, Y_2, \ldots, T)$ *for* $Y_1, Y_2, \ldots, T$ is known in the range:

$$Y_i \in \left(Y_i - \frac{1}{2}\mathrm{d}Y_i, Y_i + \frac{1}{2}\mathrm{d}Y_i\right), \quad T \in \left(T - \frac{1}{2}\mathrm{d}T, T + \frac{1}{2}\mathrm{d}T\right) \tag{5.68}$$

then the mean reaction rate can be determined as:

$$\bar{\omega} = \int\limits_{Y_1, Y_2, \ldots, T} \omega(Y_1, Y_2, \ldots, T) p(Y_1, Y_2, \ldots, T) \mathrm{d}Y_1 \mathrm{d}Y_2 \ldots \mathrm{d}T \tag{5.69}$$

The PDF-based statistical description of turbulent reactive flow has some theoretical advantages: The complex chemistry is treated exactly without applying assumptions such as "flamelet" and "fast reaction." The PDF schemes provide a closed form representation of the chemical source terms and can be applied to different combustion regimes. They can be applied equally well to non-premixed, premixed, and partially premixed flames. In general, the sub-grid PDF, the joint probability density function of all concentrations and the temperature, can be either determined by the solution of a transport equation (transported PDF method) or modeled assuming its shape (presumed PDF method). These techniques are discussed next.

Transported PDF Approach

In this method, the joint PDF is determined by the use of a balance equation [35]. Since the chemical reaction term is closed and does not include any modeling, the framework can handle any complex chemical mechanism. In this respect, the transport PDF method can have a considerable advantage over other turbulent combustion models. However, the numerical solution of the transport equation is expensive because of the high dimensionality of the PDF. Thus, application of transported PDF methods in LES generally requires high computational costs and robust solution algorithms. Current implementations therefore apply Monte Carlo methods for the solution to the PDF equation, since the memory requirements in such a scenario only depend linearly on the dimensionality. In Monte Carlo simulations, a number of stochastic particles are tracked which may be regarded of as realizations of the PDF. The histogram of the particle's properties in a cell then recovers the PDF. In order to enable proper statistics, the number of particles per cell has to be large, which may still be attractive for two-dimensional RANS calculations, but can render the technique expensive for large-scale three-dimensional applications using LES.

Presumed PDF Approach

The expensive solution to the PDF balance equation can be avoided by assuming a specific shape for the PDF. Although a PDF function can take any shape and have multiple extrema, in many combustion cases, it is observed (through experimental data and DNS studies) that probability distributions show some common features; therefore, it is reasonable to represent them with a special shape but with different controlling parameters. One of the most popular approaches is to assume the PDF as a β-function, and its controlling parameters are the mean and variance of the variable. Thus, the structure of the PDF is given by some empirical function which still contains a small number of freely selectable parameters. These parameters are then determined from the moments of the stochastic variables for which transport equations have to be solved. Although the influence of chemical reactions on the PDF shape cannot be accounted for on a physical basis as it is the case with a PDF balance equation, the method is attractive because of its much lower computational effort. The disadvantage of the presumed PDF method is that while it is relatively simple to find shape functions for one-dimensional PDFs, it is very difficult to presume or measure multi-dimensional joint PDFs. A common practice is to assume statistical independence between the parameters of the PDF and to apply one-dimensional functions to each of them. Thus, a joint PDF can be expressed as: $p(Y_1,Y_2, \ldots, T) = p_{Y1}(Y_1)\ p_{Y2}(Y_2)\ p_t(T)$. Unfortunately, this assumption may not hold in all practical combustion situations, since variables, such as temperature and species mass fraction, are closely related to flames. In other words, the influence of the PDF shape on the results may be small when the circumstances are favorable;

i.e., the reaction rates are high. In contrast, the results can become strongly dependent on the PDF shape for other scenarios (e.g., for nitric oxide formation which is very sensitive to the temperature).

5.3.2.9 Artificially Thickened Flame Approach

As previously mentioned, a key component of LES is the turbulent combustion sub-grid model, which is necessary to incorporate the effect of turbulence–chemistry interactions at the unresolved scales on the reaction rate. Various approaches have been developed [21] that can be broadly categorized into three categories: (1) finite reaction zone approaches (eddy breakup-type models, eddy dissipation concept, etc.), (2) flame front topology methods (flame surface density and flame wrinkling descriptions, the level-set flame front tracking technique, G-equation approach, etc.), and (3) statistical framework-based approaches (presumed or transported PDF method) and finite rate chemistry models (artificially thickened flame technique, partially stirred reactor concept, etc.). Evaluating the performance of these sub-models in simulating turbulence–chemistry interactions is crucial in developing higher fidelity CFD methods for predicting flame–flow dynamics and conducting combustion stability studies. In this chapter, the focus is on the thickened flame model, which has distinct advantages over some of the other approaches, in that it allows the use of suitably detailed reaction chemistry and can be applied to different combustion regimes, in general.

LES validation studies have been conducted in the past [29–31, 36] for the thickened flame approaches to assess their predictive capabilities while simulating turbulent combustion. As explained earlier, a critical aspect of the modeling technique is the determination of the efficiency function that accounts for the reduced wrinkling of the thickened flame front due to turbulence. Current models define the efficiency function using an algebraic formulation for the flame wrinkling, which is estimated based on the assumption of local equilibrium between production and destruction of SGS flame surface density. It has been suggested that in case of highly unsteady systems or in configurations prone to thermo-acoustic instabilities, deviation from local equilibrium and the lack of time-history effects in the simpler algebraic models can result in prediction inaccuracies [23, 30, 31]. Therefore, further development of the model is warranted so that the approach can be applied to the analysis of strongly unsteady systems. Some of these concerns can be addressed by incorporating a dynamic formulation for the efficiency function that explicitly incorporates the influence of strain and time-history effects on flame wrinkling by solving a transport equation. Multiple-step reaction chemistry can be appropriately used with the thickened flame approach [37].

Modeling of the filtered reaction rates in the species transport equations presents another major challenge in simulating turbulent premixed combustion using LES. This is because reaction rates are highly nonlinear functions of temperature and species mass fractions and because chemical reactions are confined to thin reacting layers at small scales that cannot be resolved on typical LES grids. The artificial flame-thickening technique, a finite rate chemistry combustion modeling approach, essentially involves artificial thickening of the flame front so that it can be resolved on the LES grid, while maintaining the same laminar flame speed and turbulence–flame interaction. From the theory of laminar premixed flames, it is known that the laminar flame speed, S_L^0, and the laminar flame thickness, δ_L^0, are related to the molecular diffusivity (D) and the mean reaction rate ($\bar{\omega}$) as follows:

$$s_L^0 \propto \sqrt{D\bar{\omega}}, \delta_L^0 \propto D/s_L^0 = \sqrt{D/\bar{\omega}} \tag{5.70}$$

Increasing the flame thickness by a factor F while maintaining a constant flame speed can be achieved by suitably modifying the diffusivity and the mean reaction rate (by replacing D with FD, and ω with ω/F). If F is sufficiently large, the thickened flame front can then be resolved on the LES computational grid. The filtered species transport equation can therefore be written as:

$$\frac{\partial\bar{\rho}\widetilde{Y}_i}{\partial t} + \frac{\partial(\bar{\rho}\widetilde{Y}_i\tilde{u}_j)}{\partial x_j} = \frac{\partial}{\partial x_j}\left(\bar{\rho}FD_i\frac{\partial\widetilde{Y}}{\partial x_j}\right) + \frac{\bar{\dot{\omega}}_i}{F} \tag{5.71}$$

where ρ is the density, Y_i is species mass fraction, $\tilde{u}$ is the filtered velocity vector, and $\bar{\dot{\omega}}_i$ is the filtered species reaction rate. The thickening of the flame front, however, leads to a modified interaction between turbulence and chemistry since the Damköhler number is decreased by the factor F. The flame becomes less sensitive to turbulence, and wrinkling of the flame front is reduced. To account for this, an efficiency function, E, is introduced [29] that recovers the underestimation of the flame front wrinkling due to the thickening approach. The balance equation for the chemical species then takes the form:

$$\frac{\partial\bar{\rho}\widetilde{Y}_i}{\partial t} + \frac{\partial(\bar{\rho}\widetilde{Y}_i\tilde{u}_j)}{\partial x_j} = \frac{\partial}{\partial x_j}\left(\bar{\rho}EFD_i\frac{\partial\widetilde{Y}}{\partial x_j}\right) + \frac{E\bar{\dot{\omega}}_i}{F} \tag{5.72}$$

It should be noted, however, that the above methodology modifies the diffusion term in the entire computational domain, which can lead to inaccuracies in the prediction of the species mass fractions. To overcome this shortcoming, a dynamic formulation has been proposed [38], wherein the thickening factor and the diffusivity are represented locally as follows:

$$F_{\text{loc}} = 1 + (F-1)\Psi(c), \quad D_{i,\text{loc}} = \frac{\mu}{Sc}EF_{\text{loc}} + (1-\Psi(c))\frac{\mu_t}{Sc_t} \tag{5.73}$$

$$\Psi(c) = 16[c(1-c)]^2, \quad c = 1 - \frac{Y_F}{Y_F^{\text{in}}} \tag{5.74}$$

where $\Psi(c)$ is a locally defined sensor function based on the reaction progress variable, c, prescribed in terms of the ratio of the fuel mass fraction in the cell (Y_F)

to that at the inlet (Y_F^{in}). The symbols μ and Sc refer to the dynamic viscosity and the Schmidt number, respectively, while F_{loc} and D_{loc} refer to the local thickening factor and local diffusivity, respectively. The filtered species transport equation can then be rewritten as:

$$\frac{\partial \bar{\rho}\widetilde{Y}_i}{\partial t} + \frac{\partial(\bar{\rho}\widetilde{Y}_i\tilde{u}_j)}{\partial x_j} = \frac{\partial}{\partial x_j}\left(\bar{\rho}EF_{\text{loc}}D_{i,\text{loc}}\frac{\partial \widetilde{Y}}{\partial x_j}\right) + \frac{E\overline{\dot{\omega}}_i}{F_{\text{loc}}} \tag{5.75}$$

As mentioned earlier, an important aspect of the thickened flame approach is the evaluation of the efficiency function E, in order to appropriately account for the reduced wrinkling of the thickened flame front due to turbulence. Based on DNS studies of flame–vortex interactions [29], different models have been proposed to define E in terms of the dimensionless wrinkling factor Ξ, which can be treated as the ratio of flame surface to its projection in the direction of propagation [29–31]. Previously, researchers have suggested modeling the efficiency function in terms of the local filter width, Δe, the unstrained laminar flame speed, S_{L}^0, the thickness of the laminar (δ_{L}^0) and thickened flames (δ_{L}^1), and the local SGS turbulent velocity, $u'_{\Delta e}$ as follows [29]:

$$E = \frac{\Xi|_{\delta_{\text{L}}=\delta_{\text{L}}^0}}{\Xi|_{\delta_{\text{L}}=\delta_{\text{L}}^1}} \geq 1, \quad \Xi|_{\delta_{\text{L}}=\delta_{\text{L}}^0} = 1 + \alpha\left(\frac{2\ln 2}{3C_{ms}(Re_{\text{t}}^{1/2} - 1)}\right)\frac{u'_{\Delta_e}}{S_{\text{L}}^0}\Gamma\left(\frac{\Delta_e}{\delta_{\text{L}}^0}, \frac{u'_{\Delta_e}}{S_{\text{L}}^0}\right) \tag{5.76}$$

$$\Gamma\left(\frac{\Delta_e}{\delta_{\text{L}}^0}, \frac{u'_{\Delta_e}}{S_{\text{L}}^0}\right) = 0.75\exp\left[-1.2\left(\frac{u'_{\Delta_e}}{S_{\text{L}}^0}\right)^{-0.3}\right]\left(\frac{\Delta_e}{\delta_{\text{L}}^0}\right)^{2/3} \tag{5.77}$$

where Re_{t} is turbulent Reynolds number, α is a model constant of order unity, and C_{ms} is a model constant, with a value of 0.28, to maintain consistency with DNS results.

A possible limitation of the above approach, as highlighted in [30], is that under highly unsteady conditions, the efficiency function model can result in inaccurate predictions, since it assumes local equilibrium between production and destruction of SGS flame surface density. In addition, the model assumes that the laminar flame speed is unaffected by strain and curvature and therefore employs the unstrained laminar flame speed while estimating the flame wrinkling. This has been shown to result in predictive inaccuracies [39], as in most cases, strain effects cannot be ignored. For example, for flames trapped in the shear layer behind a backward-facing step, the strain effects near the step are important in reducing the effective reaction rate and preserving the Kelvin–Helmholtz instability. In other words, if laminar flame is assumed independent of strain rate, the heat release in the shear layer inhibits the growth of the instability, resulting in a near smooth flame [23]. Likewise, in case of confined swirling flames, significant strain effects combined with heat loss can provoke extinction of the flame that stabilizes in the outer shear layer of the annular jet, particularly in lean conditions [39]. Further, accurately including the time-history effects of flame–vortex interaction on stretching and wrinkling of turbulent flames is of fundamental importance [40].

In order to address these concerns, a transport equation for the sub-grid flame wrinkling can be used to dynamically evaluate the magnitude of the efficiency function, by appropriately determining the flame wrinkling and subsequently updating the model parameter α during the LES computations. The transport equation for the flame wrinkling factor can be written as [23]:

$$\frac{\partial \Xi}{\partial t} + \widehat{U} \cdot \nabla \Xi = -n \cdot (\nabla U_s) \cdot n\Xi + \widehat{n} \cdot \left(\nabla \widehat{U}_\mathrm{t}\right) \cdot \widehat{n}\Xi + \left(\widehat{U}_\mathrm{t} - \widehat{U}_s\right) \cdot \frac{\nabla |\nabla \bar{b}|}{|\nabla \bar{b}|} \Xi \quad (5.78)$$

where U_t is the surface-filtered effective velocity of the flame, U_s is the local instantaneous velocity of the flame surface, n is the normal to the flame surface, and b is the reaction regress variable. The terms on the right represent the effects of strain, propagation, and differential propagation, respectively. A simplified formulation may be written, as in [23], as:

$$\frac{\partial \Xi}{\partial t} + \widehat{U}_s \cdot \nabla \Xi = G\Xi - R(\Xi - 1) + (\sigma_s - \sigma_\mathrm{t})\Xi \quad (5.79)$$

$$\widehat{U}_s = \widetilde{U} + \left(\frac{\bar{\rho}_u}{\bar{\rho}} - 1\right) s_u \Xi \tilde{\mathrm{n}} - \frac{\nabla \cdot (\bar{\rho} D \nabla \bar{\rho})}{\bar{\rho} |\nabla \tilde{b}|} \tilde{n} \quad (5.80)$$

$$G = R \frac{2c\left(\Xi^*_{\mathrm{eq}} - 1\right)}{1 + 2c\left(\Xi^*_{\mathrm{eq}} - 1\right)}, \quad R = \frac{0.28}{\tau_\eta} \frac{\Xi^*_{\mathrm{eq}}}{\Xi^*_{\mathrm{eq}} - 1}, \quad \Xi^*_{\mathrm{eq}} = 1 + 0.62 \sqrt{\frac{u'}{s_u} R_\eta} \quad (5.81)$$

where τ_η is the Kolmogorov time scale, u' is the sub-grid turbulence intensity, c is the reaction progress variable, and R_η is the Kolmogorov Reynolds number. The terms σ_s and σ_t refer to the resolved strain rate and the surface-filtered resolved strain rate, respectively, while s_u is the strained laminar burning velocity. The resolved strain rates can be obtained as follows:

$$\sigma_s = \frac{\nabla \cdot \widetilde{\mathrm{U}} - \tilde{\mathrm{n}} \cdot (\nabla \widetilde{\mathrm{U}}) \cdot \tilde{\mathrm{n}}}{\Xi} + \frac{(\Xi + 1)\{\nabla \cdot (S_u \tilde{\mathrm{n}}) - \tilde{\mathrm{n}} \cdot [\nabla (S_u \tilde{\mathrm{n}}) \cdot \tilde{\mathrm{n}}]\}}{2\Xi} \quad (5.82)$$

$$\sigma_\mathrm{t} = \nabla \cdot \left(\widetilde{U} + s_u \Xi \tilde{n}\right) - \tilde{n} \cdot \left[\nabla \left(\widetilde{U} + s_u \Xi \tilde{n}\right)\right] \cdot \tilde{n} \quad (5.83)$$

The strained laminar burning velocity can be obtained as follows:

$$\frac{\partial S_u}{\partial t} + \widehat{U}_s \nabla \cdot s_u = -\sigma_s S_u + \sigma_s s_u^\infty \frac{(s_\mathrm{L}^0 - S_u)}{(S_u - s_u^\infty)}, \quad s_u^\infty = s_\mathrm{L}^0 \max\left(1 - \frac{\sigma_s}{\sigma_{\mathrm{ext}}}, 0\right) \quad (5.84)$$

where S_L^0 is the unstrained laminar flame speed and σ_{ext} is the strain rate at extinction. Additional details on the modeling can be found in [23]. An alternative approach to estimate the laminar burning velocity, which accounts for the influence of strain and heat losses, is to use empirical relations that can be derived based on strained laminar flamelet computations [39]. However, this has not been considered in the present analysis.

Once the value of the flame wrinkling is obtained using the above formulation, it can be used to estimate the model parameter α, and the efficiency function can therefore be dynamically evaluated during the computations. Additionally, the strained flame speed can be used to replace the unstrained value. A distinct advantage of the thickened flame approach is that the reaction rates can be calculated using Arrhenius rate laws, much like a direct numerical simulation (DNS) calculation, without the need of any ad hoc sub-models, while allowing the explicit use of multi-step chemistry mechanisms as well. Moreover, phenomena such as ignition and flame–wall interactions are directly accounted for, without requiring additional sub-modeling. In addition, the proposed wrinkling-based approach is an improvement over the existing algebraic models, since deviation from local equilibrium and the absence of time-history and strain effects in the equilibrium models can result in inaccurate predictions, especially in case of lean premixed flames under highly unsteady conditions.

For the chemical reaction schemes, an appropriate multiple-step chemistry mechanism for propane combustion in air (like Jones-Lindstedt (JL) mechanism [34]) can be utilized. While simplified global schemes can provide adequate results if the main species concentrations are of interest, they cannot be expected to work as well under unconventional combustion conditions (e.g., oxy-fuel combustion) and are likely to inaccurately estimate the temperature in the absence of an adequate set of reactions. The role of reaction chemistry can also be significant in combustion systems prone to thermo-acoustic instabilities, near extinction and re-ignition phenomena, as well as influencing the flame speed. However, the reaction schemes used within the thickened flame framework should typically include a limited number of intermediate species as it can lead to difficulties for wrinkled and/or stretched flame fronts [36, 37].

5.4 Modeling of Radiation

The radiant energy is calculated through solving the radiative transfer equation (RTE). This equation is based on the conservation principle applied to a monochromatic bundle of radiation. The RTE is given by Viskanta [41] as follows:

$$\frac{1}{c}\frac{\mathrm{d}I_\nu}{\mathrm{d}t} = -(\kappa_\nu + \sigma_\nu)I_\nu + \kappa_\nu n_\nu^2 I_{b\nu} + \frac{\sigma_\nu}{4\pi}\iint_{\Delta\nu,\Omega=4\pi} \Phi_\nu\left(\overset{\rho'}{s} \rightarrow \overset{\rho}{s}; \nu' \rightarrow \nu\right) I_{\nu'}\left(\overset{\rho'}{s}\right)\mathrm{d}\Omega'\mathrm{d}\nu' \tag{5.85}$$

where the different coefficients and terms in the above equation are defined as: "I_ν" is the spectral radiation intensity, "c" is the speed of electromagnetic wave in vacuum, "κ_ν" is the spectral absorption coefficient, "σ_ν" is the scattering coefficient, "$I_{b\nu}$" is Planck's spectral blackbody intensity of radiation, "n_ν" is the spectral index of refraction of the medium, and "Φ_ν" is the phase function. The spectral radiation intensity is a function of three spatial coordinates, time, and two angles, in addition to the radiation frequency or the wavelength. Solution of radiative transfer requires models to account for the directional and spectral natures of radiation. One of the popular methods to treat the directional nature of radiation is the discrete ordinates (DO) method. This method is originally suggested by Chandrasekhar [42] for astrophysical applications.

The spectral nature of radiation is a very important aspect in the gas radiation treatment. In general, there are three different models used to define the radiative properties of combustion gases [43]. Those models are the spectral line-by-line models, the spectral band models, and the global model. The complexity of the models decreases from the line model to the global model. In the line-by-line (LBL) model, the RTE is integrated into detailed molecular spectrum for the gases. This model is used only for benchmark solutions due to the enormous number of computational requirements. The statistical narrowband (SNB) model provides the spectral transmissivity averaged over a narrowband. This model is suitable for the radiation heat transfer prediction in high-temperature mediums. The wideband model (WBM) is a simplification of the SNB model; it yields wideband absorptance and requires the knowledge of the path length in the model as well as the spectral parameters associated with the path length.

5.4.1 Simple Gray Gas (SGG) Model

The simple gray gas (SGG) model, in addition to being simple, requires low execution time. The model considers the effective absorption coefficient as the main parameter controlling the radiative properties of a gas mixture and assumes that radiant absorption and emission by gas molecules are independent of the frequency of the radiation. Under the gray gas assumption and neglecting scattering of radiation, the RTE for the radiation intensity (integrated into the entire spectrum) in 3D Cartesian coordinates is expressed as follows:

$$\zeta \frac{\partial I}{\partial_x} + \eta \frac{\partial I}{\partial_y} + \mu \frac{\partial I}{\partial_z} = -\kappa_e I + \kappa_e I_b \tag{5.86}$$

In the full modeling of a gas-fired furnace, the RTE is solved for known temperature field and species concentration that are determined by conservation equations of motion, energy, and mass of species. At the start of the solution, an initial guess is made for all of the parameters. Then, iterative methodology is used in order to get the final converged solution. The RTE is solved based on the calculations of temperature and species concentrations from the previous step. A good estimation of the effective absorption coefficient from the known properties (a mean beam length L_{m} and a characteristic gas temperature) of emission by the gas can be obtained from interpretation of the total emissivity of the gas upon the bounding surface. The characteristic temperature can be the volume-average gas temperature T_{m}, as follows.

$$\kappa_e = -\left(\frac{1}{L_{\mathrm{m}}}\right) \ln\left[1 - \varepsilon_g(T_{\mathrm{m}}, L_{\mathrm{m}})\right] \tag{5.87}$$

Or the characteristic temperature can be the local temperature, T, giving a locally varying value,

$$\kappa_e = -\left(\frac{1}{L_{\mathrm{m}}}\right) \ln\left[1 - \varepsilon_g(T, L_{\mathrm{m}})\right] \tag{5.88}$$

The mean beam length is estimated as

$$L_{\mathrm{m}} = 3.6 \frac{V}{S} \tag{5.89}$$

where V represents volume of furnace and S is the wall surface area.

5.4.2 Exponential Wideband Model (EWBM)

The exponential wideband model is based on a physical analysis of gas absorption. This model provides a set of semiempirical expressions to predict the total band absorptance of infrared active molecules. This model can be used to predict radiative properties in a wide range of temperature, total pressure range, volumetric fraction, and path length [44]. In this model, it is assumed that the total absorption of a vibration–rotation band can be approximated with correlations dependent on three parameters: the integrated band intensity, α; the bandwidth parameter, ω; and the mean line-width-to-spacing parameter, β. These parameters depend on temperature, but pressure effects are also accounted for through the equivalent broadening pressure, P_e. The parameters yield asymptotic relations for the total band absorptance A_k. These relations are known as the four-region expression, which is defined as linear, square root, log-root, and logarithmic regions. The expressions of the band transmissivity, τ_k, are derived from the relation:

$$\tau_k = \frac{\tau_H}{A_k} \frac{\partial A_k}{\partial \tau_H} \tag{5.90}$$

The optical depth at the band head τ_H and the pressure correction parameter η are calculated from:

$$\tau_H = \frac{\alpha X}{\omega} \tag{5.91}$$

$$\eta = \beta P_e \tag{5.92}$$

where X is the density path length. The parameters α and β are calculated according to the simplified relations presented by Lallemant and Weber [44], while ω is given by a correlation dependent on temperature. In order to calculate the total emissivity of H_2O–CO_2 mixtures, band energy approximation (BEA) is used. In this method, it is assumed that the blackbody emissive power is constant over each absorption band. The total emissivity is given as [45]:

$$\varepsilon_g \cong \sum_{k=1}^{k=N} \frac{\overline{E^{\circ}}_{\nu C,k}}{\sigma \cdot T^4} \cdot A_k - \Delta\varepsilon_{c+w} \tag{5.93}$$

The correction term, $\Delta\varepsilon_{c+w}$, accounts for the overlap between the 2.7 and 15 μm bands for mixtures of CO_2 and H_2O. The simplified procedure proposed by Modak [46] is used in the present study to calculate this overlapping.

5.4.3 Leckner Model

The model was developed by Leckner in 1972 [47]. The CO_2 and H_2O emissivities are treated separately. Then, the total emissivity of the mixture is determined by taking a total band overlap correction term into consideration. The advantage of this model is that it can be applied to any arbitrary partial pressures of CO_2 and H_2O. The total emissivity for a mixture of CO_2 and H_2O is calculated as follows:

$$\varepsilon_g = \varepsilon_c + \varepsilon_w - \Delta\varepsilon_{cw} \tag{5.94}$$

In this model, a zero partial pressure emissivity for either CO_2 or H_2O is given by:

$$\ln \varepsilon_o = a_o + \sum_{i=1}^{i=M} a_i \lambda^i \tag{5.95}$$

in which:

$$a_i = C_{oi} + \sum_{I=1}^{j=M} C_{ji}\tau^j \tag{5.96}$$

$$\lambda = \log_{10}(p_i L) \tag{5.97}$$

where $p_i\,L$ is the pressure path length of either H_2O or CO_2 (given in bar cm) and $\tau = T/1000$ (the gas temperature T is given in K). The coefficients C_{ji} are found in Ref. [47]. The pressure correction term is determined using the relation:

$$\frac{\left[\frac{\varepsilon_i}{\varepsilon_o} - 1\right]}{\left[\frac{\varepsilon_i}{\varepsilon_o} - 1\right]_{\max}} = \exp\left\{-\zeta(\lambda_{\max} - \lambda)^2\right\} \tag{5.98}$$

where ε_i is the emissivity of either CO_2 or H_2O. At high pressures, the emissivity follows an asymptotic behavior according to:

$$\left(\frac{\varepsilon_i}{\varepsilon_o}\right)_{\max} = \frac{A \cdot P_E + B}{P_E + A + B - 1} \tag{5.99}$$

The parameters ξ, $\lambda_{\max}$, A, and B are given in Ref. [47]. The effective pressure P_E is defined in the following two equations for H_2O and CO_2, respectively.

$$P_E = P_T\left(1 + 4.9\frac{p_w}{P_T}\sqrt{\frac{273}{T}}\right) \tag{5.100}$$

$$P_E = P_T\left(1 + 0.28\frac{p_c}{p_T}\right) \tag{5.101}$$

The total emissivity of the gas mixture of H_2O and CO_2 for the desired partial and total pressures is equal to the sum of emissivities of the two gases minus a correction term $\Delta\varepsilon_{cw}$ due to overlap in some spectral regions. The overlap between the two gases is approximated by:

$$\Delta\varepsilon_{cw} = \left\{\frac{\zeta}{10.7 + 101\zeta} - 0.0089\zeta^{10.4}\right\} \cdot \lambda^{2.76} \tag{5.102}$$

in which the partial pressure ratio ζ is given by:

$$\zeta = \frac{p_w}{p_w + p_c} \tag{5.103}$$

and the parameter λ is defined as:

$$\lambda = \log_{10}((p_w + p_c) \cdot L) \tag{5.104}$$

5.4.4 Perry Model

Empirical correlations for the emissivities of water vapor, carbon dioxide, and four mixtures of the two gases were developed [48]. They are valid to be used for pressure path length of range 0.005–10 m atm. The emissivities at three temperatures of 1000, 1500, and 2000 K at six values of partial pressure ratios of H_2O/CO_2, namely 0, 0.5, 1.0, 2.0, 3.0, and ∞, can be calculated by the following equation. Empirical constants for different partial pressure ratios of H_2O/CO_2 are given in [48]. These correlations were developed based on the data in Hottel's emissivity charts [49] and were adjusted to the more recent data from cross-handler [50]. Linear interpolation or extrapolation of the emissivities determined at 1000, 1500, and 2000 K is used in order to obtain the emissivity at different temperatures, and it is given by:

$$\varepsilon_g T_g = \frac{\varepsilon_g T_H (T_g - T_L) + \varepsilon_g T_L (T_H - T_g)}{500} \tag{5.105}$$

where T_H and T_L are the higher and the lower temperatures, respectively.

5.4.5 Weighted-Sum-of-Gray-Gas (WSGG) Model

There are several gas radiative property models, called global models; they are based on the concept of weighted sum of gray gases. Example of these models is Hottel and Sarofim's gas radiative property model [49]. Equation 5.22 is used for the evaluation of total emissivity in terms of the weighted sum of gray gases, and it is useful especially for the zonal method of analysis of radiative transfer.

$$\varepsilon = \sum_{i=o}^{i=I} a_{\varepsilon,i}(T)\left[1 - e^{-\kappa_i PL}\right] \tag{5.106}$$

where $a_{\varepsilon,i}$ is the emissivity weighting factors for the gray gas I (these weighting factors are dependent on gas temperature T), κ_i is the absorption coefficient, P is the sum of partial pressures of absorbing gases, and L is the thickness of gas layer or path length. The weighting factor for clear gas, i.e., for $i = 0$, is defined as:

$$a_{\varepsilon,o}(T) = 1 - \sum_{i=1}^{i=I} a_{\varepsilon,i} \quad (5.107)$$

A common representation of the temperature dependency of the weighting factors is a polynomial of order $J - 1$ given as:

$$a_{\varepsilon,i}(T) = \sum_{j=1}^{j=J} b_{\varepsilon,i,j} T^{j-1} \quad (5.108)$$

where $b_{\varepsilon,i,j}$ is referred to as the emissivity gas temperature polynomial coefficients. The absorption coefficients κ_i and the polynomial coefficients $b_{\varepsilon,i,j}$ are obtained by fitting Eq. 5.22 to a table of total emissivities previously calculated from the EWBM. The total emissivities for various gas mixtures can be obtained utilizing experimental measurements, by calculation based on spectral lines or bands based on Hottel's charts, etc. The total absorptivity is calculated as follows:

$$\alpha = \sum_{i=1}^{i=I} a_{\alpha,i}(T, T_s)\left[1 - \mathrm{e}^{-\kappa_i PL}\right] \quad (5.109)$$

where $a_{\alpha,i}$ is the absorptivity weighting factor, T is the gas temperature, and T_s is the surface irradiation temperature. The weighting factors for a CO_2/H_2O/clear gas mixture with $P_{co_2} = 0.1P$, $P_{H_2O} = 0.2P$, and P = 101.3 kPa have been reported by Truelove [51].

All the WSGG model parameters, i.e., weighting factors and absorption coefficients, are intended for air combustion conditions, and their validity to be used in oxy-fuel combustion is limited. Pressure path lengths and ratios of H_2O to CO_2 in oxy-fuel combustion are drastically different from that of conventional combustion. Therefore, there is a need to derive new parameters to be used in these models. Johansson et al. [52] have developed two WSGG models: The first consists of three-gray-one-clear gas and the second of four-gray-one-clear gas.

5.5 Modeling Species Transport

The mass fraction of each species m_l is predicted through the solution of a convection–diffusion equation for the ith species. The conservation equations can be expressed in the following form:

$$\frac{\partial}{\partial x_i}\left(\rho \overline{\mathrm{U}}_i m_l\right) = -\frac{\partial}{\partial x_i} J_{l,i} + R_l \quad (5.110)$$

where R_l is the mass rate of creation or depletion by chemical reaction of the species i and $J_{l,i}$ is the diffusion flux of species i, which arises due to concentration gradients which is given by:

$$J_{l,i} = -\left(\rho D_{l,m} + \frac{\mu_t}{Sc_t}\right)\frac{\partial m_l}{\partial m_i} \tag{5.111}$$

where $D_{l,m}$ is the diffusion coefficient for species l in the mixture and Sc_t is the turbulent Schmidt number; $\frac{\mu_t}{\rho D_t}$ is equal to 0.7. An eddy dissipation model [53] that relates the rate of reaction to the rate of dissipation of the reactant- and product-containing eddies is used to calculate the rate of reaction. The source of chemical species i due to reaction, R_i, is computed as the sum of the reaction sources over the N_R reactions; thus:

$$R_i = M_i \sum_{k=1}^{N_R} R_{i,k} \tag{5.112}$$

where M_i is the molecular weight of species i and $R_{i,k}$ is the molar rate of creation/destruction of species i in reaction k. The reaction rate, $R_{i,k}$, is controlled either by an Arrhenius kinetic rate expression or by the mixing of the turbulent eddies containing fluctuating species concentrations. The rate of reaction $R_{i,k}$ is given by the smaller value of the two expressions below:

$$R_{i,k} = v_{i,k} M_i A \rho \frac{\varepsilon}{k} \frac{m_R}{v_{R,k} M_R} \tag{5.113}$$

$$R_{i,k} = v_{i,k} M_i A B \rho \frac{\varepsilon}{k} \frac{\sum_p m_p}{\sum_j^N v_{j,k} M_j} \tag{5.114}$$

where m_P is the mass fraction of a product species (P), m_R is the mass fraction of a reactant (R), and R is the reactant species giving the smallest value of $R_{i,k}$. A is an empirical constant equal to 4.0, B is an empirical constant equal to 0.5, $v_{R,k}$ is the stoichiometric coefficient for reactant i in reaction k, and $v_{j,k}$ is the stoichiometric coefficient for product i in reaction k. The eddy dissipation model relates the rate of reaction to the rate of dissipation of the reactant- and product-containing eddies, and the term k/ε represents the time scale of the turbulent eddies.

5.6 Modeling Reaction Kinetics

5.6.1 Chemistry Reduction/Acceleration Techniques

Apart from implementation of the combustion models, reaction chemistry must also be appropriately employed. Chemical mechanisms usually contain many species and reactions. To compute a reacting flow using such a mechanism, an equally substantial number of transport equations would have to be solved, often involving many intermediate species. Furthermore, the involved radicals have a short lifetime and very high net production rates which depend strongly and nonlinearly on the concentrations of other species and temperature. This leads to very stiff equations and modeling problems that grow with the size of the mechanism. Therefore, both the mechanism size and its associated stiffness are key factors to be addressed, and it is highly desirable to somehow reduce the dimensionality of the chemical reaction system. Different methods have been developed in order to address these issues and are discussed briefly below.

5.6.1.1 Reduced Chemistry Mechanisms

Traditional approaches to chemistry reduction are based on the quasi-steady-state assumption (QSSA) for species and the partial equilibrium assumption (PEA) for elementary reactions [54]. Significant progress has been achieved in automating the chemistry reduction based on QSSA and PEA based on data from canonical combustion problems, such as 1D premixed and non-premixed flames and simpler 0D reactor models [54]. The final outcome of chemistry reduction is a skeletal mechanism, or a reduced mechanism made up of global steps, with a significant reduction in the number of reaction steps and species.

One-Step Global Chemistry

The most simplistic approach is to describe the entire reaction system by a single reaction, using a fitted rate constant based on experimental measurements. Only the major species are considered, and the intermediate species are not included. Moreover, the reaction, in general, may be applicable only to a specific range of combustion operating conditions. Note that if a homogeneous combustion system is described by a one-step reaction, a single combustion progress variable is sufficient to describe the chemistry, and the flame front approaches are generally applicable.

Multi-Step Reduced Chemistry

Chemistry reduction approaches, which result in a skeletal mechanism or a reduced mechanism made up of global steps, with a significant reduction in the number of reactions and species, are often adopted to account for the complexity of chemistry in turbulent flames, while ensuring that the schemes can be efficiently incorporated in numerical computations. These multiple-step reduced combustion chemistry mechanisms are often considered to include the effects of reaction chemistry and to get a better insight into the emissions while ensuring that the calculations are not computationally intensive.

5.6.1.2 Low-Dimensional Manifolds

While these simplified schemes are often used in LES and can provide adequate results if the main species concentrations are of interest, they cannot be expected to work as well under unconventional combustion. Further, predicting the flame shape, its stabilization process, and pollutant emissions in practical burners also requires the use of sufficiently detailed chemistry. Additionally, mechanism reduction does not guarantee the elimination of the shortest time scales. In a skeletal mechanism, the fastest reactions may still be retained. In global mechanisms, algebraic relations for quasi-steady-state (QSS) species or resulting from the PEA may still preserve some of the stiffness in the original detailed mechanism [54]. Since the use of detailed chemistry directly within the LES simulations can be computationally intensive, significant efforts have been made in order to integrate chemistry effects accurately and efficiently using alternative approaches, such as the low-dimensional manifold strategy. The technique involves a dynamical system-based approach to identify the fast and slow time scales of chemically reacting flows. The method is based on the premise that the system in composition space, parameterized with a limited set of composition variables, lies on or around a low-dimensional manifold, which is characterized by the slow time scales of the system [54]. By constructing databases of relevant quantities using detailed simulations of simple flames, the approach can reduce the cost of performing reacting flow computations with extensive chemical kinetic mechanisms, while still retaining the accuracy of the detailed results. The intrinsic low-dimensional manifold (ILDM) and the flamelet-generated manifold (FGM) are such examples.

Intrinsic Low-Dimensional Manifolds (ILDMs)

The general idea behind ILDM is explained using Fig. 5.10a, which displays the reaction paths of a homogeneous methane–air system in the chemical state space [54]. This space is spanned by the mass fractions of all species Y_i, the pressure p, and enthalpy h. At a certain time, the state of the system is represented by a point in this multi-dimensional state space. Chemical reactions change the species composition of the system during the reaction process, so the system state moves along a trajectory through the state space. The trajectory connects the initial state with the final state, i.e., the equilibrium state. The projections of trajectories onto a certain plane in state space are displayed in Fig. 5.10a. Lines with different symbols denote paths from different initial compositions. Since all initial compositions were chosen to have the same elementary composition, they all end up in the same equilibrium point (denoted with an empty circle). The remarkable feature of the system behavior is that all trajectories converge together toward an attracting

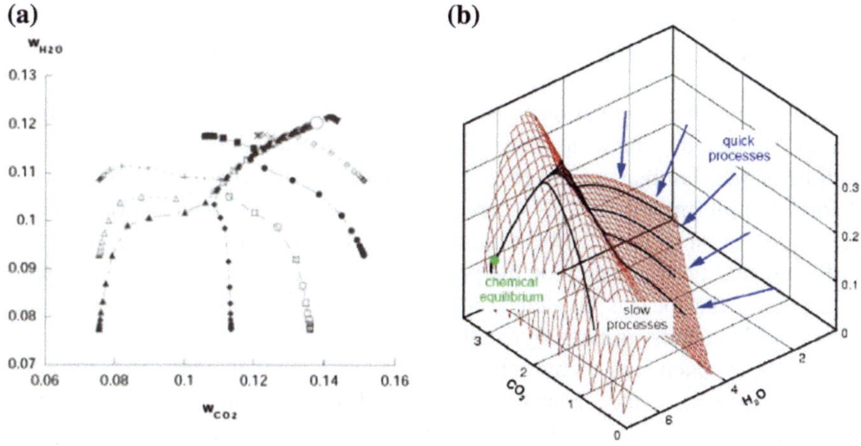

Fig. 5.10 **a** Reaction paths of a homogeneous CH_4–air system, projected onto the $Y_{CO_2}-Y_{H_2O}$ plane and **b** 3D schematic of ILDM

manifold, long before the system state reaches the equilibrium. The velocity of the system along the trajectory is not apparent from Fig. 5.10a, but it may be noted here that the movement along the common manifold is very slow, compared to movement toward this manifold. This is depicted in Fig. 5.10b, showing the multi-dimensional trajectories and the approach toward the attracting manifold.

The common manifold is referenced to as intrinsic low-dimensional manifold (ILDM) of the chemical system, and the observation leads to the following idea concerning the modeling: System states of the manifold quickly relax toward the manifold. This happens within time scales that cannot be resolved. The system state then moves slowly along the common manifold, which can be suitably resolved. It is therefore reasonably sufficient to presume that the system state is always on the manifold, due to the fast relaxation toward it. Since the manifold has a lower dimension than the entire state space, a smaller number of parameters are sufficient to represent the system state, and instead of solving equations for each species mass, equations for the manifold parameters can be solved. These parameters are usually properly chosen mass fractions or linear combinations of mass fractions.

Flamelet-Generated Manifolds (FGMs)

The FGM approach involves performing detailed simulations of laminar one-dimensional flames to generate a lookup table prior to LES computations. Specifically, the principal idea is that the thermochemical states in a laminar flamelet lie on a one-dimensional attracting manifold in state space. Thus, after computation of a laminar flamelet, a suitable parameterization has to be chosen (usually a linear combination of selected species mass fractions) and the states from the flamelet can be readily tabulated. When heat losses or a variable mixture fraction has to be considered, multiple flamelets with different initial temperatures or compositions have to be computed. The number of flamelets depends on the desired tabulation resolution. The advantage of this manifold computation method over the formerly introduced ones is that diffusion is included in a natural way. Furthermore, it is relatively easy to compute the chemistry table because reliable codes for flamelet computation exist and can be used. For example, in case of premixed flames, the chemical flame structure can be reproduced by a collection of one-dimensional laminar premixed flames calculated at various equivalence ratios and for different enthalpy levels (to account for heat losses) using detailed chemistry. A lookup table can then be generated to store the chemical information, with coordinates that typically include the progress of reaction (Y_c), the mixture fraction characterizing the local equivalence ratio (Y_z), and the enthalpy (h), thereby combining the advantages of mathematically reduced databases with a good estimation of the whole chemical flame structure. When coupled with the presumed PDF-based approach, the mean filtered values for any particular thermochemical property can then be obtained by performing integration over the corresponding PDF, thereby reducing the cost of reacting flow calculations, while retaining the accuracy of complex chemistry mechanisms.

5.6.2 Modified Two-Step Model for Oxy-combustion of Methane

It is widely recognized that numerical modeling of combustion applications is a computationally demanding process. That is why it is necessary to apply simplified reaction mechanisms to reduce the computing cost. Examples of CFD modeling in combustion application using the global mechanisms are found in the literature [55, 56]. The currently available simple computational mechanisms are cheap but not likely expected to give accurate results, especially under oxy-fuel combustion conditions. The replacement of inert N_2 with a chemically reactive compound, CO_2, has been shown to change the importance of some of the elementary reactions governing the combustion, thereby requiring a modification of the global multi-step reaction mechanisms to make them valid under oxy-fuel conditions [57]. In this section, the two-step hydrocarbon oxidation mechanism for the calculations of the reaction kinetics by Westbrook and Dryer [58], which was modified by Andersen et al. [59], is presented. The model was modified to handle the increased CO_2 concentration under oxy-fuel conditions. The Westbrook and Dryer model [58] consists of two reactions, where the last step, oxidation of CO to CO_2, is reversible. The mechanism is listed in the form of three irreversible steps. The modified reaction rate data by Andersen et al. [59] are listed in Table 5.1.

Table 5.1 Modified two-step methane–oxygen combustion mechanisms with kinetic rate data [59]

Reaction number	Reactions	*A*	β	E_a (J/kmole)	Reaction orders
Reaction 1	$CH_4 + 1.5O_2 \rightarrow CO + 2H_2O$	$1.59 * 10^{13}$	0	$1.998 * 10^{8}$	$[CH_4]^{0.7}[O_2]^{0.8}$
Reaction 2	$CO + 0.5O_2 \rightarrow CO_2$	$3.98 * 10^{8}$	0	$4.18 * 10^{7}$	$[CO][O_2]^{0.25}[H_2O]^{0.5}$
Reaction 3	$CO_2 \rightarrow CO + 0.5O_2$	$6.16 * 10^{13}$	−0.97	$3.277 * 10^{8}$	$[CO_2][H_2O]^{0.5}[O_2]^{-0.25}$

5.6.3 Modified JL Mechanism for Oxy-combustion of H_2-Enriched Methane

The H_2-enriched CH_4 reaction mechanism includes the combination of CH_4 reaction mechanism with that of H_2 with the former being the base mechanism. Ilbas et al. [60] investigated numerically the non-premixed turbulent H_2 and H_2-hydrocarbon flames where they combined two-step general reaction mechanism for hydrocarbon oxidation with global reaction mechanism for H_2 combustion. De and Acharya [61, 62] did a similar work in which two-step chemistry for CH_4 combustion was combined with single-step H_2 reaction mechanism proposed by Marinov et al. [63] in studies where large eddy simulation (LES) was used. In the modified JL mechanism for oxy-combustion of H_2-enriched methane, the modified four-step JL reaction mechanism for oxy-fuel combustion of CH_4 proposed by Frassoldati et al. [64] is combined with the single-step Marinov [63] H_2 reaction mechanism to solve chemical kinetics. The details of the reactions are presented in Table 5.2 with the first six reactions for the JL mechanism and the last one for the single-step H_2 reaction. The first reaction considers the oxidation of H_2 coming from partial oxidation of CH_4 under certain temperature. However, the second reaction considers the single-step oxidation of incoming H_2 from the inlet nozzle under certain temperature. One should notice that the conditions at which each reaction should occur are different in terms of temperature, flame speed, and molecular activity. Based on the calculations of flame speed in the work done by Frassoldati et al. [64] and Marinov et al. [63], the kinetic rates of both reactions are calculated as presented in Table 5.2. One should notice that the reaction parameters are different for the two reactions.

Table 5.2 Modified JL reaction mechanism for oxy-combustion of CH_4 [64] and the single-step mechanism for H_2 [63]; units of reaction parameters are: cal, mol, l, and s

Reaction	Order	Pre-exponential factor (A)	Temperature exponent (n)	Activation energy (E_a)
$CH_4 + 0.5O_2 \Rightarrow CO + 2H_2$	$[CH_4]^{0.5}$ $[O_2]^{1.30}$	3.06E+10	0.0	30,000
$CH_4 + H_2O \Rightarrow CO + 3H_2$	$[CH_4]^{1.0}$ $[H_2O]^{1.0}$	3.84E+09	0.0	30,000
$CO + H_2O \Longleftrightarrow CO_2 + H_2$	$[CO]^{1.0}$ $[H_2O]^{1.0}$	2.01E+09	0.0	20,000
$H_2 + 0.5O_2 \Longleftrightarrow H_2O$	$[H_2]^{0.3}$ $[O_2]^{1.55}$	8.03E+16	−1.0	40,000
$O_2 \Longleftrightarrow 2O$	$[O_2]^{1.0}$	1.5E+09	0.0	113,000
$H_2O \Longleftrightarrow H + OH$	$[H_2O]^{1.0}$	2.3E+22	−3.0	120,000
$H_2 + 0.5O_2 \Longleftrightarrow H_2O$	$[H_2]^{1.0}$ $[O_2]^{0.5}$	1.8E+16	0.0	35,002

5.7 H_2-Enriched Methane Oxy-combustion in a Model Gas Turbine Combustor: A Case Study

This study presents the results of experimental and numerical investigations of oxy-combustion H_2-enriched CH_4 diffusion flames in an atmospheric pressure swirl-stabilized gas turbine model combustor [65]. The flame characteristics in terms of temperature and species distributions, structures, and flow fields are studied experimentally and numerically over ranges of operating parameters. The effects of equivalence ratio, oxidizer composition, H_2 enrichment and swirl vane angle on flame stability, temperature distribution, and flow field are studied in detail. The swirl number considered in the experimental measurements is 1.10 with corresponding swirl vane angle of 55°. ANSYS Fluent was used to perform the numerical study, and the models adopted are: $k - \varepsilon$ (standard), discrete ordinate (DO), eddy dissipation concept (EDC) for turbulence, radiation, and species transport. Combined modified Jones-Lindstedt reaction mechanism and Marinov reaction mechanism for H_2 were considered as reaction mechanism for the numerical study. The numerical results are in good agreement with their corresponding experimental data. It was observed that utilizing H_2-enriched CH_4 improves the flame stability. The numerical results showed that oxy-fuel combustion of H_2-enriched CH_4 is not achievable at an equivalence ratio of 0.95 and above due to stability issues. The stability is highly enhanced by the corresponding formation of inner recirculation zone. A swirler with swirl vane angle of 65° achieved better stability limits of the gas turbine combustor.

5.7.1 *Boundary Conditions and Solution Technique*

A swirl-stabilized gas turbine model combustor with a power range from 4.1 to 6.2 MW/m^3/bar has been considered to conduct the recent experiments. Schematic diagram representing the experimental setup with more details can be found in Nemitallah and Habib [66]. The present study includes analysis of flame structures, blow-off limits, and measurement of flame temperature distributions in both axial and radial directions within the combustor for ranges of fuel composition (H_2 enrichment), equivalence ratio, and oxidizer mixture composition. A swirler with swirl vane angle of 55° was considered in this study with a corresponding swirl number (S) of 1.10 calculated based on:

$$S = {}^2\!/_3 \left[\frac{1 - (d_h/d_o)^3}{1 - (d_h/d_o)^2} \right] \tan\theta \tag{5.115}$$

where d_o is the outside diameter, d_h is the hub diameter, and θ is the swirl vane angle. The considered values of equivalence ratio were 0.55, 0.65, and 0.75. The combustor energy level was maintained constant at 4 MW/m^3 throughout all

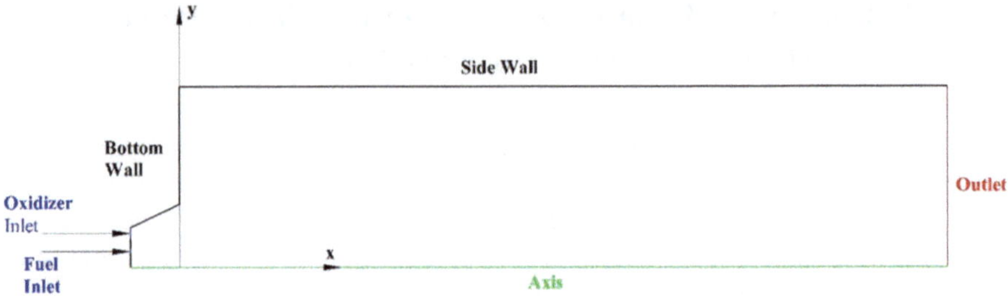

Fig. 5.11 Schematic diagram showing 2D axisymmetric of the combustor

experiments. The fuel composition was varied from 0%H_2 (i.e., 100%CH_4) to 50% H_2 (i.e., 50%CH_4), while the oxidizer mixture composition was varied from 100% (i.e., 0%CO_2) up to the value of blow-off. The above-mentioned modified JL reaction mechanism for oxy-combustion of H_2-enriched CH_4 is applied. The gas turbine model combustor was developed numerically by considering its half, axisymmetric, with appropriate consideration for the exit boundary condition as shown in Fig. 5.11. The length along the x-axis of the combustor is 300 mm, and its diameter along the y-axis is 70 mm. Computational domain was developed by discretizing the geometry into 77 × 170 grid points after conducting the grid dependency study of non-uniform mesh using Gambit 2.2. The boundary conditions were set by considering both inlets of the fuel (a mixture of CH_4 and H_2) and the oxidizer (a mixture of CO_2 and O_2) as velocity inlet conditions, and the exit of the combustor was considered as a pressure outlet boundary condition. The wall of the combustor was divided into two, the bottom-wall and the sidewall, due to differences in their materials (i.e., steel and quartz, respectively). The centerline of combustor was considered as an axis. The mesh was then exported to ANSYS Fluent R15.0 where 2D axisymmetric with gravity in negative x-direction was activated to obtain a steady-state solution.

The modified JL reaction mechanism for CH_4 combined with Marinov single-step H_2 mechanism and transport data files were imported through CHEMKIN mechanism into the ANSYS Fluent R15.0. In each of the cases considered, the fuel and oxidizer velocity (taking swirler distribution into consideration) with their respective mass fractions were accurately specified. The mixed thermal boundary condition was specified for the two types of walls, and the radiation heat transfer through the sidewall was modeled as semitransparent with a diffuse fraction of 0.9. The outlet of the combustor was modeled as pressure outlet with zero gauge pressure and backflow total temperature of 300 K, which means that the combustor was expected to discharge to the environment at atmospheric conditions. The pressure–velocity coupling was handled using SIMPLE scheme. All the partial differential equations described in the previous sections were discretized, apart from pressure discretization that was done using the standard technique. All other spatial discretization was achieved using the second-order upwind method. The criterion for convergence was set to be 10^{-3} for momentum equations and 10^{-6} for others.

Similar to the experimental conditions, oxidizer mixture composition, fuel composition, and equivalence ratio were varied so as to study their effect on flame structure, temperature field, and velocity field. A range of equivalence ratio was considered, from 0.55 to 1.00 at an interval of 0.1. Oxidizer mixture composition and fuel composition were changed from 100%O_2 to 50%O_2 and from 0%H_2 to 50%H_2, respectively. The effects of swirl vane angle on flame structure, temperature field, and velocity field were also examined numerically by considering three different swirlers with swirl vane angles of 55°, 65°, and 75°.

5.7.2 *Results and Discussions*

5.7.2.1 Flow and Flame Characteristics

The developed numerical method was used in predicting and analyzing the flow field in detail. Figure 5.12 shows the contour plots of radial and swirl velocity fields at an equivalence ratio of 0.65, 80%CH_4/20%H_2 fuel composition, and 50%O_2/50%CO_2 oxidizer mixture. Figure 5.12 shows the formation of inner and outer recirculation zones created downstream of the burner. These recirculation zones are characterized by negative values of radial velocity component. The inner recirculation zone appeared around the combustor axis and span between 0 and 20 mm along the axis, while the outer recirculation appeared in the vicinity of the reactor wall and span between 15 and 35 mm along the axis. Recirculation is due to the sudden expansion of the swirling oxidizer mixture as it exits the bluff body into the reactor which led to the rigorous positive pressure gradient and, subsequently, the recirculation zones are created. Mixing of hot flue gases with a cold fresh oxidizer is enhanced at the recirculation zones due to high turbulence intensities as a result of large fluctuations in the velocity. The resulting effects of recirculation on temperature and stability are explained in later section. The swirl effect due to swirl velocity is extended about 60 mm in the axial direction. This can be attributed to the high axial flow momentum due to high inlet fuel velocity in the axial direction. The swirl effect is degraded as the fuel is consumed in the chemical reaction process. The contour plots showing the kinetic rates of reactions also testified to this scenario as shown in Fig. 5.12.

Figure 5.13 shows the contour plots of species mass fractions at an equivalence ratio of 0.65, 80%CH_4/20%H_2 fuel composition, and 50%O_2/50%CO_2 oxidizer mixture. Figure 5.13 indicates that the mass fraction of CO is high within the flame zone and decreases in both axial and radial directions of the combustor. The high value of CO exhibited within the flame zone is due to the fact that CO is an intermediary product when burning CH_4 with O_2. It disappears down the reactor length and radius as a result of its further reaction with O_2 that escaped the reaction zone through outer recirculation zone due to sudden expansion to produce CO_2. The small fraction of CO at the exit plane is due to dissociation of CO_2 due to high flame temperature. Similarly, the concentrations of H_2O are high within the flame

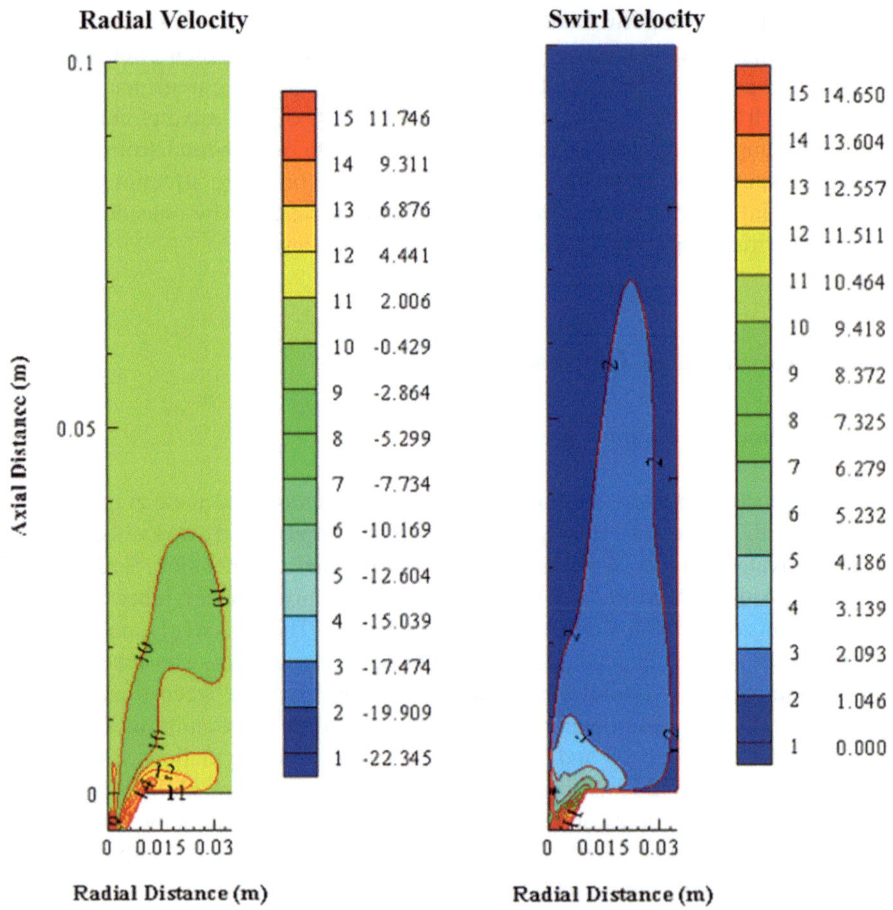

Fig. 5.12 Numerical contour plots showing velocity fields at 0.65 equivalence ratio, 80%CH_4/20%H_2 fuel composition, and 50%O_2/50%CO_2 oxidizer mixture

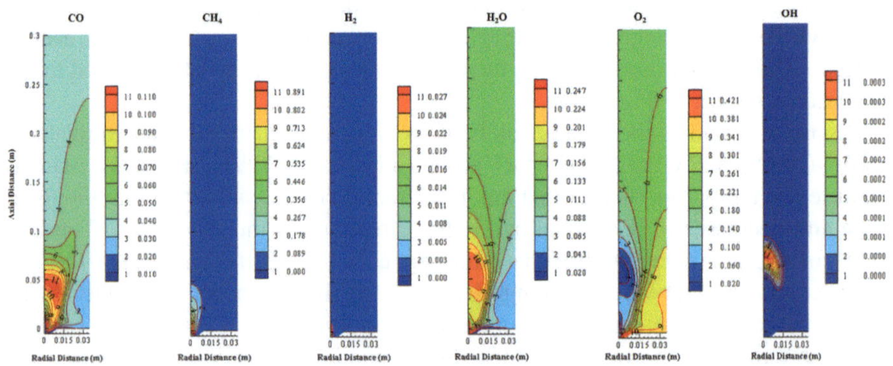

Fig. 5.13 Numerical contour plots showing species mass fraction distributions at 0.65 equivalence ratio, 80%CH_4/20%H_2 fuel composition, and 50%O_2/50%CO_2 oxidizer mixture

core as H_2O is a main product of the oxy-combustion process. However, H_2O concentrations decrease as the process proceeds in both axial and radial directions. The decrease may be attributed to the reduction in the reaction intensity and the increase in the amount of CO_2 in the reactor as a result of incoming fresh oxidizer. The results indicated that H_2O formation increases as the reaction proceeds and reaches its maximum value of 60-mm distance from the inlet section of the combustor. At this point, the static temperature attains its peak value, while the reaction driver (i.e., CH_4) approaches zero.

Unlike CO and H_2O, O_2 attains its least mass fraction values within the burning zone, while it attains its highest values in the vicinity of inner and outer recirculation zones. The attainment of least values is due to the fact that O_2 is among the major requirements to initiate and sustain the combustion process and is being consumed during the reaction process to generate the products. The attainment of highest values is due to the scape of a large amount of the supplied oxidizer at the sudden expansion through the outer recirculation zone. There was a further dilution of the fresh oxidizer with the burned gases which enables the reaction of O_2 and CO as explained previously and also reduces the static temperature. The concentrations of OH radicals remain zero everywhere in the combustor except in the flame core where the maximum temperature and reaction intensities are attained. Due to the high inlet flow velocity of fuel, the reactions are delayed and, accordingly, the core of the flame is located downstream in the combustor as shown in the contour plots of OH radicals. As expected, the mass fractions of CH_4 and H_2 were only high at the nozzle outlet and decrease downstream of the burner. The figure also indicates that H_2 appears only in the zone between the inlet section and a distance of 10 mm along the axis of the combustor due to high flame speed and high reactivity nature of hydrogen. Since the flame speed of CH_4 is lower when compared to that of H_2, it took a while for CH_4 to be completely consumed. The CH_4 species exist only between the axis and inner recirculation zone from inlet section up to 60 mm along the axis of the combustor. The mass fraction is zero elsewhere indicating that there were no cases of unburned hydrocarbons.

Figure 5.14 shows the plots of kinetic rate of reaction (KRR) along the combustor axis for two CH_4 reactions and H_2 reaction at 0.65 equivalence ratio, 80% CH_4/20%H_2 fuel composition, and 50%O_2/50%CO_2 oxidizer mixture. KRR_1 ($CH_4 + 0.5O_2 \Rightarrow CO + 2H_2$) and KRR_2 ($CH_4 + H_2O \Rightarrow CO + 3H_2$) represent the first and the second reactions of CH_4, while KRR_4 ($H_2 + 0.5O_2 \Rightarrow H_2O$) represents the oxidation reaction of H_2. Figure 5.14 shows that the peak value of KRR_2 corresponds to the peak value of CO, while the second peak value of KRR_1 corresponds to the peak value of CO_2. This indicates that the two reactions were actually responsible for the formation of the intermediate and final product. Furthermore, the CH_4 reactions take place only between the axis of the combustor and the inner recirculation zone. For the case of H_2, the reaction is fast with a high rate of reaction as shown in the secondary axis of Fig. 5.14. This can be attributed to the high reactive nature of H_2 and high rate of generation of intermediate H_2 within the flame core due to the very high flame temperature as a result of burning with pure oxygen.

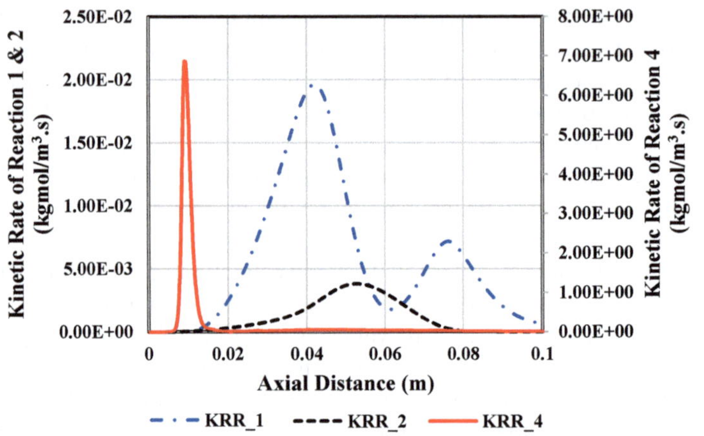

Fig. 5.14 Numerical kinetic rates of reactions (KRR) of CH_4 (1 and 2) and H_2 (4) reactions along the axis of the combustor at 0.65 equivalence ratio, 80%CH_4/20%H_2 fuel composition, and 50% O_2/50%CO_2 oxidizer mixture

5.7.2.2 Effect of Equivalence Ratio

The effects of equivalence ratio on the flame stability and structure are investigated experimentally and numerically. Figure 5.15 shows numerically the effects of equivalence ratio on flame structure/stability and temperature distribution at 80% CH_4/20%H_2 fuel composition and 50%CO_2/50%O_2 oxidizer mixture composition. The figure shows that an increase in the equivalence ratio tends to a slight increase in the flame length. This can be attributed to the reduction in the amount of O_2 available for the combustion process as the equivalence ratio is increased. As a result, the flame is extended for the combustion process to complete. Further increase in equivalence ratio (near stoichiometric) causes the flame to become completely unstable and lifts off the burner. This shows that stable oxy-combustion flames cannot be achieved near stoichiometric conditions in gas turbine combustors. As it could be observed from Fig. 5.15, the flame instability in terms of liftoff is observed at 0.95 equivalence ratio (i.e., near stoichiometric). Also, it is observed that the temperature increases with the increase of equivalence ratio in the region of stable flame operation (i.e., from 0.55 to 0.85 equivalence ratio), while it decreases with the increase of equivalence ratio when the flame becomes unstable (near stoichiometric conditions). Actually, at higher equivalence ratios, there exists low O_2 concentration and due to the higher specific heat capacity of CO_2, the radiation heat loss becomes higher; thereby, the combustion temperature is reduced.

Figure 5.15 also indicates that there is an increase in temperature along the axis of the combustor as the equivalence ratio is increased. This is expected since the increase in the equivalence ratio creates less availability of excess oxidizer to cool the combustion products; thereby, the combustion gases leave the combustor at a higher temperature. For all the equivalence ratios considered, temperature

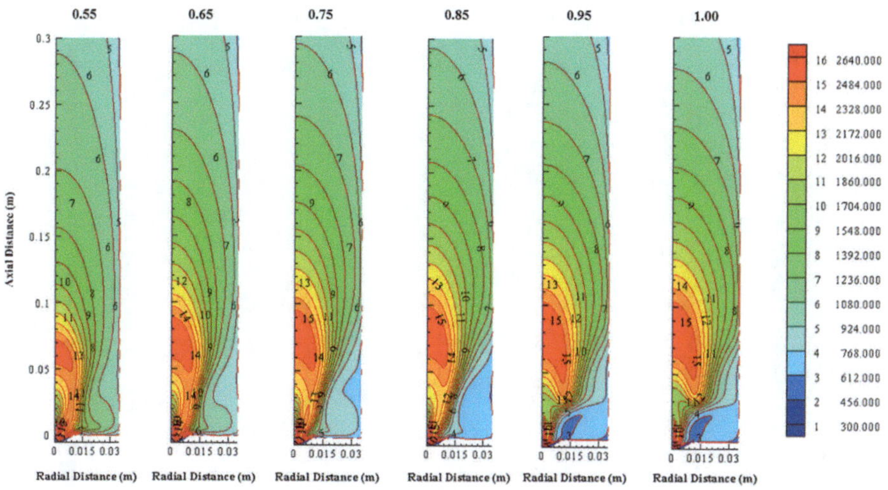

Fig. 5.15 Numerical contour plots showing the effects of equivalence ratio on flame structure/stability and temperature distribution at 50%CH_4/50%H_2 fuel composition and 50%O_2/50%CO_2 oxidizer mixture

distributions along the axis of the combustor decrease toward the exit section. In the radial zone between a radius of 0.02 m up to the combustor wall, along the radius of the combustor, there exists little irregularity in the rate at which the combustion products are being cooled. This can be attributed to the large amount of unburnt oxidizer that escapes the combustion zone through the outer recirculation zone which is responsible for the rapid cooling of the combustion products by enhancing the heat transfer through the combustor wall.

The contour plots of radial and swirl velocity components are presented in Fig. 5.16 for a range of equivalence ratio and at 80%CH_4/20%H_2 and 50%CO_2/50%O_2 fuel composition and oxidizer mixture, respectively. As shown in Fig. 5.16, the equivalence ratio significantly affects the radial and the swirl velocity fields, thereby affecting the flame stability. As the equivalence ratio increases, the sizes of both the inner and the outer recirculation zones decrease. When operating close to stoichiometric conditions, the outer recirculation zone completely disappears, while the size of the inner recirculation zone drastically reduces. This indicates that the inner recirculation zone is primarily responsible for flame stability. The swirl effect is extended for a long distance inside the combustor at low equivalence ratio due to the high oxidizer flow rate and effective mixing. As the equivalence ratio increases, the swirl effect is limited to the zone close to the inlet section resulting in a reduction in the size of the outer recirculation zone until the flame liftoff at an equivalence ratio of 0.95 as shown in Fig. 5.16.

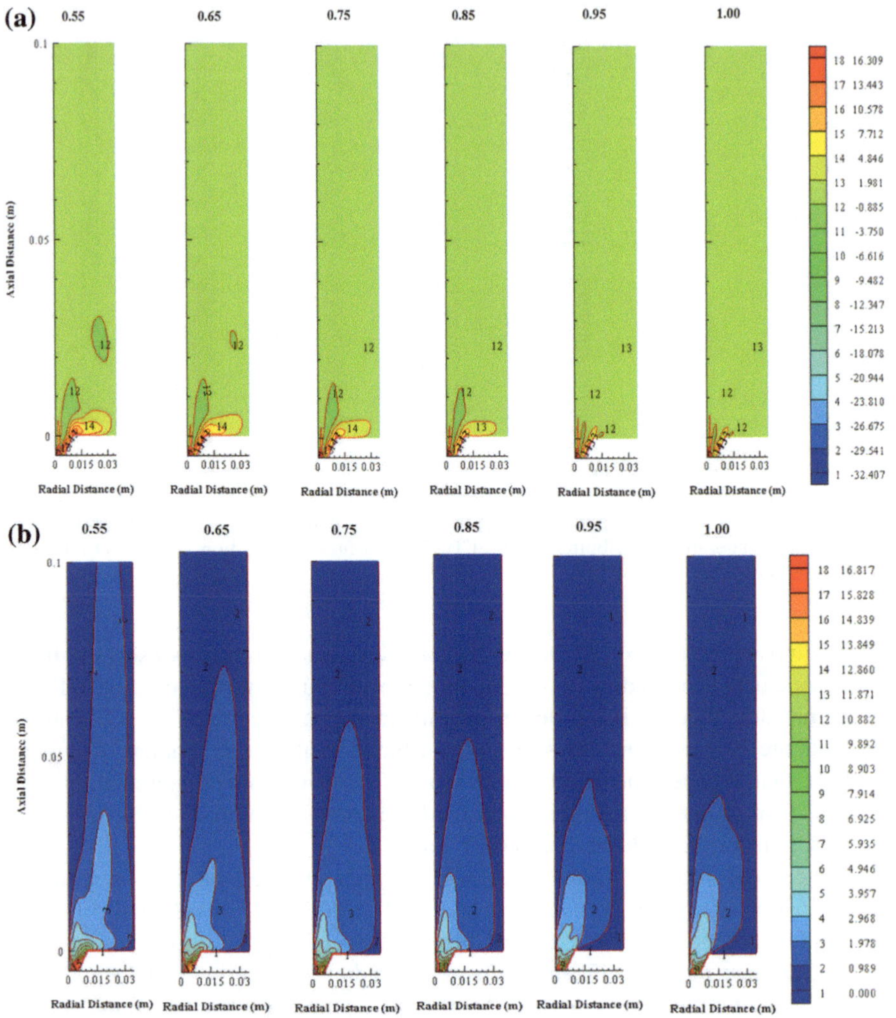

Fig. 5.16 Numerical contour plots showing radial (**a**) and swirl (**b**) velocity fields at different equivalence ratios, 50%CH_4/50%H_2 fuel composition, and 50%O_2/50%CO_2 oxidizer mixture

5.7.2.3 Effects of Oxidizer Mixture Composition

Figure 5.17 shows the effect of oxidizer mixture on the flame structure at an equivalence ratio of 0.65 and 80%CH_4/20%H_2 fuel composition. Figure 5.17 also shows a comparison between the experimental and computed flame structure. The figure indicates that the surface area of the flame decreases as the percentage of O_2 in the oxidizer mixture decreases until blow-off point is reached at a concentration of O_2 in the oxidizer mixture of 21.9%. The reason behind this is that decrease in O_2 percentage through the addition of CO_2 increases the oxidizer flow rate and

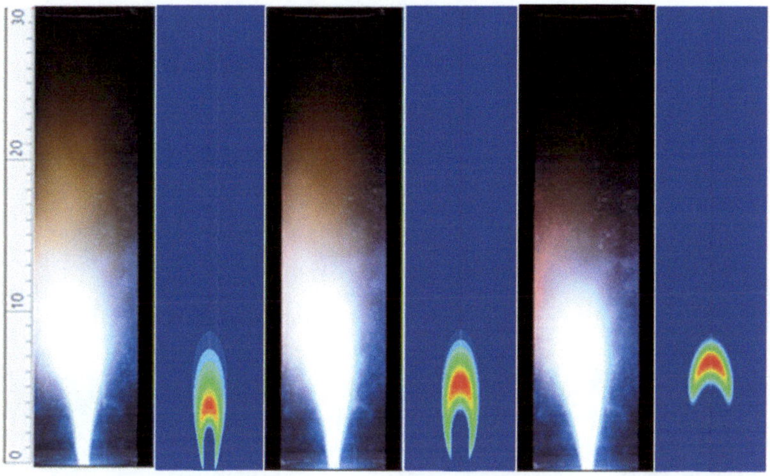

Fig. 5.17 Experimental (dark background) and numerical (blue background) flame cores; the experimental is captured using high-speed camera, and the numerical is captured by distributions of OH radicals, at different percentages of O_2, 0.65 equivalence ratio, and fuel composition of 80% CH_4/20%H_2

Reynolds number which enhanced the mixing and subsequently resulted in a reduction of residence time by decreasing the flame length. This can be explained on the basis that a high percentage of O_2 creates a low mixing rate and low Reynolds number which results in flame elongation. One can also deduce that flame is liable to blow-off at high Reynolds number. Moreover, the intensity of reactions, flame speed, and flame temperature are reduced with the reduction of O_2 (an increase of CO_2) in the oxidizer mixture. Figure 5.17 indicated a good agreement between the shapes of the flame core obtained experimentally through flame visualization using a high-speed camera and numerically through the distributions of the OH radicals. However, the experimental flame shape is not of full symmetric shape like the calculated flame shape. This can be attributed to the high-swirl intensity, and as a result, flame wobbling was experienced during the experiment. As a matter of fact, the OH radicals can represent effectively the core of the flame, where the temperature and reaction intensity are at their highest values. Examining Fig. 5.17 carefully, it could be realized that a large amount of the oxidizer supplied escaped through the outer recirculation zone to the reactor walls due to sudden expansion which was created between the bluff body and the combustor wall. The model combustor was actually designed in a manner such that a large amount of fresh oxidizer or excess oxidizer can escape the burning zone through the side walls and later mix-up with the combustion product downstream of the burner, thereby cooling the flue gases along the length of the combustor.

Figure 5.18 shows, in detail, the computed temperature field at different percentages of O_2, 0.65 equivalence ratio, and fuel composition of 80%CH_4/20%H_2. At a composition of 100%O_2 in the oxidizer mixture, a maximum value of

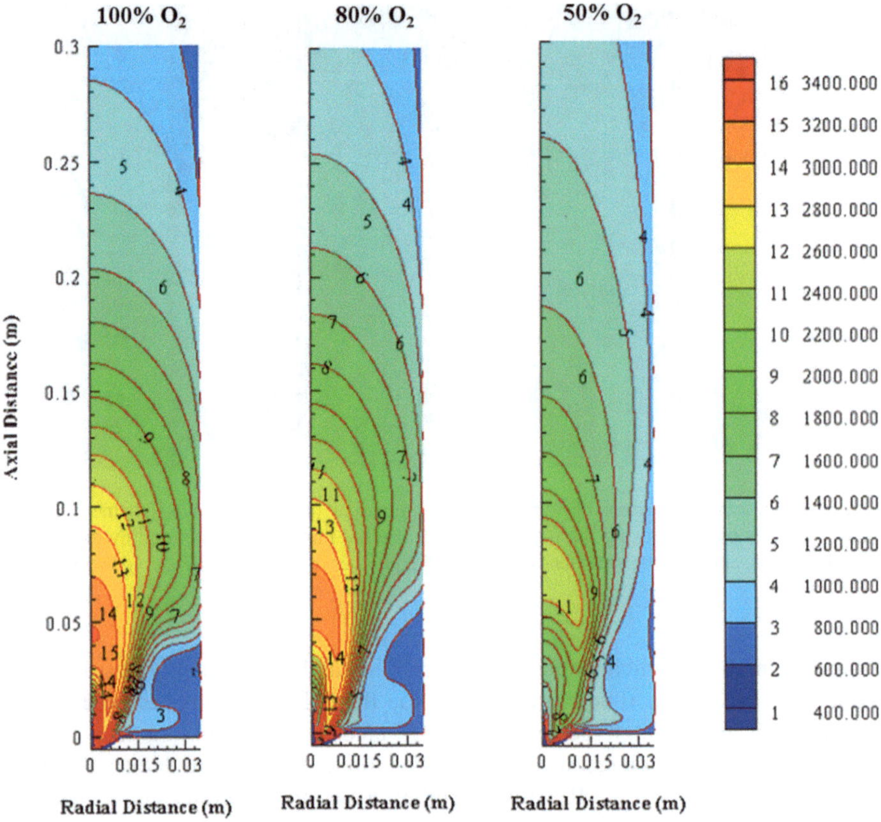

Fig. 5.18 Numerical temperature contours at different percentages of O_2, 0.65 equivalence ratio, and fuel composition of $80\%CH_4/20\%H_2$

temperature was attained in the combustor at the burning zone (along the axis), while the minimum value of temperature was observed around the outer recirculation zone. But, as the percentage of O_2 in the oxidizer is reduced due to the introduction of CO_2, the temperature in the outer recirculation zone gradually increased, while the reduction in the values of temperature was experienced at the burning zone. Such situations are due to two major reasons including flame speed or burning velocity and Reynolds number, which are the main parameters controlling the stability of the flame. At a high O_2 concentration in the oxidizer mixture, there exists high flame speed which eventually influences the combustion performance positively. Also, since O_2 has lower heat capacity as compared to CO_2 with poor radiative heat transfer properties, hence the combustor remained at a higher temperature. On the other hand, the introduction of CO_2 resulted in a reduction of O_2 percentage in the oxidizer, and hence, the Reynolds number is increased. Due to the physical (high molecular heat capacity, low thermal diffusivity, etc.) and chemical properties of CO_2, the burning velocity is reduced, and radiative heat transfer was

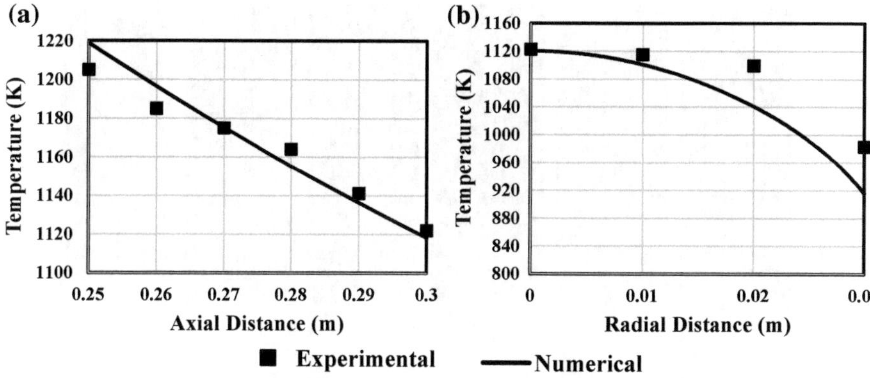

Fig. 5.19 Experimental and numerical temperature distributions at 0.65 equivalence ratio, 80% CH_4/20%H_2 fuel composition, and 50%O_2/50%CO_2 oxidizer mixture along axial (**a**) and radial (**b**) directions at the exit plane of the combustor

modified, thereby resulting in modifications in the temperature field. A large amount of heat is being absorbed by CO_2 and loses the same through radiative heat transfer. It was also noted that the flame decreases in length as the percentage of O_2 is decreased as a result of high Reynolds number.

Therefore, stability is being affected by high Reynolds number which was experienced as the CO_2 percentage in the oxidizer is increased. Seeking a specific comparison between the measured and the computed values of temperature in order to validate the applied model, Fig. 5.19 provides such comparison of temperature distributions along the axial and the radial directions within the combustor at 0.65 equivalence ratio and 80%CH_4/20%H_2 fuel composition. As it could be observed from Fig. 5.19, there exist good agreements between experimental and numerical results of temperature distributions. Due to heat transfer by convection and radiation through the wall of the combustor, the temperature is reduced in the radial direction. There exists large recirculation of fresh (excess) oxidizer at the outer recirculation zone which causes a drop in the temperature near the combustor wall. Figure 5.19b shows that the maximum temperature occurred at the axis of the combustor where the flame is located.

5.7.2.4 Effects of Hydrogen Enrichment

Figure 5.20 shows the effect of H_2 concentration on flame structures captured experimentally using a high-speed camera at 0.65 equivalence ratio and 50%CO_2/ 50%O_2 oxidizer mixture. An increase in the concentration of H_2 resulted in the extension of the flame length. This may be attributed to high reactivity, diffusivity, and combustibility nature of hydrogen. These features of H_2 enabled the resulting H_2-enriched flames to be more stable. The flames were characterized with a lot of wobbling due to the high turbulence intensities associated with the swirler.

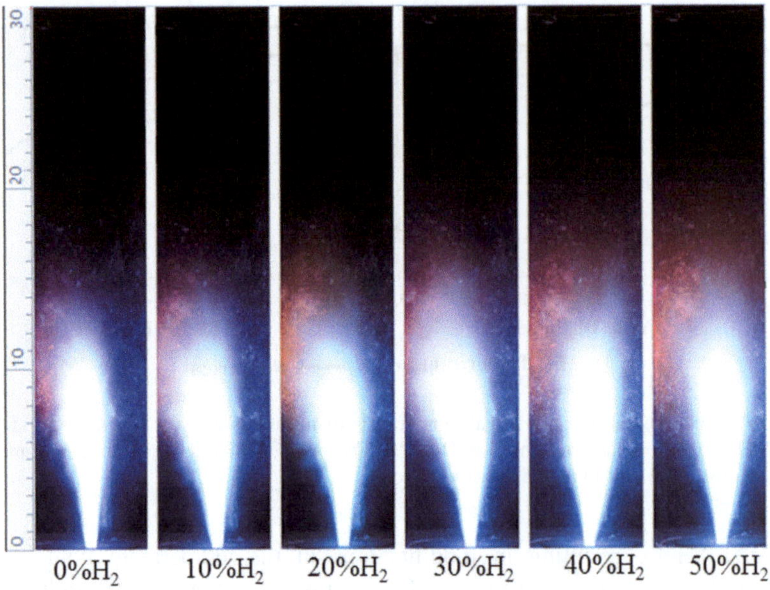

Fig. 5.20 Captured flame photographs showing the effect of H_2 concentrations on the flame structure at 0.65 equivalence ratio and 50%CO_2/50%O_2 oxidizer mixtures for 55° swirl vane angle

The effect of fuel composition (H_2 enrichment) on temperature distributions in axial and radial directions is presented in Fig. 5.21 at 0.65 equivalence ratio and constant oxidizer mixture. Three fuel compositions were considered including pure methane (0%H_2/100%CH_4) and two H_2-enriched fuels consisting of 10%H_2/90%CH_4 and 20%H_2/80%CH_4. Through the axis of the combustor, it can be observed that the higher the percentage of H_2 in the fuel composition the higher the temperature. However, the effect of H_2 enrichment is insignificant at low enrichment value, 10% H_2. This may be attributed to the considered constant energy level for all cases, where the added amount of H_2 to balance CH_4 in order to maintain the same energy level is not able to sustain higher temperature level. Further increase in the level of H_2 enrichment resulted in significant increase in temperature level due to the increased flow rate of hydrogen as shown in Fig. 5.21.

5.7.2.5 Effects of Swirl Vane Angle

Figure 5.22 shows the effects of swirl vane angle on temperature field at 0.65 equivalence ratio, 50%CH_4/50%H_2 fuel composition, and 50%O_2/50%CO_2 oxidizer mixture. As the swirl vane angle increases from 55° to 75°, the flame temperature decreases as shown in Fig. 5.22. Increasing swirl vane angle actually enhances the rate of mixing of the fresh cold oxidizer mixture with the hot burned gases, and as a result, the combustion temperature is lowered. The optimum value

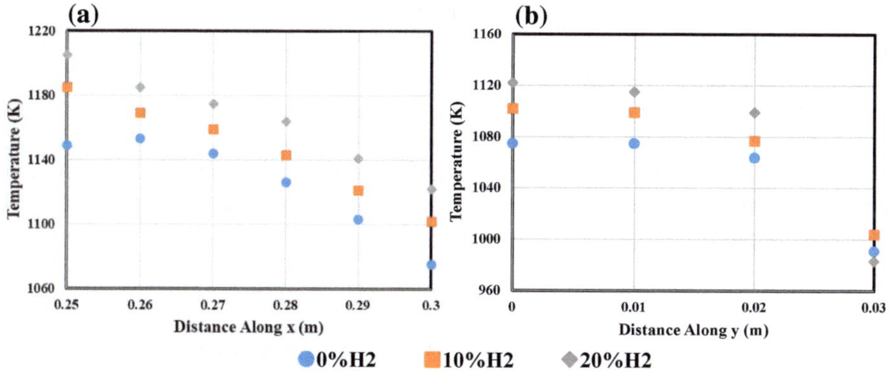

Fig. 5.21 Effect of fuel composition on axial (**a**) and radial (**b**) temperature distributions measured experimentally at fixed oxidizer mixture composition of 50%O_2/50%CO_2 and at an equivalence ratio of 0.65 for 55° swirl vane angle

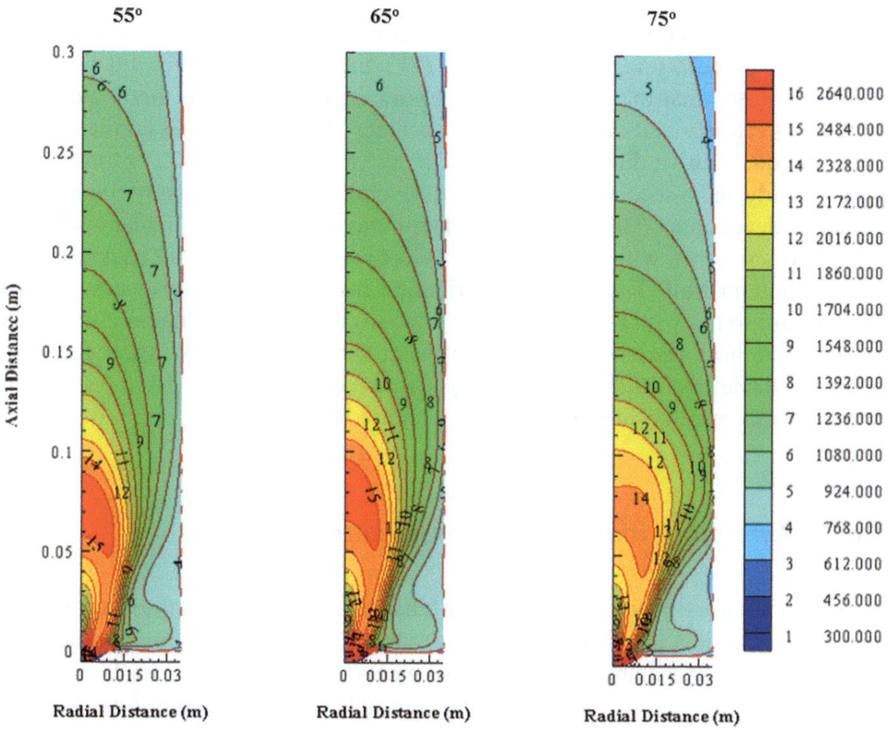

Fig. 5.22 Numerical contour plots showing the effect of swirl angle on flame stability and temperature distribution at 0.65 equivalence ratio, 50%CH_4/50%H_2 fuel composition, and 50%O_2/50%CO_2 oxidizer mixture

could lie in the range of 55°–65° or even 70° as discussed by Linck et al. [67] but not 75° because the shape of the flame exhibited by this angle is completely conical which is not practicable. The results showed that the flame is stable at all the cases due to the presence of inner recirculation zone. However, the swirl effect is stronger at lower swirl vane angles and is reduced while increasing the swirl vane angle. The swirler with swirl vane angle of 65° exhibited a lower amount of CO at the exit plane. This may be attributed to the higher mixing rate for the case of swirler with this vane angle.

5.8 Investigation of a Turbulent Premixed Combustion Flame in a Backward-Facing Step Combustor; Effect of Equivalence Ratio: A Case Study

In this study, Nemitallah et al. [68] carried out large eddy simulation (*LES*) to analyze lean premixed propane–air flame stability in a backward-step combustor over a range of equivalence ratio. The artificially thickened flame approach coupled with a reduced reaction mechanism is incorporated for modeling the turbulence–combustion interactions at small scales. Simulation results are compared to high-speed particle image velocimetry (*PIV*) measurements for validation. The results indicated that the numerical framework captures different topological flow features effectively and with reasonable accuracy, for stable flame configurations, but some quantitative differences exist. The recirculation zone (*RZ*) is formed of a primary eddy and a secondary eddy, and its overall size is significantly impacted by the equivalence ratio. The temperature distribution inside the recirculation zone is highly non-uniform, with much lower values observed close to the backward step and the bottom-wall. The mixture distribution inside the RZ is also non-uniform because of mixing with reactants and heat loss to the walls. The flame is stabilized closer to the backward step as the equivalence ratio increases. At lower fuel fractions, the flame lifts off the step starting at equivalence ratio of 0.63 and the liftoff distance is increased while the equivalence ratio is lowered.

5.8.1 Operating and Boundary Conditions

A planar combustor in which there is a step (change in plane level) in the flow direction was used to conduct the experimental work and the simulations. In this combustor, a premixed flame was stabilized near a backward-facing step. Figure 5.23a shows a schematic representation of this step combustor and the relevant dimensions. The inlet section of the combustor consists of a rectangular cross-sectional stainless steel duct of a 160-mm-spanwise width and 40-mm height. Air was fed to the combustor inlet by an Atlas Copco GA 30 FF air compressor through a flow meter, and the inflow was choked. At an axial location of 0.45 m downstream the choke plate, the cross section of the duct was contracted gradually to a height of 20 mm over an axial distance of 0.15 m, followed by a 0.4-m-long and 20-mm height duct of constant cross-sectional area. This was followed by the backward-facing step where the height expands with an expansion ratio of 2:1 back to 40 mm. This acted as the nominal anchoring point for the flame. The fuel flow rate was measured using a Sierra C100 M mass flow controller before its injection, 20 mm downstream of the choke plate, through a number of holes in the manifold. The mass flow controller allows a maximum flow rate of 2.36 g/s for propane with uncertainty in the measurement of $\pm 1\%$ of the full scale. The distance between the fuel injection point and the backward step was enough to mix the gases efficiently (see Altay et al. [69]). The length of the combustor downstream the step was 0.5 m, and the combustor was opened to the atmosphere. The combustor length was sufficiently short to prevent the coupling with the acoustics (see Hong et al. [70]). The temperature of the air–fuel mixture was measured using a K-type thermocouple located at a distance of 0.2 m upstream of the step. In order to allow for optical access to the flame, a quartz window was installed just downstream of the step.

Figure 5.23b shows a schematic representation of the particle image velocimetry (*PIV*) system used to measure the 2D velocity fields. The light source consists of a Nd:YLF laser of wavelength 527 nm capable of producing dual pulses at a rate up to 5 kHz with a peak power output of 25 mJ/pulse. A 1280 $\times$ 1024 pixel NAC GX-1 CMOS camera, with an F-mount Nikon 60 mm microlens with an aperture of *f/*8, was used for imaging at a rate of 1 kHz. The interval between the laser pulses was set to 30–150 μs depending on the flow velocity and field of view. A light sheet less than 1 mm thick in the imaged region was generated using a spherical lens with a focal length of 1000 mm and a cylindrical lens with a focal length of -20 mm. The former was used to reduce the diameter of the beam, and the latter was used to diverge the beam to generate a sheet. Bypass air, taken from the inlet pipe upstream of the choke plate, was routed through a cyclone-type seeder. Seeding particles consisting of 1.5–3-μm diameter Al_2O_3 were injected into the main air immediately downstream of the choke plate. The PIV measurements were processed using the LaVision DaVis 7.2 software. The field of view was set to 65–120 mm in the streamwise direction depending on the operating conditions, which corresponds to 0.05–0.1 mm/pixel. The LaVision DaVis 7.2 software was used to process the 1280 $\times$ 1024 or 800 $\times$ 800 pixel images of the seeding particles, using a multi-pass approach with the final pass of a 32 $\times$ 32 or 16 $\times$ 16 pixel window with 50% overlap, producing velocity fields with a spatial resolution of 0.8–1.5 mm. The backward-facing step combustor geometry was shown in Fig. 5.23c. The combustor had an inlet channel height (H) of 20 mm upstream of the step and a channel height of 40 mm downstream of the step. Upstream of the step, the inlet channel length was 50 mm which is long enough to allow for further flow development before reaching the step. The total length of the combustor channel downstream of the step was 350 mm with a fixed width of 160 mm.

A parallelized, unstructured, finite-volume *LES* code was used for solving the reactive compressible *3D* Navier–Stokes equations with second-order spatial and

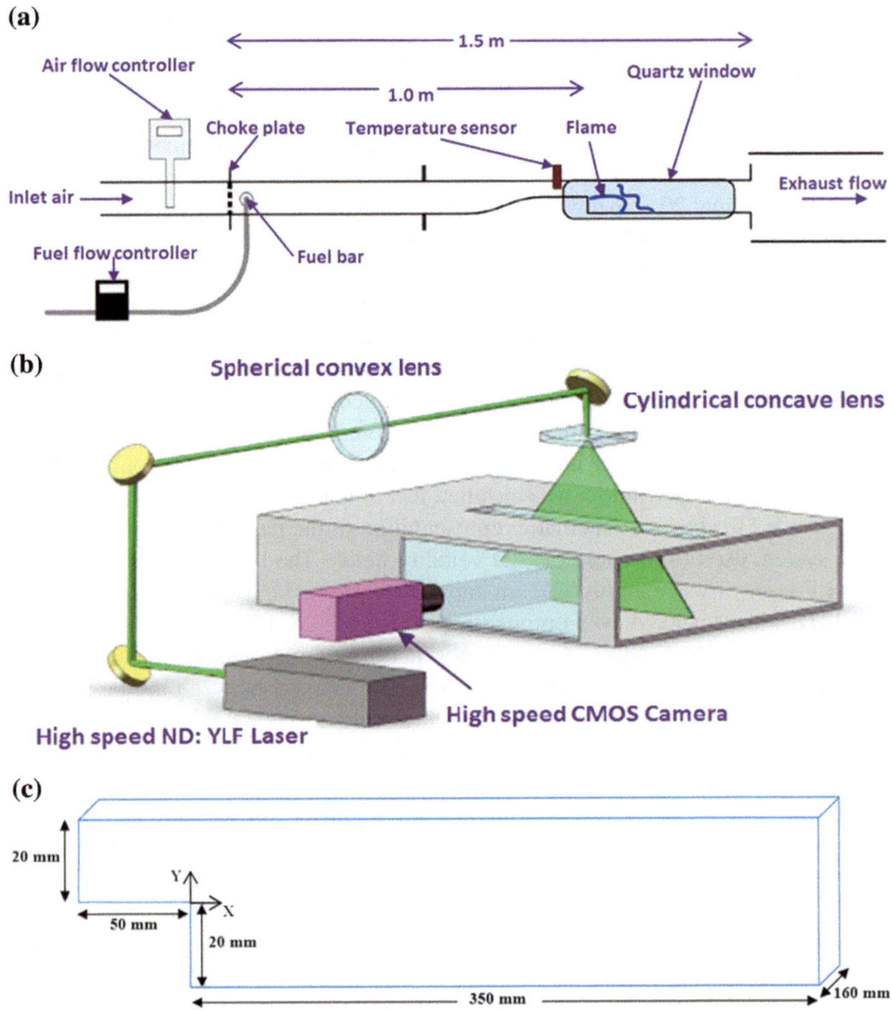

Fig. 5.23 Schematic diagrams showing: **a** the experimental setup of the present backward-facing step combustor, **b** the setup for the high-speed particle image velocimetry (*PIV*), and **c** the *3D* representation of the present combustor to be used in the *OpenFOAM* simulations

temporal accuracy. The one-equation eddy viscosity model was employed to estimate the sub-grid-scale (*SGS*) stress terms. The choice of the numerical grid was controlled by the values of the physical length scales of the flow. Pope [71] suggested a filter width (Δ) to integral length-scale (L_I) ratio of 0.083 to resolve at least 80% of the turbulent kinetic energy and capture the bulk of the energy-containing structures. Using the step height (H) as the integral length scale, the filter width was estimated to be 1.8 mm. Based on that, a non-uniform mesh was created using *OpenFOAM* consisting of approximately 16,600 cells in the *x*-*y* plane. Downstream the step, the total number of cells in the *x*- and *y*-directions was 270 and 59, respectively. The corresponding values of grid size in the *x*- and *y*-directions, Δx and Δy, were 1.0 mm and 0.5 mm, respectively. The used LES grid resolution was compatible with the PIV system resolution for easy post-processing and valid comparisons. In order to prevent excessive numerical dissipation or instability, the temporal resolution was determined based on physical time scale estimates as function of velocity and grid size. The time step for the simulations was estimated at approximately 2.8 μs; however, a value of 1 μs was used in order to adequately resolve the chemical time scales and consider local refinement and acceleration of the fluid above the bulk inlet velocity.

The dimensions of the combustor have been calculated carefully to match the flow and combustion conditions. The reactor length was short enough to avoid the interaction between the heat release field and the pressure field which results in acoustics instabilities. In all cases, a mixture of air and propane was introduced at a design bulk inlet velocity of 5.2 m/s. To achieve such design inlet velocity in parallel with the required design inlet Reynolds number of 6500, the total inlet area was calculated. While designing the inlet cross section, trials have been made to keep high combustor width to step height ratio to minimize the wall effects. Based on that and to achieve such flow conditions, the width of the reactor was calculated to be 160 mm. In all simulations, the inlet velocity was fixed to 5.2 m/s, which corresponded to a fixed Reynolds number of 6500. The inlet flow temperature was kept unchanged at 293 K. The effects of the equivalence ratio on the size of the recirculation zone and flame stability were numerically investigated over the range of 0.45–0.85. Details of the experimental study were reported in Hong et al. [70]. All simulations were performed in *3D* while considering periodic boundary conditions in the *z*-direction, and at atmospheric pressure. Inlet uniform velocity with fluctuations of around 5% of the inlet average value was considered in all simulations in order to model turbulence in the incoming flow. This value of 5% was set as boundary condition for the turbulence fluctuations at the inlet section based on the measured values by the PIV system close to the inlet section. To initiate the flame, a high-temperature pulse was applied just before the combustor step to ignite the fuel. Thus, the reacting mixture is convected downstream and eventually stabilizes the flame in the wake of the step. The no-slip conditions were applied along all walls, while the zero Neumann condition was specified for the other variables. At the exit section of the reactor, zero Neumann conditions were specified for all variables except the pressure, for which wave-transmissive conditions were used. To resolve the flow features in the wall boundary layer and to maintain reasonable computational efficiency, appropriate wall functions were utilized. In order to correctly predict the temperature and species concentrations inside the recirculation zone, the heat transfer (loss) through the combustor walls should be correctly predicted. A combined convection and radiation heat transfer mechanism were applied to reactor walls in order to correctly predict the heat transfer through the walls of the reactor and, accordingly, correctly predict the flow and combustion characteristics inside the recirculation zone. A conservative value time step of 1 μs is used in the reacting flow simulations, to take into account local refinement and acceleration of the fluid above the bulk inlet velocity and to adequately resolve the chemical time scales. In all simulation cases, the initial transients left the domain at a time of about 0.1 s. The average values were calculated based on the time range from $t = 0.2$ s to $t = 0.3$ s to confirm that the average values correspond to the steady-state conditions. Numerical computations started with quiescent conditions and the unsteady flow characteristics evolve naturally, and the total computation time was set to 0.3 s in all simulations. The effect of total consumption time was tested numerically for a range from 0.15 s up to 0.7 s, and no significant changes in the results were obtained for time more than 0.3 s. Only, the computational cost in terms of the required number of processors and total time for converged solution was significantly increased.

5.8.2 LES Model

In *LES* models, the governing equations are filtered at a scale smaller than that of the large eddies, which have significant impact on the flow by virtue of their energy. The impact of scales below the filter scales were modeled using sub-filter-scale models. This was essential for accurate prediction of the combustion characteristics, especially when the filter scale is sufficiently close to the Kolmogorov scale. In this case study, the accuracy of the *LES* models (as detailed in Sects. 5.3.2.5 and 5.3.2.6) was examined by comparing the predicted results to data of the experimental measurement and analyzed the simulation results.

5.8.3 Combustion Modeling Technique

For combustion in a turbulent environment, under most conditions, the chemical reactions are confined to thin layers at scales smaller than those resolved on the *LES* grid. In *RANS* models, the averaged chemical source term in the governing equations is modeled in terms of the mean field variables and the modeled fluctuations. In *LES* models, although finer grid resolution is used and some of the large scales are resolved, the instantaneous flame thickness is still too small to be captured by the *LES* grid. In the case of premixed combustion, several approaches have been suggested for modeling the filtered reaction rate terms [21]. In this case study, the models that incorporate finite-rate chemistry such as the artificially thickened flame approach were chosen (as detailed in Sect. 5.3.2.8).

The artificially thickened flame (*ATF*) approach allows for finite-rate kinetics as described by Arrhenius rate laws to be used in the calculations, as in the case of *DNS* model. Global single-step reaction kinetics cannot capture some of the flame characteristics, especially at conditions in which the premixed flames interact strongly with flow gradients. However, the reaction mechanisms should have a small number of reaction steps with fewer intermediate species to limit the computation complexity

Table 5.3 Propane–air reaction kinetics mechanism [72]

Jones-Lindstedt (CHEMKIN format)	*A* (pre-exponential coefficient)	*β* (temperature exponent)	*Ea* (activation energy)
C_3H_8 + 1.5 $O_2 \Rightarrow 3CO + 4H_2$	7.1E+13	0.00	3.0E+4
FORD/C_3H_8 0.5/ FORD/O_2 1.25/			
$C_3H_8 + 3H_2O \Rightarrow 3CO + 7H_2$	3.0E+11	0.00	3.0E+4
$H_2 + 0.5\ O_2 \Rightarrow H_2O$	1.21E+18	−1.0	4.0E+4
FORD/H_2 0.25/ FORD/O_2 1.5/			
$H_2O + 0\ O_2 + 0$ $H_2 \Rightarrow H_2 + 0.5\ O_2$	7.06E+17	−0.877	9.8E+4
FORD/H_2 −0.75/ FORD/O_2 1/ FORD/H_2O 1/			
$CO + H_2O = CO_2 + H_2$	2.75E+12	0.00	2.0E+4

[36]. Large eddy simulation (LES) with appropriate turbulent combustion models and reaction mechanisms is considered as one of the more promising CFD approaches, balancing computational complexity and predictive accuracy. As regards the chemical reaction schemes, appropriate multiple-step chemistry mechanism for propane combustion in air, Jones-Lindstedt (JL) mechanism [72], has been utilized. While simplified global schemes can provide adequate results if the main species concentrations are of interest, they cannot be expected to work as well under unconventional combustion conditions (e.g., oxy-fuel combustion) and are likely to inaccurately estimate the temperature in the absence of an adequate set of reactions. The role of reaction chemistry can be significant in combustion systems prone to thermo-acoustic instabilities, near extinction and re-ignition phenomena, as well as influencing the flame speed. However, the reaction schemes used within the thickened flame framework should typically include a limited number of intermediate species as it can lead to difficulties for wrinkled and/or stretched flame fronts [36, 37]. The parameters of the JL mechanism (listed in Table 5.3) have been coded in the *CHEMKIN* format and coupled to the *OpenFOAM* software.

5.8.4 Results and Discussions

The results of the experimental and numerical investigation are presented in this section. These results include the recirculation zone shape and structure in addition to the flame location and overall shape (macrostructure). The numerical results, obtained using the present *LES* model, were compared with the corresponding PIV measurements under different conditions. The impact of equivalence ratio on the flow field and flame stabilization was also examined.

5.8.4.1 Model Validation and the Flow Field

Figure 5.24 depicts a comparison of the numerical results for both average and rms of the axial velocities at an equivalence ratio of 0.63 with the *PIV* measurements. The plots are presented on a background of the streamlines of the mean flow velocity in order to highlight the important flow features such as the recirculation zones. The mean axial velocity is first reduced because of the expansion over the step, followed by the formation of recirculation zone due to unfavorable pressure gradient. The velocity increases within the shear layer and downstream of the recirculation zone as a result of combustion heat release. The axial rms velocity values are highest within the shear layer. This can be attributed to the growth of local instability within the shear layer and to the roll-up of coherent vortices. Small reduction in the turbulence intensity is observed within the mixing layer due to the heat release from the combustion process. Good agreement between the experimental measurements and the corresponding numerical results can be seen in Fig. 5.24 in terms of the overall features and average and rms axial velocity contours. The overall length of the recirculation zone, primarily impacting the size of the secondary eddy, is under-predicted.

As can be seen from Fig. 5.24, there are two eddies inside the recirculation zone at the equivalence ratio of 0.63. The first is the large primary eddy (*PE*) spinning in the clockwise direction and constituting most of the recirculation zone. The second is the secondary eddy (*SE*) spinning in the opposite direction and located between the step and the PE close to the corner. The sizes of both *PE* and *SE* are controlled by the equivalence ratio. For more detailed presentations of the *LES* model results and the *PIV* data at the equivalence ratio of 0.63, Figs. 5.25 and 5.26 present line plots of the normalized axial average and rms velocities at different axial locations. The predicted flow evolves faster than the measurements since, as mentioned above; the size of SE is under-predicted. The plots indicate overall good agreement

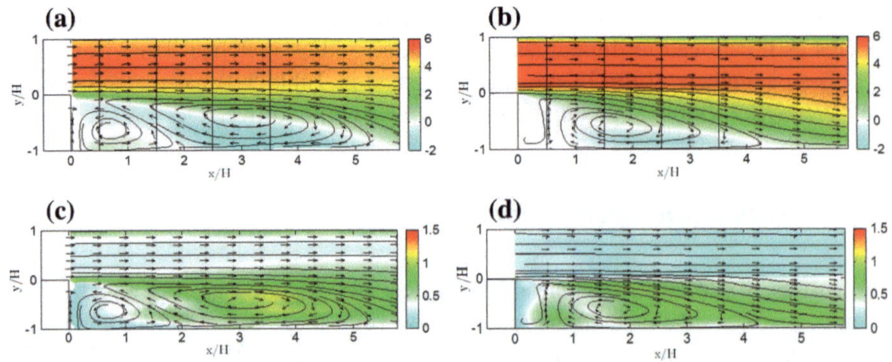

Fig. 5.24 Comparison between the *PIV* data (left) and *LES* model results (right) for the contour plots of average axial velocity (m/s, **a** and **b**) and rms axial velocity (m/s, **c** and **d**) plotted on the streamlines of the mean velocity as a background at $\Phi = 0.63$

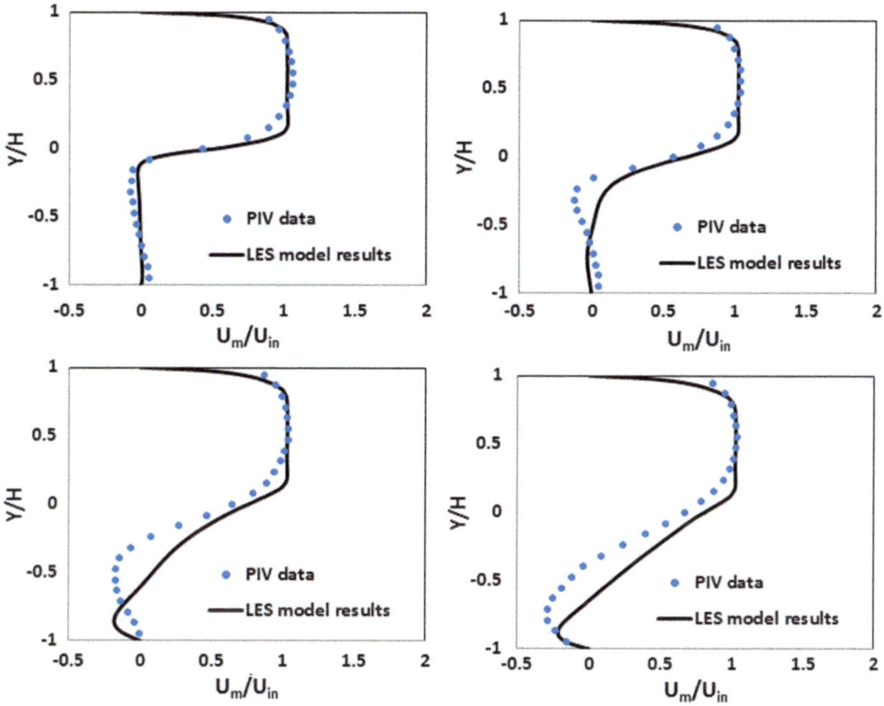

Fig. 5.25 Comparison between the *PIV* data and *LES* model results for the normalized average axial velocity at different normalized axial locations at $\Phi = 0.63$

between the *LES* model and the experimental data. Despite the heat release from the combustion process and turbulent dilatation, the flow field is modified, and the axial average velocity is increased downstream. The turbulence intensity is high in the region close to the upper wall and in the shear layer area where flow separation and shear layer instability lead to vortex shedding and associated fluctuations. As shown in Fig. 5.26, the velocity fluctuation is under-predicted in the upper half of the combustor. This may be attributed to several reasons such as the use of periodic boundary conditions and how the inlet fluctuations are modeled. Close to the reattachment zone ($x/H > 5$), the streamline curvatures are the highest. Significant unsteadiness is typically experienced by the flow around the reattachment point. As a result, the velocity fluctuations and the turbulent kinetic energy (*TKE*) increase. From the comparison of the present model results with the experimental data, the results are in good agreement with wide ranges of operating conditions. Slight under-prediction of the negative velocity is observed. This may result from uncertainties or fluctuations in the experimental conditions (equivalence ratio, reactant mixture velocity, etc.) or the use of periodic boundary conditions and the use of wall functions. It should, however, be noted that three-dimensional effects in the reacting case may be important since the chemical reaction rates can depend

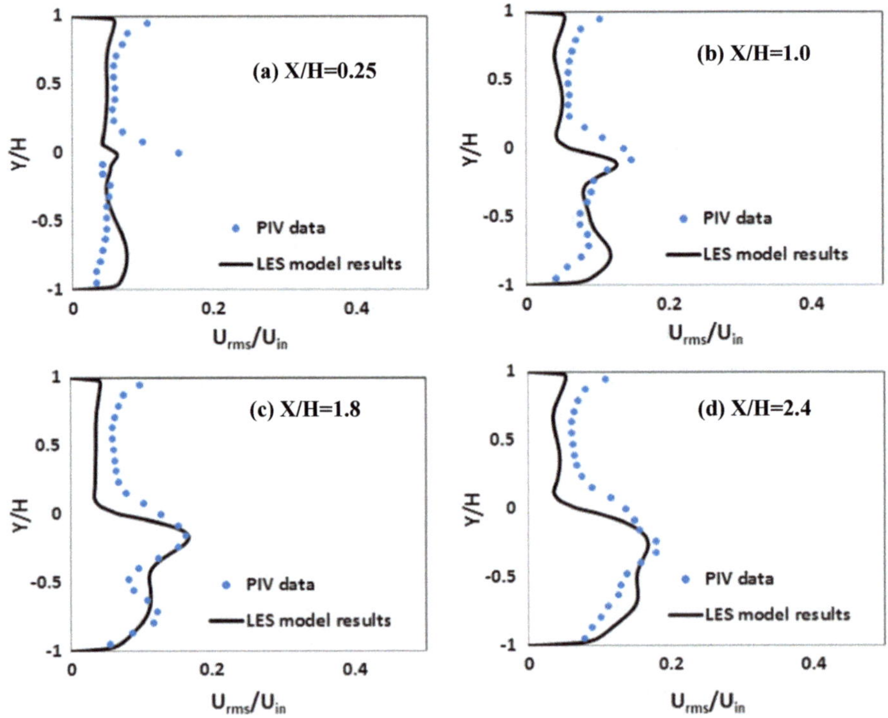

Fig. 5.26 Comparison between the *PIV* data and *LES* model results for the normalized rms axial velocity at different normalized axial locations at $\Phi = 0.63$

critically on the small-scale three-dimensional flow structures. The use of periodic boundary conditions in the spanwise direction, therefore, might explain the slight discrepancy in the simulation results and the underestimation in the rms axial velocity. Precise knowledge of experimental inlet conditions coupled with fully 3D simulations that can effectively capture the effects of all four walls (e.g., boundary layer thickening) and vortex stretching effects may result in more accurate predictions for the temperature as well as velocity fluctuations.

Figure 5.27 shows the contours of the average axial velocity at different equivalence ratios based on the *PIV* measurements. Streamline-based arrows depict the direction of the flow, and the colors shown in the color bars illustrate the mean velocity variation. As the equivalence ratio increases, the size of the PE is reduced, and its location is displaced upstream toward the step. The *SE* size is also reduced by increasing the equivalence ratio, and it almost collapses at the equivalence ratio of 0.85 as shown in Fig. 5.26c. The higher temperature ratio across the flame leads to faster acceleration of the flow and the reduction of the overall recirculation zone length. These findings show the strong dependence of the flow field and recirculation zone on the equivalence ratio in premixed combustion. More experimental data can be found in our previous work by Hong et al. [70] for the same combustor setup.

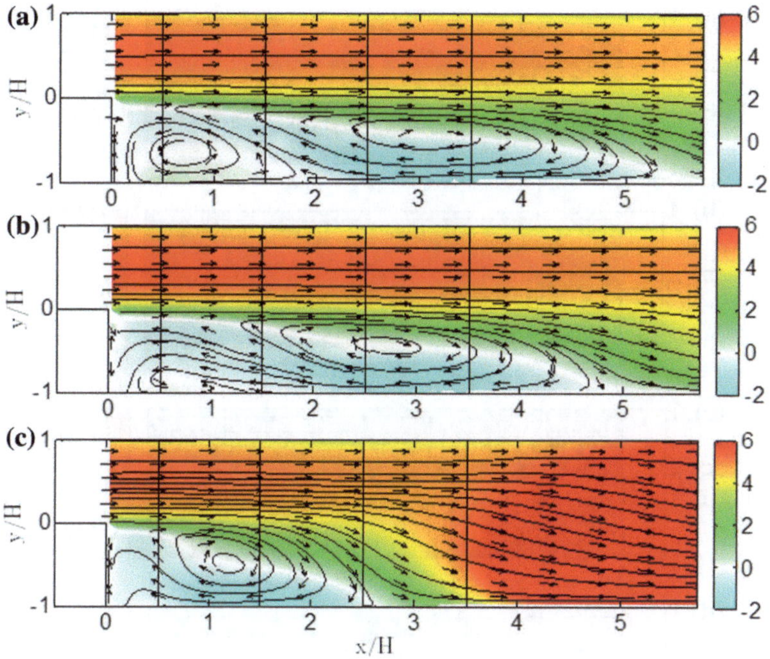

Fig. 5.27 Effect of equivalence ratio on the distribution of the average axial velocity (m/s) based on the *PIV* data: **a** $\Phi = 0.63$, **b** $\Phi = 0.69$, and **c** $\Phi = 0.85$

5.8.4.2 The Reactive Field

Figure 5.28 depicts the instantaneous temperature contours (at $t = 0.3$ s) at different equivalence ratios based on the LES results. The results are obtained at fixed inlet Reynolds number for a range of equivalence ratio of 0.5 and 0.85. As shown in the figure, the flame tip or leading edge moves toward the step as the equivalence ratio increases. At lower equivalence ratios, the flame is stabilized near the middle of the average recirculation zone (further away from the step) where it is ignited by the recirculating products. This is supported by experimental data; see Hong et al. [73]. At higher equivalence ratios, the size of the recirculation zone is reduced, the temperature levels are increased, and the flame is moved to upstream locations. The flame tip propagates upstream with respect to the recirculation zone, anchoring near the step where the primary eddy forms. The angle of the flame with respect to the flow also increases at higher equivalence ratios, consistent with the higher burning velocity of the embedded laminar flames. The flame can move further into the reactant stream and upstream toward the step. It is also interesting to see that the temperature of the recirculating gases near the step remains significantly lower than that of the products downstream of the flame.

Mixing with reactants and heat loss to the walls near the step contributes to the temperature distribution within the recirculation zone. This feature must be

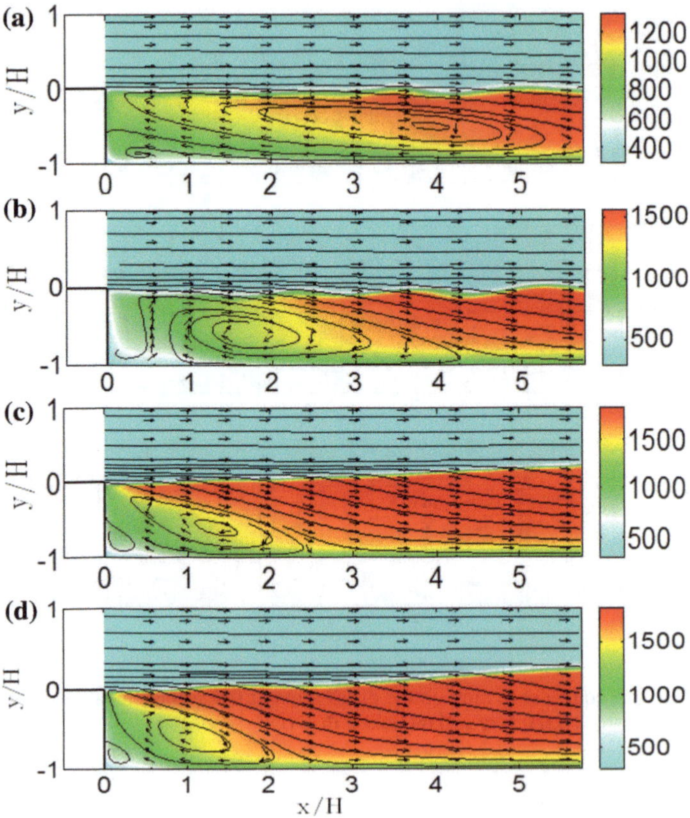

Fig. 5.28 Effect of equivalence ratio on the distribution of the instantaneous temperature (K, at $t = 0.3$ s) based on the *LES* model results: **a** $\Phi = 0.50$, **b** $\Phi = 0.63$, **c** $\Phi = 0.75$, and **d** $\Phi = 0.85$

considered when modeling flame stabilization in confined domains. Experimental results supporting these trends were reported in Hong et al. [70, 73]. The reduction in temperature in this region is partly due to the entrainment of the reactants into the recirculation zone and partly due to its proximity to the step and channel walls. The temperature in the secondary eddy is the lowest at the lower equivalence ratio where the flame is anchored a significant distance away from the step, thus making it possible for reactants to diffuse across the layer near the step. It is also interesting to see that, at low equivalence ratios, the flame is stabilized/embedded within the shear layer where significant waviness is exhibited. On the other hand, the flame propagates outward and further into the reactant stream at higher values of the equivalence ratio. As the flame propagates outward with respect to the shear layer and toward the reactants, its angle increases, and it anchors closer to the step. Less reactant survives, and the temperature in the RZ increases.

The profiles of instantaneous temperature and species concentrations (at $t = 0.3$ s) are presented in Fig. 5.29 at different axial locations within and across the

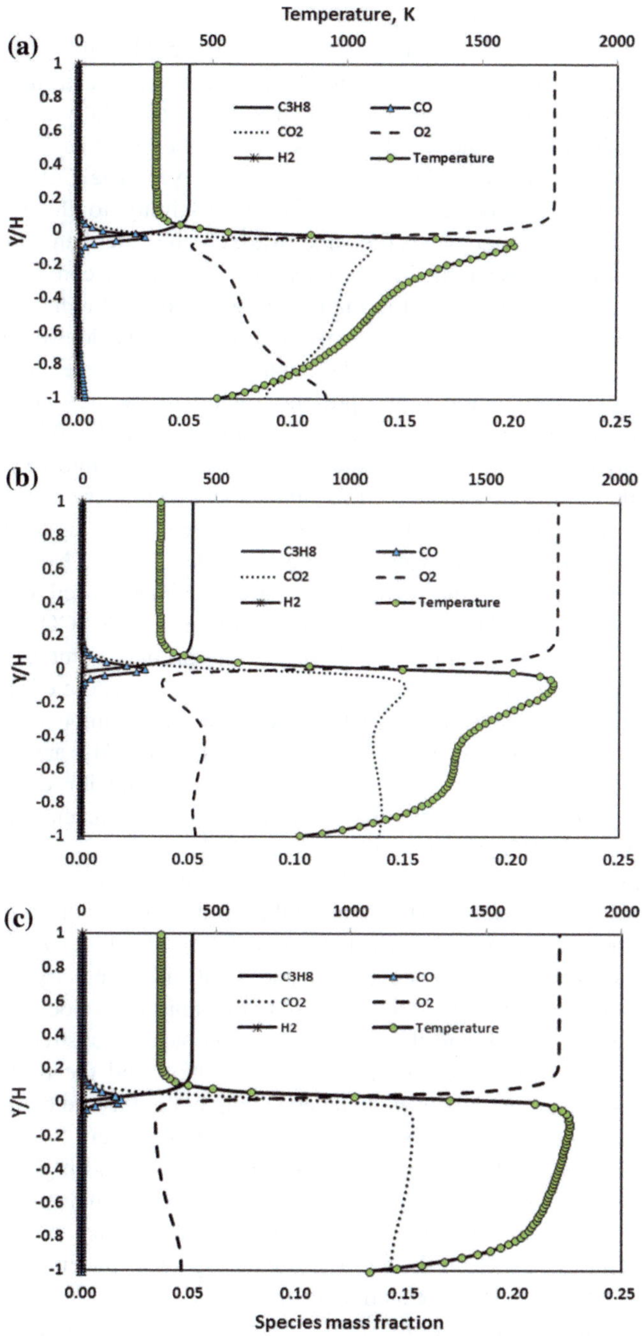

Fig. 5.29 Instantaneous temperature (K, at $t = 0.3$ s) distributions and species concentrations at different axial locations at $\Phi = 0.85$: **a** $x/H = 0.25$, **b** $x/H = 1.0$, and **c** $x/H = 2.4$

recirculation zone at equivalence ratio of 0.85. The temperature is low in the upper part of the combustor corresponding to the incoming cold flow and is high in the lower part because of the presence of the flame and the recirculating hot products. Close to the lower wall, the temperature is the lowest especially close to the step because of the heat loss to the walls and mixing with the reactants. The temperature profiles show the flame spreading along the shear layer. The CO_2 profiles show areas filled with products but cooled by their proximity to the walls. Further downstream, the flame approaches the upper wall. On the other hand, the upper wall temperature remains close to the inlet temperature. Despite the convective heat loss, oxygen concentrations are highest in the incoming stream and reduced in the lower half due to consumption and dilution by the products. The low oxygen concentration spreads into the upper half of the combustor as the flame spreads upward.

An interesting observation is the presence of relatively high concentration of oxygen in the upstream portions of the recirculation zone, $x/H = 0.25$, significantly higher than that found in the downstream regions. While finite oxygen concentrations should be expected since the mixture is lean, the higher values closer to the step indicate that reactants are transported across the shear layer and remained unburned, especially at lower values of the equivalence ratio. Note that O_2 concentration decreases within the shear layer where reactions are active and increases toward the lower wall. The concentrations of carbon dioxide (CO_2) resemble the temperature concentrations since CO_2 is one of the primary combustion products, except close to the walls where the temperature decreases because of heat loss. Like the temperature, average values of CO_2 reached a maximum within the mixing layer. The concentrations of intermediate species, CO and H_2, are highest in the region near the step, and they are reduced in the axial direction due to their oxidation to CO_2 and H_2O, respectively. Nevertheless, finite concentrations of intermediates, especially CO, persist until the exit.

Figure 5.30 shows the effect of the equivalence ratio on the instantaneous temperature distribution (at $t = 0.3$ s) along the combustor. This figure presents the axial distribution of the instantaneous temperature along the line $y/H = 0$, the lower and upper wall average temperatures, and the axial mean velocity along the line $y/H = 0$. As the equivalence ratio is raised, the burning velocity and product temperature are increased, and the flame front is propagated upstream and upward into the reactant stream. Both the product temperature and the proximity of the flame and products to the lower walls raise its temperature as the equivalence ratio increases from 0.45 to 0.85, as shown in Fig. 5.30b, c. Figure 5.30a shows clear distinction between the flame shape and position at lower and higher equivalence ratios. At lower values, Fig. 5.30a, the flame is initiated downstream (the flame "lifts off" the step) and stabilizes in the region of low velocity inside the recirculation zone. This is clear from the temperature and species concentrations shown in Figs. 5.28 and 5.29. Further reduction in the equivalence ratio results in longer flame liftoff distance. Simulations show that the flame blows out at equivalence ratio of around 0.45. These results were confirmed by the experimental measurements by Hong et al. [70] for the same combustor setup. The temperature distribution confirms that the flame tip moves upstream as the equivalence ratio is raised.

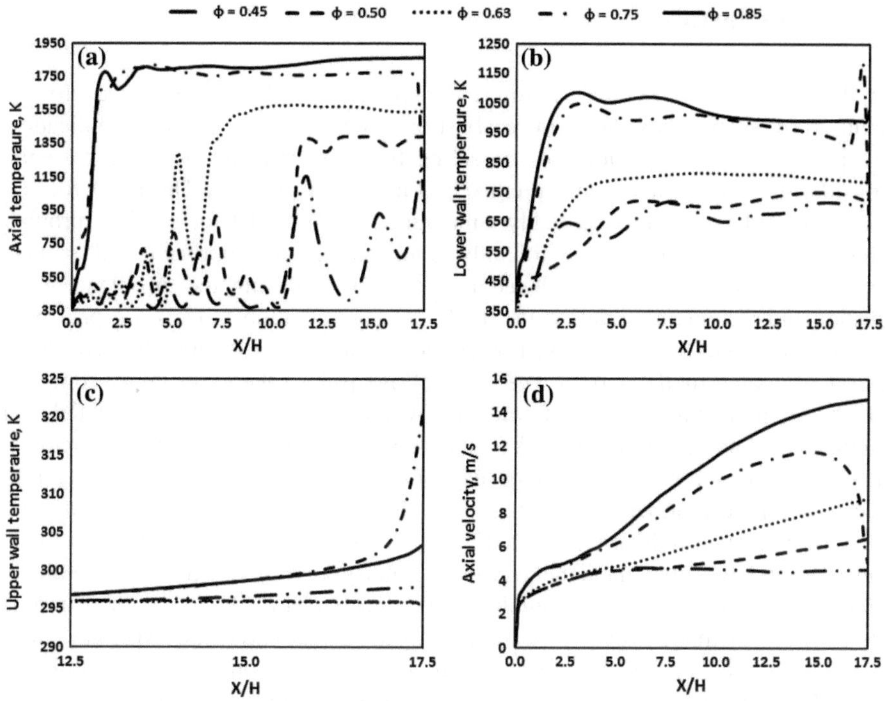

Fig. 5.30 Effect of equivalence ratio on the axial distributions of: **a** temperature (K, through the line $y/H = 0$, T $t = 0.3$ s), **b** lower wall temperature (K, at $t = 0.3$ s), **c** upper wall temperature (K, at $t = 0.3$ s), and **d** average axial velocity (m/s, through the line $y/H = 0$), based on the *LES* model results

At intermediate values, the flame is anchored close to the step. Away from the inlet section, the flame is extended to the upper half of the combustor due to the expansion of the flow and the reduction in the flow momentum due to the increase in the flow area. This may justify the increase in the temperature near the upper wall close to the combustor exit. The flame propagates outward and moves to upstream locations and further into the reactant stream at higher values of the equivalence ratio. Due to the higher density ratio across the flame at higher fuel concentration in the reactants, and the associated volumetric expansion in the products, the average axial velocity grows in the axial direction. The angle of the flame with respect to the flow also increases at higher equivalence ratios. At certain equivalence ratio, the flame reaches the upper wall and it may be reflected back to the lower wall. This may result in fluctuations in the temperature through the center of the combustor as presented in Fig. 5.30. Due to the higher density ratio across the flame at higher fuel concentration in the reactants, and the associated volumetric expansion in the products, the average axial velocity increases in the axial direction as presented in Fig. 5.30d.

5.9 Conclusions

In this chapter, the details of different numerical modeling techniques for the solution to reacting flow fields in combustion systems used in different applications are presented. The equations for modeling turbulent reacting flow are presented considering non-premixed and premixed combustion regimes. The main focus is made on understanding large eddy simulation (LES) model for solving turbulent premixed combustion. The artificially thickened flame approach and reduced reaction mechanisms are introduced in detail for modeling the turbulence–combustion interactions at small scales. Different gas radiation models are presented, and their influence on the combustion characteristics is investigated. It is concluded that the Leckner model and the Perry model overpredict the temperature compared to the exponential wideband model. Weighted-sum-of-gray-gas model can predict accurate results compared to the benchmark model; however, it requires new parameters for different ratios of H_2O and CO_2. The chemistry reduction/acceleration techniques including reduced chemistry mechanisms and low-dimensional manifolds are presented. This is followed by detailed description of the modified two-step methane–oxygen combustion mechanisms along with the kinetic rate data. Also, the modified JL mechanism for oxy-combustion of H_2-enriched methane is described in detail. The chapter ends with two case studies as direct applications of the presented numerical models on real combustion systems. The obtained numerical results are compared with the available experimental data for validation. In the first case study, detailed numerical modeling and the obtained results are presented considering H_2-enriched methane oxy-combustion in a model gas turbine combustor. The results showed that enriching CH_4 with H_2 increases the flame stability. Based on the analysis of the oxy-combustion H_2-enriched CH_4 flames, the optimum operating equivalence ratio in the gas turbine combustor was found to be around the value of 0.85. The flame was completely unstable for equivalence ratio of 0.95 (i.e., near stoichiometric condition). It was found that better performance of gas turbine combustor can be achieved by utilizing swirl vane angles of 65° or 70° and not beyond because the flame achieved at 75° appeared to be not practicable. The second numerical case study considers modeling of premixed combustion in a backward-facing step combustor. Results showed that the numerical solution is able to predict the flow with reasonable agreement, although some quantitative differences exist. The equivalence ratio significantly affects the flow field. At lower equivalence ratio, the flame is initiated downstream (flame liftoff) and stabilizes in the region of low velocity inside the recirculation zone close to the reattachment point. At higher values of equivalence ratios, the flame approaches the step and penetrates deeper into the reactants. At lower fuel fractions, the flame is embedded in the shear layer while propagating outward into the reactant stream at higher values.

References

1. Huang Y, Yang V (2009) Dynamics and stability of lean-premixed swirl-stabilized combustion. Prog Energy Combust Sci 35:293–364
2. Kashir B, Tabejamaat S, Jalalatian N (2015) A numerical study on combustion characteristics of blended methane-hydrogen bluff-body stabilized swirl diffusion flames. Int J Hydrogen Energy 40:6243–6258
3. Habib MA, Nemitallah MA, Ahmed P, Sharqawy MH, Badr HM, Muhammad I, Yaqub M (2015) Experimental analysis of oxygen-methane combustion inside a gas turbine reactor under various operating conditions. Energy 86:105–114
4. Yu B, Lee S, Lee CE (2015) Study of NOx emission characteristics in CH_4/air non-premixed flames with exhaust gas recirculation. Energy 91:119–127
5. Gao X, Duan F, Lim SC, Yip MS (2013) NOx formation in hydrogen–methane turbulent diffusion flame under the moderate or intense low-oxygen dilution conditions. Energy 59:559–569
6. Li YH, Chen GB, Lin YC, Chao YC (2015) Effects of flue gas recirculation on the premixed oxy-methane flames in atmospheric condition. Energy 89:845–857
7. Stone C, Menon S (2002) Swirl control of combustion instabilities in a gas turbine combustor. Proc Combust Inst 29(1):155–160
8. Schluter J (2004) Static control of combustion oscillations by coaxial flows: a large eddy simulation investigation. J Propuls Power 20(3):460–467
9. Franzelli B, Riber E, Gicquel L, Poinsot T (2012) Large eddy simulation of combustion instabilities in a lean partially premixed swirl flame. Combust Flame 159:621–637
10. Roux S, Lartique G, Poinsot T, Meier U, Berat C (2005) Studies of mean and unsteady flow in a swirled combustor using experiments, acoustic analysis, and large eddy simulations. Combust Flame 141(1–2):40–54
11. Huang Y, Sung H-G, Hsieh S-H, Yang V (2003) Large eddy simulation of combustion dynamics of lean premixed swirl stabilized combustor. J Propul Power 19(5):782–794
12. Angelberger C, Veynante D, Egolfopoulos F, Poinsot T (1998) Large eddy simulations of combustion instabilities in premixed flames. In: Proceedings of the summer program, Center for Turbulence Research
13. Fureby C (2012) A comparative study of flamelet and finite rate chemistry LES for a swirl stabilized flame. J Eng Gas Turbines Power 134(4)
14. Versteeg HK, Malalasekera W (1995) An introduction to computational fluid dynamics: The finite volume method. Longman Scientific and Technical
15. Fluent 6.2 User's Guide. Fluent Inc. (2003) Center Research Park, 10 CavendishCourt, Lebanon, NH 03766, USA
16. ANSYS (2013) ANSYS fluent theory guide, vol 15, Nov 2013
17. Wilcox DC (2000) Turbulence modeling for CFD. DCW Industries
18. Chen L, Ghoniem AF (2012) Simulation of oxy-coal combustion in a 100 kWth test facility using RANS and LES: a validation study. Energy Fuels 26:4783–4798
19. Garnier E, Adams N, Sagaut P (2009) Large eddy simulation for compressible flows. Springer
20. Perez FEH (2011) Sub-filter scale modeling for large eddy simulation of lean hydrogen-enriched turbulent premixed combustion. Ph.D. thesis, University of Toronto
21. Echekki T (2011) Turbulent combustion modeling: advances, new trends and perspectives. Springer, Berlin
22. Lin W (2010) Large-eddy simulation of premixed turbulent combustion using flame surface density approach", Ph.D. Thesis, University of Toronto
23. Weller HG, Tabor G, Gosman AD, Fureby C (1998) Application of a flame wrinkling LES combustion model to a turbulent mixing layer. In: 27th symposium on combustion. The Combustion Institute, pp 899–907
24. Pitsch H (2002) LES of premixed turbulent combustion using a level-set approach. In: Proceedings of Combustion Institute, vol 29, pp 2001–2008

25. Candel S, Poinsot T (1990) Flame stretch and the balance equation for the flame area. Combust Sci Technol 3:1–15
26. Durand L (2007) Development, implementation and validation of LES models for inhomogeneously premixed turbulent combustion. Ph.D. thesis, Technical University of Munich
27. Wang P (2005) Large eddy simulation of turbulent swirling flows and turbulent premixed combustion. Ph.D. thesis, Lund Institute of Technology
28. Baudoin E (2010) Large eddy simulation of turbulent premixed and partially premixed combustion. Ph.D. thesis, Lund Institute of Technology
29. Colin O, Ducros F, Veynante D, Poinsot T (2000) A thickened flame model for large eddy simulation of turbulent premixed combustion. Phys Fluids 12(7):1843–1863
30. Charlette F, Meneveau C, Veynante D (2002) A power-law flame wrinkling model for les of premixed turbulent combustion Part I: non-dynamic formulation and initial tests. Combust Flame 131:159–180
31. Charlette F, Meneveau C, Veynante D (2002) A power-law flame wrinkling model for LES of premixed turbulent combustion part ii: dynamic formulation. Combust Flame 131:181–197
32. Karlsson A, Chomiak J (1996) Flame liftoff in diesel sprays. In: 26th symposium on combustion, pp 2557–2564
33. Norsin PAN (2000) Complex chemistry modeling of diesel spray combustion. Ph.D. thesis, Imperial College of Science, Technology and Medicine
34. Jones WP, Lindstedt RP (1988) Global reaction schemes for hydrocarbon combustion. Combust Flame 73(3):233–249
35. Vaos EM, Lindstedt RP (2006) Transported PDF modeling of high-Reynolds-number premixed turbulent flames. Combust Flame 145:495–511
36. Roux S, Lartigue G, Poinsot T, Meier U, Bérat C (2005) Studies of mean and unsteady flow in a swirled combustor using experiments, acoustic analysis, and large eddy simulations. Combust Flame 141:40–54
37. Selle L, lartigue G, Poinsot T, Kaufmann P, Krebs W, Veynante D (2002) Large eddy simulation of turbulent combustion for gas turbines with reduced chemistry. In: Proceedings of summer program 2002, Center for Turbulence Research
38. Durand L, Polifke W (2007) Implementation of the thickened flame model for large eddy simulation of turbulent premixed combustion in a commercial solver. ASME Paper No. GT2007–28188
39. Polifke W, Hirsch C, Zellhuber M, Komarek T, Chong L (20069) Influence of strain and heat loss on flame stabilization in a non-adiabatic combustor. In: Proceedings of the European combustion meeting
40. Steinberg AM, Driscoll JF (2009) Straining and wrinkling processes during turbulence-premixed flame interaction measured using temporally-resolved diagnostics. Combust Flame 156(12):2285–2306
41. Viskanta R (2005) Radiative transfer in combustion systems: fundamentals & applications. Purdue University, USA
42. Chandrasekhar S (1960) Radiative transfer. Dover Publications, New York
43. Modest MF, Zhang H (2000) The full-spectrum correlated-k distribution and its relationship to the weighted-sum-of gray-gas method. In: Proceedings of the 2000 IMECE, vol HTD-366–1. ASME, Orlando, pp 75–84
44. Lallemant N, Weber R (1996) A computationally efficient procedure for calculating gas radiative properties using the exponential wide band model. Int J Heat Mass Transfer 39:3273–3286
45. Siegel R, Howell J (2002) Thermal radiation heat transfer, 4th edn. Taylor and Francis, New York
46. Modak AT (1979) Radiation from products of combustion. Fire Res 1:339–361
47. Leckner B (1972) Spectral and total emissivity of water vapor and carbon dioxide. Combust Flame 19:33–48

48. Hottel HC, Noble JJ, Sarofim AF (2007) Heat and mass transfer (Chap. 5). In: Perry's chemical engineers' handbook, 8th edn. McGraw-Hill, New York
49. Hottel H, Sarofim A (1967) Radiative transfer. McGraw-Hill Inc., New York
50. Grosshandler WL (1993) RADCAL: a narrow-band model for radiation calculations in a combustion environment 1993. NIST Technical Note 1402
51. Truelove JS (1976) A mixed grey gas model for flame radiation. United Kingdom Atomic Energy Authority Report AERER-8494, Harwell
52. Johansson R, Andersson K, Leckner B, Thunman H (2010) Models for gaseous radiative heat transfer applied to oxy-fuel conditions in boilers. Int J Heat Mass Transf 53:220–230
53. Magnussen BF, Hjertager BH (1976) On mathematical models of turbulent combustion with special emphasis on soot formation and combustion. In: 16th Symposium (Int'l) on combustion, The Combustion Institute
54. Echekki T, Mastorakos E (2011) Turbulent combustion modeling: advances, new trends and perspectives. Springer, Berlin
55. Brink A, Mueller C, Kilpinen P, Hupa M (2000) Possibilities and limitations of the eddy break-up model. Combust Flame 123:275–279
56. Saario A, Oksanen A (2008) Comparison of global ammonia chemistry mechanisms in biomass combustion and selective noncatalytic reduction process conditions. Energy Fuels 22:297–305
57. Glarborg P, Bentzen LL (2007) Chemical effects of a high CO_2 concentration in oxy-fuel combustion of methane. Energy Fuels 22:291–296
58. Westbrook CK, Dryer FL (1981) Simplified reaction mechanisms for the oxidation of hydrocarbon fuels in flames. Combust Sci Technol 27:31–43
59. Andersen J, Rasmussen CL, Giselsson T, Glarborg P (2009) Global combustion mechanisms for use in CFD modeling under oxy-fuel conditions. Energy Fuels 23:1379–1389
60. Ilbas M, Yilmaz I, Veziroglu T, Kaplan Y (2005) Hydrogen as burner fuel: modelling of hydrogen-hydrocarbon composite fuel combustion and NOx formation in a small burner. Int J Energy Res 29:973–990
61. De A, Acharya S (2012) Parametric study of upstream flame propagation in hydrogen-enriched premixed combustion: effects of swirl, geometry and premixedness. Int J Hydrogen Energy 37:14649–14668
62. De A, Acharya S (2012) Dynamics of upstream flame propagation in a hydrogen-enriched premixed flam. Int J Hydrogen Energy 37:17294–17309
63. Marinov N, Westbrook C, Pitz W (1996) Transport phenomena in combustion. In: Eighth international symposium on transport processes, vol 8, pp 118–141
64. Frassoldati A, Cuoci A, Faravelli T (2009) Simplified kinetic schemes for oxy-fuel combustion. Sustain Foss 1–14
65. Aliyu M, Nemitallah MA, Said SA, Habib MA (2016) Characteristics of H_2-enriched CH_4–O_2 diffusion flames in a swirl-stabilized gas turbine combustor: experimental and numerical study. Int J Hydrogen Energy 41(44):20418–20432
66. Nemitallah M, Habib M (2013) Experimental and numerical investigations of an atmospheric diffusion oxy-combustion flame in a gas turbine model combustor. Appl Energy 111:401–415
67. Linck M, Gupta A, Bourhis G, Yu K (2006) Combustion characteristics of pressurized swirling spray flame and unsteady two-phase exhaust jet. In: American Institute of aeronautics and astronautics, Jan 2006, pp 1–15
68. Nemitallah Medhat A, Kewlani Gaurav, Hong Seunghyuck, Shanbhogue Santosh J, Habib Mohamed A, Ghoniem Ahmed F (2016) Investigation of a turbulent premixed combustion flame in a backward-facing step combustor; effect of equivalence ratio. Energy 95:211–222
69. Altay HM, Speth RL, Hudgins DE, Ghoniem AF (2009) Flame-vortex interaction driven combustion dynamics in a backward-facing step combustor. Combust Flame 156:1111–1125
70. Hong S, Shanbhogue SJ, Ghoniem AF (2015) Impact of fuel composition on the recirculation zone structure and its role in lean premixed flame anchoring. Proc Combust Inst 35: 1493–1500

71. Pope SB (2000) Turbulent flows. Cambridge University Press
72. Jones WP, Lindstedt RP (1988) Global reaction schemes for hydrocarbon combustion. Combust Flame 73:233–249
73. Hong S, Speth RL, Shanbhogue SJ, Ghoniem AF (2013) Examining flow-flame interaction and the characteristic stretch rate in vortex-driven combustion dynamics using PIV and numerical simulation. Combust Flame 160:1381–1397

Chapter 6
Applications of OTRs in Gas Turbines and Boilers

6.1 Introduction

The growing levels of carbon dioxide (CO_2) emission in the atmosphere as a result of combustion of fossil fuel and the dissolved CO_2 in oceans represent critical environmental concerns as they lead to global warming and ocean acidification [1]. Power plants using fossil fuel for the production of electrical energy are the major contributor to greenhouse gas emissions with 41% [2]. Due to recent sharp reduction in oil prices, the conversion to renewable energy sources is expected to take longer time. As well, there are many safety issues and technical problems associated with the abandoned spread of nuclear energy production systems [3]. In parallel, the world's energy demand is tremendously growing which requires the use of fossil fuels at least at the current time. This necessitates the handling of combustion processes so as to capture CO_2 before it spreads in the atmosphere [4]. Three carbon capture technologies (CCTs) are available and can be applied. The choice of the carbon capture technology depends on the CO_2 capture process in association with the combustion process. These technologies include: (1) pre-combustion carbon capture in which CO_2 is captured before the combustion process takes place; (2) oxy-combustion carbon capture in which the fuel is oxidized using pure oxygen instead of air, and (3) post-combustion carbon capture in which CO_2 is captured downstream of the combustion process [5]. In pre-combustion CO_2 capture, carbon is removed from the hydrocarbon fuel before combustion. In post-combustion, the exiting CO_2 is removed from the flue gases containing mainly nitrogen. In oxy-combustion technology, the exhaust gases are basically water vapor (H_2O) and carbon dioxide (CO_2), where the water vapor can be simply separated via cooling and condensation while carbon dioxide is then being captured.

The oxygen required for the oxy-fuel combustion process is normally obtained through the use of cryogenic oxygen separation units. Because of the high efficiency penalty associated with using these oxygen separation units in the

M. A. Nemitallah et al., *Oxyfuel Combustion for Clean Energy Applications*, Green Energy and Technology, https://doi.org/10.1007/978-3-030-10588-4_6

oxy-combustion systems, researchers were striving to find alternative solutions for O_2 separation. One of these solutions is the use of ion transport membranes (ITMs) for oxygen production. These ITMs have the capability of extracting oxygen from air at elevated temperatures (above 650 °C) [6–8]. The permeation of oxygen through the ion transport membranes depends on the membrane type, thickness, operating temperature, and the partial pressure of oxygen across the membrane [9]. It is expected that utilization of membranes in gas separation processes would increase five times by the year 2020 [10]. The integration of these membranes in oxy-combustion reactors led to the design of oxygen transport reactors (OTRs) where oxygen–air separation occurs at one side while the fuel combustion occurring at the other membrane side [11, 12]. Currently, several studies are performed for improving the membrane performance and chemical stability under more challenging operational conditions [13–15]. In this regard, dense mixed ionic–electronic-conducting ceramic (MIEC) membrane has shown a good potential for use in oxy-fuel technology when operating at temperatures in the range 700–900 °C [16, 17]. One main challenge in scheming an ITM reactor to be used in power plants is the low oxygen flux that makes the required surface area of the membrane to be very large [18, 19]. Accordingly, improving oxygen permeation in ITMs is an essential step toward the efficient design of oxygen transport reactors (OTRs). Also, the mechanical and chemical membrane stability performance under actual operating conditions needs detailed experimental investigation. A comprehensive review of the work done on ITM membranes developed for oxy-fuel combustion up to 2010 was reported by Habib et al. [20]. Currently, numerous investigators are working on the development of various material syntheses that can be used for building oxygen separation membranes. Any promising material for oxygen transport membrane (OTM) must show good oxygen permeability in addition to being thermally and chemically stable. Among all oxygen separation membranes, the BSCF ($Ba_{0.5}Sr_{0.5}Co_{0.8}Fe_{0.2}O_{3-\delta}$) has shown the most promising oxygen permeation characteristics and high-temperature chemical stability at [21].

Nemitallah [22] investigated the oxy-combustion characteristics of methane in different designs of button-cell BSCF-ITM reactors. In this study, a modified equation for the oxygen permeation flux was developed in order to account for the effect of combustion reactions in the permeate side of the membrane on oxygen permeation flux. The results showed significant improvements in oxygen permeation flux and combustion temperature utilizing the modified reactor design due to the creation of recirculation zones for flame stabilization. In their recent work, Kirchen et al. [23] developed an ion transport membrane reactor to study the processes of permeation of oxygen in non-reacting conditions (using inert gas in place of fuel) and oxy-fuel combustion under reacting conditions. Their results indicated that replacing the inert sweep gas by a reactive gas tends to an increase in the flux of oxygen due to the reduction of oxygen partial pressure near the membrane surface in the permeate side. The interaction between oxygen permeation and fuel oxidation in the sweep side of ITM was studied by Hong et al. [24]. The results indicated increase of the permeation of oxygen with the increase of the sweep gas inlet temperature and fuel concentration. The features of oxy-combustion of methane in cylindrical oxygen transport reactors (OTRs) were numerically investigated by Ben-Mansour et al. [25]. The study revealed that combining oxygen separation with reaction in OTRs results in a significant increase in the oxygen permeation rate by a factor of approximately 2.5 in comparison with O_2 separation-only units. Moreover, it was found that most of the combustion heat transfers to the air side while a part of it heats the O_2-permeating flux. Ahmed et al. [26] studied numerically the performance of an ITM reactor under oxy-combustion conditions. The results revealed that the gradual combustion of the fuel can result in gradual temperature increase and uniform temperature distribution along the membrane surface. Wang et al. [27] investigated the characteristics of oxygen permeation in a tubular BSCF membrane. Their results showed increase of the oxygen permeation flux while increasing oxygen concentration in the shell side at a fixed temperature. The idea of using ITM reactors for zero-emission power plant (ZEPP) applications has been proposed in a set of literature works as discussed later in this chapter.

In this chapter, the most recent advancement in the OTRs' technology with more analysis is related to the membrane separation mechanism, and the available permeation equations in the literature are presented. The detailed approaches for CFD modeling of OTRs, in terms of oxygen separation and oxy-fuel combustion, are presented. The optimized techniques for modeling radiation inside OTRs are presented for optimum performance of different OTR designs. The application of the OTR technology into large-scale power plants aiming at real application of the zero emission power plant (ZEPP) concept is discussed. Novel designs of OTRs are presented for the substitution of conventional gas combustion systems, including gas turbine combustors and boiler furnaces. The performances of such OTRs in different combustion systems are optimized for maximum power output per minimum combustor size, and the results are presented in detail for each combustor design.

6.2 Development of Oxygen Permeation Model

In this chapter, we focus on oxy-fuel combustion using ion transport membranes for the separation of oxygen from air. Then, this oxygen that permeates through the membrane is used in the combustion process in the other side of the membrane as shown in Fig. 6.1. More recently, strong demand for tonnage quantities of oxygen is encouraged by the steady growth in chemical process operations. For instance, the oxy-fuel combustion process and oxygen-blown gasification to convert coal and natural gas into an intermediate synthesis gas can be further processed to produce electricity, chemicals, and transportation fuels [28]. Two fundamental approaches to air separation are currently available. These approaches are cryogenic and non-cryogenic distillation. The cryogenic distillation is typically reserved for applications that require very large (in tons) quantity of oxygen at ultra-low temperature. The latter involves the separation of air at ambient temperatures using

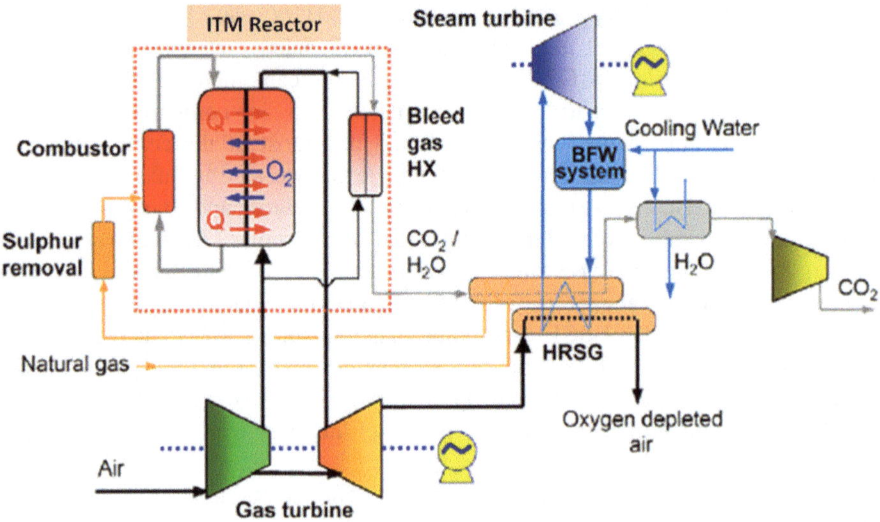

Fig. 6.1 Illustrative flow sheet for oxy-fuel combustion process using membrane reactor technology, with additional unit operations for carbon capture [37]

either molecular sieve adsorbents via pressure swing adsorption (PSA), or membrane separation process using the polymeric membranes. Recently, a third category of air separation has emerged, which is based on specialized ceramic membranes that separate oxygen from air at elevated temperatures, in contrast to the super-cooled temperatures required by conventional cryogenic distillation. This novel technique is based on dense ceramic membranes, which carry out the separation of oxygen from air at elevated temperatures, typically 800–900 °C. Ion transport membranes (ITMs), oxygen transport membranes (OTMs), and mixed ionic–electronic-conducting membranes (MIEC) are acronyms that are used to refer to high-temperature ceramic membranes [29]. These terms will be used throughout this work.

6.2.1 Concept of Operation of Ceramic-Based Membranes

Ceramic-based membranes that are used in oxygen separation systems can be categorized into pure oxygen-conducting membranes and mixed ionic–electronic-conducting membranes. The solid electrolytes are pure oxygen-conducting membranes, where electrodes are provided for the electron pathway [30]. The main advantage of this system is the control over the amount of oxygen separated via the application of an electric current. Compared to solid electrolytes, mixed ionic–electronic-conducting membranes require neither electrodes nor an external circuit to operate. The electronic conductivity itself performs as an internal short circuit

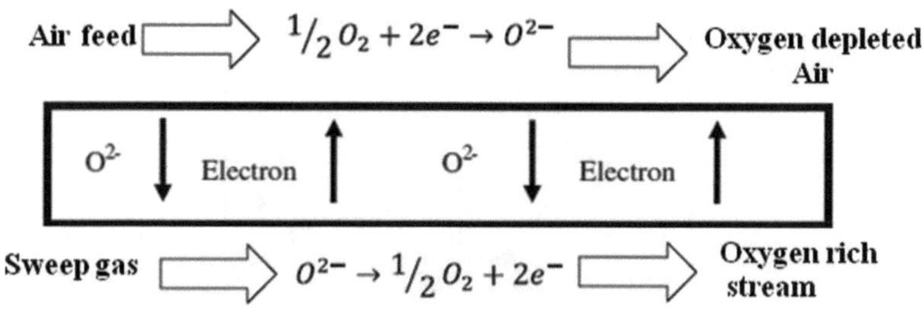

Fig. 6.2 Schematic diagram of a dense ceramic membrane based on conduction mechanism

involving oxygen partial pressure gradient. Oxygen ions permeate from the high oxygen partial pressure side (normally called feed side) to the low oxygen partial pressure side (normally referred to as permeate or sweep side). The overall charge neutrality is maintained by a counterbalancing flux of electrons, as idealized schematically in Fig. 6.2. It should be noted that oxygen separation through this process has the advantage of producing very high-purity oxygen.

It is worth noting that ceramic materials with mixed ionic–electronic-conducting characteristics typically have defined phase structures that can be derived from perovskite, fluorite, brownmillerite, and other similar types of materials [31–34]. Among ceramic membranes with mixed ionic–electronic-conducting characteristics, perovskite-type and fluorite-type are the best structures in regard to the case of oxygen permeation properties. However, the perovskite-type ceramic membranes have higher permeability and have more promising potential for improvement [35, 36].

6.2.2 *Oxygen Transport Mechanism*

Based on the difference in the oxygen chemical potentials between the feed side and the permeate side, the membrane temperature and its ambipolar conductivity, oxygen migrates from the high-pressure (feed) side to the low-pressure (permeate) side. This occurs according to the overall transport processes summarized as follows [38–41]:

(1) Mass transfer (advection and diffusion) of gaseous oxygen from the feed stream to the membrane surface, adsorption onto the membrane surface, dissociation and ionization of oxygen molecules, and subsequent incorporation of the ions into the lattice vacancies (feed-side surface exchange),
(2) Transport of lattice oxygen ions through the membrane (bulk diffusion),
(3) Association of lattice oxygen ions to oxygen molecules and desorption from the membrane surface into the gas phase (permeate side surface exchange), gaseous oxygen mass transfer (advection and diffusion) from the membrane surface to the permeate stream. As shown from Fig. 6.3, oxygen permeation through a

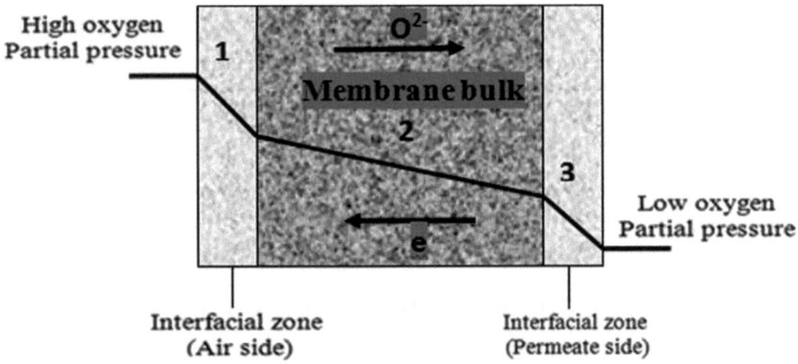

Fig. 6.3 Schematic diagram of oxygen permeation through mixed ionic-conducting membrane [30]

dense mixed ionic–electronic-conducting material is limited by surface exchange resistance, bulk diffusion limitations, or both [42–44]. It may be noted that the bulk diffusion will become the controlling step when the membrane is relatively thick.

Figure 6.3 envisions the oxygen permeation mechanism through a mixed ionic–electronic-conducting membrane. It can be seen that the permeation process from the high oxygen partial pressure (feed) side to the low oxygen partial pressure (permeate) side can be divided into three zones: (1) an interfacial zone on the high partial pressure or feed side (normally air or oxygen-enriched air); (2) a central bulk zone; and (3) an interfacial zone on the low oxygen partial pressure or sweep gas side. The demonstration of integrating both bulk diffusion and surface exchange kinetics into a single unambiguous equation has been done by few research groups [45–47]. For example, the following general assumptions were used for the derivation of Eq. (6.1) by Tan and Li [47] in the formation of mathematical models for the perovskite systems: (1) the oxygen permeation flux is controlled by the surface exchange reactions. (2) The operation is under steady-state isothermal operation. (3) The radial diffusion of gases is neglected. (4) Ideal gas law is applied to the gas phase. (5) The mass transfer resistance of gas phase to oxygen permeation is negligible. As well, the oxygen partial pressures on both shell side and tube side of the membrane surfaces are the same.

$$\frac{dN_{O_2}}{dl} = \frac{k_f\left[\left(P'_{O_2}\right)^{0.5} - \left(P''_{O_2}\right)^{0.5}\right]}{\frac{k_f \ln(r_2/r_1)(P'_{O_2})^{0.5}(P''_{O_2})^{0.5}}{\pi D_v} + \frac{(P'_{O_2})^{0.5}}{2\pi r_1} + \frac{(P''_{O_2})^{0.5}}{2\pi r_2}} \quad (6.1)$$

where N is the molar flow rate, l is the variable length of hollow-fiber membrane, k_f is the forward reaction rate constant, D_v is the oxygen vacancy diffusion coefficient, P'_{O_2} and P''_{O_2} are partial pressures of oxygen at the feed and permeate sides, respectively, and r_2 and r_1 are the outer and inner diameter radius of the membrane tube.

Equation (6.1) is specially derived for tubular membranes and has been applied successfully in the hollow-fiber systems of $Ba_{0.5}Sr_{0.5}Co_{0.8}Fe_{0.2}O_{3-\delta}$ (BSCF) [45] and $La_{0.6}Sr_{0.4}Co_{0.2}Fe_{0.8}O_{3-\delta}$ (LSCF) [48]. The detailed derivation of Eq. (6.1) can be found elsewhere [49–51]. The use of Eq. (6.1) for the perovskite systems is very useful in the scaled-up engineering calculations, wherein the oxygen permeation flux performance inside the hollow-fiber perovskite membrane modules can be estimated. Consequently, the oxygen permeation flux with respect to log-mean membrane area, $dAm = 2\pi r_m dl$ where $r_m = (r_2 - r_1)/\ln(r_2/r_1)$ can be expressed by:

$$J_{O_2} = \frac{k_f D_v\left(\left(P'_{O_2}\right)^{0.5} - \left(P''_{O_2}\right)^{0.5}\right)}{2(r_2 - r_1)k_f(P'_{O_2}P''_{O_2})^{0.5} + \frac{r_m}{r_1}D_v\left(P'_{O_2}\right)^{0.5} + \frac{r_m}{r_1}D_v\left(P''_{O_2}\right)^{0.5}} \tag{6.2}$$

For the sake of analyses, it is assumed that the oxygen permeation in Eq. (6.2) is still applicable to other mixed conducting membranes such as $SrCo_{0.9}Nb_{0.1}O_{3-\delta}$ [52] and $La_{0.2}Ba_{0.8}Co_{0.8}Fe_{0.2-x}Zr_xO_{3-\delta}$ [33]. L_c is defined by the membrane thickness at which the oxygen permeation resistance by bulk diffusion equals by the surface exchange reactions. L_c is expressed as:

$$L_c = \frac{D_v}{2k_f}\left(\frac{1}{P'^{0.5}_{O_2}} + \frac{1}{P''^{0.5}_{O_2}}\right) \tag{6.3}$$

It is noted in Eq. (6.3) that the value of membrane thickness depends on oxygen partial pressure and temperature. A larger characteristic thickness can result from operating at lower temperature or lower oxygen partial pressure. When the membrane has a thickness far less than the critical thickness, L_c, the resistance by bulk diffusion can be neglected; therefore, the surface exchange reaction becomes the rate-limiting step. Lee et al. [48] indicated that the oxygen permeation in membranes in the thickness range of 1–2.6 mm was controlled both by the bulk diffusion of oxide ions and by surface exchange. In this case, Eq. (6.1) can be further simplified as [53]:

$$\frac{dN_{O_2}}{dl} = \frac{k_f\left[\left(P'_{O_2}\right)^{0.5} - \left(P''_{O_2}\right)^{0.5}\right]}{\frac{(P'_{O_2})^{0.5}}{2\pi r_1} + \frac{(P''_{O_2})^{0.5}}{2\pi r_2}} \tag{6.4}$$

Surface exchange reactions may involve many sub-steps, which are oxygen transfer from the gas phase to membrane phase, physical adsorption on surface, dissociation with electronic transfer to yield chemisorbed oxygen species, and the incorporation in surface layer including the reverse reactions [54].

The oxygen permeation flux is inversely proportional to the membrane thickness. Thus, reducing the membrane thickness is expected to increase the oxygen permeation as long as bulk diffusion prevails. Kim et al. [43] derived an equation that is applicable in tubular perovskite membranes to explain the oxygen permeation data in $Sr_{0.5}Co_{0.8}Fe_{0.2}O_{3-\delta}$ and $Sm_{0.5}Sr_{0.5}CoO_{3-\delta}$ in the surface exchange reaction limited. Equation (6.5) can be used for surface exchange limited region in tubular perovskite membranes.

$$J_{O_2} = \frac{\pi r_1 r_2 k}{(r_1 + r_2)S}\left(\frac{\sqrt{P'_{O_2}}}{\sqrt{P^{o}_{O_2}}} - \frac{\sqrt{P''_{O_2}}}{\sqrt{P^{o}_{O_2}}}\right) \tag{6.5}$$

where k is the surface exchange coefficient unit and S is the effective membrane area. When the membrane is relatively thick, the overall oxygen permeation is mainly controlled by the bulk diffusion. In this case, the oxygen flux, J_{O_2}, is generally described by Wagner's theory [50] and is given by Eq. (6.6):

$$J_{O_2} = \frac{1}{4^2F^2L}\int_{\mu_{O_2}(2)}^{\mu_{O_2}(1)} t_i t_e \sigma_t \mathrm{d}\mu_{O_2} \tag{6.6}$$

where F (C/mol) is the Faraday constant and L (m) is the membrane thickness. t is the product of the electronic transference number t_e, ionic transference number t_i, and total conductivity σ_t. Lin et al. [44] and Qi et al. [55] derived the oxygen permeation flux equations for ionic- or mixed-conducting ceramic membranes in terms of electrical conductivity. In this case, the oxygen permeation flux can be expressed as:

$$J_{O_2} = \frac{RT}{4F^2L}\left[\sigma_h(P'_{O_2})\left(1 - \left(\frac{P''_{O_2}}{P'_{O_2}}\right)^{1/4}\right) + \sigma_e(P'_{O_2})\left(\left(\frac{P''_{O_2}}{P'_{O_2}}\right)^{1/4} - 1\right)\right] \tag{6.7}$$

where R (J/mol K) is the gas constant and T (K) is the temperature. Equation (6.7) has been used successfully by Rui et al. [56] to describe the oxygen permeation through dense ceramic membranes with finite rate of chemical reaction. Oxygen permeation flux within $Bi_{1.5}Y_{0.3}Sm_{0.3}O_{3-\delta}$ (t = 1.2 mm, T = 873–953 K) and $La_{0.6}Sr_{0.4}Co_{0.2}Fe_{0.8}O_{3-\delta}$ (t = 1.12 mm, T = 1273 K) is well explained by this equation. Although Eq. (6.7) is originally an empirical equation, it has been applied in the literature [57] with reasonable predictions.

Wang et al. [50] revealed that the controlling step of the oxygen permeation for the 1.5 mm thickness BSCF tubular membrane was bulk diffusion at the temperature range of 700–900 °C; hence, the surface exchange reaction does not favor the oxygen permeation flux. The oxygen permeation flux is described in Eq. (6.8).

$$J_{O_2} = \frac{\pi L C_i D_a}{2S\ln\left(\frac{r_1}{r_2}\right)}\ln\left(\frac{P''_{O_2}}{P''_{O_2}}\right) \tag{6.8}$$

where C_i stands for density of oxygen ions and D_a for ambipolar oxygen ion-hole diffusion coefficient. D_a could be determined from oxygen permeation flux data. Ge et al. [51] theoretically indicated that the oxygen permeation flux through $La_{0.4}Sr_{0.6}CoO_{3-\delta}$ disk-shaped membrane at 950 °C was controlled by both the bulk diffusion and oxygen surface exchange reactions; therefore, the oxygen permeation flux is given as:

$$J_{O_2} = \frac{D_v}{2L} \frac{\dfrac{k_f}{\dfrac{1}{\left(P'_{O_2}\right)^{0.5}} + \dfrac{1}{\left(P''_{O_2}\right)^{0.5}}}}{\dfrac{k_f}{\dfrac{1}{\left(P'_{O_2}\right)^{0.5}} + \dfrac{1}{\left(P''_{O_2}\right)^{0.5}}} + \dfrac{D_v}{2L}} \left(C'^{e}_{v} - C^{e}_{v}\right) \tag{6.9}$$

where C^e and C'^e are the oxygen vacancy concentrations in the material under thermal equilibrium with the atmosphere surrounded by oxygen partial pressure at feed side and oxygen partial pressure at the permeate side and (P'_{O_2} and P''_{O_2}), respectively. D_v and k_f can be determined by proper fitting of the experimental data into Eq. (6.9), which requires prior knowledge of (a) C^e and C'^e values and (b) the oxygen permeation fluxes under the applied oxygen partial pressure gradient across the membrane at selected temperatures by oxygen permeation experiments [51].

Chang et al. [58] thoroughly compared the performance of symmetric (1.5 mm thick) and asymmetric mixed-conducting membranes (200 μm-thick thin dense layer and 1.3-mm-thick support) with correlation of the overall oxygen permeation resistance across the membrane. The authors prepared the asymmetric membrane consisting of the support and thin dense layer from the same composition $SrCo_{0.4}Fe_{0.5}Zr_{0.1}O_{3-\delta}$ (SCFZ) perovskite-type oxide. The oxygen permeation flux in both membrane architectures is exemplified as:

$$J_{O_2} = \frac{1}{S} \frac{RT}{R_p + R_d} \ln\left(\frac{P'_{O_2}}{P''_{O_2}}\right) \tag{6.10}$$

In this equation, R_p and R_d are being the resistance in the porous support and thin dense layer, respectively. For an asymmetric membrane, the overall resistance is the sum of the resistance in the porous support and in the thin dense layer ($R_{overall} = R_p + R_d$). In the symmetric membrane, $R_{overall} = R_p$. The authors concluded that the oxygen permeation flux on the asymmetric membrane was higher than the symmetric membrane due to the significant decrease of bulk diffusion resistance in the thin dense layer of the asymmetric membrane. Chang et al. [58] reported detailed calculations of the overall oxygen permeation resistance and modeling conducted to the transport resistance through a membrane.

6.2.3 Oxygen Permeation with Chemical Reactions

The performance of electronic-conducting ceramic membranes in permeating oxygen at high temperatures led to the idea of integrating the two processes of oxygen permeation and combustion in one piece of equipment. To achieve that, a major industrial effort is presently devoted to the development of the mixed-conducting ceramic membrane reactor technology for combustion of hydrocarbons. In this reactor, oxygen permeates through the ionic- or mixed-conducting ceramic membrane via a complex mechanism. This mechanism usually includes adsorption of oxygen and charge transfer reaction on the membrane surface exposed to air, oxygen vacancy and electron diffusion in the membrane bulk, and charge transfer and chemical reaction on the membrane surface exposed to a reducing gas. The detailed mathematical formulation for oxygen permeation through mixed-conducting ceramic membranes is fairly complex [44]. The oxygen permeation through oxygen ionic- or mixed-conducting ceramic membranes under reaction conditions was also examined by Rui et al. [56]. The authors considered a model that takes into account the different electrical transport mechanisms (*p*-type and *n*-type transports) and finite reaction rate. In their work that includes reaction consuming oxygen in one side of the membrane, it was demonstrated that the oxygen partial pressure in the reaction side decreases and the oxygen permeation flux increases with the increase in the reaction rate for both the *p*-type and the *n*-type transport-dominated mechanism. The authors also reported that the increase in the reaction rate results in a transition of the transport mechanism from *p*-type to *n*-type. This transition leads to an increase in the permeation flux by up to 30 times. This effect offers one explanation for the large discrepancies in published permeation data for membrane reactors of partial oxidation reaction employing an oxygen-permeable dense ceramic membrane [56]. The authors reported also, for a membrane with a specific transport mechanism, the increase in the reactant partial pressure results in an increase in the reaction rate and, consequently, the oxygen permeation flux. However, the increase in the inlet inert gas amount has a complex effect on the oxygen permeation flux because it lowers both oxygen partial pressure and the reaction rate at the same time.

The oxidative coupling of methane (OCM) and selective oxidation of ethane (SOE) reactions involve oxidative reactions of methane or ethane to form ethylene as the intermediate (desired) product. The final (equilibrium) product is carbon dioxide (and water). Both the OCM and SOE reaction mechanisms are very complex and may involve over than hundreds of steps [59, 60]. Akin and Jerry [57] demonstrated how the extent of the reaction (or reactivity) and reactant flow rate affect the oxygen permeation flux. They used the following simple reaction to exemplify the complex oxidative reactions of methane or ethane to ethylene and finally to carbon dioxide:

$$CO + 1/2O_2 \rightarrow CO_2 \tag{A}$$

In this equation, carbon monoxide represents a hydrocarbon reactant. The use of this simple reaction can help in obtaining semi-analytical solution for the combustion process with oxygen permeation. The basic principle illustrated with this rather simple reaction can be extended for application to the more complex reaction systems if detailed reaction mechanism and kinetic equations are known. In order to obtain analytical expressions for the oxygen partial pressure in the reaction chamber, Akin and Jerry [57] used a simple reactor model, continuously stirred tank reactor (CSTR), to describe the reaction chamber in their work. Since in most laboratory studies, membrane reactor experiments were conducted on disk-shaped or short tubular membranes, such as the BYS used in their work, the CSTR model can catch the major characteristics of the oxidation reaction in the membrane reaction chamber. Instead of setting specific reaction kinetics in the model to account for the reaction rate, they only considered two extreme cases in the modeling and analysis: (1) extremely fast reaction rate, which reflects complete conversion of the reactant (CO in this work) with oxygen permeating to the reaction chamber, (2) extremely slow reaction rate which implies no reaction between the oxygen permeating with the reactant (CO). In the second case, the reactant fed into the reaction side behaves like an inert gas, such as the case of oxygen permeation experiments with helium as purge. The real case would lie between these two extreme cases. Oxygen permeation flux through dense ionic- or mixed-conducting ceramic membranes can be related to air- and reaction-side oxygen partial pressures as proposed in their model [57]. For ionic- or mixed-conductors with ionic transference number close to 1 and electron conduction dominated by the transport of the electron holes (such as yttria-stabilized zirconia and doped bismuth oxide) [61–63], the oxygen permeation flux can be estimated by the following equation:

$$J_{O_2} = k({P'}_{O_2}^{1/n} - {P''}_{O_2}^{1/n}) \tag{6.11}$$

In this case, it should be noted that the driving force is the difference between the oxygen partial pressures in the air side and the reaction side with a positive value for constant n (i.e., $n > 0$). For convenient notations, this group of membranes with oxygen permeation equation in the form of Eq. (6.11) is referred to in this paper as membranes with p-type flux equation.

For ionic conductors with electron conduction dominated by the electrons, such as yttria-doped zirconia at low oxygen partial pressure range [64], the permeation flux equation can be expressed by

$$J_{O_2} = k({P''}_{O_2}^{1/n} - {P'}_{O_2}^{1/n}) \tag{6.12}$$

This also applies for mixed conductors with ionic transference number close to zero, such as lanthanum cobaltite.

Table 6.1 Summary of parameters used in the parametric study [57]

Parameter	Range
Q^0_{feed} (ml/min)	100–675
T (°C)	850
A (cm^2)	1.8
k (ml/min cm^2 atm$^{1/n}$)	
BYS (ml/min cm^2 atm$^{1/n}$)	0.035
LSCF (ml/min cm^2 atm$^{1/n}$)	0.16
n	
BYS	3.34
LSCF	−8.06

It should be noted, in this case, that the driving force is the oxygen partial pressure in the reaction side minus the oxygen partial pressure in the air side with a negative value for constant n ($n < 1$). For convenience in notation, these groups of membranes are referred to here as membranes with n-type flux equation. Table 6.1 summarizes the parameters used in their study. The values for constants, k and n in Eq. (6.11), for a membrane with p-type flux equation are chosen for the bismuth-based oxide $Bi_{1.5}Y_{0.3}Sm_{0.2}O_{3-\delta}$ (BYS) [65]. The k and n values for a membrane with n-type flux equation in Eq. (2.12) are calculated from the literature oxygen permeation data given for $La_{0.6}Sr_{0.4}Co_{0.2}Fe_{0.8}O_{3-\delta}$ (LSCF) [45]. As proposed by Xu and Thomson [45], the concentration of oxygen vacancies at both surfaces of the membrane (C'_V and C''_V) is also governed by surface exchange kinetics for the following two surface reactions (forward reaction in the feed side and backward reaction in the permeate side):

$$\frac{1}{2}O_2 + V^{-}o \overset{k_f/k_r}{\leftrightarrow} O_o^x + 2h^{-} \tag{B}$$

O_o^x represents lattice oxygen in the perovskite crystal structure. k_f and k_r are, respectively, the forward and reverse reaction rate constants for the forward reaction (or the reverse and forward rate constants for backward reaction). It may be noted that, because of the high electronic conductivity, the electron holes are essentially constant at both surfaces of the membrane and, thus, the reverse reaction rate of Reaction B and the forward reaction rate of Reaction C are pseudo-zero-order at steady state under isothermal conditions. According to this feature, they have correlated the steady-state oxygen permeation flux as a function of P'_{O_2}, P''_{O_2} and membrane temperature and thickness:

$$J_{O_2} = \frac{D_v k_r (P'^{0.5}_{O_2} - P''^{0.5}_{O_2})}{2Lk_f (P'_{O_2} P''_{O_2})^{0.5} + D_v (P'^{0.5}_{O_2} + P''^{0.5}_{O_2})} \tag{6.13}$$

where D_v, k_r and k_f are being functions of the specific properties of the membrane as well as the operating temperature. The values of these coefficients have been

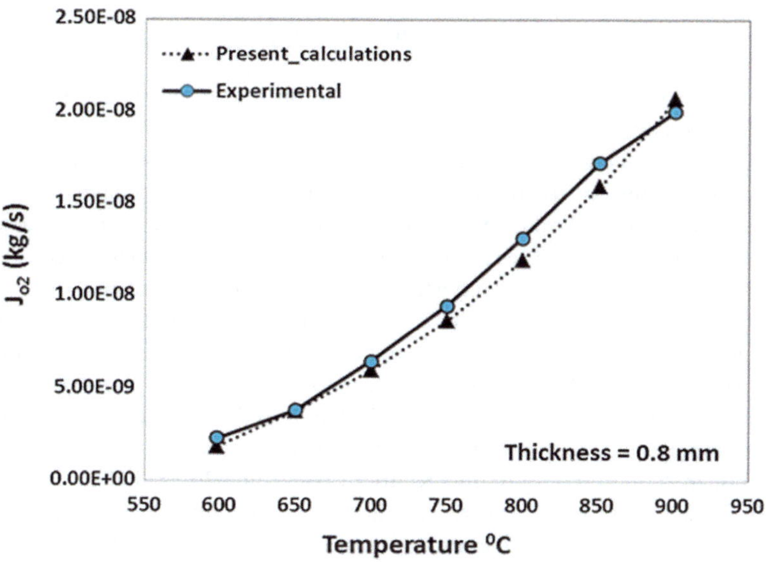

Fig. 6.4 Fitting of the experimental data of oxygen permeation through LSCF-1991 membrane having a thickness of 0.8 mm [65] with our present permeation model

Table 6.2 Obtained pre-exponential coefficients and activation energies of D_v, k_f, and k_r for LSCF-1991 membrane from our work [3] through the fitting of experimental data of [65]

Expression	Pre-exponential coefficients		Activation energy (kJ/mol)
	Unit	Value	
$D_v = D_v^o \exp(-E_D/RT)$	m^2/s	1.58×10^{-5}	73.6
$k_f = k_f^o \exp(-E_f/RT)$	$m/atm^{0.5}$ s	1.11×10^{10}	226.9
$k_r = k_r^o \exp(-E_r/RT)$	mol/m^2 s	3.85×10^{11}	241.3

determined by Habib et al. [3] through fitting the experimental oxygen flux data in the work done by Kusaba et al. [65] as a function of temperatures as shown in Fig. 6.4. As revealed from this figure, both the experimental data and our present *Dkk* model (D_v, k_r and k_f model) are in a very good agreement. Also, it is indicated that the temperature of the membrane surface increases as the oxygen permeation flux across the membrane increases due to the reduced surface resistance of the membrane because of temperature increase. The fitted values of the coefficients D_v, k_r and k_f are listed in Table 6.2 and the values of the activation energies.

A semi-empirical form found in the literature also was used extensively to determine the local oxygen flux as a function of the membrane temperature, the feed and permeate oxygen partial pressures, and empirical constants that depend on the specific material. This form allows for interchangeable oxygen flux mechanisms to be implemented quickly and effectively within the model in order to explore the impact of different ITM membrane materials [30].

$$J_{O_2} = A^* \exp(-B/T_M) * (P'^{\,n}_{O_2} - P''^{\,n}_{O_2}) \tag{6.14}$$

The pre-exponential term A accounts, in some sense, mostly for the diffusion coefficient and the membrane thickness dependence. The term B represents the effective activation energy or Arrhenius dependence due to both surface exchange kinetics and diffusion coefficient activation energy. This mechanism is chosen because it is simple and relatively accurate with respect to experimental data. Also, this model reasonably captures the impact of both surface exchange kinetics and the temperature dependence of the oxygen vacancy diffusion coefficient. However, it is limited in the sense that it applies to a specific membrane thickness and also cannot be extrapolated too far from the experimental conditions used to obtain the fitted values for A and B [30]. The functional dependence on partial pressure is assumed to be $n = 0.25$ for LNO and $n = 0.5$ for LSCF, based on global surface exchange kinetics theory and experimental results and the values of A and B are 2.011 mol m^{-2} s^{-1} pa^{-n}, 10,240 K for LNO and 26.75 mol m^{-2} s^{-1} pa^{-n}, 16,510 K for LSCF, respectively. This is consistent with the mixed control, i.e., both diffusion and surface kinetics, in contrast to diffusion dominant where n is typically less than zero [66].

6.3 CFD Modeling of OTR

The mathematical model consists of the conservation equations for mass, momentum, and energy, and transport equations for scalar variables. These equations were solved for providing predictions of the flow pattern in addition to thermal and emission characteristics of the reacting medium inside the membrane reactor. The equations which govern the conservation of mass, momentum, and energy as well as the equations for species transport can be stated as [67]:

$$\nabla \cdot (\rho U) = S_i \tag{6.15}$$

$$\nabla \cdot (\rho U U) = -\nabla p + \mu \nabla^2 U \tag{6.16}$$

$$(\rho C_p) U \cdot \nabla T = \nabla \cdot (\lambda \nabla T) \tag{6.17}$$

$$\nabla \cdot (\rho U Y_i) = \nabla \cdot (\rho D_{i,m} \nabla Y_i) = S_i \tag{6.18}$$

The oxygen moves across the membrane surface from feed side to permeate side, and a source-sink term is added to those equations. The conservation equations were modified and listed in previous works [3, 68, 69]. In all cases, the source/sink term (S_i) permits considering the mass flow of the species across the membrane. Clearly, the method requires preliminary knowledge of the permeability characteristics of the membrane for proper formulation of the source term. This source term only takes into account the transfer of oxygen molecules across the membrane.

Depending on each specific membrane, any mechanism can be included in the model. From the above-mentioned equations for oxygen permeation, the following form for the source term can be implemented in the model:

$$S_i = \pm \frac{J_{O_2}.A_{\text{cell}}.\text{MW}_{O_2}}{V_{\text{cell}}} \tag{6.19}$$

where A_C and V_C are the computational cell area (m^2) and volume (m^3), respectively, MW_{O_2} is the molecular weight of O_2 in kg/mol, and J_{O_2} is the oxygen permeation flux in mol/m^2 s. The positive sign is used in the permeate side cells at low partial pressure side of oxygen, and the negative sign is used in the feed-side cells at high partial pressure side of Oxygen. The source/sink S_i term (kg/m^3/s) is modeled in such a way to vanish except for the case when $i = O_2$ when the computational cell is adjacent to the membrane surface. The emissivity of the membrane surface is assumed to be equal to 0.8 for the majority of membrane materials.

Coupling of heat transfer through the membrane surface between the feed and permeate sides is made through a heat balance equation which considers also the radiant heat absorbed by the membrane:

$$Q^{\text{o}}_{\text{sweep}} - Q^{\text{o}}_{\text{feed}} - 2\sigma\varepsilon(T^4_{\text{mem}} - T^4_{\infty}) = 0 \tag{6.20}$$

where $Q^{\text{o}}_{\text{sweep}}$ is the heat transfer from the reacting sweep gases to the membrane and $Q^{\text{o}}_{\text{feed}}$ is the heat released from the membrane surface to the feed side. The emissivity of the membrane surface is assumed to be equal to 0.8, and T_{mem} and T_{∞} are the membrane and surrounding wall temperature, respectively. The diffusion coefficient is determined by specifying the binary mass diffusion coefficient of the component i in the component j. The corresponding diffusion coefficient is:

$$D_{i,m} = \frac{1 - X_i}{\sum\limits_{j,i \neq i} \left(\frac{X_i}{D_{i,j}}\right)} \tag{6.21}$$

Gambit software, or any other mesh builder software, can be used to construct the mesh with very small boundary control volumes near to both sides of the membranes in order to correctly predict the permeation process. In order to reduce the numerical errors, structured meshing can be performed to divide the flow domain into sub-domains (feed sides and permeate sides). The membrane cell computational geometry consisted of two mass flow inlets, boundaries for introducing the feed and sweep gas mixture, and two pressure outlets for retentate and permeate flows. Grid refinement should be performed to achieve the grid independence. The general-purpose commercial CFD code ANSYS Fluent software, OpenFOAM or any other software can be used for the solution of the conservation equations. Transport of O_2 across the membranes can be achieved through using series of user-defined functions (UDFs) that are written in VC++ and compiled and hooked to the Fluent software. The convective terms can be discretized by a second-order upwind scheme, while the SIMPLE algorithm can be used in all cases to couple the continuity and Navier–Stokes equations, due to its stability and accurateness. The solution convergence was carefully checked by monitoring the residuals of all variables as well as the species mass flow rate.

6.4 Modeling of Reaction Kinetics and Radiation

Modeling of reaction kinetics in integrated systems such as gas turbine combustors utilizing OTR or fire tube furnaces required careful selection of the computational model to cope with the very large number of computational cells contained in the 3D computational domain. In the case of ITM reactor simulation, the large number of computational cells requires large time for each run, especially where a more detailed reaction kinetics mechanism is considered. Based on this fact, the general simplest oxidation reaction of methane is considered ($CH_4 + 2O_2 \rightarrow CO_2 + 2H_2O$). The main source of error in this model is neglecting important intermediate species like hydrogen and carbon monoxide. However, this simple model is expected to produce close results to those obtained using the detailed reactions mechanisms, and it can give reasonable distributions of temperature and species concentrations. This model was considered in the recent work by Nemitallah et al. [70], which was validated against the model data of Mancini and Mitsos [71], and the results showed a good agreement. The combustion starts once oxygen encounters fuel in the permeate side of the membrane. It was assumed that the combustion products are only CO_2 and H_2O. Once the combustion products are formed, they should be carried by the main stream. It was assumed that the reactions are very fast so that chemical equilibrium is attained everywhere in the reaction zone. No catalysts are included in that work so that no reactions can occur on the membrane surface. So, the permeated oxygen diffuses in the permeate flow till it reaches CH_4 molecules so that the combustion occurs at these locations. The values of species concentrations are averaged at each cell. The combustion model just uses the permeated oxygen in the combustion process away from the membrane surface. However, it affects the permeation process through the consumption of oxygen which results in a reduction of the oxygen partial pressure in the permeate side. The consumed oxygen allows the newly permeated oxygen molecules to move toward the combustion region and move away from the membrane surface. As a result, the partial pressure gradient force is increased and accordingly oxygen flux across the membrane is increased. Another effect of the combustion process on oxygen permeation is coming from the effect of the elevated temperature due to combustion on the permeation process. This should result in increased oxygen flux due to the reduction in the resistance of the membrane to oxygen permeation with increased temperature. This is clearly defined in the oxygen permeation equations as mentioned above.

In fire tube boiler applications, substantial portion of heat are transferred through the radiation mechanism from the combustion gases to the surrounding water.

Integrating ceramic membranes into a reactor based on the operational concept of fire tube boilers necessitates proper calculations of the temperature distribution inside the domain. This is due to the dependence of oxygen permeation flux on the operating temperature. Based on these facts, the radiative transfer equation (RTE) was solved through the whole domain for applications in gas turbine combustors and fire tube boiler furnaces. The RTE is given as follows [72]:

$$\frac{1}{c}\frac{\mathrm{d}I_\nu}{\mathrm{d}t} = -(\kappa_\nu + \sigma_\nu)I_\nu + \kappa_\nu n_\nu^2 I_{b\nu} + \frac{\sigma_\nu}{4\pi}\int_{\Delta\nu}\int_{\Omega=4\pi} \Phi_\nu(\vec{s}\,' \rightarrow \vec{s}; \nu' \rightarrow \nu) I_{\nu'}(\vec{s}\,')\mathrm{d}\Omega'\mathrm{d}\nu' \tag{6.22}$$

where c is the speed of electromagnetic wave in vacuum, I_ν is the spectral radiation intensity, and κ_ν is the spectral absorption coefficient. σ_ν is the scattering coefficient, n_ν is the spectral index of refraction of the medium, $I_{b\nu}$ is the Planck's spectral blackbody intensity of radiation, in addition to Φ_ν is the phase function.

The solution of radiative transfer entails models to account for the directional and spectral natures of radiation. Based on the calculations of the radiative intensity, the radiative heat flux vector gradient can be determined and used in the enthalpy equation in order to account for heat sources/sinks as a result of heat transfer by radiation. In the work of Nemitallah et al. [85], the scattering has been neglected in the radiative transfer equation (RTE). This can be attributed to its low significance in case of combustion of light gaseous fuels such as methane and propane. Through the application of such assumption, the RTE becomes a differential equation, which can be solved easily as compared to its integral form. One of the popular methods to treat the directional nature of radiation is the discrete ordinate (DO) method. This method is originally suggested by Chandrasekhar [72] for astrophysical applications. As a matter of fact, effect and applicability of the used radiation model especially in case of oxy-combustion applications are very crucial. In the recent research work by Rajhi et al. [73], a study has been performed targeting the evaluation of most of the available gas radiation models in oxy-combustion environment. The study has been performed for a range of temperature from 900 up to 2000 K and for high values of CO_2 concentrations, up to 0.9. The results using all considered radiation models were compared with the available experimental data. Among all radiation models, the exponential wideband model (EWBM) has shown the highest agreement with the experimental data. Additionally, the EWBM was validated using the recorded data for oxy-combustion in a gas turbine model combustor [74], and it has proved its applicability. The exponential wideband model (EWBM) is based on a physical analysis of gas absorption and provides a set of semi-empirical expressions to predict the total band absorptance of infrared active molecules. This model can be used to predict radiative properties in a wide range of temperature, total pressure range, volumetric fraction, and pass length [75].

6.5 Integration of OTRs with Conventional Combustors for ZEPP Applications

Recently, several studies investigated the integration of gas separation membranes in power cycles [70, 76–78] aiming at clean energy production while capturing CO_2 at lower cost [79]. Membranes can be used in all CCTs including pre-combustion [80], post-combustion [81], and oxy-combustion [82]. In case of oxy-combustion technology, the idea is to replace the currently used conventional combustor by multi-membrane reactor system. Air is fed to the feed side of each membrane, and a mixture of gaseous fuel plus exhaust recycled gases (consists mainly of H_2O plus CO_2) is fed to the permeate side where oxy-combustion occurs. A monolith structure design of an ITM reactor was proposed by Mancini and Mitsos [83] to produce power in the range of 300–500 MWe based on the cycle first law efficiency. To produce such power, 100,000 square membrane channels were utilized (50,000 channels per each stream) resulting in reactor volume of 1000 m^3, total surface area of 266,700 m^2, reactor height of 4.75 m, and reactor length of 44.44 m. Nemitallah et al. [70] developed, based on three-dimensional (3D) numerical simulations, a monolith structure design of an ITM reactor that can produce power of 5 MWe to replace a conventional gas turbine combustor. To produce such power, the ITM reactor was designed to have a volume of 10 m^3, membrane surface area of 2700 m^2, height of 3.35 m, and length of 0.9 m. Also, Habib and Nemitallah [84] presented a design of a reactor consisting of multi-separated-ITM units for application in a carbon-free fire tube boiler. The results showed impossible operation under counter-current flow configuration, and the large drop in reactor efficiency was encountered because of the large drop in membrane surface temperature due to heat transfer to the surrounding water inside the boiler. In another study, Nemitallah et al. [85] presented a design of a multi-can carbon-free gas turbine combustor utilizing multiple oxygen transport reactors (OTRs) of the shell and tube type for zero-emission power plant (ZEPP) applications. The design of the gas turbine combustor is calculated based on optimizations of flow configuration (co- and counter-current), shell-side and tube-side (feed and sweep) flow rates, inlet fuel concentration in the sweep flow (CH_4 plus CO_2), and membrane tube diameter, pitch (spacing) and length. Based on required power output in the range from 10 to 15 MWe, the final design of the gas turbine combustor is calculated to have 16 cans (OTRs) with 3000 membrane tubes per can and volume of 5.2272 m^3 per can. Next, the above-mentioned studies for designing OTRs for different applications are presented along with the detailed results and analyses.

6.6 Application of OTR into Gas Turbine Combustor

In this section, two detailed numerical studies are presented on the integration of the membrane-based OTRs into gas turbine combustors. The first study presents a monolith structure design of an ion transport membrane reactor for gas turbine combustion application [70]. The second study presents distinctive design for the application of OTRs into gas turbine combustors. In the second study, a design of a multi-can carbon-free gas turbine combustor utilizing multiple shell-and-tube OTRs for ZEPP applications is presented [85]. Both studies are presented successively in the next sections.

6.6.1 Monolith Structure Design OTR for Replacement of a Gas Turbine Combustor

The use of ion transport membrane reactors to substitute the conventional gas turbine combustors is a promising technology for the applications of ZEPP. An ITM monolith structure reactor design is introduced in this study for substituting a conventional gas turbine combustor. Due to reactor symmetry, only 3D four quarters of four adjacent channels sharing one common edge are considered in all simulations using LSCF-1991 membranes. Effect of feed and sweep flow rates have been considered, and it was calculated in order to meet the power required for the reactor and keeping the reactor size as compact as possible. Effects of flow configurations, channel width, and percentage of CH_4 in the permeate side flow are introduced under constant inlet gas temperature of 1173 K and fixed operating pressure of 10 bars. The reactor geometry has been calculated based on the calculations of the minimum possible channel width. Counter-current flow configuration design resulted in improved oxygen permeation flux and improved heat transfer characteristics. However, this flow configuration resulted in unacceptable increase in the membrane temperature. It was found that any reduction in the channel width below 15 mm results in large increase in the viscous pressure drop. Also, increasing the amount of CH_4 in the permeate side over 5% was found to be non-applicable because of oxygen permeation flux limitation. The reactor length was fixed to 0.9 m to be similar to that of real gas turbine combustors with 25,000 channels for each stream. The present reactor design resulted in a reactor height of 3.35 m and an overall volume and membrane surface area of 10 m^3 and 2700 m^2, respectively. The reactor is capable of delivering power ranging from 5 to 8 MWe based on cycle first law efficiency.

6.6.1.1 Numerical Modeling

A monolith structure ITM reactor design is used in the present study with total number of feed and permeate channels of 50,000, 25,000 for each stream. Each channel has a square cross section of width of 15 mm based on the calculations of the minimum possible channel width as discussed in the coming section. The length of the reactor was fixed to 0.9 m, similar to that of industrial gas turbines. Due to symmetry in the monolith structure design, only four quarters of four adjacent channels sharing one common edge are considered in the simulations as shown in Fig. 6.5. Figure 6.5 shows a traverse cross section, and LSCF-1991 membranes were used to separate the oxygen from the stream containing the fuel, methane, and the sweep gas, CO_2 plus H_2O. More details about membrane specifications are listed in Table 6.3 and also in the work by Nemitallah et al. [69]. The present reactor design results in reactor height of 3.35 m and overall volume and membrane surface area of 10.1 m^3 and 2700 m^2, respectively. In order to understand how an ITM reactor performance depends on the flow configuration, calculations are performed for co-current versus counter-current at the same operating conditions. Equilibrium is assumed for simplicity because it provides upper-bound estimates on the wall temperature and reactive ITM performance in general.

The channel width was varied in this study in order to investigate the effect of reactor geometry on oxygen permeation flux and combustion process. The temperature of streams in all feed and permeate channels of the reactor was fixed to 900 °C, and the pressure for all streams was maintained at ten times the atmospheric pressure, 10 bars. Simulations for the effect of the volume flow rate of feed and permeate flows were done in order to get maximum oxygen permeation flux and improving the combustion process in the permeate channels. The pre-exponential coefficients of the *Dkk* (D_v, k_f and k_r coefficients model) oxygen

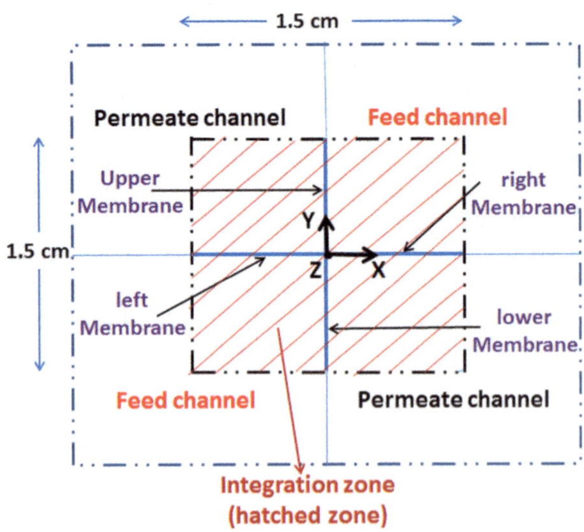

Fig. 6.5 Schematic diagram of a traverse cross section in four adjacent channels in the 3D membrane reactor showing the membranes and the considered integration zone

Table 6.3 Membrane specifications

Parameter	Value
Membrane thickness	0.9 mm
Membrane material	LSCF-1991
Density	6000 kg/m^3 [86, 87]
Thermal conductivity	4 W/m/K [86, 87]

flux permeation equation model were calculated by fitting the available experimental data in the literature, where D_v is the diffusion coefficient of oxygen vacancies and k_f and k_r are, respectively, the forward and reverse reaction rate constants as discussed below.

6.6.1.2 The Oxygen Permeation Model

In the ceramic membrane reactor, oxygen permeates through a mixed-conducting ceramic membrane, which includes adsorption of oxygen and charge transfer reaction on the membrane surface exposed to air, oxygen vacancy, and electron diffusion in the membrane bulk and charge transfer and chemical reaction on the membrane surface exposed to a reducing gas [3]. As proposed by Xu and Thomson [45], at steady state, an expression that combines surface exchange on the feed and permeate sides and bulk diffusion in terms of the oxygen partial pressures can be derived in the form:

$$J_{O_2} = \frac{D_v k_r \left(P'^{\,0.5}_{O_2} - P''^{\,0.5}_{O_2}\right)}{2Lk_f \left(P'_{O_2} P''_{O_2}\right)^{0.5} + D_v \left(P'^{\,0.5}_{O_2} + P''^{\,0.5}_{O_2}\right)} \tag{6.23}$$

where

$$D_v = D_v^o \exp(-E_D/RT) \tag{6.24}$$

$$k_f = k_f^o \exp(-E_f/RT) \tag{6.25}$$

$$k_r = k_r^o \exp(-E_r/RT) \tag{6.26}$$

The value of the pre-exponential coefficient of D_v is in m^2/s, of k_f is in m/atm$^{0.5}$ s, and that of k_r is in mol/m^2 s. The activation energy for LSCF-1991 membrane has been determined by fitting the experimental oxygen flux data in the work done by Kusaba et al. [65] as function of temperatures (see [68, 69]) for three different membrane thicknesses of 0.8, 1, and 2 mm. The fitted values of the pre-exponential coefficients of D_v, k_r, and k_f are listed in Table 6.4 and the values of the activation energies are listed there in the table.

Gambit 2.2 was used to construct the mesh with more than 800,000 finite volumes beside very small boundary control volumes near to both sides of the

Table 6.4 Pre-exponential coefficient and activation energy of D_v, k_f, and k_r for LSCF-1991 membrane comes from the fitting of experimental data of Kusaba et al. [65]

Expression	Pre-exponential coefficients		Activation energy (kJ/mol)
	Unit	Value	
$D_v = D_v^o \exp(-E_D/RT)$	m^2/s	1.58×10^{-5}	73.6
$k_f = k_f^o \exp(-E_f/RT)$	$m/atm^{0.5}$ s	1.11×10^{10}	226.9
$k_r = k_r^o \exp(-E_r/RT)$	mol/m^2 s	3.85×10^{11}	241.3

membranes in order to correctly predict the permeation process. In order to reduce the numerical errors, structured meshing was performed to divide the flow domain into sub-domains (feed sides and permeate sides). The membrane cell computational geometry consisted of two mass flow inlet boundaries for introducing the feed and sweep gas mixture and two pressure outlets for retentate and permeate flows. The membrane thickness considered in the calculations of the oxygen permeation equation is 0.9 mm, and also it was considered in all of the simulations with the properties of the LSCF material. The computational grid consisted of 30 cells for each channel side normal to the flow direction and 900 cells along the axis of the membrane reactor. Grid refinement was performed to achieve the grid independence by analyzing the mass fractions within the geometrical domain. The refinement has been done for all computational cell sizes from 1×1 to 0.5×1 mm^2 because of a difference around 4% in the results. Further reduction in the cell size below 0.5×1 mm^2 did not result in any considerable variations in the results. The general-purpose commercial CFD code Fluent 12.1 was selected for the solution of the steady-state conservation equations adopting the laminar formulation. The equations were numerically solved in a Cartesian coordinate system; transport of O_2 across the membranes was achieved using series of user-defined functions (UDF) that are written in VC++ and compiled and hooked to the Fluent software. The convective terms were discretized by a second-order upwind scheme, while the SIMPLE algorithm was used in all cases to couple the continuity and Navier–Stokes equations, due to its stability and accurateness. The solution convergence was carefully checked by monitoring the residuals of all variables as well as the species mass flow rate. Residuals were dropped to the order of 10^{-5} or less, which is at least two orders of magnitude tighter than Fluent default criteria.

6.6.1.3 OTR Design Specifications

Several designs are possible for the ITM reactor that can replace the gas turbine combustor; however, the optimum design requires a parametric study for many parameters affecting the reactor performance. These parameters include the volume flow rates in feed and permeate sides, channel width and the percentage of CH_4 in the permeate side flow. The permeate stream has an upper-bound on diluent flow rate because the inlet methane concentration should not fall below roughly 5% for mass transfer and combustion stability reasons. For this reason, the percentage of CH_4 in the permeate channels was kept at 5% vol. and the flow rates in both sides of the membrane were changed in order to get maximum oxygen permeation flux and complete conversion of CH_4 in the permeate channels. It was found from the simulations that any reduction in the channel width below 15 mm results in large increases in the viscous pressure drop. Basically, as the channel width decreases, the flow velocity must increase to maintain the same mass flow rate. The oxygen permeation at high flow velocities when reducing the channel width below 15 mm is not proportional to the fuel flow, and the combustion temperature was reduced sharply. The channel length was considered fixed in all simulations to 0.9 m to be similar to the operating length of a real gas turbine combustor, and the operating pressure for both streams was fixed to 10 bars. The total number of channels was calculated in order to get as compact design as possible for the ITM reactor and gives a reasonable operating power output. A monolith ITM reactor structure design is used in this study with total number of feed and permeate channels of 50,000, 25,000 for each stream. The membranes separate oxygen-containing streams (typically air) from streams containing the fuel, methane, and the sweep gas, CO_2 plus H_2O. This reactor design results in reactor height of 3.35 m and overall volume and membrane surface area of 10 m^3 and 2700 m^2, respectively. The reactor is able to deliver power ranging from 5 to 8 MWe based on cycle first law efficiency. In order to understand how an ITM reactor depends on the flow configuration, calculations are performed for co-current versus counter-current at the same operating conditions. The ITM reactor specifications for replacing a gas turbine combustor are listed in Table 6.5.

Table 6.5 Membrane reactor specifications for substituting a gas turbine combustor

Parameter	Reactive co- and counter-current
Permeate T_{in} (K)	1173
Feed T_{in} (K)	1173
Feed $m^{\bullet}_{air}$ (kg/s/cell)	1.4336×10^{-3}
Permeate $m^{\bullet}_{CH_4}$ (kg/s/cell) 5% vol.	1.54944×10^{-5}
Permeate $m^{\bullet}_{CO_2,in}$ (kg/s/cell)	4.289×10^{-4}
Permeate $m^{\bullet}_{H_2O,in}$ (kg/s/cell)	1.754×10^{-4}
Feed P_{tot} (bar)	10
Permeate P_{tot} (bar)	10
Number of cells per stream	25,000
Power (MW), bases on 1st law efficiency	5:8

6.6.1.4 Model Validation

Figure 6.6 compares the present model calculation results with the work done by Mancini and Mitsos [71]. For this validation, the counter-current flow configuration cases considered in their work were conducted using the present model, and the data for oxygen permeation flux and oxygen partial pressure in both sides of the membrane were plotted as shown in the figure. Due to the reactor symmetry, only four adjacent quarters of four adjacent cells are considered in this simulation as shown in Fig. 6.5. The calculations were performed at fixed pressure of 10 bars in both sides of each membrane and feed and permeate inlet temperatures of 973 and 1173 K, respectively. The total flow rates of feed gases for all reactor cells

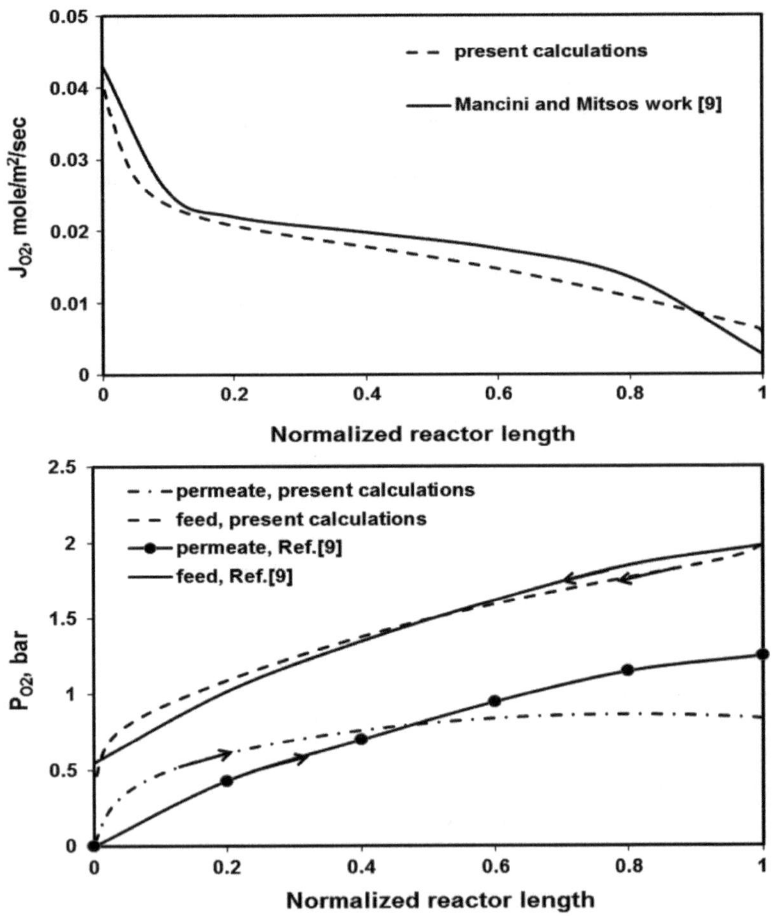

Fig. 6.6 Counter-current, non-reactive: axial dependence of oxygen permeation flux and oxygen partial pressure at fixed pressure of 10 bars in both sides of the membrane for the present calculations and the work done by Mancini and Mitsos [71]

(100,000 cells, 50,000 cells per stream) are O_2 (3.3 kmol/s), N_2 (12.41 kmol/s) and H_2O (1 kmol/s). Also, the total permeate gases' flow rates for all reactor cells are CO_2 (8.34 kmol/s) and H_2O (8.34 kmol/s). For this comparison and validation of the present work with Mancini and Mitsos work [71], the same oxygen permeation model was used with increasing the oxygen flux one order of magnitude as in their work. The considered comparison case is for counter-current flow regime and non-reactive, only oxygen separation was considered. For non-reactive separation-only case, one can see that the oxygen permeation flux profiles for the present calculations and the reference work calculations are in a good agreement and the flux drops from inlet to exit direction. This flux drop is due to the increased oxygen partial pressure in the permeate side of the membrane as shown in the same Fig. 6.6b. Also, the oxygen mole fraction of the feed stream drops sharply in the case of counter-current configuration indicating a high recovery ratio, and an effective ITM configuration as shown in Fig. 6.6.

6.6.1.5 Co-current OTR Design

Most of the analysis in the work of Nemitallah et al. [70] is related to the co-current flow configuration because of membrane temperature rise problems. In case of counter-current flow, the membrane temperature is increased exceeding the permissible operating range as discussed in the coming section. The counter-current has the lowest average permeation driving force (partial pressure) in contradiction to heat exchangers in which the counter-current heat exchangers has higher efficiency than the co-current exchangers. The overall mass transfer coefficient depends on both the average wall temperature and average partial pressure driving force. Thus, since the average wall temperature is higher, the average partial pressure driving force must be higher if the same operating conditions are applied. Figure 6.7 shows the co-current temperature line plots on a plane $y = 0.00375$ m and at different axial locations, in the X-direction, normal to the upper membrane and crossing the upper feed and permeate channels. This figure shows the development of temperature profile in the axial direction and the location of the flame zone. The combustion starts close to the membrane surface at axial location close to 0.1 m, and the sudden increase in temperature at this region causes a jump in the oxygen permeation flux. The permeate gas temperature rises slowly at first due to slow oxygen transport starting from 1173 K at inlet and then increases gradually to reach 1400 K at the exit plane. To see the flame shape and its development in the axial direction, Fig. 6.8 shows the temperature contour plots at different axial locations in planes normal to the flow direction. The flame starts in a small region in the corners between membranes in the permeate channels. The size of the flame starts to increase in the axial direction until filling the whole permeate channels at the axial location of 0.8 m, close to the rector exit plane, as shown in the figure.

Figure 6.9a shows the axial average temperature distribution of feed gas, membrane, and permeate gas, while Fig. 6.9b shows the axial average species mole fractions in the permeate side of the membrane. As shown in Fig. 6.9a, the

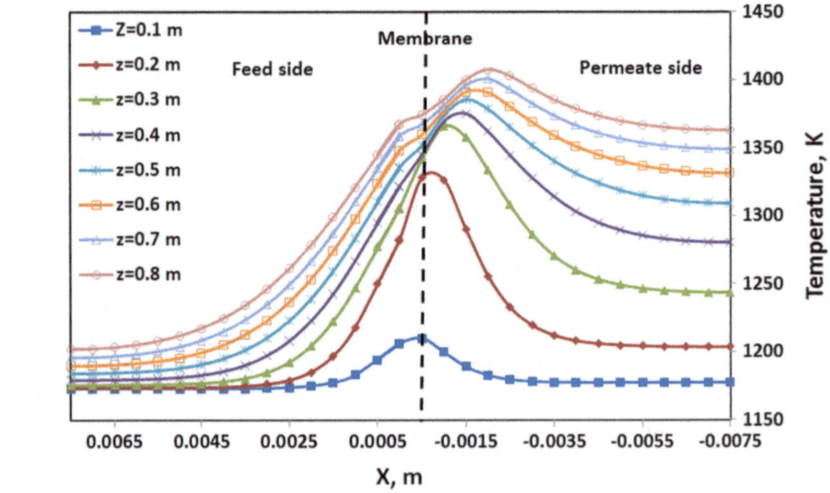

Fig. 6.7 Co-current temperature line plots on a plane $y = 0.00375$ m and at different axial locations, in the X-direction, normal to the upper membrane and crossing the upper feed and permeate channels

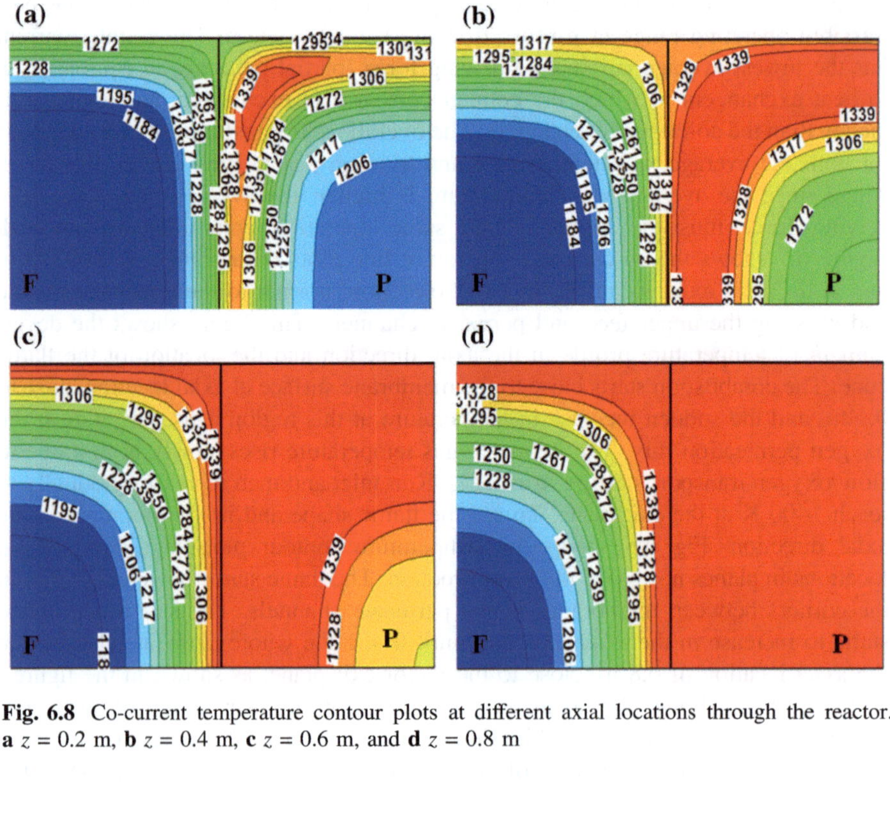

Fig. 6.8 Co-current temperature contour plots at different axial locations through the reactor. **a** $z = 0.2$ m, **b** $z = 0.4$ m, **c** $z = 0.6$ m, and **d** $z = 0.8$ m

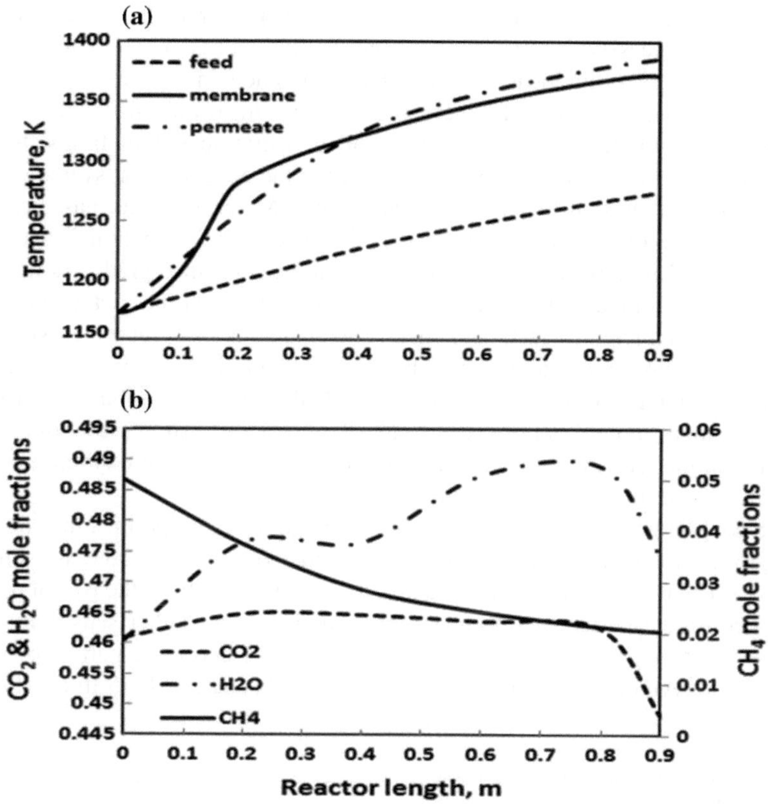

Fig. 6.9 Co-current: **a** axial average temperature distribution of feed gas, membrane, and permeate gas and **b** axial average species mole fractions in the permeate side of the membrane

membrane temperature rises slowly at the first part of the reactor due to slow oxygen transport and the resulting gradual combustion of the fuel. The permeate stream remains relatively close to the membrane temperature in this region because the convective heat transfer coefficient is high enough to accommodate the local heat release without a noticeable temperature difference. Due to the higher air flow rates and thus heat capacity in the feed channels, the average temperature of the feed flow remains low and little bit away from the membrane temperature. This increase in air flow rates has double merits enhancing the reactor operation, one is to cool down the membrane surface temperature because the combustion zone is very close to the membrane wall and the second is to increase the oxygen permeation flux [3, 68]. The membrane temperature begins to increase faster than those of permeate and feed streams as the Arrhenius term in the flux constitutive relation increases rapidly, and the chemical reactions accelerate, and the fuel is rapidly consumed. The implications of these results are that the local heat transfer coefficient is quite important because it dictates how much additional diluent is required, or equivalently, how much smaller the reactor channels must be to accommodate the localized excessive temperature. Accordingly, unless local cooling can be provided precisely at the location where the membrane temperature overshoots, the heat transfer coefficient must be increased. As a result, if only the transverse heat convection is sufficient to maintain the wall temperature within limits, then the reactor is likely to meet this criterion easily when the effects of transverse radiative heat transfer are present and the application of ITM reactors into gas turbines is possible. On the other hand, the average axial species mole fractions' distribution shown in Fig. 6.9b indicates that the fuel starts to be consumed gradually from inlet to exit and the combustion is uniformly distributed through the rector length. The reactions are sufficiently fast, and that is why no fuel is present in significant amounts at reactor exit. The concentrations of H_2O and CO_2 are increased through the reactor length due to combustion as both of them are combustion products. However, their mole fractions are reduced at reactor exit section due to the reduction in the reaction rates and also due to the permeation of oxygen that is not consumed close to the rector exit plane.

The longitudinal distributions of the average oxygen permeation flux and average local oxygen partial pressure in both sides of the membrane are shown in Fig. 6.10. The results indicate that the flux drops significantly as the partial pressure of oxygen on the permeate side increases. This is due to higher sensitivity to the permeate partial pressures at large ratios of feed to permeate partial pressures. A jump in oxygen permeation flux occurs at axial distance of 0.15 m at which the

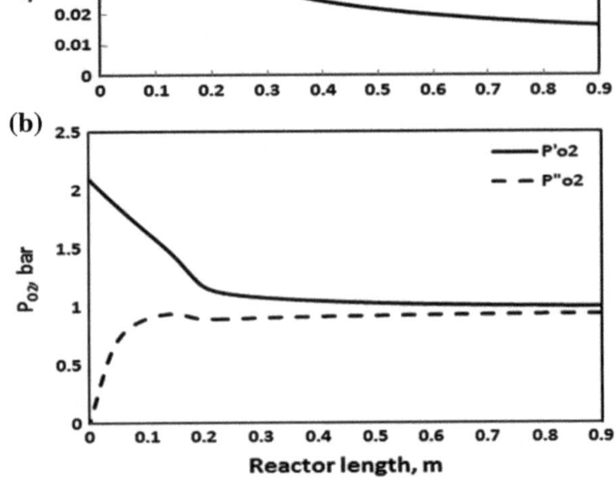

Fig. 6.10 Co-current: **a** axial oxygen permeation flux distribution and **b** axial local oxygen partial pressure distribution in both sides of the membrane

combustion starts, and the membrane temperature starts to increase sharply at this point. This behavior is a direct consequence of a dominant oxygen flux dependence on temperature than on partial pressure differences and has serious implications for the feasibility of a reactive ITM. The increase in membrane temperature reduces the surface resistance of the membrane to oxygen permeation [88]. Further, the sudden acceleration of the oxygen flux as the temperature increases in a region of high partial pressure difference is the primary cause of the temperature overshoot and could also have implications for materials stability. The increase in the oxygen partial pressure through the axial direction is due to the increased mass fractions of O_2 in the axial direction which can be seen easily in Fig. 6.11 at different axial locations on planes normal to the flow direction. As shown in the figure, the amount of O_2 increases in the axial direction in the permeate side and occupies larger area, and it is reduced in the feed side. To capture a complete 3D view about the other parameters, Fig. 6.12 shows the contour plots of CH_4, H_2O, CO_2 mole fractions' distribution and axial velocity distribution, respectively, on a plane normal to the flow direction at axial location of $z = 0.6$ m. In the permeate channels, the species concentrations are reduced close to the membrane surface due to oxygen permeation and the velocity is increased in the direction from the membrane surface to the channel centerline as shown in the figure.

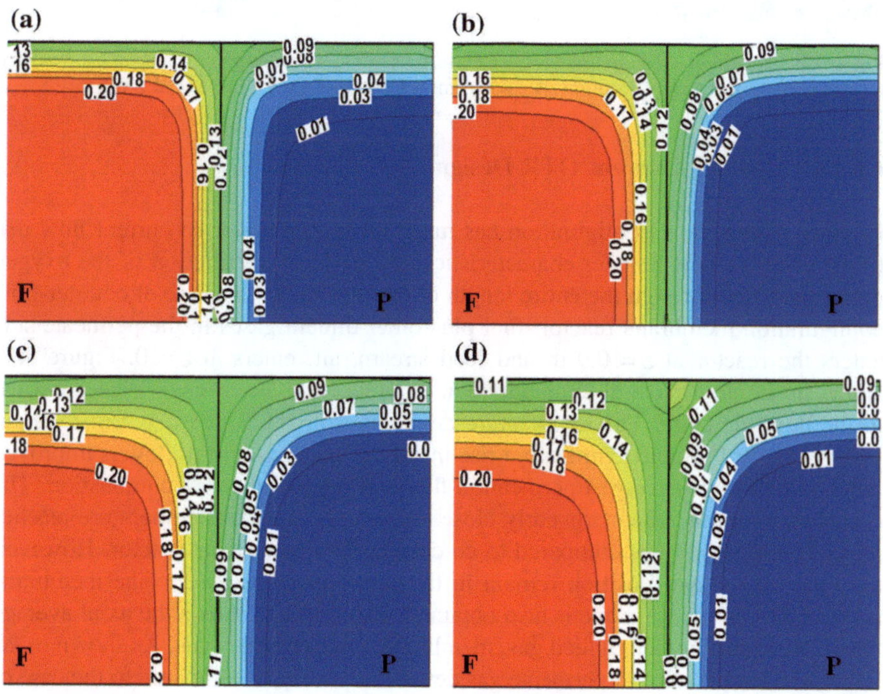

Fig. 6.11 Co-current oxygen mole fraction distribution at different axial locations through the reactor. **a** $z = 0.2$ m, **b** $z = 0.4$ m, **c** $z = 0.6$ m, and **d** $z = 0.8$ m

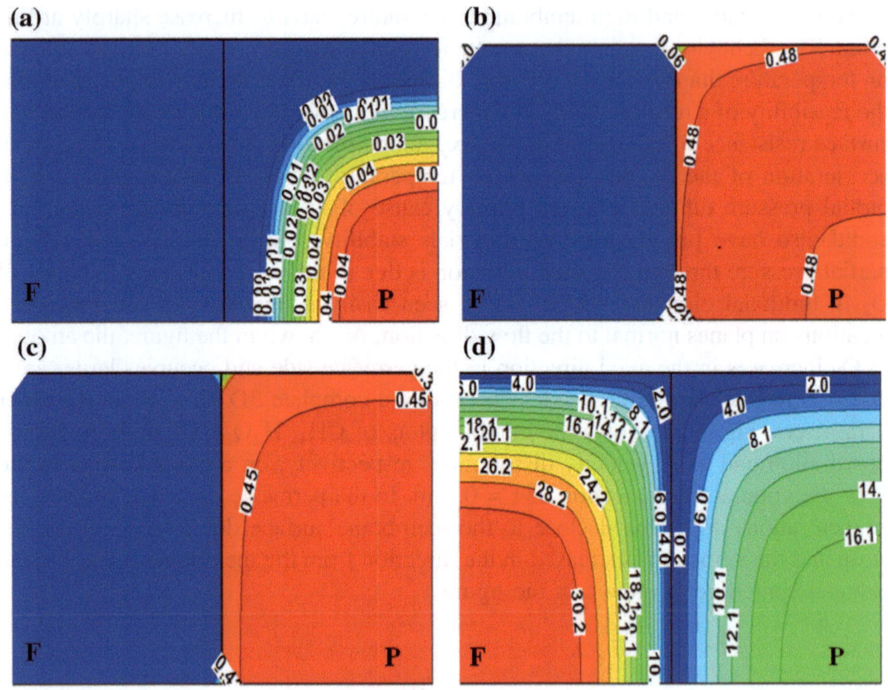

Fig. 6.12 Co-current contour plots of CH_4, H_2O, CO_2 concentrations and axial velocity, respectively, at fixed axial location of $z = 0.6$ m. **a** CH_4, **b** H_2O, **c** CO_2, and **d** z-velocity

6.6.1.6 Counter-Current OTR Design

Counter-current flow configuration has many advantages over co-current flow due to the effective heat transfer characteristics and uniform distribution of the oxygen permeation flux through the entire length of the reactor. In this case of counter-flow configuration membrane reactor, fuel plus other diluent gases in the permeate side enters the reactor at $z = 0.9$ m and feed stream, air, enters at $z = 0$. Figure 6.13 shows the contour plots of temperature at different axial locations on planes normal to the flow direction. The temperature develops from 1173 at $z = 0.9$ m to about 1440 K at $z = 0$ m. the temperature rise in case of counter-current flow is improved much due to higher oxygen permeation flux and improvement in combustion. The combustion in this case starts early close to fuel inlet, and the temperature reaches its maximum rapidly as compared to co-current flow membrane reactor. However, the flame shape and position remain in the same position close to the membrane surface, and the flame shape is also obtained. Figure 6.14a shows the axial average temperature distribution of feed gas, membrane, and permeates gas. As shown in the figure, the temperature of permeate gas develops quickly as compared to the case of co-current and reaches its maximum value faster. The membrane temperature is uniform through the reactor length except at inlet and exit regions. Due to high air

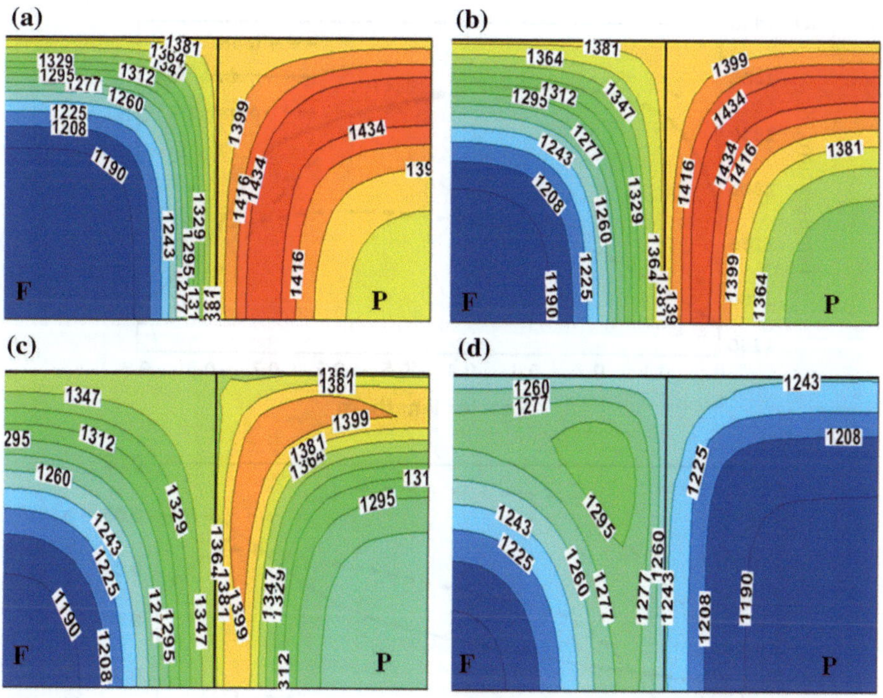

Fig. 6.13 Counter-current temperature contour plots at different axial locations through the reactor. **a** $z = 0.2$ m, **b** $z = 0.4$ m, **c** $z = 0.6$ m, and **d** $z = 0.8$ m

flow rates in the permeate sides and thus its specific heat, the feed stream temperature is lower than the temperature of membranes and permeate streams. The temperature profiles are quite important, and careful selection of the inlet conditions could lead to large improvements in ITM performance. The oxygen flux is a strong function of the local membrane temperature, and so the average membrane temperature is given for each simulation. Interestingly, the counter-current separation-only ITM has the highest average temperature due to the well-balanced heat exchange between the streams. In Fig. 6.14b, the average axial mole fraction distributions of CH_4, CO_2, and H_2O are presented. As shown, CH_4 mole fractions start to decline sharply after small distance from fuel inlet indicating higher oxygen permeation flux and improved combustion as compared to the smooth reduction of CH_4 through the reactor length in case of co-current reactor design. CO_2 and H_2O concentrations are increased due to combustion, and after that their concentrations starts to decline due to the permeation of oxygen.

Due to effective heat transfer and oxygen permeation mechanisms in case of counter-current flow, the combustion is improved, and the membrane temperature is higher as compared to co-current. As shown in Fig. 6.15a, the average flux through the reactor is higher than in case of co-current flow configuration. The oxygen permeation flux is uniform through the reactor length with a jump close to fuel inlet.

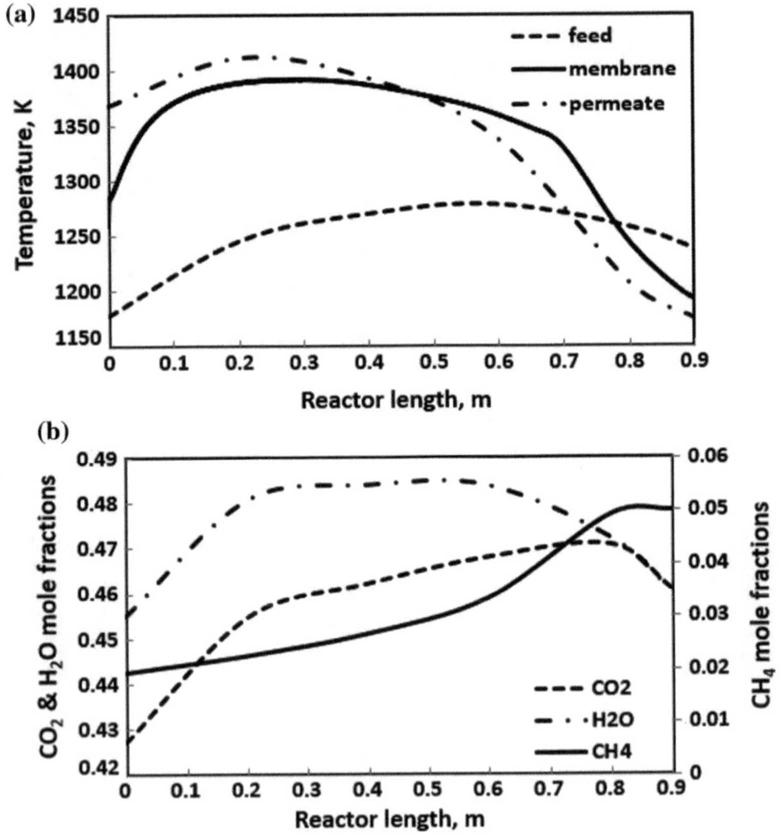

Fig. 6.14 Counter-current: **a** axial average temperature distribution of feed gas, membrane, and permeate gas and **b** axial average species mole fractions in the permeate side of the membrane

This jump in oxygen flux is indicating the start of combustion and an increased membrane temperature. Examination of the partial pressure profile reveals that the partial pressure difference is essentially constant along the reactor length, indicating good material stability potential by minimizing chemical expansion stress as shown in Fig. 6.15b. It is important to note that the oxygen mole fraction of the feed stream drops by 72.5%, indicating a high recovery ratio and an effective ITM configuration. Further, the more sensitive region where the permeate partial pressure is low coincides with the region where the feed partial pressure is low, which appears to be a better match-up than the co-current case, where the high feed matches up with the low permeate. Figure 6.16 shows the oxygen mole fractions' distribution at different axial locations of the reactor on planes normal to the flow direction. As shown, the amount of O_2 increases in the axial direction in the permeate side and occupies larger area than the case of co-current flow which indicates higher recovery ratio.

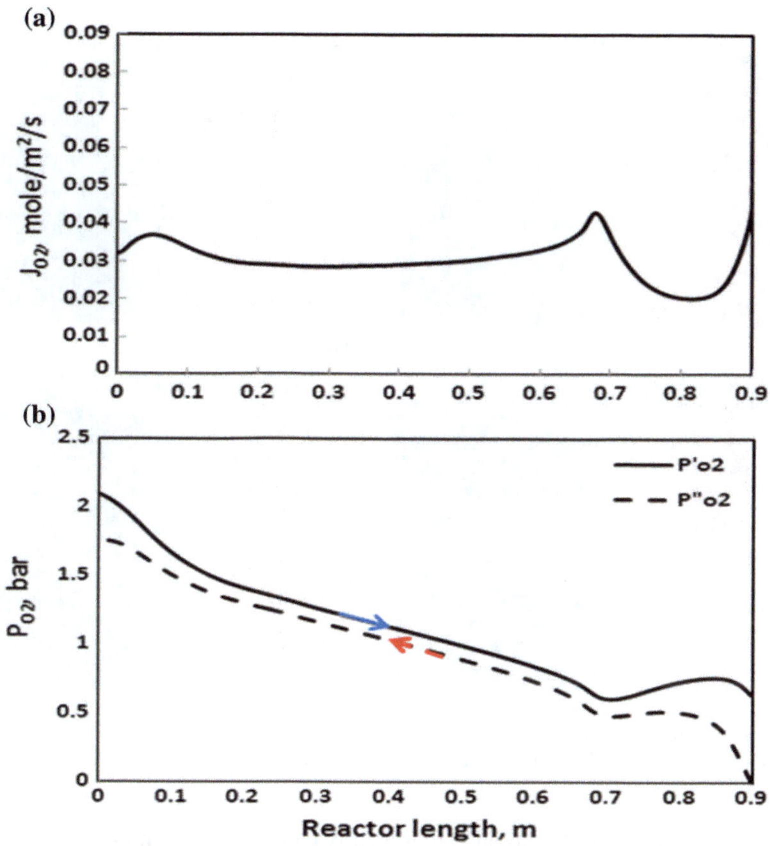

Fig. 6.15 Counter-current: **a** axial oxygen permeation flux distribution and **b** axial local oxygen partial pressure distribution in both sides of the membrane

6.6.1.7 Effect of Channel Width

The individual monolith channel width and accordingly the reactor volume is a very important parameter in designing the ion transport membranes (ITM). Variations in the reactor volume have similar effects like varying the volume flow rates. Larger reactor size may result in increased oxygen permeation flux. However, increasing the ITM reactor size is too costly and, consequently, there may be an optimal channel width in order to achieve a certain power output according to the limitations provided for specific applications. In the work by Nemitallah et al. [70], a design of an ITM reactor was presented in order to substitute a gas turbine combustor. In this case, the reactor length was limited to 0.9 m to occupy the same length as a real gas turbine combustor. The total number of channels is limited by the channel width and the available space for a real application. Modifications applied to the channel width can directly impact the surface area-to-volume ratio as

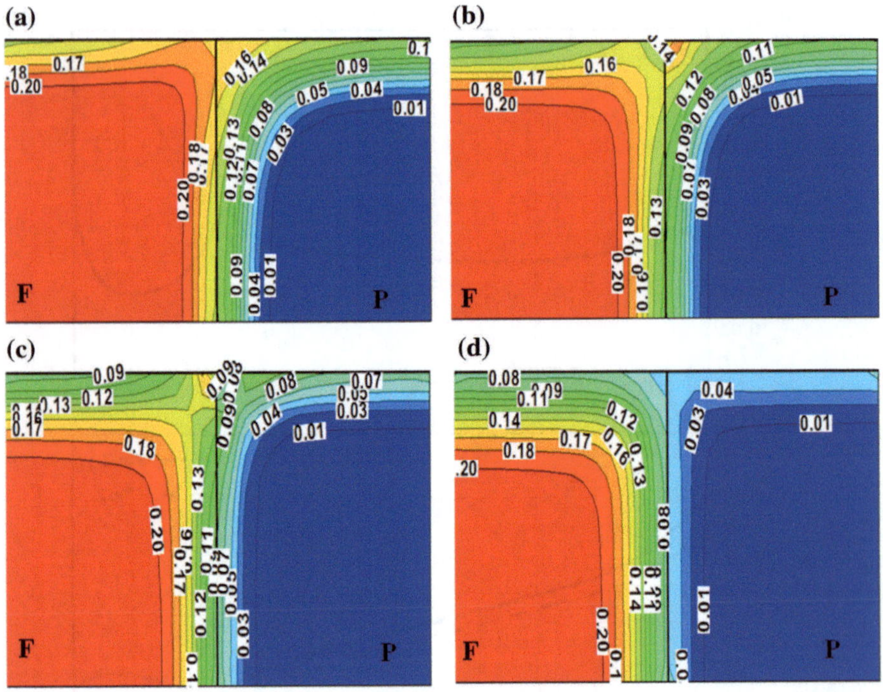

Fig. 6.16 Counter-current oxygen mole fraction distribution at different axial locations through the reactor. **a** z = 0.2 m, **b** z = 0.4 m, **c** z = 0.6 m, and **d** z = 0.8 m

well as the reactor aspect ratio and significantly cause changes in pressure drop for a fixed number of channels and inlet flow rates. The channel width has direct effect on many aspects of the ITM performance. These aspects include mainly the heat transfer coefficient, the surface area-to-volume ratio, and the local flow conditions among others. A larger channel width reactor near the inlet would result in a relatively low heat transfer coefficient, and so even if the reaction rate is small, the wall temperature could still be significantly high. As a matter of fact, the amount of oxygen permeation flux is increased as the channel width is reduced. Based on this fact, the simulations [70] were done in order to calculate the minimum operating channel width as it is explored later in detail. It was found from the simulations that any reduction in the channel width below 15 mm results in a large increase in the viscous pressure drop. Basically, as the channel width decreases, the flow velocity must increase to maintain the same mass flow rate. The oxygen permeation at high flow velocities when reducing the channel width below 15 mm is found not to be proportional to the fuel flow and the combustion temperature was sharply reduced. This analysis can be confirmed from the results in Fig. 6.17 that compares the results of the minimum allowable operating channel width, 15 mm, with 20 mm channel width. Large enhancement in oxygen permeation flux is obtained by reducing the channel width from 20 to 15 mm as shown in Fig. 6.17a.

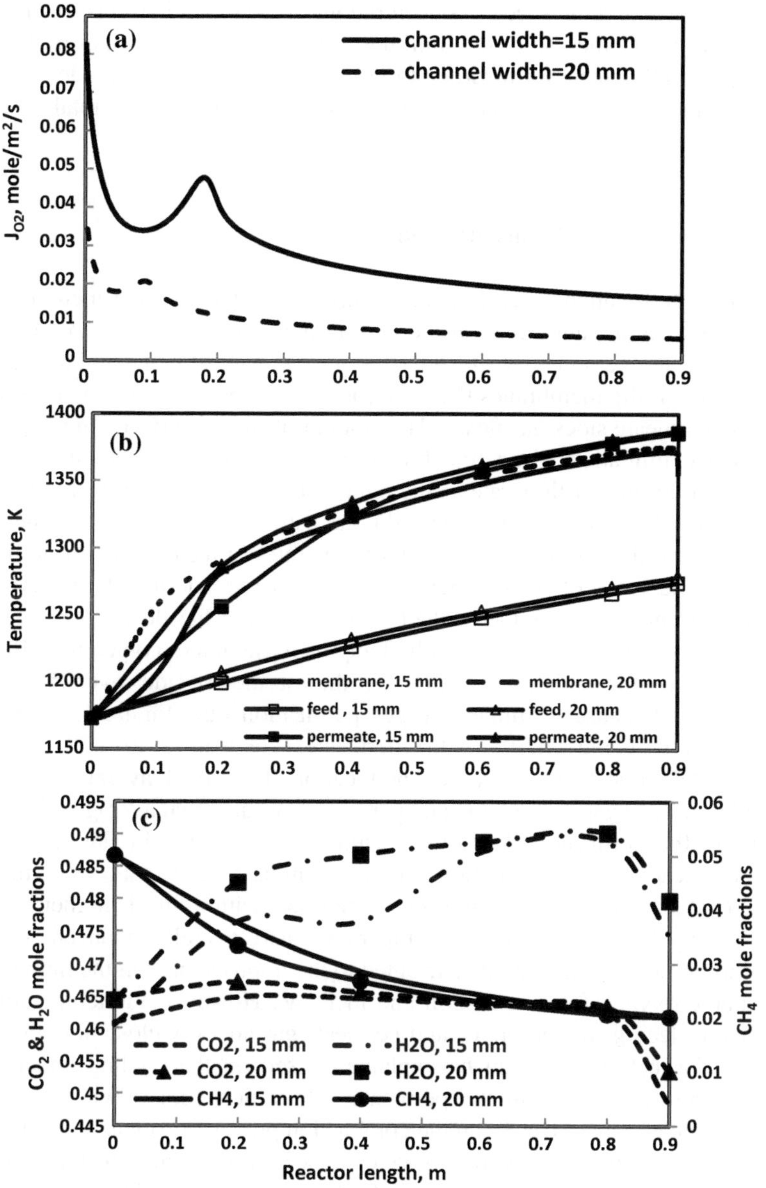

Fig. 6.17 Co-current: effect of channel width on **a** axial oxygen permeation flux distribution; **b** axial average temperature distribution of feed gas, membrane, and permeate gas; **c** axial average species mole fractions in the permeate side of the membrane

The temperatures and the species concentrations are very similar in both cases. This is enough to justify the use of the minimum permitted channel width of 15 mm in the design of ITM reactors in order to substitute a gas turbine model. By using 25,000 channels for each stream, this will result in a reactor with total volume of 10 m^3.

6.6.1.8 Effect of Fuel Concentration

In his work, Nemitallah et al. [70] have shown that the average membrane temperature has a significant effect on the performance of an ITM reactor. The membrane surface temperature greatly depends on the intensity of combustion in the permeate side of the membranes that is accordingly dependent on the percentage of fuel in the permeate sides gas flows. The amount of methane (CH_4) in the permeate flow has two limitations. The first limitation is that the permeate stream has an upper-bound on diluent flow rate as the inlet methane concentration should not fall below roughly 5% for mass transfer and combustion stability reasons. The second limitation, according to the optimization of flow rates in feed and permeate sides, is imposed by the value of oxygen permeation flux across the membrane. At certain flow circumstances such as the studied case, increasing the flow rates in feed and permeate sides is not expected to result in any increase in the oxygen flux. Based on this, increasing the amount of CH_4 in the permeate side over 5% is also non-applicable because of limited oxygen permeation flux. Figure 6.18 shows the influences of reducing CH_4% in the permeate side from 5 to 2.5% vol. while keeping H_2O% constant on oxygen permeation flux, axial average temperature distribution of feed gas, membrane and permeate gas and axial average species mole fractions in the permeate side of the membrane. As shown in Fig. 6.18a, better oxygen permeation flux distribution is obtained in case of 5% CH_4 as compared to the case of 2.5%. As well, the jump in oxygen permeation flux is higher in case of 5% indicating higher combustion temperature and overall enhancement in the combustion process. This may be attributed to the improvement in the ratio between the available oxygen for combustion and methane. However, the temperature distributions for feed gas, membrane, and permeate gas are very close in both cases as shown in Fig. 6.18b. This can be explained based on the cooling effects of the excess permeated oxygen in the permeate side which cool down the gases. In Fig. 6.18c, the effects of CH_4% on the combustion process are clear and can be seen straightforwardly from the average species concentrations through the rector. For both cases, 5 and 2.5%, the amount of H_2O at inlet is the same. However, in case of 5% CH_4, the concentrations of H_2O are much higher as compared to the case of 2.5% CH_4. This indicates an improvement in combustion process as H_2O is a combustion product. Further, CO_2 concentrations are also improved.

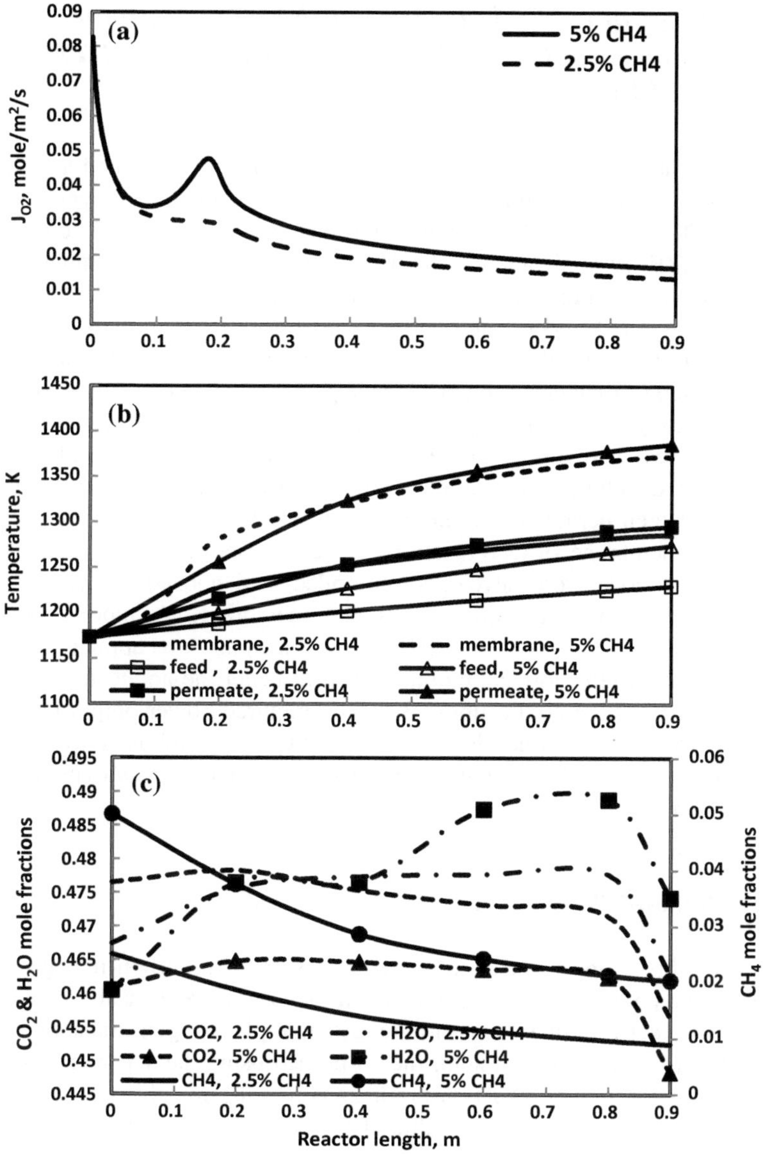

Fig. 6.18 Co-current: effect of CH_4% on **a** axial oxygen permeation flux distribution; **b** axial average temperature distribution of feed gas, membrane, and permeate gas; **c** axial average species mole fractions in the permeate side of the membrane

6.6.2 Design of a Multi-can Carbon-Free Gas Turbine Combustor Utilizing Multiple Shell-and-Tube OTRs for ZEPP Applications

Based on 3D numerical simulations, Nemitallah et al. [85] presented a design of a multi-can carbon-free gas turbine combustor utilizing multiple oxygen transport reactors (OTRs) of the shell and tube type that can be applied for zero-emission power plant (ZEPP) applications. The design of the gas turbine combustor is calculated based on optimizations of different parameters. These include flow configuration (co-current and counter-current), inlet fuel concentration in the sweep flow (CH_4 plus CO_2), shell-side and tube-side (feed and sweep) flow rates as well as membrane tube diameter, pitch (spacing), and length. High-temperature mixed-conducting perovskite-type BSCF ($Ba_{0.5}Sr_{0.5}Co_{0.8}Fe_{0.2}O_{3-\delta}$) ceramic membrane tubes were used. These consider square arrangement of the tubes. A mesh was developed and used by the ANSYS Fluent software. As well, a modified oxygen permeation equation was utilized accounting for reacting flow and sub-step membrane surface reactions. Series of user-defined functions (UDFs) written in C++ code were developed, compiled, and hooked to the software to account for oxygen permeation across the membrane. A modified two-step oxy-combustion reaction kinetics mechanism for methane was applied. The results showed that co-current flow configuration can fit better the application of OTR in gas turbine combustion applications. Based on required power output in the range from 10 to 15 MWe, the final design of the gas turbine combustor was calculated [85] to have 16 cans (OTRs) with 3000 membrane tubes per can and volume of 5.2272 m^3 per can.

6.6.2.1 Proposed Power Cycle

The aim of this proposal is to introduce a design of a carbon-free gas turbine combustor for power cycle applications. The idea is to replace the conventional combustor in conventional gas turbine by multiple oxygen transport reactors (OTRs). In this combustion system, fuel is burned using the separated oxygen by membranes to produce combustion products consisting mainly of CO_2 and H_2O. H_2O can be condensed easily and, thus, CO_2 is captured. Figure 6.19 shows the schematic diagram of the proposed multi-can carbon-free gas turbine utilizing OTRs for power cycle applications. The proposed cycle consists of multi-can OTRs combustion system, high-pressure turbine (T_1), low-pressure turbine (T_2), in addition to, two compressors. The combustion system consists of multi-can combustors; inside each can, there are a set of OTRs. Combustion occurs inside all OTRs in all cans, and the elevated temperature combustion products (CO_2 plus H_2O) are passed through the turbine (T_1) to generate power. One compressor (C_1) is used to compress air to the combustion system, and the other (C_2) compresses a portion of the exhaust gases leaving the high-pressure turbine to be recirculated to

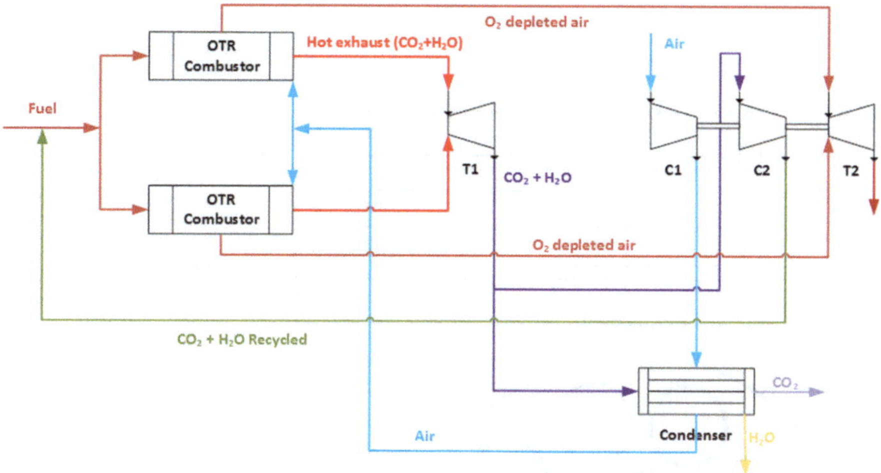

Fig. 6.19 Representation of the present multi-can carbon-free gas turbine utilizing OTRs for power generation

the OTRs within the combustion system. A condenser is used to separate water from the exhaust stream for CO_2 capture, and the rejected heat is used to heat the air before being introduced to the combustion system. The low-pressure turbine utilizes the hot oxygen-depleted air mixture leaving the OTRs to generate additional power.

The can surface and the OTRs are forming like a shell-and-tube design. Air is fed to the OTRs in the shell side from a compressor (C_1), and oxygen is separated from air through the membranes. Another compressor (C_2) is used to recirculate part of the separated CO_2 to be mixed with the fuel before introducing them to the OTRs. The recirculated CO_2 acts as a diluent inside the combustor to control the flame temperature. The stream of the mixture of fuel and recirculated carbon dioxide enters into the fuel box (fuel distributor) and then flows inside the oxygen transport reactors as presented in Fig. 6.20. Air enters the annulus air box surrounding the fuel box and flows into the volumes surrounding the oxygen transport membranes. The pressurized air is to fill the volume around the membranes inside each can combustor as presented in Fig. 6.20, and the oxygen penetrates from the outside surface to the inside surface of the cylindrical membranes to react with the fuel. The fuel flows inside the cylindrical membrane of each of the oxygen transport reactors and burns using the transported oxygen. The oxygen-depleted air mixture (mainly nitrogen) is collected in an annulus (nitrogen box) at combustor outlet. The oxygen-depleted air mixture is used to drive the second turbine (T_2) providing the work required for compressing the air and the recirculated CO_2.

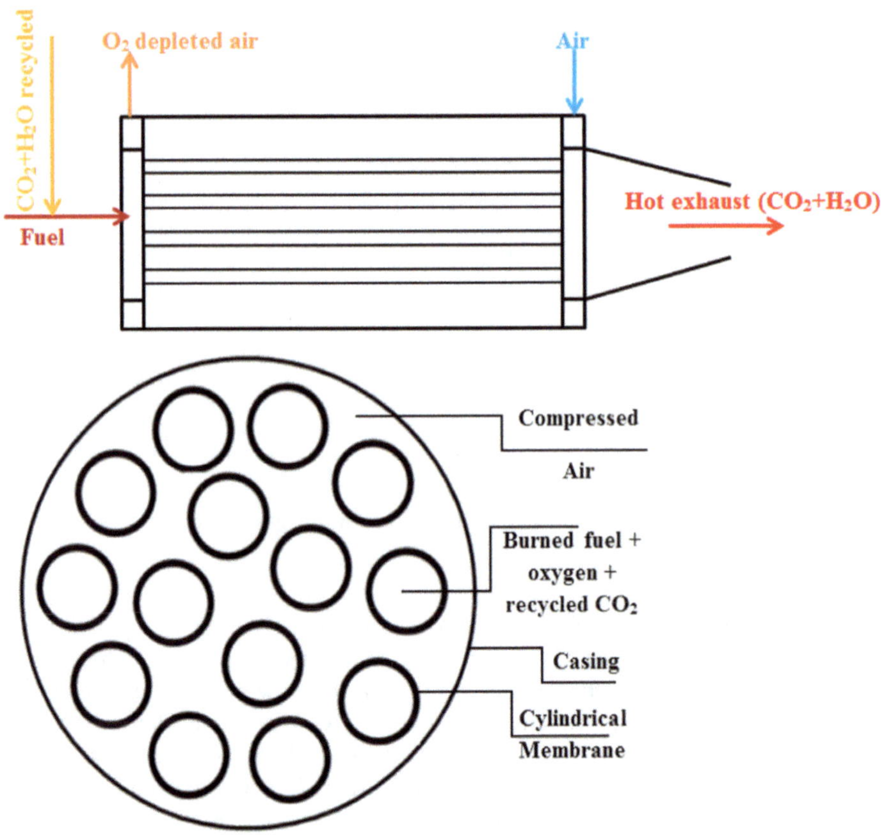

Fig. 6.20 Representation of a single combustor (single can) of the present gas turbine having multiple OTRs: (top) combustor details and (bottom) sectional view of the combustor

6.6.2.2 OTR Design and Flow Conditions

In this section, a new design of a multi-can gas turbine combustor is introduced. Each can consists of multiple tubular oxygen transport membrane reactors, based on detailed 3D numerical simulations for zero emission power cycle applications. The proposed system is of a shell-and-tube design, where air is being passed in the shell side and fuel plus recycled gases are being passed inside the membrane tubes. Oxygen transports across the membrane tubes to the permeate side to react with fuel and produce combustion products consisting mainly of H_2O plus CO_2. In the present design, square arrangement of membrane tubes is considered as shown in Fig. 6.21. Based on required power output in the range from 10 to 15 MWe, the number of membrane tubes within each shell is expected to be huge making the simulations very difficult to be performed in the full 3D for the whole shell. Generating power in this range requires high amount of oxygen for combustion and, accordingly, vast number of membrane tubes having small diameter (high area-to-volume ratio) is required to satisfy the power requirement. The diameter ratio between the shell and membrane tube is very huge and, thus, the wall effect can be neglected. Based on that, the present design can be simplified in the numerical simulations, as per Fig. 6.21, considering only quarter portion of four adjacent membrane tubes of square arrangement to be solved for the flow field and reactions while oxygen permeation in the 3D mode. Due to symmetry of the square arrangement, a quarter-section of four adjacent membrane tubes for the entire length of the reactor is considered to represent the entire system in the full 3D numerical simulations as shown in Fig. 6.21. The case considered in the present study of shell-and-tube design is not axisymmetric to reduce the model from 3D to 2D or even 1D. If we tried to represent the model in the 2D domain, then we should consider the axial direction in addition to another direction. If we look at Fig. 6.21 trying to select the other direction (in addition to the axial direction for 2D representation), we can observe that there is no symmetry in any radial direction and, accordingly, the results will not be similar in any radial direction normal to the axial direction. Based on that, 3D representation of the model is considered instead of 2D or even 1D for accuracy reasons. Symmetric boundary conditions are considered in all longitudinal sides of the considered membrane reactor unit. The OTR is constructed of BSCF perovskite-type ceramic membranes having thickness of 1 mm. More details regarding the membrane material can be found in our previous work, Nemitallah [22], and in the work done by Behrouzifar et al. [89].

Optimizations of the OTR in terms of design and flow conditions are performed to calculate the optimum design of the reactor in terms of the high oxygen permeation flux, small size, and high combustion temperature. In the considered OTR, normal air (21% O_2 plus 79% N_2) is fed to the shell (feed) side of the membrane, and a mixture of fuel (methane) and recycled gases (H_2O and CO_2), with different fractions, is fed to the sweep (tube) side of the membrane. Table 6.6 represents the different considered simulation cases for the optimization study. Nine simulation cases were performed under unique design and operating conditions. The values of the flow rates and gas temperature and concentrations are calculated based on

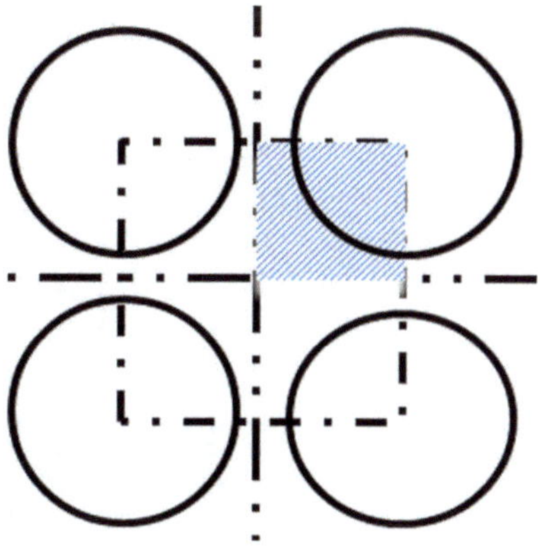

Fig. 6.21 Schematic diagram of the considered square arrangement showing the integration (hatched) zone

Table 6.6 Representation of flow and design conditions for all simulation cases

Case	1	2	3 (base)	4	5	6	7	8	9
Flow configuration	Counter-current		Co-current						
Inlet temperature (K)	1173								
Feed/permeate pressure (bar)	10								
Feed air flow rate (kg/s/cell)	2.960619×10^{-3}								
Pitch (mm)	22						18	22	
Membrane tube diameter (mm)	17							10	24
Reactor length (m)	1.8				0.9	3.6	1.8		
Total permeate flow rate (kg/s/cell)	2.57606×10^{-4}	5.15212×10^{-4}	2.57606×10^{-4}	5.15212×10^{-4}	2.57606×10^{-4}				
Volume fractions of permeate flow	10% CH_4	5% CH_4	10% CH_4	5% CH_4	10% CH_4				
	45% CO_2	47.5% CO_2	45% CO_2	47.5% CO_2	45% CO_2				
	45% H_2O	47.5% H_2O	45% H_2O	47.5% H_2O	45% H_2O				

limitations for operation of such kind of membranes in terms of expected oxygen permeation flux and minimum and maximum operating temperatures of the membrane [3]. Fuel flow rate and inlet concentrations are calculated based on the expected oxygen permeation flux to allow for complete conversion of the fuel [70]. The size of the membrane tubes is selected based on allowable pressure drop limits [83]. In all cases, inlet flow temperature in both feed and sweep sides is considered fixed at 900 °C (1173 K), total pressure in both sides of the membrane is fixed at 10 bars, and the feed flow rate is also kept constant at 2.960619×10^{-3} kg/s. The flow rates inside the shell (feed) side and tube (permeate) side are calculated based on the expected average oxygen permeation flux of BSCF membranes under reacting flow conditions based on previous studies [22, 89]. Based on the expected average oxygen permeation flux based on certain membrane tube diameter, fuel flow rate is calculated and accordingly a range for sweep gas flow rate is specified. Effect of fuel concentration is investigated in this study considering different values of species concentrations in tube side. Membrane tube length is a flexible parameter, and it depends on the specific application of the OTR. In this study, different membrane tube lengths are considered to study its effect on combustion inside the OTR. Membrane tube diameter is one of the most critical parameters affecting surface-to-volume ratio of the OTR, heat transfer rate, and flow characteristics within the OTR. As a matter of fact, reducing membrane tube diameter results in better oxygen permeation flux and higher surface-to-volume ratio (compact size OTR) of the OTR. However, there is a limit for the minimum membrane tube diameter that can be used depending on the values of viscous pressure drop. Effect of membrane tube diameter is also investigated in the present study. Effect of flow configuration is studied by considering the simulations under co-current and counter-current flow configurations. Also, effect of pitch distance between adjacent membrane tubes on OTR performance is investigated.

6.6.2.3 CFD Modeling

A 3D geometry has been constructed using the commercial ANSYS 15.0 Workbench software to be fed to the commercial ANSYS Fluent 15.0 code for the solution of the domain under reacting flow conditions. Due to symmetry, a quarter, of four adjacent membrane tubes, for the whole reactor length is selected for the simulations as shown in Fig. 6.21. The ANSYS Fluent code solves mass, momentum, energy, and species transport equations in the full 3D domain. The ANSYS Fluent code, in its normal programing, does not consider any additional mass source terms. Based on that, we developed a visual C++ code accounting for the oxygen transfer across the membrane by adding a source/sink term to the conservations of mass and species transport. The code is supplied to ANSYS Fluent 15.0 software in the form of user-defined functions (UDFs), which can be compiled and hooked to the software. To do such integration of the UDF with the ANSYS Fluent software, Microsoft Visual Studio 2008 was first coupled with ANSYS Fluent and, then, the UDF is integrated with the code. The heat

Table 6.7 Summary of geometrical and boundary conditions

Boundary conditions	Geometrical conditions	Parameter
Inlet conditions	– Membrane tube diameter in the range from 10 to 24 mm – Pitch distance in the range from 18 to 22 mm	– Flow rate – Species concentrations – Temperature – Inlet pressure of 10 bar
Outlet conditions	– At membrane tube length in the range from 0.9 to 3.6 m – Membrane tube diameter in the range from 10 to 24 mm – Pitch distance in the range from 18 to 22 mm	– Pressure outlet – Outlet pressure of 10 bar – Back flow condition of air consisting of 21% O_2 and 79% N_2
Membrane	– Thickness of 1.0 mm – Membrane tube diameter in the range from 10 to 24 mm	– BSCF solid wall conditions – Oxygen permeation $S_i = \pm\frac{J_{O_2}.A_{cell}.MW_{O_2}}{V_{cell}}$ – Coupled heat transfer $Q''_{sweep} - Q''_{air} - 2\sigma\varepsilon_{mem}(T^4_{mem} - T^4_{\infty}) = 0$ Q^o_{sweep} is the heat transfer from the reacting sweep gases Q^o_{feed} is the heat released from the membrane surface to the feed side T_{mem} and T_{∞} are the membrane and surrounding wall temperature, respectively – Emissivity of 0.8 from both sides

transfer across the membrane from the permeate side (where combustion occurs) to the feed side is considered through coupling between the two sides based on the thermal properties of the membrane material [22]. Table 6.7 provides a summary of the main geometrical and boundary conditions considered in the numerical simulations.

6.6.2.4 Solution Procedures

A grid independence study was performed to avoid the effect of computational cell size on the calculated results by calculating the difference in oxygen permeation flux for different grids with different cell size. The cell sizes considered in the grid refinement study include 2 (axial) × 1 (radial), 1 × 1, 1 × 0.5, and 1 × 0.25 mm^2. For the three grids, simulations were performed at the same conditions of case (3) and oxygen permeation flux was calculated and compared for all cases as presented in Fig. 6.22. Refining the cell size from 1 × 1 to 1 × 0.5 mm^2 resulted in increase of oxygen permeation flux of about 11%, while refining the cell size from 1 × 0.5 to 1 × 0.25 mm^2 resulted in changes of oxygen flux of less than 3%. Based on that, the grid with cell size of 1 × 0.5 mm^2 (Grid 2) has been considered in all simulations as it resulted in best results in terms of accuracy and computational time.

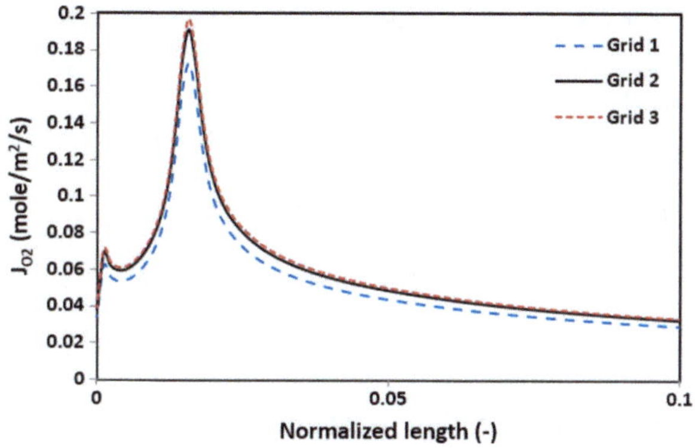

Fig. 6.22 Comparison of oxygen permeation flux at the conditions of case (3) considering three grids with different cell sizes: Grid 1 (1×1 mm^2), Grid 2 (1×0.5 mm^2), and Grid 1 (1×0.25 mm^2)

Software based on finite-volume approach was employed for the discretization of the conservation equations for the numerical calculations. Pressure–velocity coupling was achieved using a built-in double-precision solver semi-implicitly. Second-order upwind schemes were used for the species and momentum discretization, while standard scheme was applied for the pressure. The energy, momentum, and species transport as well as the radiative transfer equations were solved using steady pressure-based solver. The energy and radiation models are implemented one after the other (uncoupled) sequentially using the conservative finite-volume scheme [90]. The used pressure–velocity segregate algorithm was the Semi-Implicit Method for Pressure Linked Equations (SIMPLE) with higher order of the relaxation terms. Thermal conductivity, viscosity, and specific heat of the gas were determined as a mass fraction-weighted average of all the species. For each species, the specific heat was determined through the application of a temperature piecewise polynomial fit. Advanced Multi-Grid (AMG) solver was used to ensure convergence using the following values for under-relaxation factor: 0.3 for pressure, 1.0 for density, 0.7 for momentum, 1.0 for body force, 0.8 for energy, and 1.0 for species mass fractions. Recursive *f*-cycle process was set, and gradient-stabilized bi-conjugate method was used to prevent convergence irregularities. An absolute convergence check criterion was set to 10^{-5} for all the residuals, which implies convergence after all the residuals falls below the set criterion [73].

Seeking more accurate calculations of reaction rates and species distributions within the OTR, the two-step reaction mechanism for methane, by Westbrook and Dryer [91] modified by Andersen et al. [92] for oxy-combustion conditions (medium with high CO_2 concentrations), is applied in the present study. The mechanism combines three irreversible reaction steps as per Table 6.8. The modified reaction rates by Andersen et al. [92] are presented in Table 6.8.

Table 6.8 Modified reaction rates by Andersen et al. [92]

Reaction number	Reactions	A	$\beta(-)$	E_a (kJ/mol)	Reaction orders
Reaction 1	$CH_4 + 1.5O_2 \leftrightarrow CO + 2H_2O$	1.59×10^{13}	0	1.998×10^2	$[CH_4]^{0.7}[O_2]^{0.8}$
Reaction 2	$CO + 0.5O_2 \leftrightarrow CO_2$	3.98×10^8	0	41.8	$[CO][O_2]^{0.25}[H_2O]^{0.5}$
Reaction 3	$CO_2 \leftrightarrow CO + 0.5O_2$	6.16×10^{13}	−0.97	3.277×10^2	$[CO_2][H_2O]^{0.5}[O_2]^{-0.25}$

6.6.2.5 Permeation Mechanism

The mechanism of oxygen permeation depends on the selectivity of dense perovskite ceramic membranes to oxygen, and the flux of oxygen across the membrane depends mainly on differences in partial pressures of oxygen on both sides of the membrane and surface temperature of the membrane. The permeation process from feed side to permeate side involves three steps, namely surface exchange of the oxygen ions between air and the membrane in feed side; diffusion of oxygen ions and released electrons through the membrane lattice and surface exchange, and recombination of oxygen molecules at the permeate side of the membrane. In this study, high-temperature mixed-conducting perovskite-type BSCF ceramic membranes are used due to their high oxygen permeation flux [89]. The most commonly used equation in the literature for oxygen permeation flux is the one developed by Xu and Thomson [45]:

$$J_{O_2} = \frac{(k_r/k_f)\left[\left(1/P''^{\,0.5}_{O_2}\right) - \left(1/P'^{\,0.5}_{O_2}\right)\right]}{\left[1/\left(k_f P'^{\,0.5}_{O_2}\right)\right] + [2L/D_v] + \left[1/\left(k_f P''^{\,0.5}_{O_2}\right)\right]} \tag{6.27}$$

The parameters P'_{O_2} and P''_{O_2} represent oxygen partial pressure in feed and permeate sides, respectively. The parameters D_v, k_f, and k_r are, respectively, the diffusion coefficient, the forward reaction rate constant, and the reverse reaction rate constant, and are presented as [45]:

$$D_v = D_v^o \exp\left(\frac{-E_D}{\bar{R}T}\right) \tag{6.28}$$

$$k_f = k_f^o \exp\left(\frac{-E_f}{\bar{R}T}\right) \tag{6.29}$$

$$k_r = k_r^o \exp\left(\frac{-E_r}{\bar{R}T}\right) \tag{6.30}$$

where the parameters E_D, E_f, and E_r represent the activation energies and the parameters D_v^o, k_f^o, and k_r^o represent the pre-exponential coefficients. The above equation for oxygen permeation assumes two elementary (non-sub-step surface reactions) surface reactions [22]:

$$\frac{1}{2}O_2 + V^{\bullet\bullet}o \overset{k_f/k_r}{\leftrightarrow} O_o^x + 2h^{\bullet} \quad (6.31)$$

$$O_o^x + 2h^{\bullet} \overset{k_r/k_f}{\leftrightarrow} \frac{1}{2}O_2 + V^{\bullet\bullet}o \quad (6.32)$$

Actually, the above two surface reactions include many sub-step reactions including oxygen adsorption, dissociation, bulk diffusion, charge transfer, and recombination reactions. Behrouzifar et al. [89] proposed a modified equation for oxygen permeation to account for sub-step reactions as follows:

$$J_{O_2} = \frac{(k_r/k_f)\left[\left(1/P''^{\,n}_{O_2}\right) - \left(1/P'^{\,n}_{O_2}\right)\right]}{\left[1/(k_f P'^{\,n}_{O_2})\right] + [2L/D_v] + \left[1/(k_f P''^{\,n}_{O_2})\right]} \quad (6.33)$$

The equation in this form can be used to predict more accurate values of oxygen permeation flux under no-reacting flow conditions. For reacting flow conditions, flow fluctuations are encountered near membrane surface due to heat release of combustion. Further modification was done by Behrouzifar et al. [89] on the calculations of oxygen partial pressure in both sides of the membrane to account for the variations in flow Reynolds number near membrane surface as follows [89]:

$$P'^{*}_{O_2} = (a' + b'\mathrm{Re}'^{c'})P'_{O_2} \quad (6.34)$$

$$P''^{*}_{O_2} = (a'' + b''\mathrm{Re}''^{c''})P''_{O_2} \quad (6.35)$$

where a, b, c are constants and Re′ and Re″ are the Reynolds numbers in feed side and permeate side, respectively, and they can be expressed as follows:

$$\mathrm{Re}' = q'/(\pi v' \lambda') \quad (6.36)$$

$$\mathrm{Re}'' = q''/(\pi v'' \lambda'') \quad (6.37)$$

The parameters q, v, and λ represent, respectively, volumetric flow rate, kinematic viscosity, and distance between membrane surface and air inlet. The above expressions can be combined to come up with the modified equation for oxygen permeation as follows [89]:

Table 6.9 Oxygen permeation model parameters [22, 89]

Parameter	Unit	Value
$D_{\mathrm{v}}^{\mathrm{o}}$	m^2/s	5.9807×10^{-5}
$k_{\mathrm{f}}^{\mathrm{o}}$	$m/atm^n/s$	41.688
$k_{\mathrm{r}}^{\mathrm{o}}$	$mol/m^2/s$	1.166×10^4
E_{D}	J/mol	9.2709×10^4
E_{f}	J/mol	1.4668×10^5
E_{r}	J/mol	1.0291×10^5
n	–	0.25
a'	–	0.1015
b'	–	1.8687
c'	–	0.4525
a''	–	0.1891
b''	–	9.3439
c''	–	0.132

$$J_{O_2} = \frac{(k_{\mathrm{r}}/k_{\mathrm{f}})\left[\left(1/P_{O_2}^{\prime\prime * n}\right) - \left(1/P_{O_2}^{\prime * n}\right)\right]}{\left[1/\left(k_{\mathrm{f}} P_{O_2}^{\prime * n}\right)\right] + [2L/D_{\mathrm{v}}] + \left[1/\left(k_{\mathrm{f}} P_{O_2}^{\prime\prime * n}\right)\right]} \tag{6.38}$$

All parameters in this equation are listed in Table 6.9, and the equations have been validated against experimental data in the literature and have been proved high accuracy [22, 89].

6.6.2.6 Model Validation

The experimental work done, by Behrouzifar et al. [89] on $Ba_{0.5}Sr_{0.5}Co_{0.8}Fe_{0.2}O_{3-\delta}$ (BSCF) ceramic membranes utilizing a button-cell ITM reactor, is used to validate the present model results in terms of oxygen permeation flux. The considered button-cell ITM reactor in their work was developed numerically by constructing a mesh with typical dimensions. The simulations, using the commercial CFD ANSYS Fluent 15.0 code, were performed in the 2D domain due to symmetry around the axis of the reactor. The considered membrane in the experiment has a diameter of 100 and a thickness of 1 mm. The present CFD model for oxygen permeation was applied, and the results are compared with the measured experimental data as shown in Fig. 6.23. Different simulation cases were performed over a range of pure helium flux as sweeping gas and at fixed inlet temperature and fixed feed air flux of 750 °C and 150 sccm, respectively. Both experimental and numerical results showed increase of oxygen permeation flux with the increase of sweep gas flux, and both results are in a good agreement as presented in Fig. 6.23. This can be attributed to the effective purging process of the permeated oxygen in the permeate side while increasing the sweep gas flux and, thus, oxygen partial pressure is reduced in the permeate side resulting in higher partial pressure driving force and higher oxygen permeation flux across the membrane.

To make sure that the flow characteristics are accurately captured in the considered 3D numerical model, a comparison is made of the present model results and

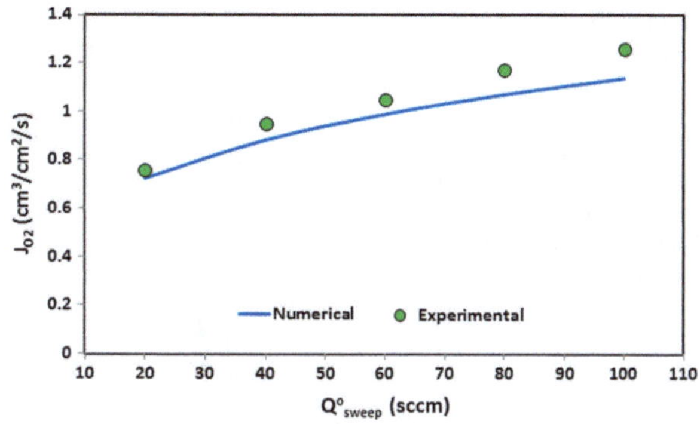

Fig. 6.23 Comparison of the present model results of oxygen permeation flux as function of sweep gas flux and the corresponding experimental data by Behrouzifar et al. [89]

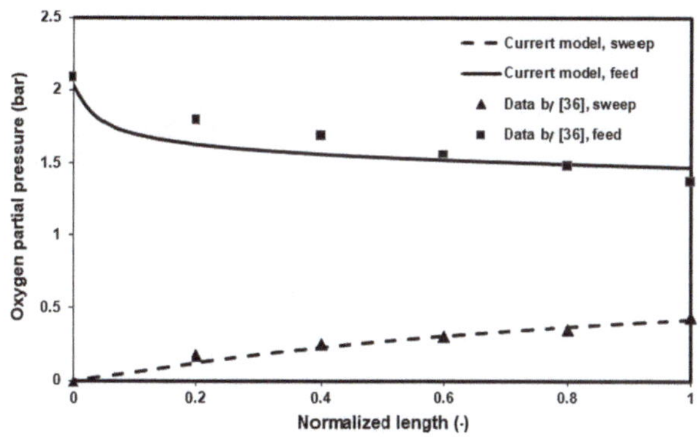

Fig. 6.24 Comparison of the present 3D numerical model results and the data by Mancini and Mitsos [83] in terms of local distributions of oxygen partial pressure in feed and sweep sides of the membrane

the data by Mancini and Mitsos [83]. The comparison is performed in terms of local distributions of oxygen partial pressure in feed and sweep sides of the membrane considering monolith structure oxygen transport reactor. As shown in Fig. 6.24, the local values of oxygen partial pressure on membrane surface in feed and sweep sides are captured accurately by the present model.

6.6.2.7 Co-current vs. Counter-Current Flow Configurations

Before proceeding with the optimization of OTR based on design and flow conditions, the influences of flow configuration, including co-current (case 3) and counter-current (case 1), are first examined. Distributions of axial oxygen permeation flux and oxygen partial pressure, axial and radial temperatures, and axial species concentrations are presented considering co-current and counter-current flow configurations. The plots showing the axial distributions of oxygen permeation flux and oxygen partial pressure are presented in Fig. 6.25 [85]. The arrows on the figure show the directions of feed and sweep flows for co-current and counter-current flow configurations. Air flow direction is reversed in case of counter-current flow configuration. Figure 6.26 shows the axial distributions of feed flow, permeate

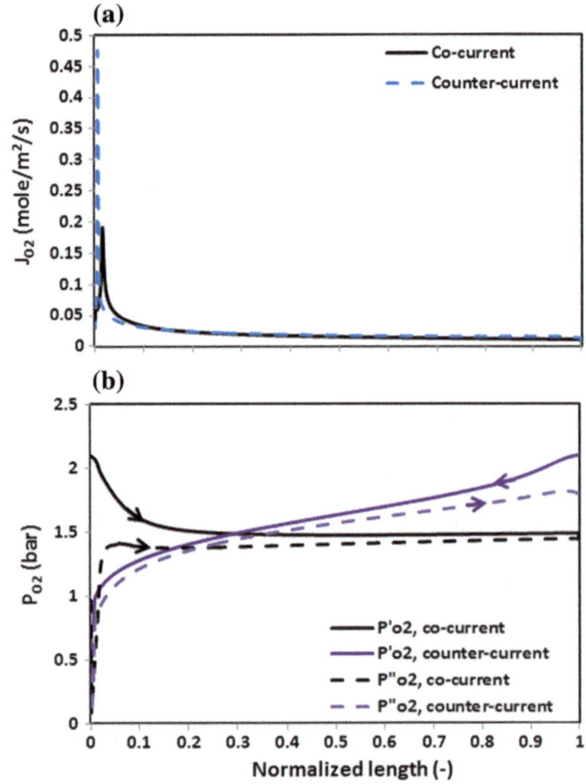

Fig. 6.25 Influence of flow configuration, co-current (case 3) and counter-current (case 1), on the axial distributions of: **a** oxygen permeation flux and **b** oxygen partial pressure in feed and permeate sides of the membrane

flow and, membrane temperatures, in addition to the axial distributions of mass fractions of CH_4 and CO. Figure 6.27 presents the radial distributions of temperature at different axial locations on a plane 45° passing through the center of the membrane tube. Based on Eq. (6.12), oxygen permeation flux is direct function of membrane temperature and oxygen partial pressure differences across the membrane. In the zone where the flame core exists, membrane temperature is the highest and accordingly oxygen permeation flux is expected to be high in this zone. Also, in the zone where partial pressure differences are the highest, oxygen permeation flux is expected to be high. It seems for the plots in Figs. 6.25a and b and 6.26a that oxygen permeation flux is much affected by differences in oxygen partial pressure in case of co-current flow and is much affected by temperature in case of counter-current flow configuration. This may be attributed to the effective heat transfer mechanism in case of counter-current flow, which results in higher membrane surface temperature for the counter-current flow case as compared to the case of co-current flow as shown in Fig. 6.26a. Also, effective diffusion of permeated oxygen in the sweeping flow is obtained for counter-current as compared to co-current. This enables better mixing between fuel and oxygen and, as a result, combustion starts early and intensively for counter-current as compared to

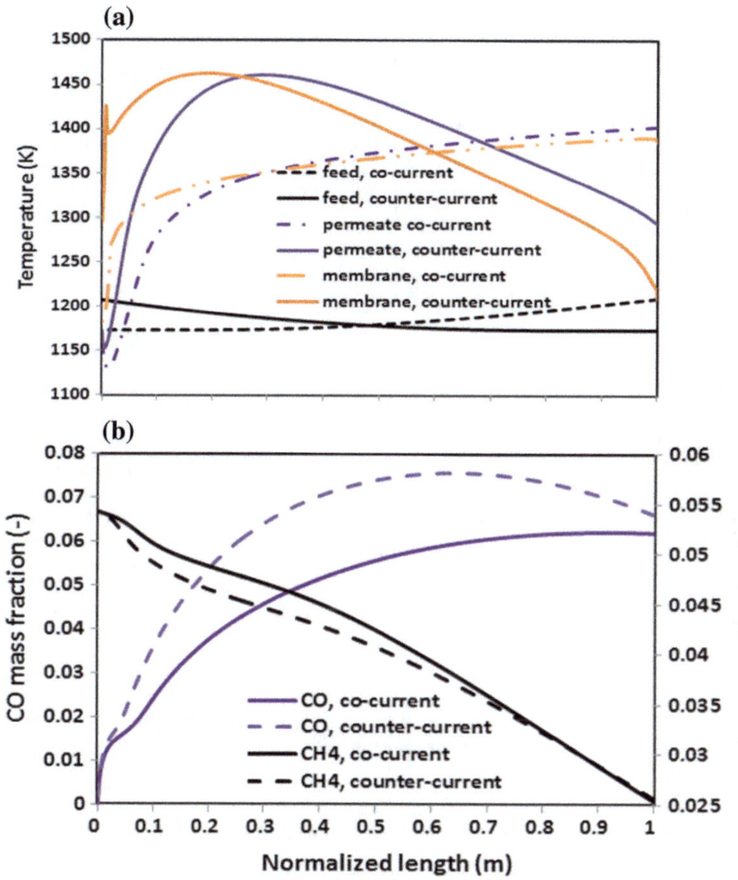

Fig. 6.26 Influence of flow configuration, co-current (case 3) and counter-current (case 1), on the axial distributions of: **a** feed flow, permeate flow and membrane temperatures and **b** mass fractions of CH_4 and CO

co-current. Based on that, fuel is consumed faster and CO species, as an intermediate combustion product, are produced faster in case of counter-current as compared to co-current. This is clearly shown in Fig. 6.26 from the axial distributions of temperatures and species concentrations. For the case of co-current, temperature increases due to combustion gradually and the lines representing permeate flow and membrane temperatures remain below those of counter-current case up to a distance from 0.65 to 0.75 of total length of the reactor. In the last quarter of the reactor, the reaction rates of co-current flow are higher than those of counter-current flow and the temperature curves flip to have higher temperature at exit section in case of co-current as compared to the counter-current flow as shown in Fig. 6.26a. This may be attributed to the effective heat transfer mechanism in case of counter-current and, as a result, heat transfer to the feed flow in higher rates resulting in reduction in

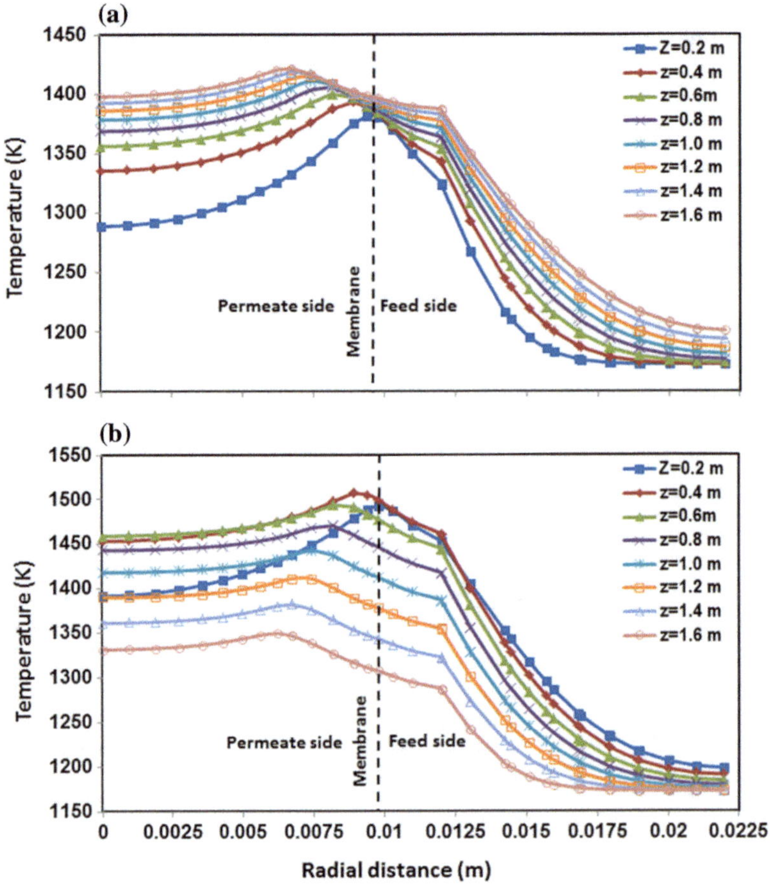

Fig. 6.27 Influence of flow configuration, **a** co-current (case 3) and **b** counter-current (case 1), on the radial distributions of temperature at different axial locations on a plane 45° passing through the center of the membrane tube

temperature close to the exit section of the reactor. Also, for co-current flow, oxygen flux distribution is affected by oxygen partial pressure differences which are the highest near the inlet section, as shown in Fig. 6.25b, resulting in a peak of oxygen flux close to the inlet section, as shown in Fig. 6.25a.

In comparison between co-current and counter-current flow configurations, the counter-current flow resulted in high peak combustion temperature in the flame core zone which may not be withstood by the membrane material. On the other hand, the surface temperature of the membrane increases gradually from inlet to exit which reduces the potential for thermal fracture of the membrane material. The hot exhaust gases leaving the OTR should be used in the expansion inside the turbine. The exhaust gas temperature is much better in case of co-current as compared to counter-current, which results in the generation of higher power in the turbine.

Both co-current and counter-current flows resulted in similar recovery ratio of the available fuel for combustion as they resulted in similar CH_4 mass fractions at the exit section as shown in Fig. 6.26b. More details about the temperature distributions within the reactor can be obtained from the plots in Fig. 6.27 for radial temperature distributions at different axial locations for both flow configurations. The peak value of temperature, about 1510 K, is obtained at axial distance of 0.4 m, close to the inlet section of the sweep flow, for the case of counter-current flow as shown in Fig. 6.27b, while a peak temperature of about 1420 K is obtained at axial distance of 1.6 m, close to the exit section of the sweep flow, for the case of co-current flow as shown in Fig. 6.27a. Sharp reduction in temperature beside the membrane in the feed side is obtained in case of counter-current flow, as shown in Fig. 6.27b, indicating effective heat transfer from permeate (combustion) zone to the feed zone across the membrane. Based on the above discussion, it is clear that the co-current flow configuration has better characteristics when compared to the counter-current flow configuration in terms of leaving exhaust gas temperature, gradual temperature distribution, and gradual oxygen permeation flux. So co-current flow configuration can fit better the application of OTR in gas turbine combustion applications. Based on that, the optimizations of the operating and design conditions are performed considering co-current flow configuration.

6.6.2.8 Effect of Inlet Fuel Concentration

The concentration of fuel in the inlet sweep gas is controlled by the oxygen permeation flux which is, in turn, controlled by the membrane material, membrane surface area, and gas flow rates. Increasing membrane surface area results in larger size OTR which may not be suitable for an application like gas turbine combustion system. In designing an OTR for application in gas turbine, Nemitallah et al. [85] tried to make the size of the OTR as compact as possible while extracting the highest possible oxygen permeation flux and generating the design power. Based on the required output power, fuel flow rate is calculated and, then, optimizations to the total sweep and permeate flows and species concentrations are performed. Lowering the fuel concentration below 5% (by vol.) may not be acceptable for such OTR application due to the associated severe combustion instabilities at such operating conditions [83, 84]. Also, increasing the inlet fuel concentration is limited by the expected oxygen permeation to complete the combustion of the fuel to prevent the creation of rich mixture zones. While designing this OTR, Nemitallah et al. [85] tried to secure enough oxygen permeation flux for complete conversion of the fuel, lean combustion, through the control of the OTR size. Based on these limitations, they examined numerically different inlet fuel concentrations, from 5 to 10%, and the OTR performance in terms of distributions of oxygen permeation flux and oxygen partial pressure, temperature, and species concentrations are presented in Figs. 6.28 and 6.29.

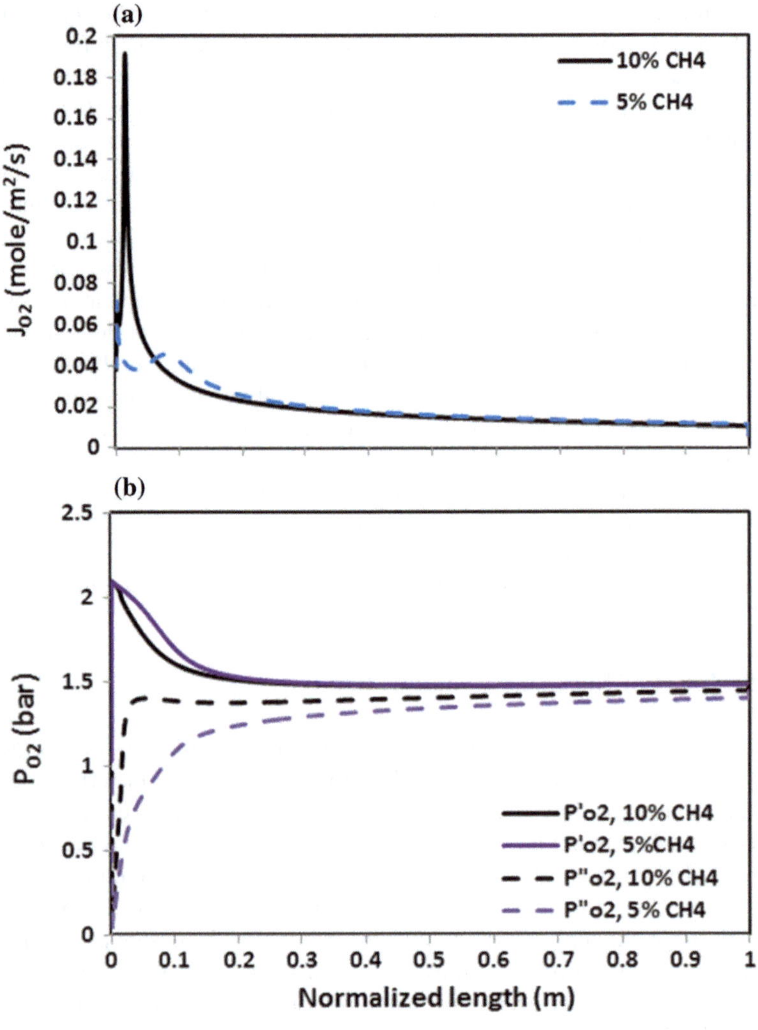

Fig. 6.28 Influence of inlet fuel concentration, 5% CH_4 (case 4) and 10% CH_4 (case 3), on the axial distributions of: **a** oxygen permeation flux and **b** oxygen partial pressure in feed and permeate sides of the membrane

Better oxygen permeation flux is obtained in the case of 10% CH_4 as compared to the case of 5% CH_4 as shown in Fig. 6.28a. This can be attributed to earlier and better mixing between permeated oxygen and fuel in case of 10% CH_4 due to increased fuel concentration in the sweeping gas. Thus, combustion starts earlier near the inlet section and the temperature increases faster for the case of 10% CH_4 as presented in Fig. 6.29a. The sudden jump in temperature due to combustion results in a jump in oxygen permeation at the flame zone near the inlet section for the case of 10% CH_4 as shown in Fig. 6.28a. Due to higher oxygen permeation flux

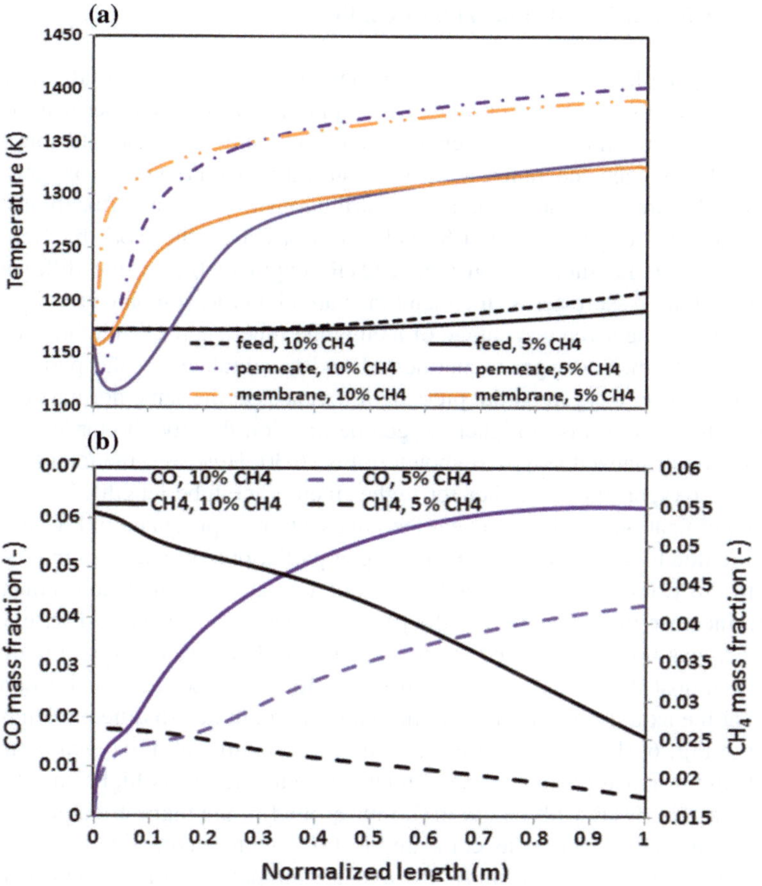

Fig. 6.29 Influence of inlet fuel concentration, 5% CH_4 (case 4) and 10% CH_4 (case 3), on the axial distributions of: **a** feed flow, permeate flow and membrane temperatures and **b** mass fractions of CH_4 and CO

in case of 10% CH_4, oxygen partial pressure is reduced faster in feed side and increased faster in permeate side as compared to the case of 5% CH_4 as presented in Fig. 6.28b. As compared to the case of 5% CH_4, higher oxygen permeation flux for the case of 10% CH_4 resulted in faster burning rate of the fuel and higher production rate of CO as an intermediate combustion species. This is clearly indicated from the slope of the curves of axial distributions of CH_4 and CO for both cases of 10% CH_4 and 5% CH_4 presented in Fig. 6.29b. However, full conversion of the fuel is not obtained for both cases which may require an additional after burning stage of the leaving exhaust gases to fully burn the fuel.

6.6.2.9 Effect of Membrane Tube Length

Membrane tube length is an important parameter controlling the total surface area of the membrane and, accordingly, the amount of oxygen permeation flux. Increasing the membrane tube length results in increase in the membrane area available for oxygen separation and, as a result, the total amount of oxygen flux is increased. Figure 6.30 shows the axial distributions of oxygen permeation flux for membrane tube lengths of 0.9, 1.8, and 3.6 m based on the work by Nemitallah et al. [85]. For the shortest considered OTR length of 0.9 m, the differences in oxygen partial pressure across the membrane are higher as compared to the cases of longer OTR at the same percentage of membrane tube length. While increasing the reactor length, more oxygen is permeated and accumulated in the permeate side resulting in lower oxygen partial pressure differences at the same normalized length of the OTR. This results in higher oxygen permeation flux for shorter length OTRs at the same normalized length as shown in Fig. 6.30. However, the total amount of permeated oxygen flux is higher for higher membrane tube lengths. The peaks of oxygen permeation flux happen near the inlet section as presented in Fig. 6.30. The peaks are due to higher differences in oxygen partial pressure across the membrane close to the OTR inlet section and, also, due to start of combustion indicated by sudden increase in temperature in the permeate side as presented in Fig. 6.31a. As the sweep gas flow rate is fixed in all cases, the peak values of oxygen permeation flux are almost the same, as shown in Fig. 6.30, and, also, the combustion starts almost at the axial location (different normalized length due to different lengths), as shown in Fig. 6.31a. Due to the increase in the total amount of oxygen permeation flux at higher reactor lengths, the combustion temperature is higher as shown in Fig. 6.31a. The OTR with length of 3.6 m resulted in the highest temperature, and the reaction rates are faster as compared to OTR with length of 0.9 and 1.8 m as shown in Fig. 6.31a. Accordingly, the average temperature of feed gas and membrane is higher for the case of 3.6 m long reactor. This can be attributed directly to

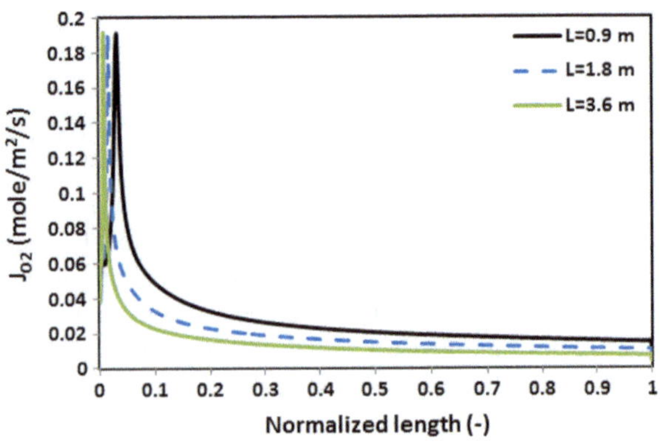

Fig. 6.30 Influence of membrane tube length, $L = 0.9$ m (case 5), $L = 1.8$ m (case 3), and $L = 3.6$ m (case 6), on the axial distributions of oxygen permeation flux

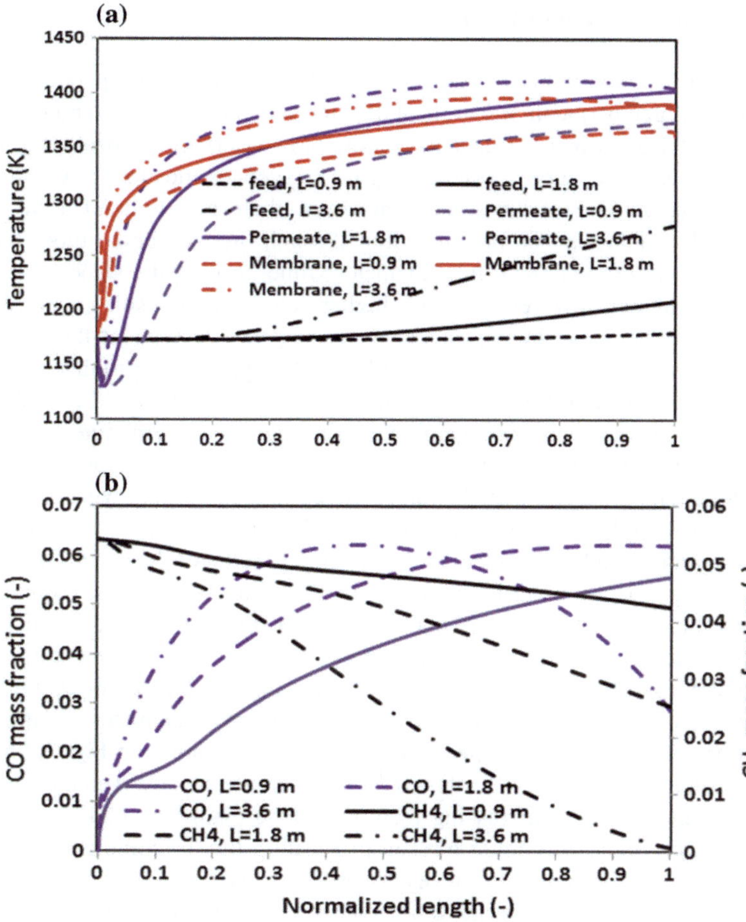

Fig. 6.31 Influence of membrane tube length, $L = 0.9$ m (case 5), $L = 1.8$ m (case 3), and $L = 3.6$ m (case 6), on the axial distributions of: **a** feed flow, permeate flow, and membrane temperatures and **b** mass fractions of CH_4 and CO

the obtained highest amount of permeated oxygen for reactor length of 3.6 m. Due to higher reaction rates for the case of 3.6 m long reactor, the production rate of CO as an intermediate combustion product is higher, at early stages of combustion, as compared to the cases of shorter reactors as per the plots in Fig. 6.31b. Near reactor exit section, higher concentrations of CO are obtained for shorter reactors indicating incomplete combustion of the fuel as shown in Fig. 6.31b. Also, the increase of total amount of permeated oxygen for the case of 3.6 m long reactor results in full conversion of the fuel as shown in Fig. 6.31b from the axial distributions of fuel mass fractions. OTRs with shorter length resulted in almost 50% conversion of the available fuel.

6.6.2.10 Effect of Membrane Tube Diameter

The diameter of the membrane tube is one of the most important parameters controlling the oxygen permeation flux, heat transfer rate, pressure drop, and the overall size of the OTR in terms of surface to volume ratio and reactor aspect ratio. Lower size of the membrane tube results in higher oxygen permeation flux and better surface-to-volume ratio of the reactor; however, there is a limit in terms of the maximum allowable viscous pressure drop through the membrane tube [83, 84]. The decrease in the membrane tube diameter while keeping the same sweep flow rate results in higher flow velocity. This higher velocity may result in severe flame instabilities. As per plots in Fig. 6.32a, increasing the diameter of the membrane tube resulted in lower oxygen permeation flux with a peak closer to the sweep inlet section. This may be attributed to the reduction in flow velocity and better diffusion of permeated oxygen at higher membrane tube diameter. Membrane tube with a diameter of 17 mm resulted in the highest peak value close to the inlet section. The lowest membrane tube diameter, 10 mm, resulted in the highest oxygen permeation flux as shown in Fig. 6.32a. The encountered increase in flow velocity while reducing membrane tube diameter results in better purging of the permeated oxygen and, accordingly, oxygen partial pressure is reduced in the permeate side at lower membrane tube diameter as shown in Fig. 6.32b. Due to the increased oxygen permeation flux while reducing the membrane tube diameter, combustion temperature is increased in the permeate side as shown in Fig. 6.33a. However, the start of combustion is delayed for 10 mm membrane tube diameter resulting in non-uniform distribution of temperature for this case as shown in Fig. 6.33a. This delay in combustion may be attributed to the increase in flow velocity at lower membrane tube diameter. Increasing membrane tube diameter to 24 mm does not have significant impact on either oxygen permeation flux or combustion temperature, as shown in Fig. 6.33a, but it results in larger OTR size. Due to combustion delay in case 10 mm membrane tube diameter, lower fuel conversion ratio is encountered in this case as presented in Fig. 6.33b from the axial distributions of fuel mass fractions. Due to lower oxygen permeation flux at higher membrane tube diameters, incomplete combustion is encountered and, accordingly, the concentrations of CO are increased as shown in Fig. 6.33b. Reducing membrane tube diameter below 10 mm resulted in unstable calculations and divergence of the solution due to higher values of viscous pressure drop and flame instabilities. Based on the above discussion, a membrane tube with a diameter of 17 mm is considered as the best choice for the present OTR design.

6.6.2.11 Effect of Membrane Tube Pitch

Membrane tube pitch affects directly the velocity of the air flow and the overall volume of the OTR. In the current OTR design, the cross-sectional area for air flow in the shell side is considered to be very similar to the area for the sweep flow through inside membrane tubes. Based on that, the calculations were stated with a

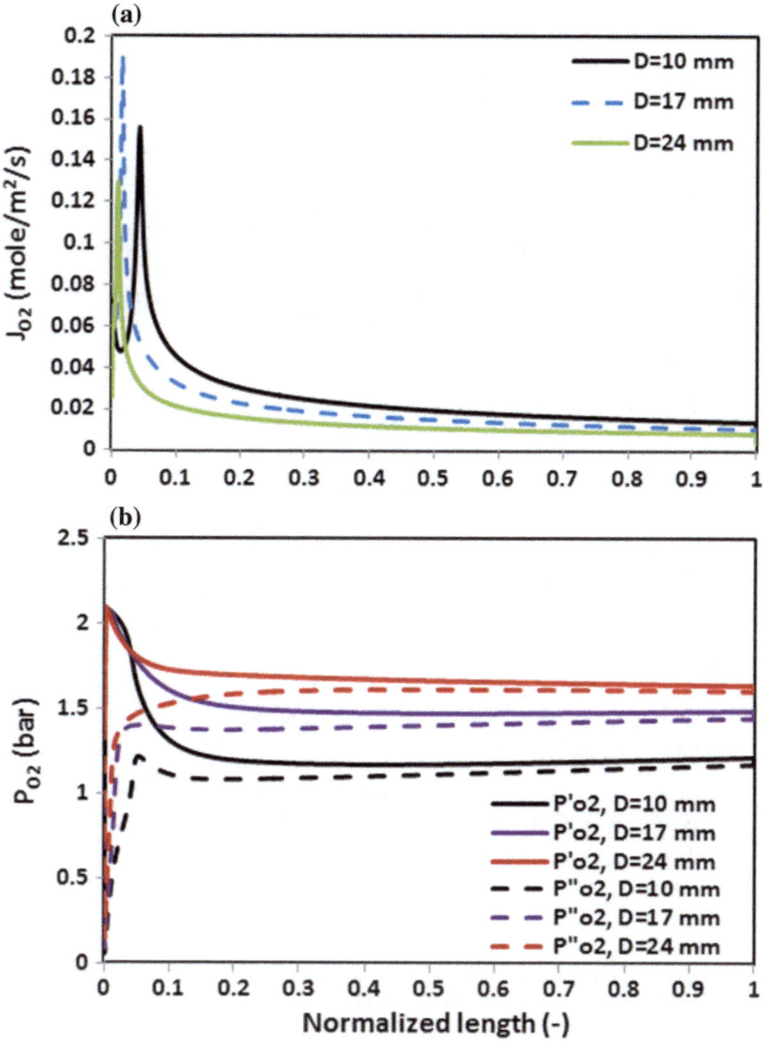

Fig. 6.32 Influence of membrane tube diameter, D = 10 mm (case 8), D = 17 mm (case 9), and D = 24 mm (case 3), on the axial distributions of: **a** oxygen permeation flux and **b** oxygen partial pressure in feed and permeate sides of the membrane

membrane tube pitch of 22 mm which results in an area for air flow in the shell side equal to that of sweep flow inside membrane tubes. Increasing the membrane tube pitch above 22 mm should result in bigger size OTR, and more air flow will be required to keep fixed flow velocity in the feed side. Usually, it is preferred to increase air flow rate to get high oxygen permeation flux, and this is the case in the present study. In this study, simulations were performed also at membrane tube pitch of 18 mm, and the results in terms of oxygen permeation flux and species

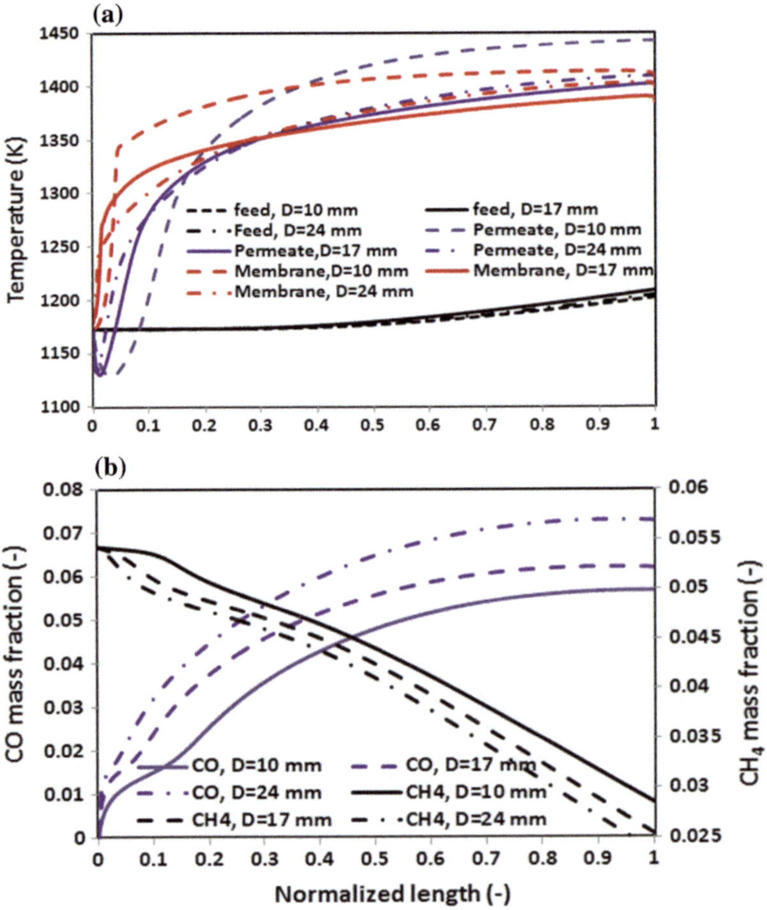

Fig. 6.33 Influence of membrane tube diameter, D = 10 mm (case 8), D = 17 mm (case 9), and D = 24 mm (case 3), on the axial distributions of: **a** feed flow, permeate flow and membrane temperatures and **b** mass fractions of CH_4 and CO

concentrations are presented in Fig. 6.34. Reducing membrane tube pitch, from 22 to 18 mm, resulted in higher viscous pressure drop in feed side as compared to the viscous pressure drop in the sweep side. This can be attributed to the expected higher flow rate in the feed side as compared to the sweep flow rate. Thus, some flow and combustion instabilities are encountered and, accordingly, fluctuations in oxygen permeation flux across the membrane are also encountered as shown in Fig. 6.34a. For 22 and 18 mm membrane tube pitches, the peak of oxygen permeation flux occurs at the same axial distance; however, it is stronger in case of 22 mm pitch. This may be attributed to the stability of the combustion flame in case of 22 mm pitch as compared to the case of 18 mm pitch. Very similar species concentrations and fuel conversion ratio are obtained for the two membrane tube

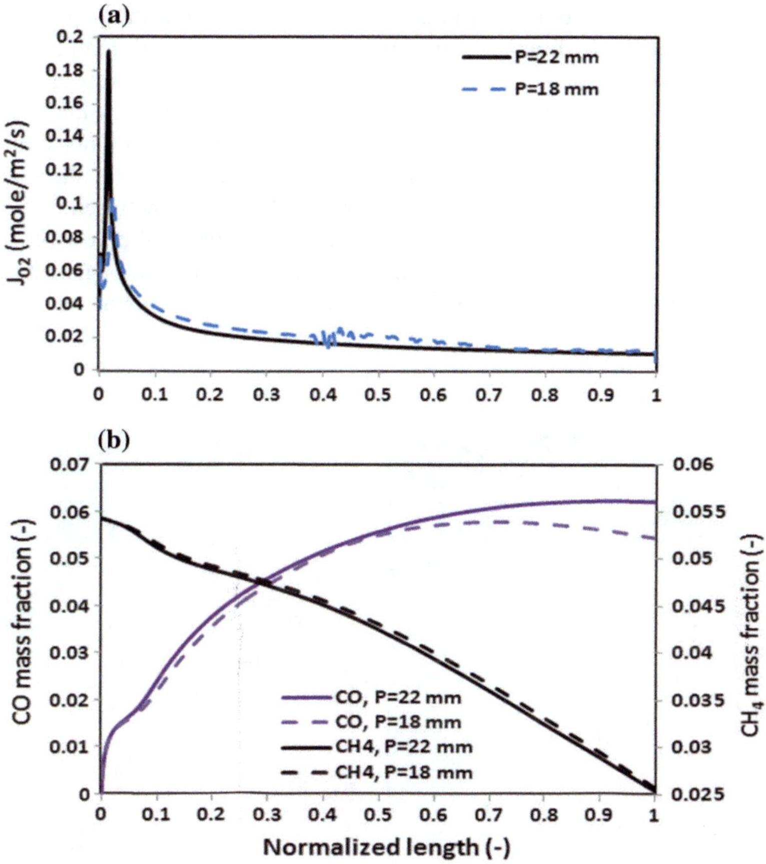

Fig. 6.34 Influence of pitch distance between membrane tubes, $P = 18$ mm (case 7) and $P = 22$ mm (case 3), on the axial distributions of: **a** oxygen permeation flux and **b** mass fractions of CH_4 and CO

pitches. Based on the above discussion, the membrane tube pitch of 22 mm is considered to be the optimum value for the present OTR design based on the considered flow rates.

6.6.2.12 Final Design of the Gas Turbine Combustor

Based on the above detailed results considering distinctive design and flow conditions, the optimum design of the OTR can be identified. The final design of the OTR has the following specifications—length of 3.6 m, membrane tube diameter of 17 mm, and membrane tube pitch of 22 mm—which present the simulation case (6); see Table 6.6. This design showed the best performance in terms of fuel conversion efficiency and power output under co-current flow conditions.

Figures 6.35, 6.36, and 6.37 present detailed contours of temperature, oxygen mass fraction, and fuel mass fraction distributions, respectively, at different axial locations for the optimum design case of the carbon-free gas turbine combustor (case 6). The membrane is presented in these figures as a bold circular black line separating feed stream from the sweep stream. Near the inlet section, the combustion starts, and the flame is suspended beside the membrane surface where the source of oxygen exists. The flame occupies an annular cylindrical zone between the center of the membrane tune and its surface as shown in Fig. 6.35a. As the flow proceeds in the axial direction, more oxygen is extracted, and the flame propagates to fill the whole membrane tube at axial location of 1.6 m as shown in Fig. 6.35.

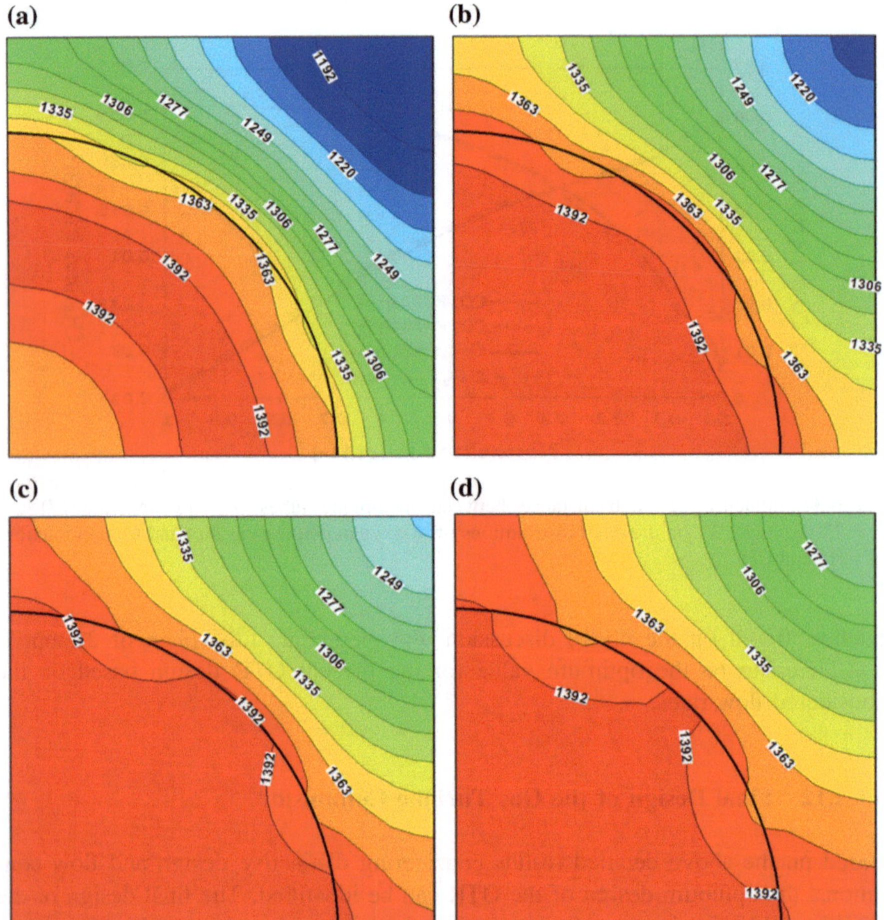

Fig. 6.35 Temperature contour plots on planes normal to flow direction at different axial locations, **a** $z = 0.8$ m, **b** $z = 1.6$ m, **c** $z = 2.4$ m, and **d** $z = 3.2$ m, for the optimum design case of the carbon-free gas turbine combustor (case 6)

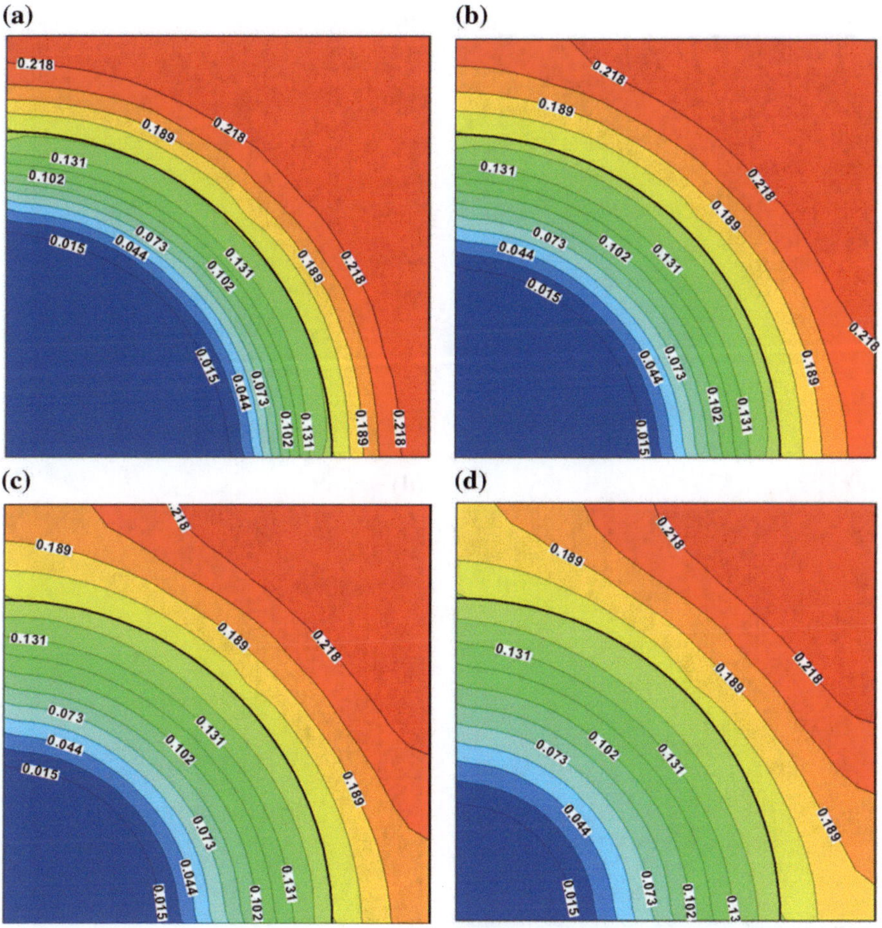

Fig. 6.36 Contour plots of oxygen mass fractions on planes normal to flow direction at different axial locations, **a** z = 0.8 m, **b** z = 1.6 m, **c** z = 2.4 m, and **d** z = 3.2 m, for the optimum design case of the carbon-free gas turbine combustor (case 6)

Heat transfer from the flame zone to the feed air and the permeation of oxygen across the membrane cools down the membrane surface temperature from both feed and permeate sides. This is clearly indicated in Fig. 6.35 from the reduced contour levels in the zone beside the membrane. Figure 6.36 shows the contours of oxygen mass fraction distributions at different axial locations. In the vicinity of the membrane, the concentrations of oxygen are reduced in the feed side and increased in the permeate side due to transfer of oxygen molecules from the feed to the permeate side. As the flow proceeds in the axial direction, more oxygen is permeated and, as a result, oxygen concentrations are increased in the permeate side and better diffusion of oxygen in the sweep flow is obtained. Figure 6.37 shows the contours of fuel mass fraction distributions at different axial locations. Fuel concentrations are

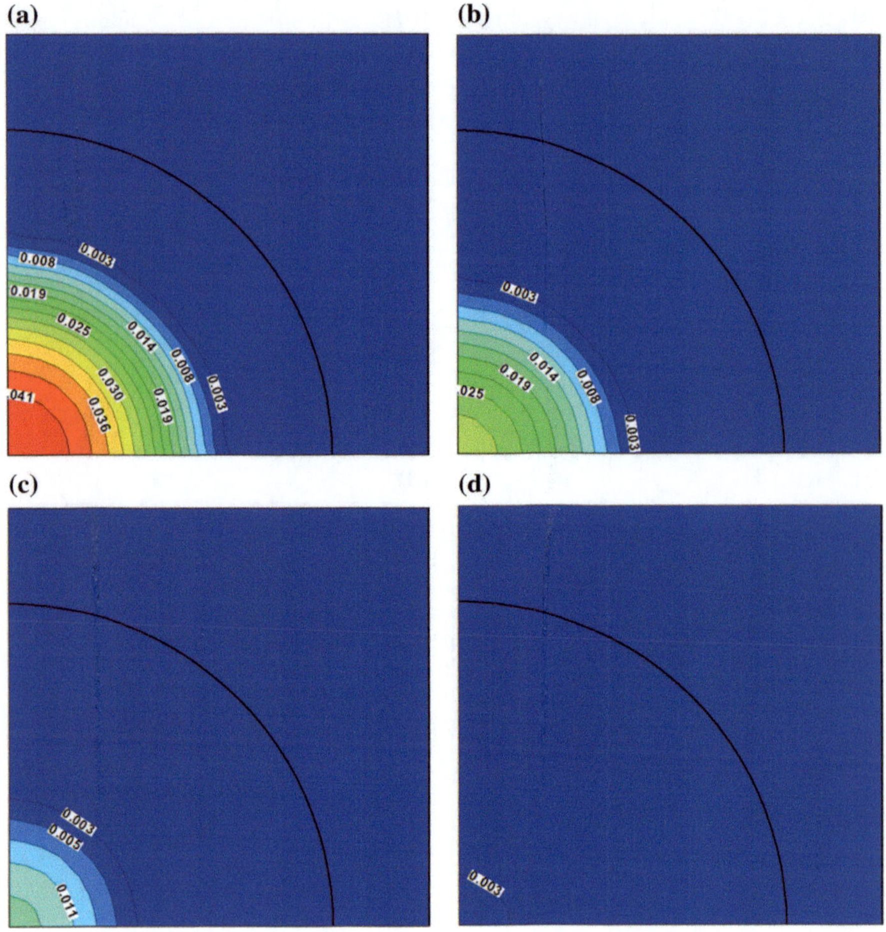

Fig. 6.37 Contour plots of CH_4 mass fractions on planes normal to flow direction at different axial locations, **a** $z = 0.8$ m, **b** $z = 1.6$ m, **c** $z = 2.4$ m, and **d** $z = 3.2$ m, for the optimum membrane reactor design (case 6)

high near the inlet section around the center of the membrane tube. Fuel concentrations are reduced in the radial direction toward the membrane surface due to the permeation of oxygen and the consumption of fuel in the combustion process. As the flow proceeds downstream, the fuel is consumed in the combustion process until it vanishes near the exit section as shown in Fig. 6.37.

Based on the above detailed optimization study, the best design of the OTR to be used in carbon-free gas turbine applications is obtained and the flow and design parameters are presented in Table 6.10. The design considers co-current flow configuration, inlet temperature of 1173 K, and operating pressure of 10 bars. The OTR has a length of 3.6 m, membrane tube diameter and pitch of 17 and

Table 6.10 Design of the present gas turbine combustor for substitution of conventional gas turbine combustor

Parameter	Value
Flow configuration	Co-current
Inlet temperature (K)	1173
Feed/permeate pressure (bar)	10
Feed air flow rate (kg/s/cell)	2.960619×10^{-3}
Pitch (mm)	22
Membrane tube diameter (mm)	17
Reactor length (m)	3.6
Total permeate flow rate (kg/s/cell)	2.57606×10^{-4}
Volume fractions of permeate flow	10% CH_4
	45% CO_2
	45% H_2O
Number of cans (combustor)	16
Number of membrane tubes per can	3000
Volume per can (m^3)	5.2272
Total output power (MWe)	10–15

22 mm, respectively, feed and sweep flow rates of 2.960619×10^{-3} kg/s and 2.57606×10^{-4} kg/s per cell, respectively, and fuel concentration in the sweep flow of 10% (by vol.). Based on output required power of the gas turbine in the range from 10 to 15 MWe based on the cycle first law efficiency, the total number of membrane tubes is calculated. The final design considers 16 cans (OTRs) to be fitted with the gas turbine; each can combustor has a total number of 3000 membrane tubes. This should result in a total volume of 5.2272 m^3 for each can (OTR).

6.7 Application of OTR into Fire Tube Boilers

In this section, a design of an ITM reactor is introduced in a two-pass fire tube boiler furnace to produce steam for power generation toward ZEPP applications [84]. Oxygen separation, combustion, and heat exchange occur in the first pass containing the multiple-units ITM reactor. In the second pass, heat exchange between the combustion gases and the surrounding water at 485 K (P_{sat} = 20 bar) occurs mainly by convection. The emphasis is to extract sufficient oxygen for combustion while maintaining the reactor size as compact as possible. Based on a required power in the range of 5–8 MWe, the fuel and gases flow rates were calculated. Accordingly, the channel width was determined to maximize oxygen permeation flux and keep the viscous pressure drop within a safe range for a fixed reactor length of 1.8 m. Three-dimensional simulations were conducted for both counter- and co-current flow configurations. Counter-current flow configuration proved its suitability in fire tube boilers for steam generation over the co-current flow configuration. The resultant reactor consists of 12,500 ITM units with a height of 5 m, membrane surface area of 2700 m^2, and a total volume of 45.45 m^3.

6.7.1 Reactor Features and Boundary Conditions

In this study, the main goal is to design an ITM reactor in order to be able to substitute a conventional fire tube boiler furnace toward the application of oxy-fuel combustion technology for carbon capture. In this design, multiple membrane reactor units are used in place of the fire tube in the conventional fire tube boiler. Each ITM reactor unit consists of two crossing membranes, creating four channels with the same dimensions as shown in Fig. 6.38. Two of these channels are feed channels while the other two are permeate channels, separated by the crossing membranes. $La_{0.1}Sr_{0.9}Co_{0.9}Fe_{0.1}O_{3-\delta}$ dense ceramic membrane type was used in all simulations. The membrane has a thickness of 0.9 mm, thermal conductivity of 4 W/m K, specific heat at constant pressure of 450 J/kg K, and a density of 6000 kg/m^3 [45, 65]. Each membrane unit is surrounded by walls which separate the ITM unit from the surrounding water in the boiler first pass. In Fig. 6.38, two

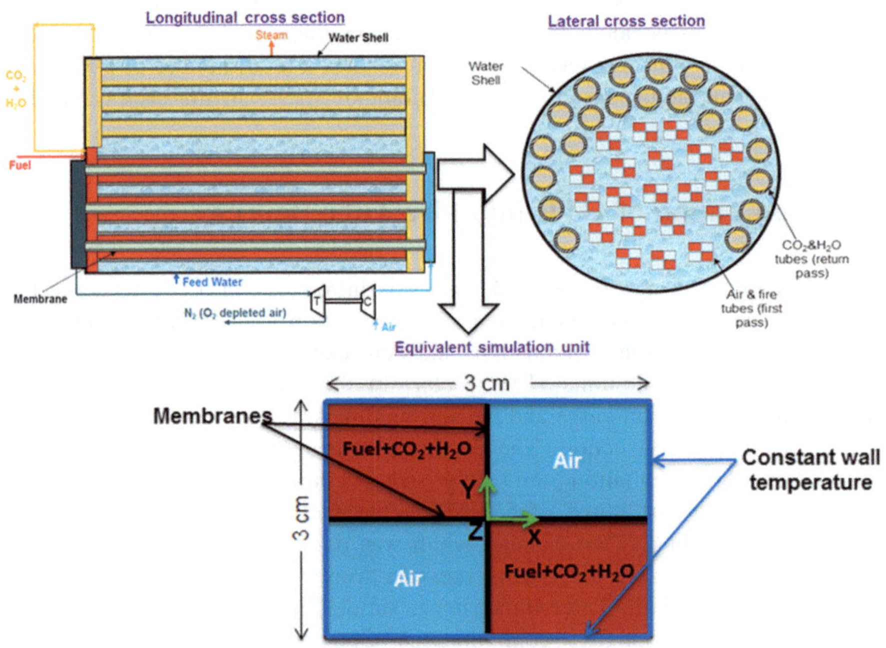

Fig. 6.38 Schematic diagram of two sections in a two-path fire tube boiler utilizing ITM reactor in the first path, and the corresponding integration zone for an ITM reactor unit

cross sections are presented for the targeted two-pass fire tube boiler in the longitudinal and lateral directions. In the first pass, oxygen permeates from the air contained feed channels to the permeate channels in order to react with the fuel in a medium of recycled CO_2 and H_2O of equal volume percentages. In this section, the oxy-combustion process occurs in the permeate channels inside each ITM reactor unit and part of the produced heat is exchanged with the surrounding water through the ITM unit walls. In the second pass, only the remained energy in the combustion gases is exchanged with the surrounding water. To make use of the energy contained in the feed channels gas and improve boiler efficiency, the oxygen-depleted air out from the feed channels is used in order to drive a turbine which is connected to a compressor. This compressor is used to raise the air pressure to 20 bars before entering the ITM reactor as shown in Fig. 6.38.

Due to fixed design and operating conditions of all ITM reactor units, all the 3D numerical simulations were conducted considering only one unit consisting of two feed and two permeate channels as shown in the bottom drawing in Fig. 6.38. The operating pressure of the boiler is set to 20 bars, and accordingly, the surrounding walls temperatures were set to the saturation temperature of water at this pressure, 485 K. The gas temperature at the inlet sections of both feed and permeate channels was set to 1173 K. The inlet boundary conditions, in feed and permeate channels, were specified in terms of the values of the mass flow rates, kg/s. The outlet boundary conditions for all channels were set to pressure outlet conditions. In all simulations, the length of each ITM unit was set to 1.8 m in order to collect more oxygen for combustion, increase the heat exchange surface area, and to be similar to commercial fire tube boilers length. The channel width has a great effect on the overall ITM reactor performance. It directly affects the pressure drop through each channel, the total surface-to-volume ratio of the reactor, and the coefficient of heat transfer. As the channel width is reduced, more oxygen permeation flux is collected due to the increased flow velocity and reduced oxygen partial pressure in the membrane permeate side. Also, reducing the channel width should result in improvement in the heat transfer coefficient, which is very essential in the current application. Thus, the main goal is to operate under the minimum possible channel width in order to obtain the best performance of the ITM reactor. According to the optimization which was done in the work by Nemitallah et al. [70], for the channel width under oxy-combustion conditions, 15 mm channel width has been considered in the present work. Further reduction in the channel width resulted in a sharp increase in the viscous pressure drop and a sudden drop in the combustion temperature due to encountered flame instabilities [70, 71]. 3D simulations were conducted in order to optimize the flow rate and accordingly the oxygen permeation flux through the ITM reactor. In addition, the influences of flow configurations have been considered in this study. For simplicity and reduction in the calculation time for the 3D simulations, equilibrium combustion conditions were applied. The total number of ITM reactor units in the boiler first pass has been calculated in order to deliver a thermal power output in the range of 5–8 MWe.

6.7.2 Methodology of the Numerical Solution

In order to optimize the reactor design and flow configuration, 3D simulations were conducted for the ITM reactor that substitutes the conventional two-pass fire tube boiler considering a fixed operating pressure of 20 bars. The details of the numerical solution are given below.

6.7.2.1 Oxygen Flux Mechanism

It was proven, from the literature work, that the most accurate model for oxygen permeation, especially in reacting medium applications, is the one proposed by Xu and Thomson [45]. This model accounts for three resistances for oxygen permeation, surface exchange resistance from both sides of the membrane, in addition to the bulk diffusion resistance across the membrane bulk. The surface exchange kinetics of the two surface reactions (*A*, *B*) control the oxygen vacancies concentrations (C'_V and C''_V) on both sides of the membrane.

$$\frac{1}{2}O^2 + V^{-}o \overset{k_f/k_r}{\leftrightarrow} O_o^x + 2h^- \tag{A}$$

$$O_o^x + 2h^- \overset{k_r/k_f}{\leftrightarrow} \frac{1}{2}O^2 + V^{-}o \tag{B}$$

where k_f and k_r are the forward and the reverse reaction rate constants for reaction *A* and O_o^x is the lattice oxygen in the perovskite crystal structure. At steady-state isothermal conditions, due to high electronic conductivity of LSCF membranes, the electron holes on feed and permeate surfaces of the membrane are constant [3]. So, the reverse and the forward reaction rates can be considered as pseudo-zero-order. Based on these assumptions, Xu and Thomson [45] presented a model for oxygen permeation equation supported by their experimental results. This model considers the effect of resistances from both sides of the membrane, in addition to the bulk diffusion resistance. This model is function of oxygen partial pressures in feed and permeate sides of the membrane (P'_{O_2} and P''_{O_2}, respectively), membrane thickness (L), the diffusion coefficient of oxygen vacancies (D_v), the forward and the reverse reaction rate constants for reaction *A* (k_f and k_r, respectively):

$$J_{O_2} = \frac{D_v k_r \left(P'^{\,0.5}_{O_2} - P''^{\,0.5}_{O_2}\right)}{2Lk_f\left(P'_{O_2} P''_{O_2}\right)^{0.5} + D_v\left(P'^{\,0.5}_{O_2} + P''^{\,0.5}_{O_2}\right)} \tag{6.39}$$

The values of the coefficients D_v, k_f, and k_r are calculated for a specific type of membranes based on the experimental data. In the present study, LSCF-1991 membrane type is used in the reactor design with a thickness of 0.9 mm. By fitting the available experimental data in the literature for such specific type of

Table 6.11 Fitted data of the oxygen permeation model using the available experimental data [65] in the literature

Expression	Pre-exponential coefficients	Activation energy (kJ/mol)
$D_v = D_v^o \exp(-E_D/RT)$	1.58×10^{-5} (m^2/s)	73.6
$k_f = k_f^o \exp(-E_f/RT)$	1.11×10^{10} ($m/atm^{0.5}$ s)	226.9
$k_r = k_r^o \exp(-E_r/RT)$	3.85×10^{11} (mol/m^2 s)	241.3

membranes [65], the values of the activation energies and pre-exponential coefficients of the parameters D_v, k_f, and k_r were calculated. Fittings of the experimental data of oxygen permeation flux were performed for three experimental data sets at different operating temperatures. The fitted values are listed in Table 6.11; more details can be found in our previous works [3, 69, 70].

6.7.2.2 Methodology of the Numerical Solution

The required mesh was generated using Gambit 2.2 for a 3D ITM reactor unit consisting of four adjacent channels surrounded by water. The steady-state form of the conservation equations was solved using Fluent 12.1 adopting laminar flow formulations. In order to account for oxygen transport across the membrane, Microsoft Visual Studio 2008 was compiled with fluent in order to calculate the source/sink term each time. The final number of considered cells in the axial direction was 1800. In the normal direction of the membrane, 15 cells were considered for each channel. This should result in a total number of cells of 1800 × 30 × 30 in the whole domain. A grid independence study has been performed based on the volumetric size of all cells including the cells in both sides of the membrane. Grid independence tests were performed for the 3D domain utilizing different total number of cells including 15 × 15 × 900 cells, 30 × 30 × 900 cells, and 30 × 30 × 1800 cells. Further reduction below the size of 30 × 30 × 1800 cells results in insignificant change in the calculated oxygen permeation flux. However, there is a misunderstanding of the effect of cell size (beside the membrane) on the amount of oxygen flux based on the oxygen permeation equation. As a matter of fact, the oxygen permeation equation calculates the oxygen permeation flux for each couple of cells (in feed and permeate sides) based on the cell contact area with the membrane. As a result, if the contact area of a membrane cell is reduced, the amount of oxygen flux will also be reduced. However, the total number of cells for the same membrane length will be increased. As a result, the total value of oxygen permeation flux will be fixed whatever the size of the membrane cell; however, the accuracy of the calculations will be affected while increasing the cell size. A semi-implicit method for pressure linked equations (SIMPLE) algorithm was applied in order to couple both the pressure and velocity fields [93]. To control the convergence of the solution, the summation of all residuals was set to 10^{-6} as a condition for the solution to be converged. In this study, a 3D domain consisting of large number of computational cells was considered in order to perform the ITM reactor simulations. Accordingly, the required time for each run is expected to be high, especially where a more detailed reaction kinetics mechanism is considered. Based on this fact, the general simplest oxidation reaction of methane was considered ($CH_4 + 2O_2 \rightarrow CO_2 + 2H_2O$).

6.7.3 Model Validation

The present numerical model including all of the above-mentioned details has been validated with both the experimental data of Xu and Thomson [45] and with the numerical model results of Mancini and Mitsos [71]. For the purpose of validation of the present model with the experimental data, a mesh was developed for the same button-cell reactor setup in their work [45] and the present model was applied. Figure 6.39 shows the comparison between the present model results and the experimental data of Xu and Thomson [45] for a button-cell ITM reactor. This figure shows the distribution of oxygen flux across the membrane surface against the partial pressure of oxygen in the permeate side of the membrane. The partial pressure of oxygen was controlled through the control of nitrogen flux in the sweep side of the membrane. The values of nitrogen flux were altered in the range of 4–120 ml/cm^2/min. As presented in the figure, the oxygen flux across the membrane is improved as a result of the reduction in oxygen partial pressure in the sweep side. This may be attributed to the increased partial pressure driving force across the membrane where oxygen flux is a direct function of oxygen partial pressure difference. It can be seen from the figure that both the experimental data and the numerical results are in an acceptable agreement.

Also, the results of the CFD model have been compared with the work done by Mancini and Mitsos [71] for similar reactor simulations of general power plant applications. They used a black-box model in order to predict the oxygen permeation flux, oxygen partial pressure, in addition to flow field characteristics. However, the black-box model could not capture the 3D features of the flow field and species concentrations. In their model, a monolith structure design ITM reactor consisting of many feed and permeate channels was considered. Due to symmetry,

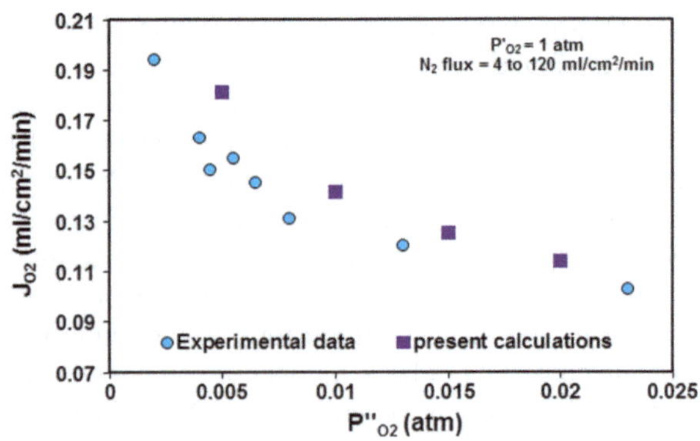

Fig. 6.39 Comparison between the present numerical model results and the experimental data of Xu and Thomson [45] in terms of oxygen flux as function of its partial pressure

3D four quarters of four adjacent cells were considered in this validation considering the whole reactor length for the case of co-current flow configuration [94]. The channel width was 15 mm with a total reactor length of around 90 m. The inlet temperature of both feed and sweep gases was 900 °C, and the operating pressure was 10 bars. For the considered case for the present model validation, the total flow rates were divided by the number of channels (50,000 channels per each stream) in order to calculate the flow rates. The feed sided gas consisted of air (15.71 kmol/s) plus water vapor (1 kmol/s). The sweep gases consisted of a mixture of CO_2 and H_2O with equal molar ratio and total flow rate of 16.68 kmol/s. The same oxygen permeation model [71] was applied for this validation with increasing the oxygen permeation flux one order of magnitude same as what have done in the referred work. Figure 6.40 compares the results of the two models regarding oxygen

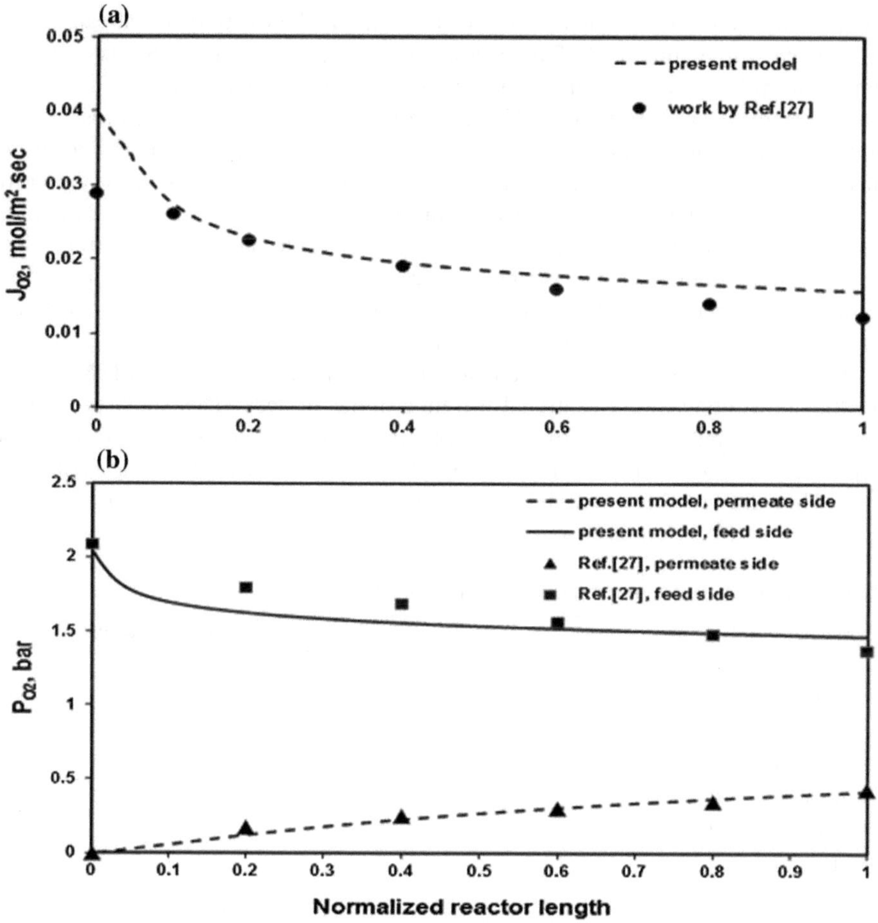

Fig. 6.40 Validation of the present model using the data by Ref. [71] for non-reactive co-current flow inside an ITM reactor at fixed operating pressure of 10 bars: **a** axial oxygen permeation flux and **b** axial local oxygen partial pressure in feed and permeate sides of the membrane

permeation flux and partial pressure distribution along the reactor. Both results are in a good agreement with common trends and mostly common values of oxygen flux and partial pressure. More about the validity of the present 3D model can be found in our previous works [70, 94]. As shown in the figure, oxygen permeation flux is reduced in the axial direction due to the increase in the oxygen partial pressure in the permeate side of the membrane. This may be attributed to the accumulation of the permeated oxygen toward the reactor exit, as there is no combustion in this case and the system is isothermal.

6.7.4 OTR Design for Boiler Furnace Substitution

Two channels for each stream, air feed stream and sweep gases stream, were considered in the simulated ITM reactor unit as shown in Fig. 6.38. Oxygen permeates from the feed channels to the permeate channels containing the fuel. In the present study, CH_4 is used as fuel and is mixed with equal volume ratios of H_2O and CO_2. Oxy-combustion process occurs in the two permeate channels, resulting in increased gases temperature and, as a result, heat is transferred from the combustion gases to the surrounding water. In this section, results for both co-current and counter-current flow configurations are presented. For the co-current flow configuration, both feed and permeate streams flow in the same axial direction. However, for the case of counter-current flow configuration, feed streams are supplied in a reverse direction to the permeate stream flows.

Table 6.12 summarizes the specifications of the proposed design of ITM reactor for the substitution of conventional fire tube boiler furnace. This study provides a design for a unit for both oxygen separation and oxy-combustion toward the application of oxy-fuel combustion technology for carbon capture and sequestration. This unit is able to derive the required oxygen for combustion from the

Table 6.12 ITM reactor specifications for a fire tube boiler substitution

Parameter	Counter-current reacting
Permeate T_{in} (K)	1173
Feed T_{in} (K)	1173
Feed $m^{\bullet}_{air}$ (kg/s)	35.84
Permeate $m^{\bullet}_{CH_4}$ (kg/s) based on 5% by vol.	0.39
Permeate $m^{\bullet}_{CO_2}$ (kg/s)	10.72
Permeate $m^{\bullet}_{H_2O}$ (kg/s)	4.39
Feed P_{tot} (bar)	20 ($T_{(sat)water}$ = 485 K)
Permeate P_{tot} (bar)	20 ($T_{(sat)water}$ = 485 K)
Total number of channels for both streams	50,000
Reactor length (m)	1.8
Power (MW), bases on cycle first law efficiency	5:8

flowing air to be burned with the fuel in the permeate channels. Toward the application of the concept of zero-emission power plants (ZEPP), the combustion products in this case are mainly CO_2 and H_2O which are recycled at equal volumes to act as sweep gases for the ITM reactor. Based on the applied reaction mechanism, each mole of fuel will produce two moles of H_2O and one mole of CO_2 after the combustion process. This means that the ratio between H_2O and CO_2 in the exhaust stream is 2:1. However, this ratio is reduced to 1:1 in this study in order to reduce the amount of CO_2 in the permeate side due to its bad effects on the combustion process. The aim and application of the present work help in applying the required ratio of H_2O and CO_2. The target is the capture of CO_2 for either storage or reuse. This necessitates the separation of CO_2 from H_2O in the exhaust stream after the energy conversion process through H_2O condensation. Also, the study presents a design of an ITM reactor for the application in fire tube boilers for steam production. The source of H_2O at the inlet section of the permeate side is coming through the extraction of a part of the generated steam by the membrane reactor to the reactor inlet section. In order to make the power per unit volume of the resultant ITM reactor close to the size of the commercial fire tube boilers for the same power range, a total of 50,000 channels (25,000 per stream) were calculated based on the optimization of channel width. As a matter of fact, when the channel width is reduced, the oxygen permeation flux is increased and, the heat transfer coefficient is improved. However, the viscous pressure drop is increased and the residence time for combustion is reduced at a certain value of channel width. It was found that any reduction in channel width below 15 mm should result in a sharp increase in pressure drop and loss of combustion [70, 71, 94]. Based on this, the channel width was maintained at 15 mm.

Due to the limitations of the amount of oxygen permeation flux, the volume flow rate of fuel should be minimized in order to reach stoichiometric combustion and increase the fuel conversion ratio. There are two limiting parameters affecting the calculation of a unique fuel percent in the sweep gas mixture. One parameter is the required ITM reactor size in order to get enough oxygen for complete fuel conversion. The second parameter is that the fuel volume percent in the sweep gas mixture should not fall below 5% due to combustion stability and mass transfer issues [45]. Accordingly, the fuel volume percent was set to the minimum possible operating value, 5%, in order to obtain the most possible compact design for the resent reactor. Based on these limitations, the total volume flow rate in the permeate side was optimized for maximum oxygen permeation. Accordingly, the required reactor length (membrane surface area) in order to collect enough oxygen for complete fuel conversion was calculated. The sweep ratio, the ratio between the sweep flow rates to the feed flow rate, is a very critical parameter for stable oxygen permeation. In the present design, there is a big heat sink around each ITM reactor unit; this necessitates increasing the amount of the feed flow rate as compared to the permeate flow rate in order to maintain the membrane temperature at a reasonable value. The specifications and the operating conditions of the resultant ITM reactor are listed in Table 6.12.

The determined length of the reactor was 1.8 m, and accordingly, the resultant reactor height is 5 m. This should result in a reactor with total volume of 45.45 m^3, membrane surface area of 2700 m^2, and thermal power output of 5–8 MWe. Comparison between the reactor performance and oxygen permeation rates for co-current and counter-current flow configurations was performed. It was found that the membrane temperature and the oxygen permeation flux are low for the case of co-current flow configuration. In this case, the membrane temperature was below its activation temperature through a long distance of the reactor and that is why the oxygen flux was limited. On the other hand, the counter-current flow configuration proved better heat transfer and accordingly oxygen permeation characteristics as discussed below. Based on that, the counter-current flow configuration was considered to be the better flow configuration for such application in fire tube boilers.

6.7.5 Operation Under Co-current Flow Configuration

In this section, analysis for the operation of the considered ITM reactor unit under co-current flow configuration is presented. In the application of ITM reactor in gas turbines, the temperature increase in case of counter-current flow is excessive and cannot be absorbed by the membrane material [70, 71]. This is due to the existence of the flame close to the membrane surface without any heat sink to absorb the resultant heat of combustion around the membrane unit. Based on that, it is preferred to apply co-current flow configuration in case of gas turbine applications. However, in the present application of ITM reactor in a fire tube boiler, such heat sink exists due to the water surrounding each membrane unit. This resulted in a sharp reduction in the membrane temperature close to the inlet section as shown in Fig. 6.41a. Once the combustion starts at axial distance of 0.5 m, the membrane temperature is improved. However, due to high radiation heat transfer as shown in Fig. 6.41c, the membrane temperature was kept low in this case of co-current flow configuration. This resulted in a sharp reduction in the oxygen permeation flux as shown in Fig. 6.42b. Close to the exit section, the membrane temperature falls below its activation temperature and accordingly the oxygen flux is zero as shown in Fig. 6.42b. In the present design, air flow rates in the feed sides are increased in order to increase the heat capacity of the feed flows aiming at maintaining almost constant temperature of the feed gas; see Fig. 6.41a. This should result in improvement in the membrane temperature and oxygen partial pressure in the feed side, and accordingly, the oxygen permeation flux is improved. The reactions in all simulations were assumed fast, and this may justify high fuel conversion rate as shown in Fig. 6.41b. In the same figure, the concentrations of CO_2 and H_2O are increased little bit due to combustion after the reaction zone, and then, their concentrations are reduced due to the reduction in the reaction rates and the permeated oxygen. The average gases' temperature in the permeate side of the membrane is reduced, and this resulted in a sharp reduction in the total heat transfer through the reactor as shown in Fig. 6.41c. Close to the inlet section of the ITM reactor, the

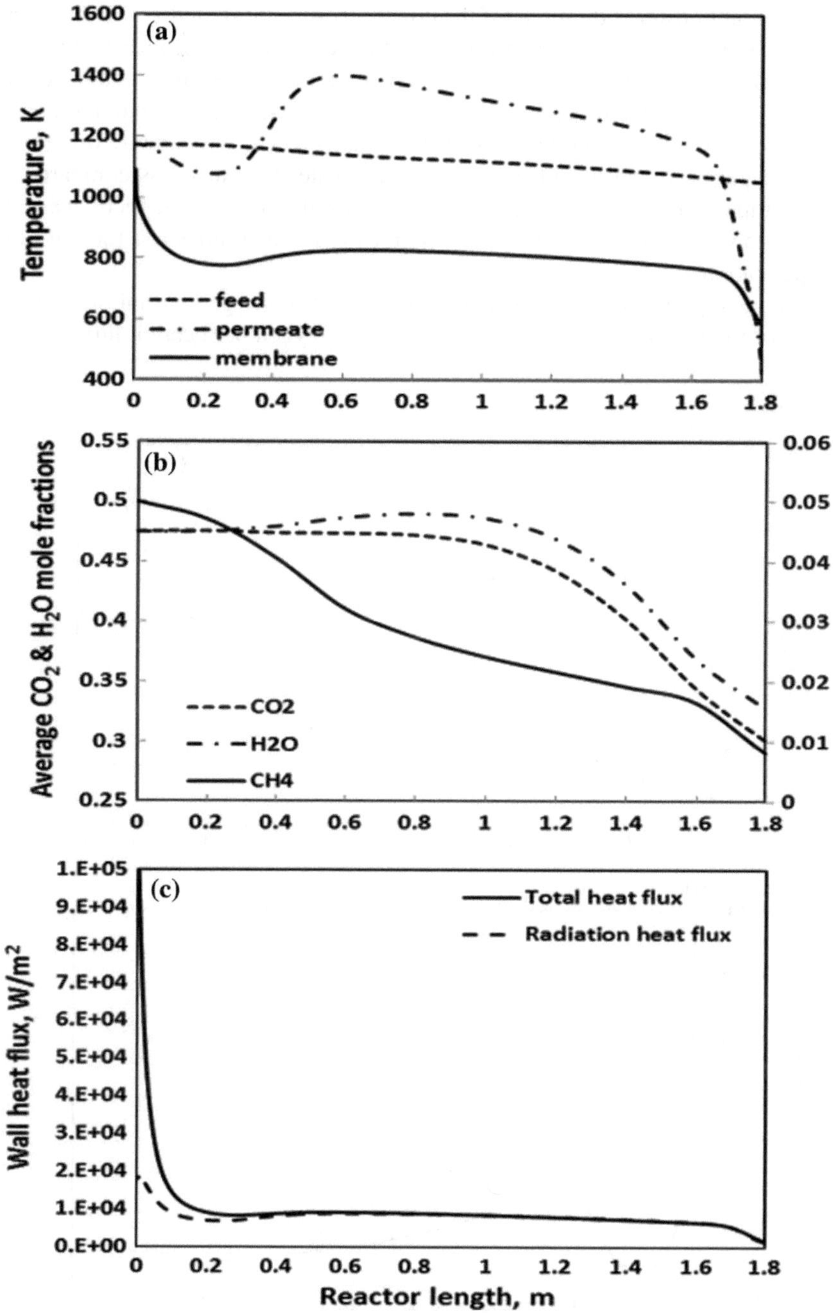

Fig. 6.41 Reactive co-current flow operation: **a** axial membrane temperature and the temperature of feed and sweep gases through the center of each channel, **b** axial average species concentrations in the permeate channel, and **c** axial average heat flux through the ITM unit wall

convection heat transfer is dominant if the combustion is not started yet. This can be attributed to the high rate of oxygen flux which is permeated to the sweep zone and the small temperature difference between the membrane and the free stream. As the combustion starts, the temperature difference increases, and the radiation heat transfer becomes the dominant heat transfer mechanism. The temperature difference between the permeate gases and the surrounding water is reduced as compared to the inlet temperature differences of the gases. This implies the ineffectiveness of co-current flow configuration for oxygen permeation and combustion for fire tube boiler applications.

Also, oxygen permeation flux is mainly a function of oxygen partial pressure in both sides of the membrane. Figure 6.42 shows the oxygen permeation flux in view

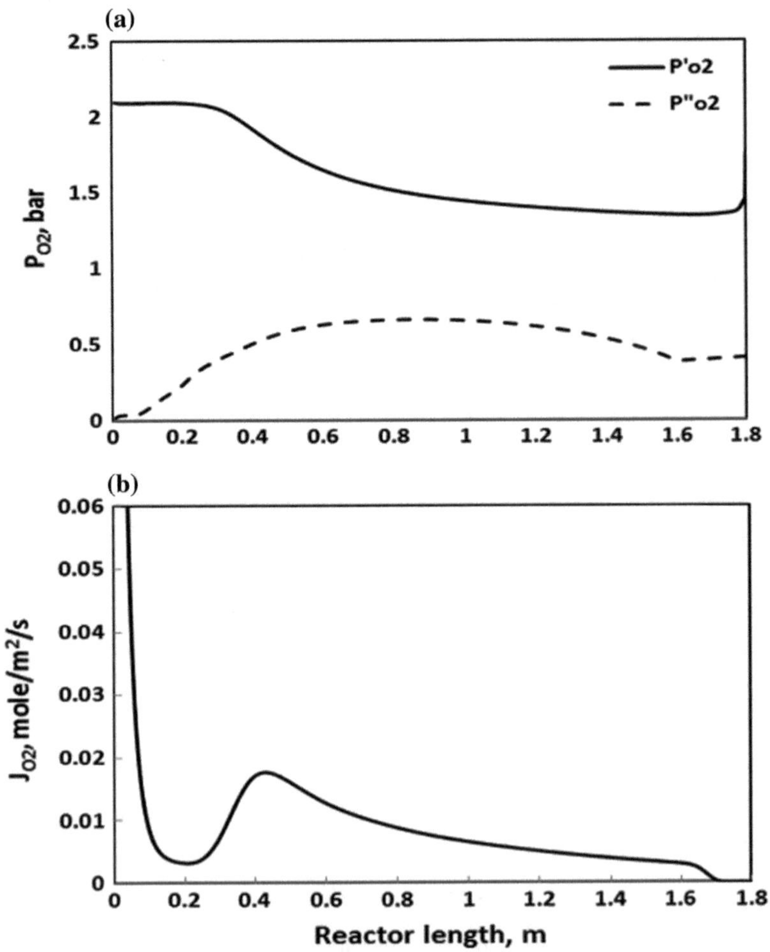

Fig. 6.42 Reactive co-current flow operation: **a** axial dependence of local partial pressure of oxygen in both sides of the membrane and **b** axial oxygen permeation flux

of oxygen partial pressure distribution. The oxygen partial pressure in feed side is reduced indicating the transfer of oxygen through the whole length of the reactor, except close to the exit section. Also, oxygen partial pressure in the permeate side is increased up to the point of the start of combustion. Then, due to the consumption of oxygen in the combustion process, oxygen partial pressure is reduced. At axial location between 1.6 and 1.7 m, the flame is quenched due to high heat transfer rate to the surrounding water and the membrane temperature falls below its activation temperature for oxygen permeation (see Fig. 6.41a). This resulted in accumulation of the available oxygen in the permeate side and, as a result, the oxygen partial pressure is increased in the permeate side as shown in Fig. 6.42a. Close to the inlet section, the membrane temperature is high and the partial pressure driving force is the highest. This resulted in high oxygen permeation flux. As the membrane temperature is reduced, the partial pressure difference is then reduced leading to a reduction in the oxygen permeation flux. At the start of combustion, a jump in membrane temperature is encountered at axial location of 0.5 m, and as a result, the oxygen flux is increased as shown in Fig. 6.42b.

A 3D representation of both the temperature and the oxygen partial pressure distributions is shown in Figs. 6.43 and 6.44 at different axial locations on planes normal to the flow direction. The permeate gases' temperature is reduced after the inlet section, and as the combustion intensity increases, the temperature is increased. The combustion starts close to the membrane surface where the oxygen source exists. The gas temperature rises sharply close to the reactor inlet to reach 1440 K at axial location of 0.6 m and then decreases gradually due to the heat loss to the surrounding water. The flame size increases reaching its maximum at $z = 0.6$ and then decreases in the axial direction until it vanishes at $z = 1.6$ m as shown in Fig. 6.43. This can be attributed to the high rate of heat transfer to the water and reduction in reaction rates because of reduced oxygen permeation flux. In order to investigate the 3D view for the other factors affecting the oxygen permeation flux, oxygen mole fraction distribution is presented in Fig. 6.44 at different axial locations. A continuous reduction in the oxygen concentrations from the inlet to the exit section is encountered in the feed sides of the membranes, indicating oxygen permeation. In the permeate side, oxygen mole fractions are increased from inlet till the combustion occurs, and after that, the concentrations are reduced due to oxygen consumption in the combustion process. Also, the steep reduction in membrane temperature due high heat transfer rate resulted in a steep reduction in the oxygen partial pressure away from the combustion zone as shown in Fig. 6.44.

In other applications of ITM reactors, like in gas turbines (see Refs. [70, 94]), the heat of combustion gases out from the ITM reactor is converted to power inside a gas turbine. So, there is no heat sink (like surrounding water in case of boiler application) close to the membrane unit. This maintains high membrane surface temperature which should result in high oxygen permeation flux. However, in the application of ITM reactor in fire tube boiler, there is a significant heat sink formed by the water walls surrounding each ITM unit. In this case, care should be taken while the calculations of the mass flow rates in both feed and permeate sides in order to maintain the flame inside the ITM unit and keep the membrane surface

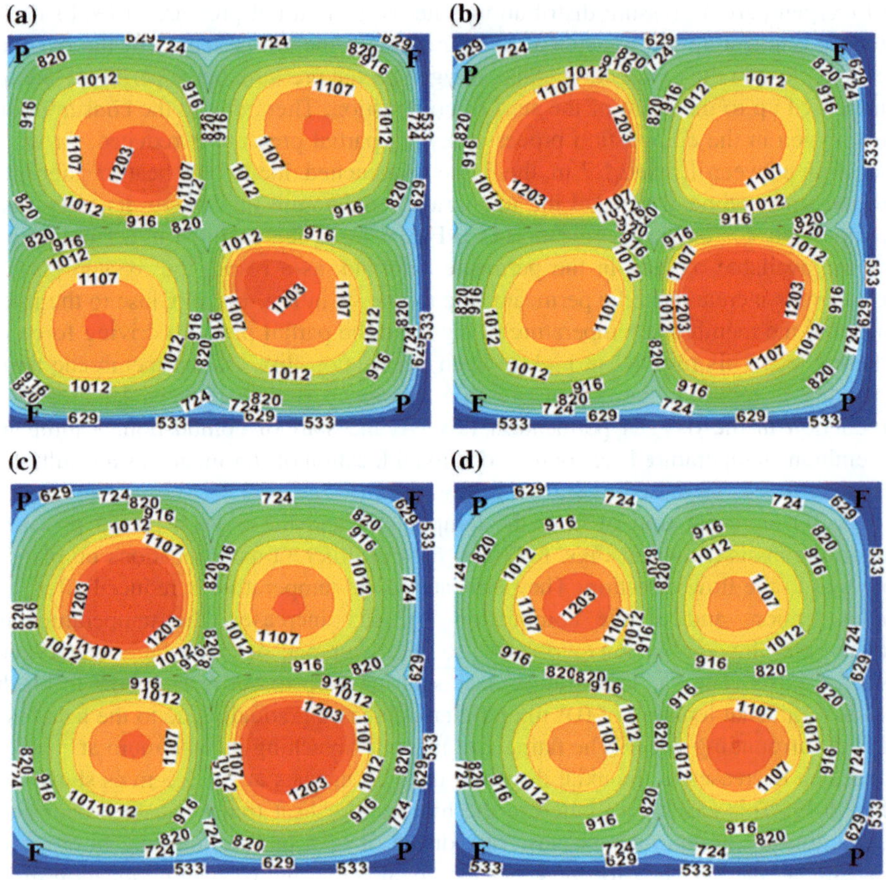

Fig. 6.43 Reactive co-current flow temperature contour plots at different axial locations: **a** $z = 0.4$ m, **b** $z = 0.8$ m, **c** $z = 1.2$ m, and **d** $z = 1.6$ m

temperature above its activation temperature for oxygen permeation. As a matter of fact, the counter-current flow configuration for heat exchange applications is much more effective than the co-current flow configuration for the same flow conditions. This should result in much improvement in the amount of oxygen permeation flux, and as a result, the combustion temperature is increased.

6.7.6 Operation Under Counter-Current Flow Configuration

In the case of counter-current flow regime, the gases in the permeate side are introduced to the reactor at axial distance of $z = 0.0$ and the feed airflow is

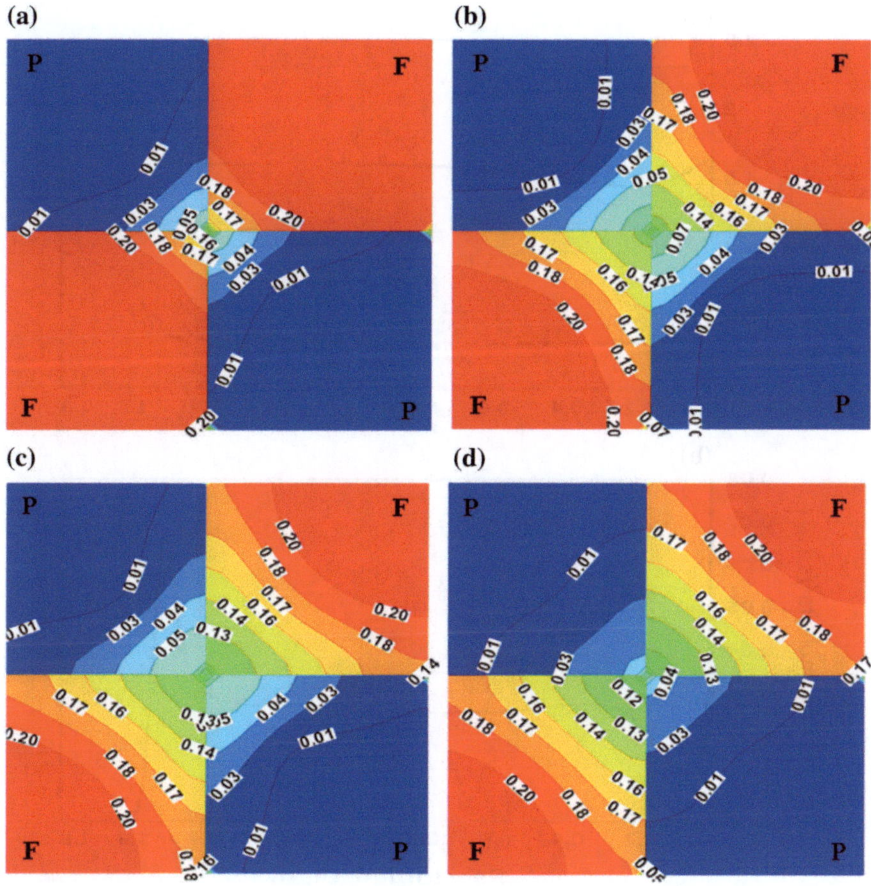

Fig. 6.44 Reactive co-current flow oxygen mole fraction contour plots at different axial locations: **a** z = 0.4 m, **b** z = 0.8 m, **c** z = 1.2 m, and **d** z = 1.6 m

introduced at axial distance of z = 1.8. In this case, the partial pressure difference between the feed and the permeate sides is uniformly distributed along the reactor length as shown in Fig. 6.45a, i.e., the regions of high oxygen partial pressure in feed and permeate sides are matching with each other, and the same for low partial pressure zones. This should result in reduction in mechanical stresses on the membrane surface. This is unlike the case of co-current, where the partial pressure differences are not uniformly distributed. This should result in a uniformly distributed oxygen permeation flux through most of the reactor length as shown in Fig. 6.45b. As a result, the combustion and heat release are also uniformly distributed along the reactor and higher amount of heat is absorbed by the surrounding water as compared to the case of co-current flow as shown in Fig. 6.46c. Both the membrane surface temperature and the feed air temperature are also uniform through most of the reactor length as shown in Fig. 6.46a.

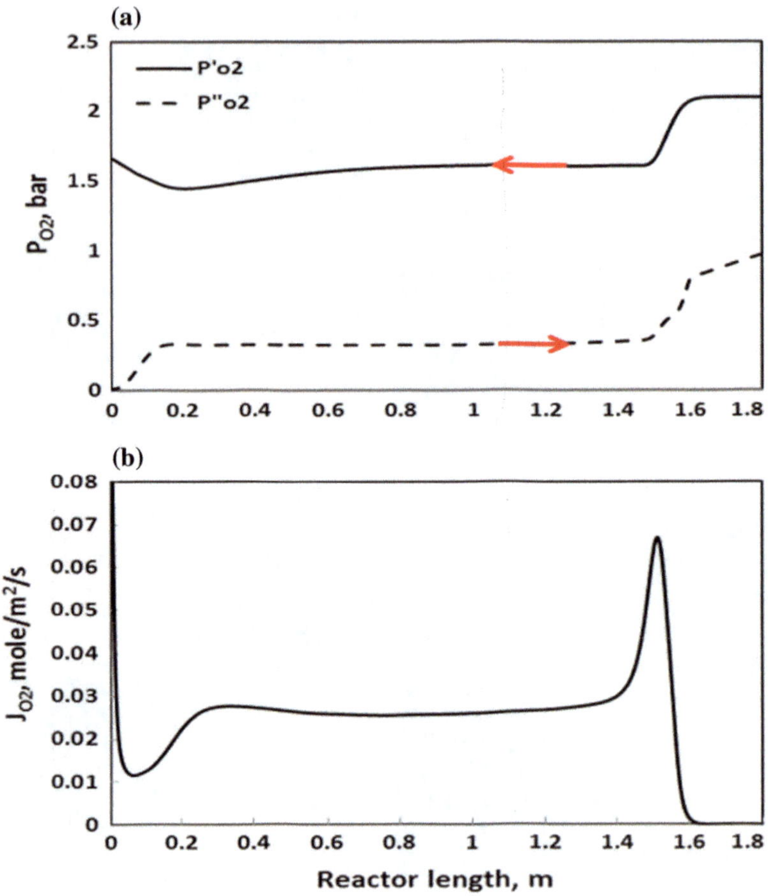

Fig. 6.45 Reactive counter-current flow operation: **a** axial dependence of local partial pressure of oxygen in both sides of the membrane (lines are showing the flow direction) and **b** axial oxygen permeation flux

The membrane temperature due to continuous combustion is elevated above the membrane activation temperature for oxygen permeation through most of the reactor length, unlike the case of co-current flow, as can be seen in Figs. 6.41a and 6.46a. There is no CH_4 in significant amount at the rector exit in case of counter-current flow as shown in Fig. 6.46b, which indicates a good fuel conversion ratio. However, there is around 5–7% of the inlet fuel which is not consumed in the reaction zone. This can be attributed to the reduced temperature in the corner regions due to the heat transfer to the surrounding water. This reduced temperature prevents any kind of reactions to occur, and as a result, part of the fuel escapes. Concentrations of CO_2 and H_2O are reduced toward the reactor exit section as shown in Fig. 6.46b. Their concentrations are mainly controlled by oxygen permeation and reaction rates. The concentrations of H_2O and CO_2 as combustion

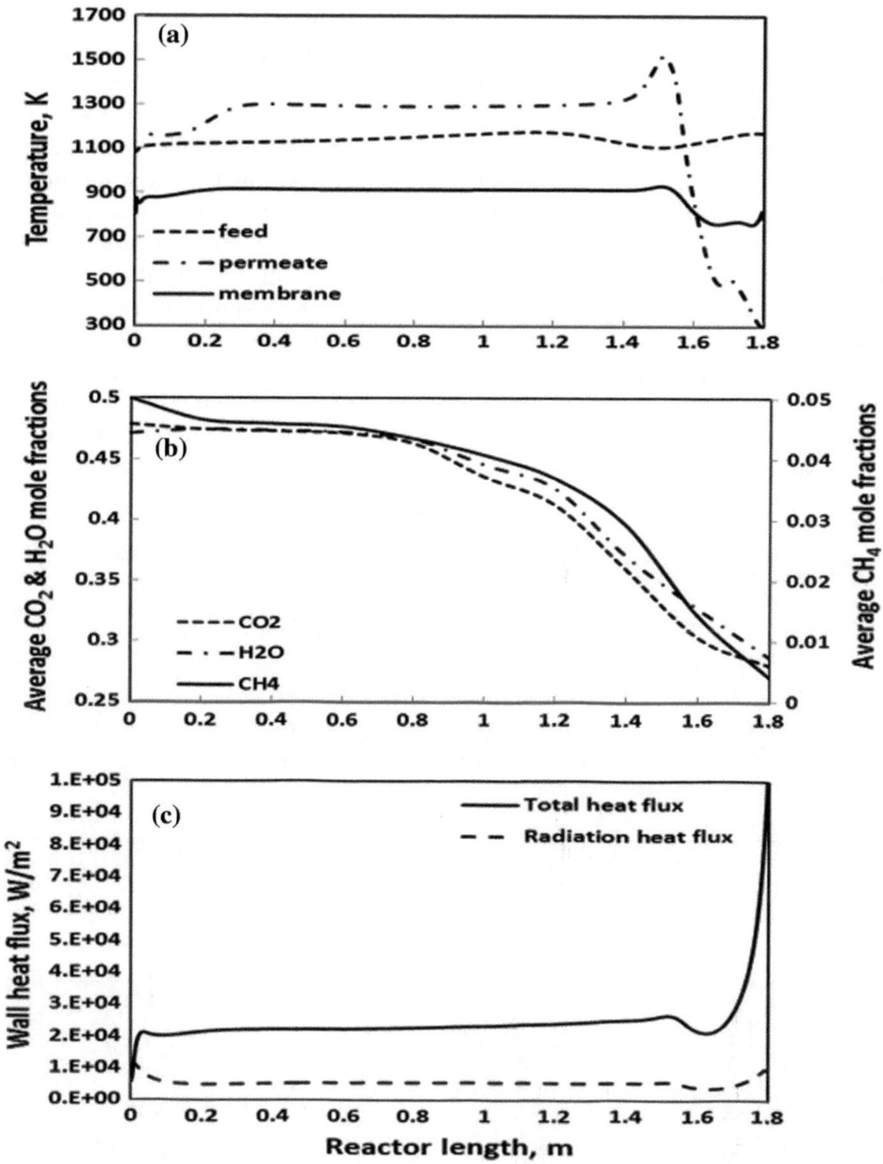

Fig. 6.46 Reactive counter-current flow operation: **a** axial membrane temperature and the temperature of feed and sweep gases through the center of each channel, **b** axial average species concentrations in the permeate channel, and **c** axial average heat flux through the ITM unit wall

products are slightly increased in the reaction zone, balanced by the oxygen permeation flux. However, their concentrations are reduced after that toward the reactor exit section. This can be attributed to the increased concentrations of permeated oxygen in the permeate side that is not consumed in the combustion process due to the sharp reduction in the reaction rates as most of the fuel is consumed. Partial oxidation of methane starts earlier at axial distance of 0.3 m and continues till the ratio of oxygen to fuel is stoichiometric at axial distance of 1.5 m as shown in Fig. 6.46a from the permeate gases' temperature plot. The fuel burns wherever oxygen exists; however, at this region, oxygen concentration is the highest due to permeation and accumulation through the channel from inlet to axial location of 1.5 m. This should result in a steep change in temperature, and as a result, the membrane resistance to oxygen permeation is reduced which causes a steep increase in the oxygen permeation flux at that location. This may justify the increase in the total and radiation heat transfer at this location as shown in Fig. 6.46c. In addition, a steep change in the oxygen partial pressure and accordingly the oxygen permeation flux is encountered at this location; see Fig. 6.45. The contour plots of oxygen mole fractions are presented in Fig. 6.47 at different axial location on planes normal to the flow direction. Air enters the reactor at z = 1.8 and leaves at z = 0, so that the oxygen concentrations are reduced in the feed sides from z = 1.8

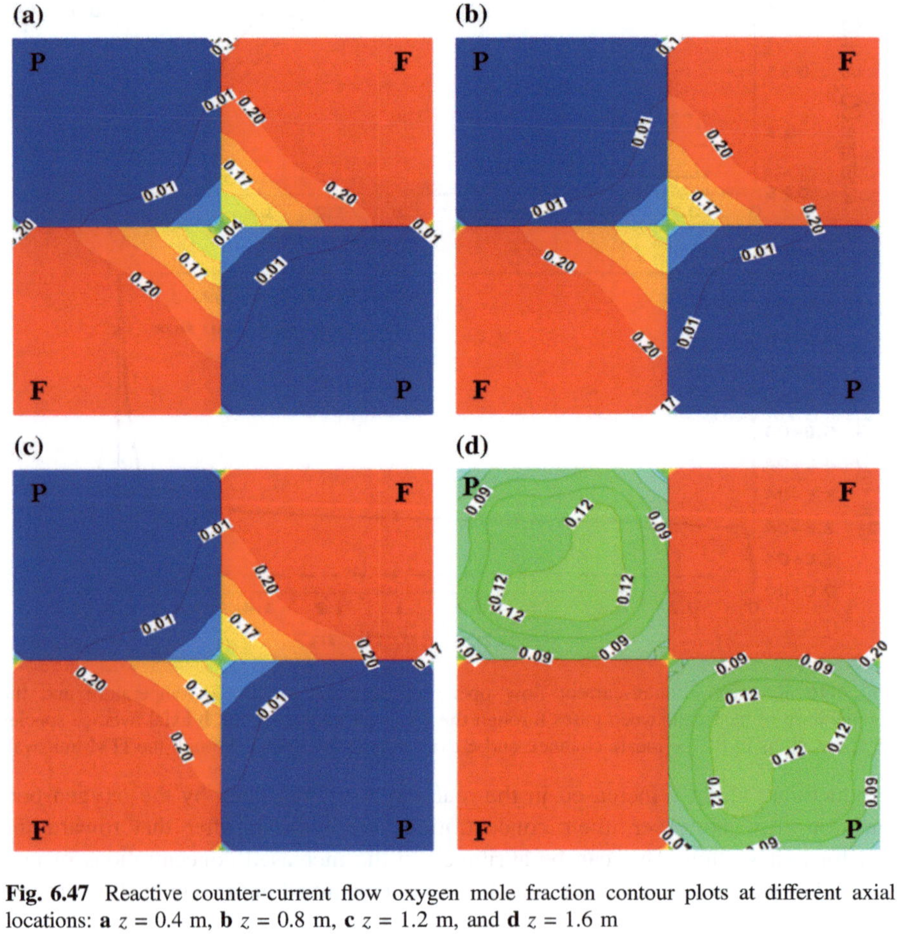

Fig. 6.47 Reactive counter-current flow oxygen mole fraction contour plots at different axial locations: **a** z = 0.4 m, **b** z = 0.8 m, **c** z = 1.2 m, and **d** z = 1.6 m

to $z = 0$ due to oxygen permeation. However, the permeated oxygen is accumulated in the reverse direction, permeate flow direction, so that the maximum oxygen concentrations are close to the exit section of the permeate gases.

To further investigate the temperature distribution in both feed and permeate sides, Fig. 6.48 represents the temperature line plots in a direction normal to the flow, *x*-direction, and crossing the upper feed and permeate channels at different axial locations. All of these lines were plotted on the same *y*-plane, $y = 0.0075$ m. As shown in the figure, at the end locations (at −0.015 and 0.015 m), the temperature is the saturation temperature of water. This temperature is increased at all axial locations in the direction from the wall to the membrane. Then, the temperature is reduced around the membrane from both sides due to oxygen permeation and heat transfer. The temperature is gradually increasing from inlet till the location of the intense combustion (stoichiometric) at axial distance of 1.5 m as shown in Fig. 6.48. The flame is located about 4 mm parallel to the membrane surface with maximum temperature of 1560 K. The curvature in temperature distribution in feed side is due to the differences between the inlet air temperature, 1173 K, and the wall temperature, 485 K. More details about the flame shape and temperature distributions at different axial locations are presented in Fig. 6.49. Combustion started early as compared to the co-current flow operation, and the size of the flame is much larger. The flame length covers most of the reactor length as shown in the figure. Close to the exit section at $z = 1.6$ m, the temperature is reduced due to heat transfer to the surrounding water and the reactions are vanished. Figure 6.50 shows some of the 3D features for the mole fractions distribution of CH_4, H_2O, and CO_2, in addition to the contour plot of the axial velocity distribution at the location of intense reactions, $z = 1.5$ m. For the mole fraction concentrations, the red zones indicate higher concentrations and the blue zones indicate zero concentrations.

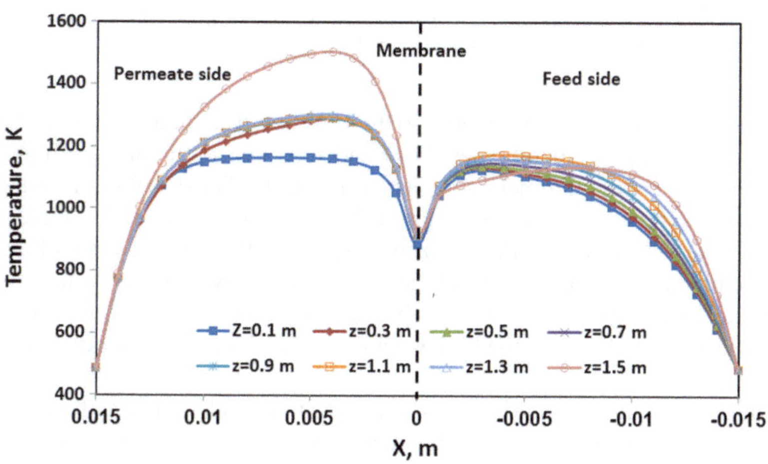

Fig. 6.48 Reactive counter-current flow temperature distributions at different axial locations through lines passing through the upper feed and permeate channels and crossing plane $y = 0.015$ m in *x*-direction

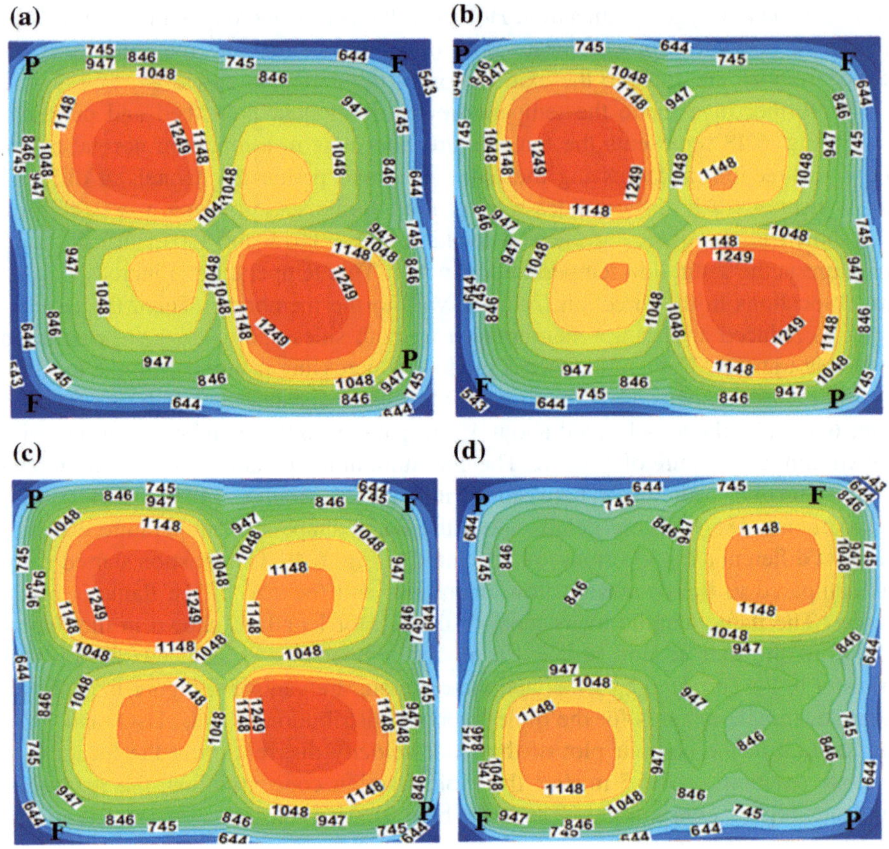

Fig. 6.49 Reactive counter-current flow temperature contour plots at different axial locations: **a** z = 0.4 m, **b** z = 0.8 m, **c** z = 1.2 m, and **d** z = 1.6 m

CH_4 concentration is the highest close to the ITM unit wall, and it is reduced toward the flame, close to the membranes. H_2O and CO_2 concentrations are high in the flame zone as they are the main combustion products. The distribution of the axial velocity of the flows is presented in Fig. 6.50d, and the negative sign indicates the flow direction. That is why the red zones represent the lower values of velocity, while the blue zones represent the higher values. The flow velocity in feed zones is higher than that in the permeate zone, because of higher flow rate in the feed side. In both sides, the velocity is the highest close to the channels centerlines due to the wall effects.

In such application of ITM reactors into fire tube boilers, it was found that the counter-current flow configuration is much more effective for oxygen permeation than the co-current one. In case of co-current, the permeated oxygen is accumulated in the direction of the feed air flow till the combustion starts. This should result in a delay in the combustion process, because the accumulated oxygen increases the

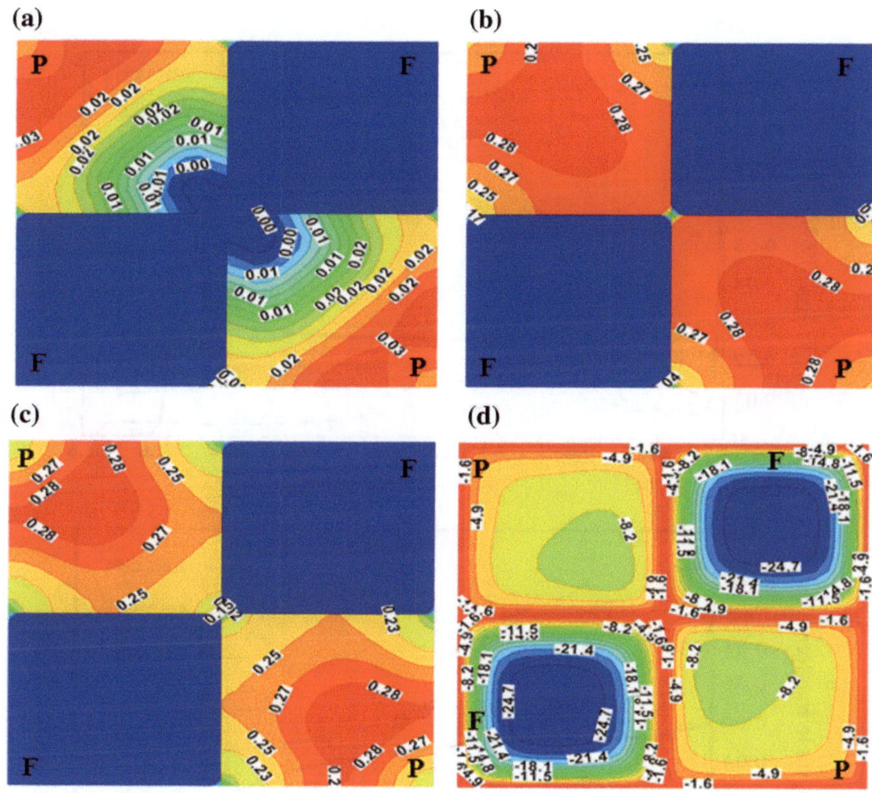

Fig. 6.50 Reactive counter-current flow contour plots on plane = 1.5 (crossing the flame zone) of: **a** CH_4 mole fractions **b** H_2O mole fractions, **c** CO_2 mole fractions, and **d** axial velocity

oxygen partial pressure in the permeate side and reduces the oxygen flux. Downstream of the combustion zone, the oxygen partial pressure in the feed side is reduced and the membrane temperature is lowered due to heat transfer to the surrounding water. This should result in a steep reduction in the oxygen permeation flux.

6.7.7 *Influence of Fuel Concentration*

From the above analysis, counter-current flow configuration has proved its suitability in fire tube boilers for steam production over the co-current flow configuration. So, optimization of the operating percentage of fuel in the sweep gases is performed in this section for the case of counter-current flow. As mentioned above, if more fuel is supplied to the reactor, this means higher requirements of oxygen permeation flux and accordingly the reactor size (more membrane surface area is

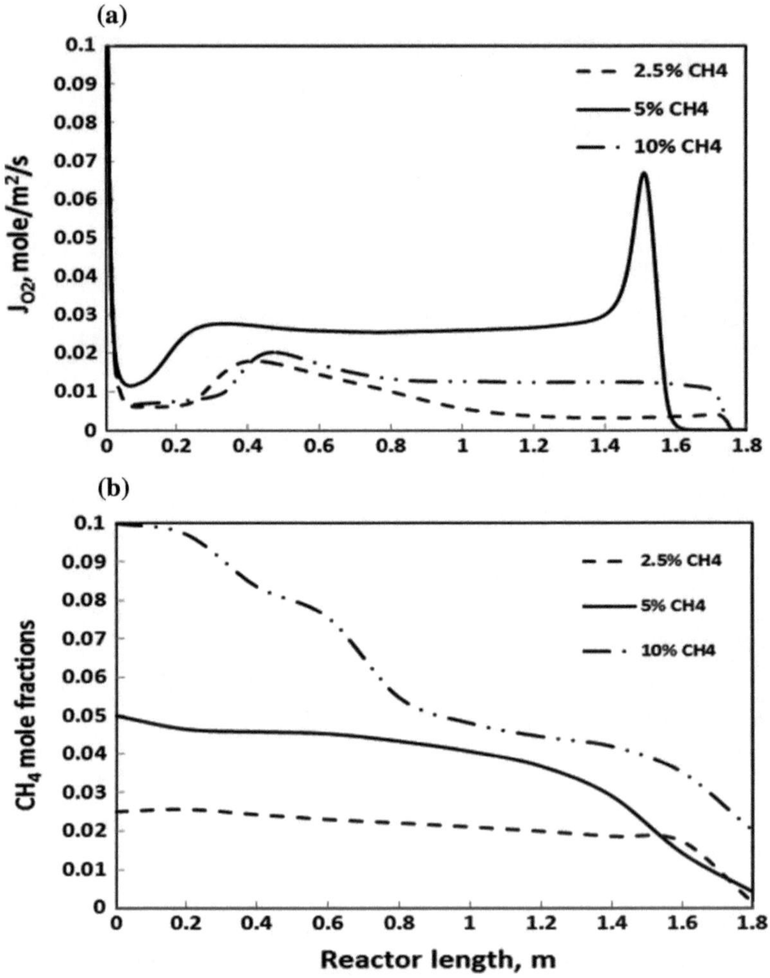

Fig. 6.51 Reactive counter-current flow operation under different fuel percentages in the permeate side flow: **a** axial oxygen permeation flux and **b** CH_4 mole fraction distributions

required). On the other hand, any reduction in the percent of the fuel below 5% (by vol.) should result in unstable combustion and accordingly reduction in oxygen flux as shown in Fig. 6.51a. This reduction in oxygen permeation flux in case of 2.5% CH_4 as compared to 5% is mainly due to the reduction in combustion and membrane temperatures as shown in Fig. 6.52a and b. This reduction in combustion temperature may be attributed to the instabilities in the combustion process due to the lower fuel concentration in the oxidizer mixture.

When the percentage of CH_4 is increased to the level of 10%, a rich mixture is formed through the entire length of the permeate channel. This reduces the possibility of creating stoichiometric fuel-oxygen zones, and the membrane and

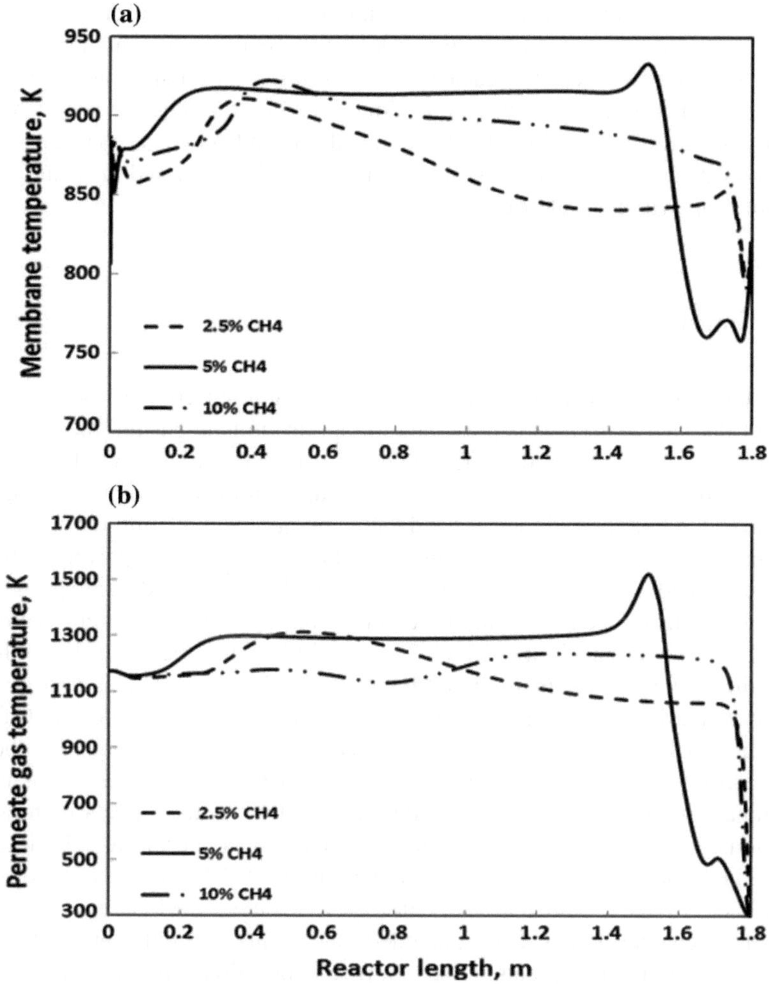

Fig. 6.52 Reactive counter-current flow operation under different fuel percentages in the permeate side flow: **a** axial membrane temperature and **b** axial permeate gas temperature through the center of the permeate channel

combustion temperatures are reduced, as shown in Fig. 6.52, due to lack of enough oxygen for combustion. This reduction in membrane temperature resulted in less oxygen permeation flux as compared to the case of 5% CH_4 as shown in Fig. 6.51a. In addition, the case of 10% CH_4 suffers from the reduction of fuel conversion and more fuel is available at the exit section of the reactor as shown in Fig. 6.51b. This discussion should not lead to the conclusion that the 5% CH_4 in the sweep gases is a unique value for all applications of ITM reactors. This is mainly because the operating fuel concentration for complete conversion requires a specified amount of oxygen. This amount of oxygen depends on the total membrane surface area, reactor design, and the flow rates in both sides of the membrane. Those parameters can vary depending on the application. One of the main constraints in this reactor design process is to keep the reactor size as compact as possible and at the same time extract enough oxygen for combustion. Based on the required power for such present application, the flow rates were calculated and based on that, the channel width was determined to maximize oxygen permeation flux and keep the viscous pressure drop within a safe range.

6.8 Conclusions

In this chapter, the idea of replacement of conventional combustion systems by OTRs of distinctive designs is presented for applications in gas turbine combustors and furnaces of fire tube boilers. Three designs of OTRs are presented, two for gas turbine applications and one for fire tube boiler applications. The reactor designs are presented based on three-dimensional detailed numerical modeling for optimization of geometry and flow rates through the OTR for maximum power output per minimum reactor size. The detailed results for the three case studies are presented in detail. Different oxygen permeation equations are presented for non-reacting and reacting flow fields inside the OTRs. The possibility of integrating the OTR with conventional combustors is discussed for ZEPP applications. This is followed by a detailed numerical optimization study of a monolith structure design OTR for replacement of a conventional gas turbine combustor. Optimizations for the feed and sweep flow rates were performed in order to meet the power requirements, and all simulations were conducted considering 3D flow fields. Effects of flow configurations, channel width, and percentage of CH_4 in the permeate side flow were investigated under constant inlet gas temperature of 1173 K and a fixed operating pressure of 10 bars. The reactor geometry was determined based on the optimization of channel width. It is concluded that counter-current flow configuration design results in improvements in oxygen permeation flux and overall heat transfer characteristics. However, it has a limitation as a result of increased membrane temperature. Any reduction in the channel width below 15 mm resulted in large increase in the viscous pressure drop, and the combustion was lost due to increased flow velocity. As well, increasing the amount of CH_4 in the permeate side over 5% was found to be non-applicable because of limited oxygen permeation flux to complete the conversion of CH_4. The designed ITM reactor has the capability of producing power output in the range of 5–8 MWe based on cycle first law efficiency.

A second detailed numerical study for the application of OTR in gas turbine combustors is also presented. The study considers a design of a multi-can carbon-free gas turbine combustor utilizing multiple shell-and-tube OTRs for ZEPP applications. In this study, a novel design of a carbon-free gas turbine combustor utilizing oxygen transport reactors (OTRs) is presented based on simulations in the full 3D domain. The design considers multi-can gas turbine combustor; each can consist of multiple oxygen transport membrane tubes of shell and tube type. Optimizations of design and flow conditions are performed to calculate the best design of the OTRs. The optimization parameters include flow configuration (co- and counter-current), shell-side and tube-side (feed and sweep) flow rates, inlet fuel concentration in the sweep flow (CH_4 plus CO_2), and membrane tube diameter, pitch (spacing) and length. BSCF ceramic tubular membranes are used in square arrangement inside each OTR (can) unit. A modified oxygen permeation equation accounting for reacting flow and sub-step membrane surface reactions is utilized. A modified two-step reaction kinetics mechanism of methane for oxy-combustion conditions is used. The co-current flow configuration has shown better characteristics when compared to the counter-current flow configuration in terms of leaving exhaust gas temperature, gradual temperature distribution, and gradual oxygen permeation flux. Better oxygen permeation flux, faster burning, and higher fuel conversion are obtained in the case of 10% CH_4 as compared to the case of 5% CH_4. The increase of total amount of permeated oxygen for the case of 3.6 m long reactor resulted in full conversion of the fuel, while OTRs with shorter length resulted in almost 50% conversion of the available fuel. Reducing membrane tube diameter below 10 mm resulted in unstable calculations and divergence of the solution due to higher values of viscous pressure drop and flame instabilities. Reducing membrane tube pitch, from 22 to 18 mm, resulted in higher viscous pressure drop in feed side as compared to the viscous pressure drop in the sweep side. The resultant gas turbine combustor can produce power in the range of 10–15 MWe based on cycle first law efficiency and consists of 16 cans. Each can consist of 3000 membrane tubes occupying a volume of 5.2272 m^3.

The final detailed numerical case study presents a novel design of an OTR for application in fire tube boilers. In this study, a design of an ITM reactor is introduced to produce steam in a two-pass fire tube boiler for steam generation. Combustion and heat exchange occurred in the first pass, and only heat was exchanged in the second one. In order to meet the power requirement in the range of 5–8 MWe, calculations of the best flow conditions are performed. The target was to introduce a compact design while extracting maximum amount of oxygen. A fixed length of 1.8 m of the reactor was considered at an operating pressure of 20 bars. Effects of flow configuration and CH_4 concentration in the sweep gases were investigated. It was found that the counter-current flow configuration has much better performance in boiler applications than the co-current flow. Based on the required power, the resultant reactor design has total number of ITM units of 12,500, height of 5 m, and total volume of 45.5 m^3.

References

1. Babu P, Linga P, Kumar R, Englezos P (2015) A review of the hydrate-based gas separation (HBGS) process for carbon dioxide pre-combustion capture. Energy 85:261–279
2. International Energy Agency (2011) World energy outlook

3. Habib MA, Nemitallah M, Ben-Mansour R (2013) Recent development in oxy-combustion technology and its applications to gas turbine combustors and ITM reactors. Energy Fuels 27:2–19. https://doi.org/10.1021/ef301266j
4. Yang H, Xu Z, Fan M, Gupta R, Slimane RB, Bland AE et al (2008) Progress in carbon dioxide separation and capture: a review. J Environ Sci 20(1):14–27
5. Mondal MK, Balsora HK, Varshney P (2012) Progress and trends in CO_2 capture/separation technologies: a review. Energy 46(1):431–441
6. Habib MA, Salaudeen SA, Nemitallah MA, Ben-Mansour R, Mokheimer EMA (2016) Numerical investigation of syngas oxy-combustion inside a LSCF-6428 oxygen transport membrane reactor. Energy 96:654–665
7. Mezghani K, Hamza A (2016) Application of $Ba_{0.5}Sr_{0.5}Co_{0.8}Fe_{0.2}O_{3-\delta}$ membranes in an oxy-fuel combustion reactor. J Membr Sci 518:254–262
8. Turi DM, Chiesa P, Macchi E, Ghoniem AF (2016) High fidelity model of the oxygen flux across ion transport membrane reactor: mechanism characterization using experimental data. Energy 96:127–141
9. Balachandran U, Kleefisch MS, Kobylinski TP, Morissette SL, Pei S (1997) Oxygen ion-conducting dense ceramic membranes (Assigned to Amoco Co.). US patent 5,639,437
10. Bernardo P, Drioli E, Golemme G (2009) Membrane gas separation: a review of state of the art. Ind Chem Eng 48(1):4638–4663
11. Farooqui AE, Badr HM, Habib MA, Ben-Mansour R (2014) Numerical investigation of combustion characteristics in an oxygen transport reactor. Int J Energy Res 38(5):638–651
12. Habib MA, Ahmed P, Ben-Mansour R, Badr HM, Kirchen P, Ghoniem AF (2013) Modeling of a combined ion transport and porous membrane reactor for oxy-combustion. J Membr Sci 446:230–243
13. Farooqui AE, Habib MA, Badr HM, Ben-Mansour R (2013) Modeling of ion transport reactor for oxy-fuel combustion. Int J Energy Res 37(11):1265–1279
14. Ben-Mansour R, Habib MA, Badr HM, Nemitallah M (2012) Characteristics of oxy-fuel combustion in an oxygen transport reactor. Energy Fuels 26(7):4599–4606
15. Nemitallah MA, Habib MA, Mezghani K (2015) Experimental and numerical study of oxygen separation and oxy-combustion characteristics inside a button-cell LNO-ITM reactor. Energy 84:600–611
16. Mezghani K, Hamza A, Habib MA, Lee D, Shao-Horn Y (2015) Effect of microstructure and thickness on oxygen permeation of $La_2NiO_{4+\delta}$ membranes. Ceram Int 42(1):666–672
17. Habib MA, Ahmed P, Ben-Mansour R, Mezghani K, Alam Z, Shao-Horn Y, Ghoniem AF (2015) Experimental and numerical investigation of la_2NiO_4 membranes for oxygen separation: geometry optimization and model validation. J Energy Resour Technol Trans ASME 137(3):03110
18. Wang L, Imashuku S, Grimaud A, Lee D, Mezghani K, Habib MA, Shao-Horn Y (2013) Enhancing oxygen permeation of electronically short-circuited oxygen-ion conductors by decorating with mixed ionic-electronic conducting oxides. ECS Electrochem Lett 2(11): F77–F81
19. Imashuku S, Wang L, Mezghani K, Habib MA, Shao-Horn Y (2013) Oxygen permeation from oxygen ion-conducting membranes coated with porous metals or mixed ionic and electronic conducting oxides. J Electrochem Soc 160(11):E148–E153
20. Habib MA, Badr HM, Ahmed SF, Ben-Mansour R, Mazghani K, Imashuku GJ, Shao-Horn Y, Mancini N, Mitsos A, Kirchen P, Ghoneim A (2011) A review of recent developments in carbon capture utilizing oxy-fuel combustion in conventional and ion transport membrane systems. Int J Energy Res 35(9):741–764
21. Salehi M, Pfaff EM, Morkis Junior R, Bergmann CP, Diethelm S, Neururer C, Graule T, Grobety B, Clemens FJ (2013) $Ba_{0.5}Sr_{0.5}Co_{0.8}Fe_{0.2}O_{3-\delta}$ (BSCF) feedstock development and optimization for thermoplastic forming of thin planar and tubular oxygen separation membranes. J Membr Sci 443:237–245

22. Nemitallah MA (2016) A study of methane oxy-combustion characteristics inside a modified design button-cell membrane reactor utilizing a modified oxygen permeation model for reacting flows. J Nat Gas Sci Eng 28:61–73
23. Kirchen P, Apo DJ, Hunt A, Ghoniem AF (2013) A novel ion transport membrane reactor for fundamental investigations of oxygen permeation and oxy-combustion under reactive flow conditions. Proc Combust Inst 34:3463–3470
24. Hong J, Kirchen P, Ghoniem AF (2013) Interactions between oxygen permeation and homogeneous-phase fuel conversion on the sweep side of an ion transport membrane. J Membr Sci 428:309–322
25. Ben-Mansour R, Nemitallah MA, Habib MA (2013) Numerical investigation of oxygen permeation and methane oxy-combustion in a stagnation flow ion transport membrane reactor. Energy 54:322–332
26. Ahmed P, Habib MA, Ben-Mansour R, Kirchen P, Ghoniem AF (2014) CFD (computational fluid dynamics) analysis of a novel reactor design using ion transport membranes for oxy-fuel combustion. Energy 77:932–944
27. Wang H, Cong Y, Yang W (2002) Oxygen permeation study in a tubular $Ba_{0.5}Sr_{0.5}Co_{0.8}Fe_{0.2}O_{3-\delta}$ oxygen permeable membrane. J Membr Sci 210:259–271
28. Kvamsdal HM, Jordal K, Bolland O (2007) A quantitative comparison of gas turbine cycles with CO_2 capture. Energy 32:10–24
29. Hashim SM, Mohamed A, Bhatia S (2010) Current status of ceramic-based membranes for oxygen separation from air. Adv Coll Interface Sci 160:88–100
30. Sunarso J, Baumann S, Serra JM, Meulenberg WA, Liu S, Lin YS, Diniz da Costa JC (2008) Mixed ionic–electronic conducting (MIEC) ceramic-based membranes for oxygen separation. J Membr Sci 320:13–41
31. Shao ZP, Yang WS, Cong Y, Dong H, Tong JH, Xiong GX (2000) Investigation of the permeation behavior and stability of a $Ba_{0.5}Sr_{0.5}Co_{0.8}Fe_{0.2}O_{3-\delta}$ oxygen membrane. J Membr Sci 172:177–188
32. Kharton VV, Viskup AP, Kovalevsky AV, Naumovic EN, Marques FMB (2001) Ionic transport in oxygen-hyperstoichiometric phases with K_2NiF_4-type structure. Solid State Ionics 143:337–353
33. Ting C, Hailei Z, Nansheng X, Yuan L, Xionggang LU, Weizhong D, Fushen L (2011) Synthesis and oxygen permeation properties of a $Ce_{0.8}Sm_{0.2}O_{2-\delta}$_$LaBaCO_2O_{5+\delta}$. J Membr Sci 370:158–165
34. Wiik K, Aasland S, Hansen HL, Tangen LL, Odegard R (2002) Oxygen permeation in the system $SrFeO_{3-x}$–$SrCoO_{3-y}$. Solid State Ionics 152–153:675–680
35. Fan CG, Zuo YB, Li JT, Lu JQ, Chen CS, Bae DS (2007) Highly permeable $La_{0.2}Ba_{0.8}Co_{0.8}Fe_{0.2-x}Zr_xO_{3-\delta}$ membranes for oxygen separation. Sep Purif Technol 55:35
36. Ishihara T, Yamada T, Arikawa H, Nishiguchi H, Takita Y (2000) mixed electronic-oxide ionic conductivity and oxygen permeating property of Fe-, Co- or Ni-doped $LaGaO_3$ provskite oxide. Solid State Ionics 135:631–636
37. Habib MA, Badr HM, Ahmed SF, Ben-Mansour R, Mezghani K, Imashuku S, lao GJ, Shao-Horn Y, Mancini ND, Mitsos A, Kirchen P, Ghoniem AF (2011) A review of recent developments in carbon capture utilizing oxy-fuel combustion in conventional and ion transport membrane systems. Int J Energy Res 35:741–764
38. Manning PS, Sirman JD, Kilner JA (1996) Oxygen self-diffusion and surface exchange studies of oxide electrolytes having the fluorite structure. Solid State Ionics 93(1–2):125–132
39. Ishihara T, Kilner JA, Honda M, Sakai N, Harumi Y, Yusaku T (1998) Oxygen surface exchange and diffusion in $LaGaO_3$ based perovskite type oxides. Solid State Ionics 113–115:593–600
40. Ruiz-Trejo E, Sirman JD, Baikov YM, Kilner JA (1998) Oxygen ion diffusivity surface exchange and ionic conductivity in single crystal gadolinia doped ceria. Solid State Ionics 113–115:565–569
41. Lane JA, Kilner JA (2000) Oxygen surface exchange on gadolinia doped ceria. Solid State Ionics 136–137:927–932

42. Tan X, Liu Y, Li K (2005) Mixed conducting ceramic hollow fibre membranes for air separation. AIChE J 71:1991
43. Kim S, Yang YL, Jacobson AJ, Abeles B (1998) Diffusion and surface exchange coefficients in mixed ionic electronic conducting oxides from the pressure dependence of oxygen permeation. Solid State Ionics 106:189–195
44. Lin YS, Wang Y, Han J (1994) Oxygen permeation through thin mixed-conducting solid oxide membranes. AIChE J 40:786–798
45. Xu SJ, Thomson WJ (1999) Oxygen permeation rates through ion-conducting perovskite membranes. Chem Eng Sci 54(17):3839–3850
46. Liu S, Tan X, Shao Z, Diniz da Costa J (2006) $Ba_{0.5}Sr_{0.5}Co_{0.8}Fe_{0.2}O_{3-\delta}$ ceramic hollow-fiber membranes for oxygen permeation. AIChE J 52:3452
47. Tan X, Li K (2002) Modeling of air separation in a LSCF hollow-fibre membrane module. AIChE J 48:1469
48. Lee T, Yang Y, Jacobson A, Abelesa B, Zhou M (1997) Oxygen permeation in dense $SrCo_{0.8}Fe_{0.2}O_{3-\delta}$ membranes: surface exchange kinetics versus bulk diffusion. Solid State Ionics 100:77–85
49. Shao Z, Xiong G, Tong J, Dong H, Yang W (2001) Ba effect in doped $Sr(Co_{0.8}Fe_{0.2})O_{3-\delta}$ on the phase structure and oxygen permeation properties of the dense ceramic membranes. Sep Purif Technol 25:419–429
50. Wang H, Wang R, Liang D, Yang W (2004) Experimental and modeling studies on $Ba_{0.5}Sr_{0.5}Co_{0.8}Fe_{0.2}O_{3-\delta}$ (BSCF) tubular membranes for air separation. J Membr Sci 243:405
51. Ge L, Shao Z, Zhang K, Ran R, Diniz da Costa J, Liu S (2009) Evaluation of mixed-conducting lanthanum-strontium-cobaltite ceramic membrane for oxygen separation. AIChE J 55:2603
52. Ito W, Nagai T, Sakon T (2007) Oxygen separation from compressed air using a mixed conducting perovskite-type oxide membrane. Solid State Ionics 178:809
53. Zhu X, Sun S, Cong Y, Yang W (2009) Operation of perovskite membrane under vacuum and elevated pressures for high-purity oxygen production. J Membr Sci 345:47–52
54. Zhu X, Cong Y, Yang W (2006) Oxygen permeability and structural stability of $BaCe_{0.15}Fe_{0.85}O_{3-\delta}$ membranes. J Membr Sci 283:38–44
55. Qi X, Lin Y, Swartz S (2000) Electrical transport and oxygen permeation properties of lanthanum cobaltite membranes synthesized by different methods. Ind Eng Chem Res 39:646
56. Rui Z, Li Y, Lin Y (2009) Analysis of oxygen permeation through dense ceramic membranes with chemical reactions of finite rate. Chem Eng Sci 64:172–179
57. Akin FT, Jerry LYS (2004) Oxygen permeation through oxygen ionic or mixed-conducting ceramic membranes with chemical reactions. J Membr Sci 231:133–146
58. Chang X, Zhang C, He Y, Dong X, Jin W, Xu N (2009) A comparative study of the performance of symmetric and asymmetric mixed-conducting membranes. Chin J Chem Eng 17:562
59. Akin F, Lin Y (2002) Oxidative coupling of methane in dense ceramic membrane reactor with high yields. AIChE J 48:2298–2306
60. Akin FT, Lin YS (2002) Selective oxidation of ethane to ethylene in a dense tubular membrane reactor. J Membr Sci 209:457–467
61. Bouwmeester HJM, Burggraaf AJ (1997) Dense ceramic membranes for oxygen separation. In: Gellings PJ, Bouwmeester HJM (eds) The CRC handbook of solid state electrochemistry. CRC Press, Boca Raton, FL Chapter 14
62. Han J, Xomeritakis G, Lin YS (1997) Oxygen permeation through thin zirconia/yttria membranes prepared by EVD. Solid State Ionics 93:263–272
63. Zeng Y, Lin YS (2000) Oxygen permeation and oxidative coupling of methane in yttria doped bismuth oxide membrane reactor. J Catal 193:58–64
64. Park JH, Blumenthal RN (1989) Electronic transport in 8 mole percent Y_2O_3–ZrO_2. J Electrochem Soc 136:2867

65. Kusaba H, Shibata Y, Sasaki K, Teraoka Y (2006) Surface effect on oxygen permeation through dense membrane of mixed-conductive LSCF perovskite-type oxide. Solid State Ionics 177:2249–2253
66. Mancini ND, Mitsos A (2011) Ion transport membrane reactors for oxy-combustion part II: analysis and comparison of alternatives. Energy 36(8):4721–4739
67. Chen L, Ghoniem AF (2012) Simulation of oxy-coal combustion in a 100 kW_{th} test facility using RANS and LES: a validation study. Energy Fuels 26:4783–4798. https://doi.org/10.1021/ef3006993
68. Ben-Mansour R, Habib MA, Badr HM, Nemitallah MA (2012) Characteristics of oxyfuel combustion in an oxygen transport reactor. Energy Fuels 26:4599–4606. https://doi.org/10.1021/ef300539c
69. Nemitallah MA, Habib MA, Ben Mansour R (2012) Investigations of oxy-fuel combustion and oxygen permeation in an ITM reactor using a two-step oxy-combustion reaction kinetics model. J Membr Sci 432:1–12
70. Nemitallah MA, Habib MA, Ben Mansour R, Ghoniem AF (2014) Design of an ion transport membrane reactor for gas turbine combustion application. J Membr Sci 450:60–71
71. Mancini ND, Mitsos A (2011) Ion transport membrane reactors for oxy-combustion part II: analysis and comparison of alternatives. Energy 36:4721–4739
72. Chandrasekhar S (1960) Radiative transfer. Dover Publications, New York, NY
73. Rajhi MA, Ben-Mansour R, Habib MA, Nemitallah MA, Andersson K (2014) Evaluation of gas radiation models in CFD modeling of oxy-combustion. Energy Convers Manag 81:83–97
74. Nemitallah MA, Habib MA (2013) Experimental and numerical investigations of an atmospheric diffusion oxy-combustion flame in a gas turbine model combustor: oxy-combustion and emission characterization, flame stabilization and model validation. Appl Energy 11:401–415
75. Lallemant N, Weber R (1996) A computationally efficient procedure for calculating gas radiative properties using the exponential wide band model. Int J Heat Mass Transfer 39:3273–3286
76. Puig-Arnavat M, Søgaard M, Hjuler K, Ahrenfeldt J, Henriksen UB, Hendriksen PV (2015) Integration of oxygen membranes for oxygen production in cement plants. Energy 91:852–865
77. Duan L, Yue L, Qu W, Yang Y (2015) Study on CO_2 capture from molten carbonate fuel cell hybrid system integrated with oxygen ion transfer membrane. Energy 93:20–30
78. Hwang KR, Park JW, Lee SW, Hong S, Lee C-B, Oh DK et al (2015) Catalytic combustion of the retentate gas from a CO_2/H_2 separation membrane reactor for further CO_2 enrichment and energy recovery. Energy 90:1192–1198
79. Chiesa P, Romano MC, Spallina V, Turi DM, Mancuso L (2013) Efficient low CO_2 emissions power generation by mixed conducting membranes. Energy Procedia 37:905–913
80. Manzolini G, Gazzani M, Turi DM, Macchi E (2013) Application of hydrogen selective membranes to IGCC. Energy Procedia 37:2274–2283
81. Voleno A, Romano MC, Turi DM, Chiesa P, Ho MT, Wiley DE (2014) Post-combustion CO_2 capture from natural gas combined cycles by solvent supported membranes. Energy Procedia 63:7389–7397
82. Romano MC (2013) Ultra-high CO_2 capture efficiency in CFB oxyfuel power plants by calcium looping process for CO_2 recovery from purification units vent gas. Int J Greenh Gas Control 18:57–67
83. Mancini ND, Mitsos A (2011) Ion transport membrane reactors for oxy-combustion part II: analysis and comparison of alternatives. Energy 36:4721e39
84. Habib MA, Nemitallah MA (2015) Design of an ion transport membrane reactor for application in fire tube boilers. Energy 81:787–801
85. Nemitallah MA, Habib MA, Badr HM (2017) Design of a multi-can carbon-free gas turbine combustor utilizing multiple shell-and-tube OTRs for ZEPP applications. J Nat Gas Sci Eng 46:172–187

86. Petric A, Huang P, Tietz F (2000) Evaluation of La–Sr–Co–Fe–O perovskites for solid oxide fuel cells and gas separation membranes. Solid State Ionics 135:719–725
87. Li S, Jin W, Xu N, Shi J (1999) Synthesis and oxygen permeation properties of $La_{0.2}Sr_{0.8}Co_{0.2}Fe_{0.8}O_{3-\delta}$ membranes. Solid State Ionics 124:161–170
88. Hong J, Kirchen P, Ghoniem AF (2012) Numerical simulation of ion transport membrane reactors: oxygen permeation and transport and fuel conversion. J Membr Sci 85:407–408
89. Behrouzifar A, Atabak AA, Mohammadi T, Pak A (2012) Experimental investigation and mathematical modeling of oxygen permeation through dense $Ba_{0.5}Sr_{0.5}Co_{0.8}Fe_{0.2}O_{3-\delta}$ (BSCF) perovskite-type ceramic membranes. Ceram Int 38:4797–4811
90. Chui E, Raithby G (1993) Computation of radiant heat transfer on a nonorthogonal mesh using the finite-volume method. Numer Heat Transfer 23(3):269–288
91. Glarborg P, Bentzen LL (2007) Chemical effects of a high CO_2 concentration in oxy-fuel combustion of methane. Energy Fuels 22:291–296
92. Andersen J, Rasmussen CL, Giselsson T, Glarborg P (2009) Global combustion mechanisms for use in CFD modeling under oxy-fuel conditions. Energy Fuels 23:1379–1389
93. Versteeg HK, Malalasekera W (1995) An introduction to computational fluid dynamics—the finite volume method. Longman Scientific and Technical
94. Nemitallah MA, Habib MA, Ben Mansour R (2013) Investigations of an ion transport membrane reactor specially designed for a power cycle. Appl Mech Mater 302:440–446

Correction to: Oxyfuel Combustion for Clean Energy Applications

Correction to:
M. A. Nemitallah et al., *Oxyfuel Combustion for Clean Energy Applications*, Green Energy and Technology, https://doi.org/10.1007/978-3-030-10588-4

In the original version of the book, the preface contained an incorrect grant number "IN171005". It should be read as "BW171005". The book has been updated with the change.

The updated version of the book can be found at https://doi.org/10.1007/978-3-030-10588-4

M. A. Nemitallah et al., *Oxyfuel Combustion for Clean Energy Applications*, Green Energy and Technology, https://doi.org/10.1007/978-3-030-10588-4_7

Correction to: Oxyfuel Combustion

MIX
Papier aus verantwortungsvollen Quellen
Paper from responsible sources
FSC® C105338

If you have any concerns about our products,
you can contact us on
ProductSafety@springernature.com

In case Publisher is established outside the EU,
the EU authorized representative is:
Springer Nature Customer Service Center GmbH
Europaplatz 3, 69115 Heidelberg, Germany

Printed by Libri Plureos GmbH
in Hamburg, Germany